HISTORY OF LIFE

FOURTH EDITION

HISTORY OF LIFE

RICHARD COWEN
University of California, Davis

BLACKWELL PUBLISHING
350 Main Street, Malden, MA 02148-5020, USA
9600 Garsington Road, Oxford OX4 2DQ, UK
550 Swanston Street, Carlton, Victoria 3053, Australia

First edition published 1990
Second edition published 1995
Third edition published 2000
Fourth edition published 2005 by Blackwell Publishing Ltd

5 2006

Library of Congress Cataloging-in-Publication Data

Cowen, Richard, 1940–
History of life / Richard Cowen.—4th ed.
p. cm.
Includes index.
ISBN 1-4051-1756-7 (pbk. : alk. paper)
1. Paleontology. I. Title.
QE711.2.C68 2005
560—dc22
2003027993

ISBN-13: 978-1-4051-1756-2 (pbk. : alk. paper)

A catalogue record for this title is available from the British Library.

Set in 10½ on 12½ pt Adobe Garamond
by SNP Best-set Typesetter Ltd., Hong Kong
Printed and bound in the United Kingdom
by TJ International Ltd, Padstow, Cornwall

The publisher's policy is to use permanent paper from mills that operate a sustainable forestry policy, and which has been manufactured from pulp processed using acid-free and elementary chlorine-free practices. Furthermore, the publisher ensures that the text paper and cover board used have met acceptable environmental accreditation standards.

For further information on
Blackwell Publishing, visit our website:
www.blackwellpublishing.com

Contents

CHAPTER EIGHTEEN: Geography and Evolution 255

CHAPTER NINETEEN: Primates 267

CHAPTER TWENTY: Evolving Toward Humans 278

CHAPTER TWENTY-ONE: Life in the Ice Age 297

Preface

FOR EVERYONE

For 36 years I taught a course called "History of Life" at the University of California, Davis. This book, now entering its fourth edition, was written for that course. However, it is meant not just for students, but for everyone interested in the history of life on our planet. Fortunately, paleontology (= paleobiology), is accessible to the average person without deep scientific training. My aim is ambitious: I try to take you to the edges of our knowledge in paleontology, showing you how life has evolved on Earth, and how we have constructed the history of that evolution from the record of rocks and fossils.

However, there is a snag. Human history is never simple, even when we try to describe events that happened last year. It's even worse when we ask why events happened. It's not likely that any account of the history of life is going to be simple either. The living world today contains all kinds of creatures that do unexpected things. There are frogs that fly and birds that can't. There are mammals that lay eggs, reptiles that have live birth, and amphibians that suckle their young. There are fishes that breathe air and mammals that never touch the land. We have to expect that there were complex and unusual ways of life in the past, and that evolution took some unexpected turns at times.

The challenge of teaching paleontology, and the challenge of writing a book like this, is to present a complex story in a way that is simple enough to grasp, yet true enough to real events that it paints a reasonable picture of what happened and why. I believe it can be done, and done so that you can learn enough to appreciate what's going on in current research projects.

Paleontologists can identify how evolution happened and how the creatures of the past lived. We can't prove it, any more than we can prove what really motivated George Washington. But we can state clearly what we know and don't know, we can suggest why certain events happened, and we can describe the evidence we used and the thoughts behind our suggestions. Then people can accept the ideas or not, as they wish.

Paleontologists have been collecting fossils, studying the rocks they came from, and assembling their data into a coherent framework, for over 200 years. At this point I think we are limited more by a lack of good ideas than by the facts available about the fossil record. I have not been shy about offering explanations of events as well as descriptions of them. Mostly they are other people's explanations, but now and again I've suggested some of my own. You can accept these or not, as you wish. The question you face is that of a jury member: is this idea sound "beyond all reasonable doubt?" If you don't accept an explanation for an event, you can leave it as an abominable mystery, with no explanation at all, or you can suggest a better explanation yourself.

There is one caution, however. No-one is allowed to dream up any old explanation for past events. A scientific suggestion (a hypothesis) has to fit the available evidence, and it has to fit with the laws of physics and chemistry, and with the principles of biology, ecology, and engineering that have been pieced together over the past 200 years of scientific investigation.

There's yet another wrinkle. A jury decides on a case, once and for all, with the evidence available. But in science, the jury is always out, and new evidence comes in all the time. You may have to change a verdict—without regret, because you made the best (wrong) decision you could based on the old evidence. Some of the ideas in the earlier editions are wrong, and you won't find them here; you'll find better ones. Sometimes the new answer is more complex, sometimes it is simpler. Always, however, the new idea fits the evidence better. That's simply the way science works: not on belief, not on emotional clinging to a favorite idea (even if it is your own), but on evidence.

I never expect to be able to write the final solution to the major questions about the history of life, but I do expect to be able to provide better answers this year than I could last year. If my lectures are the same this year as they were last year, then something is wrong with our science, or something is wrong with me. Paleontology is exciting because it is advancing so quickly.

Since paleontology is so fast-moving as a science, this book has changed too. I have radically rewritten the sections on the origin of life, on the origin and radiation of metazoan animals, on dinosaurs, on birds, and on human evolution. I have recognized more clearly that life has evolved on an evolving planet, with changing chemistry, changing geography, and changing climate, and have tried to weave these threads into the tapestry as well. Every other section has been fine-tuned to reflect new research.

As I write, the full development of the Internet in publishing and educating and informing is in the early stages of a revolution. The web site for this book, at www.geology.ucdavis.edu/cowen/HistoryofLife/ (also see www.blackwellpublishing.com/cowen), allows me to add enriching material for which I could not find space in the printed page. But more than that, it allows me to connect you to other Internet sites for colored illustrations, more detailed accounts of research projects, and snappy news articles. The web site is meant to add real background and further details and perspectives for those who are interested. The essays and mini-essays are written in the same style as the book; the trivia are irresistible; and the web links lead you to sites that are often richly illustrated in a way that a book cannot match.

I have written the book so that it stands alone without the web site; but the web pages are extensively linked back to the book. The web page will also contain updates of material, new references, and new information.

TO MY TEACHING COLLEAGUES

The course for which this book was written serves three audiences at the same time: it is an introduction to paleontology; it is a "general education" course to introduce nonspecialists to science and scientific thought; and it can serve as an introduction to the history of life to biologists who know a lot about the present and little about the past. Therefore, the style and language of this book are aimed at accessibility. I do not use scientific jargon unless it is useful. I have tried to show how we reason out our conclusions—how we choose between bad ideas and good ones. I have not diluted the English language down to pidgin to make my points. In short, I have aimed this book at the intelligent nonspecialist.

I have not covered the fossil record evenly. I've tried to write compact essays on what I think are the most important events and processes that have molded the history of life. They illustrate the most important ways we go about reconstructing the life of the past. I've used case studies from vertebrates more than from other groups simply because those are the animals with which we are most familiar. Most fossils are marine invertebrates, and most paleontologists, including myself, are invertebrate specialists. I have tried to write briefly about invertebrates at an introductory level. They are not the easiest vehicles to use at this level, and that thought has controlled my choice of subject matter.

I'm pleased with the text of this book: I believe it communicates a lot about our science in the space available. But it's impossible to communicate paleontology well without a much greater visual component than can be included in a relatively inexpensive book format. So this book is under-illustrated. I use a lot of images in my classes in an attempt to bring fossil and living organisms into the classroom, and to give life to the words and names. The web pages contains many sources for online illustrations that can be downloaded into your favorite presentation medium.

The references are a careful mixture of important books, primary literature, news reports, and review articles that bring the latest work into this edition as it went to press. I have deliberately skewed the lists to include items likely to be found in small college and city libraries.

If this book contained nothing controversial, it would be very dull and far from representing the state of paleontology as it stands today. I have tried to present arguments for and against particular ideas in case studies that are presented in some detail, such as the K–T extinction. Often, however, space or conviction has led me to present only one side of an argument. Please share your dissatisfaction and/or more complete knowledge with your students, and tell them why my treatment is one-sided or just plain wrong. That way everyone wins by exposure to the give and take of scientific argument as it ought to be practiced between colleagues.

TO STUDENTS

Several thousand people like you have voted with their comments, questions, body language, and formal written evaluations on the content of my course. Those people have had more influence on the style and content of this book than anyone else. So you and your peers at the University of California, Davis, can take whatever credit is due for the style in which the material is presented.

After all the thanks, however, I do have another point to make. You don't have to take any of the interpretations in this book at face value. Facts are facts, but ideas are only suggestions. If you can come up with a better idea than one of those I've included here, then work on it, starting with the literature references. It would make a great term paper, and (more importantly) you might be right. The 1960's slogan "Question Authority!" is still valid. Your suggestion wouldn't be the first time that a student found a new and better idea for interpreting the fossil record.

Why do I, and why should you, bother with the past? If we don't understand the past, how can we deal intelligently with the present? We and our environment are reaching such a state of crisis that we need all the help we can get. Nature has run a series of experiments over the last 3.5 billion years on this planet, changing climate and geography, and introducing new kinds of organisms. If we can read the results of those experiments from the fossil record, we can perhaps define the limits to which we can stretch our present biosphere before a biological disaster happens.

The real pay-off from paleontology for me is the fun involved in reconstructing extinct organisms and ancient communities, but if one needs a concrete reason for looking at the fossil record, the future of the human race is surely important enough for anyone.

THIS BOOK

I begin this book with the formation of Earth and the great unsolved problem of the origin of life. Then I describe an early Earth populated entirely by bacteria, so strange in its chemistry and ecology that it might well be another planet. Eventually, living things so alter their world that we begin to recognize environments and organisms that seem much more familiar. I describe the evolution of animals and begin to worry about their physical and ecological environment. By now, we are dealing with a world whose geography we can begin to reconstruct, which leads to chapters on plate tectonics and the climates of the past, and how they might have affected living things.

The vast record of invertebrates allows us to measure the diversity of life through time, which shows that there have been times of high diversity, and times of dramatic extinctions. I deal with extinction, mainly to look at the crises or "mass extinctions" that have occurred sporadically through time. Then I turn largely to the history of vertebrates, following some of the great anatomical, physiological, and ecological innovations by which fishes gradually evolved into the major classes of tetrapods on the land, including ourselves.

Some of my colleagues are dubious about evolutionary "progress," but I regard the evidence for it as overwhelming, and present many examples. I have not tried to write a simple historical catalog of fossils. Instead, I have tried to set interesting episodes in Earth history into a global picture. For example, the tragedy that overtook Mesozoic communities 65 million years ago has to be seen in terms of their success until that time, and the radiation of the mammals can only be appreciated against a background of changing planetary geography and climate. Finally, the rise to ecological dominance of humans has its counterpart in the massive changes in land faunas that accompanied it, all set in the context of the ice ages.

FURTHER READING

I have tried to list widely sold paperbacks and articles in journals such as *Nature, Science, Discover, Scientific American, National Geographic Magazine,* and *American Scientist,* perhaps the six most widely distributed journals that deal with all aspects of science. I also list books and articles in specialized journals: generally, but not always, the writing is more detailed and more technical in such journals.

Important earlier work is often summarized in more recent articles I have selected. Always, however, you should be able to work quickly backward to older papers from the references in recent articles.

MATERIAL AVAILABLE ON THE WEB SITE FOR THIS BOOK

For more references, references available freely on the Web, further reading, notes, extra stories and mini-essays by RC, sources for classroom images, and updates, see the web site for this book: http://www.blackwellpublishing.com/cowen, mirrored at http://www.geology.ucdavis.edu/~cowen/HistoryofLife/.

If you find the icon, it implies that there is a significant comment, explanation, or expanded mini-essay on that topic on the web site.

THANKS

I thank all those reviewers who gave careful and calm advice: Norm Gilinsky and Ken McKinney early on, and this time around Paul Koch and other anonymous colleagues. I hope they recognize their contribution, and I apologize for churlishly ignoring some of them. I was inspired by the web site of Doug Eernisse at California State University, Fullerton, as I began to construct my own at Davis.

I owe a great deal to the people at Blackwell who have encouraged and helped me over the years. I would like to thank especially Simon Rallison and Jane Humphreys for past help, and Nancy Whilton and Elizabeth Wald for this edition.

None of these people should be blamed for deficiencies: please complain directly to me at rcowen@ucdavis.edu.

Finally, I thank once again my wife Jo and my daughters Claire and Alexandra for tolerating my neglect of them while all this was in process. If you buy this book, you may well contribute toward their further deprivation by encouraging another new edition.

Richard Cowen
Winters, California

There is an astonishing contrast in the image on the cover. The creatures preserved in the world-renowned La Brea tar pits lived in a fertile plain, an oasis in the southern California desert. Their remains are abundant, beautiful, and evocative of a time and place that no longer exists. In its place rises one of the world's most famous cities, La Cuidad de Nuestra Señora de Los Angeles, one of the most altered places on Earth, where hardly any living thing or any inanimate object one can see would occur here naturally.

CHAPTER ONE

The Origin of Life on Earth

HOW GEOLOGY WORKS

Geology is the study of Earth we live on. It draws on methods and principles from many sciences: physics, chemistry, biology, mathematics, and statistics are just a few. Geologists have to know at least a little of several sciences: they cannot be narrow specialists. Geology is a broad science that works best for people who think broadly. So most geologists cannot be successful if they are geeks (though a few seem to manage it). Above all, geology deals with the reality of Earth: its rocks, minerals, its rivers, lakes and oceans, its surface, and its deep structure. Always, the reality of evidence from fieldwork controls what can and what cannot be said about Earth. Geological hypotheses are tested against evidence from rocks, and many beautiful theories have failed those demanding tests.

Some geologists deal with Earth as it is now: they don't need to look at the past. Earth history doesn't matter much to a geologist trying to deal with ecological repair to an abandoned gold mine. But many geologists have to deal with the history of Earth, and they find that they are studying a planet that changes, at all scales of space and time, sometimes in the most surprising ways. We have known for 200 years that life on Earth has changed: we can collect fossils as direct and solid proof of that. But gradually, geologists have come to realize that life has evolved on a planet that is changing too.

Ideas about changing geography, changing climate, and changing chemistry have become much more important recently in discussing Earth history. And our best sources of insight into those changes come from the fossil evidence of the creatures that survived them (or not). So paleontology is not just a fascinating side branch of geology, but a vital component of it.

As they run their life processes, organisms take in, alter, and release chemicals. Given enough organisms and enough time, biological processes can change the chemical and physical world. Photosynthesis, which provides the oxygen in our atmosphere, is only one of these processes. In turn, physical processes of Earth such as continental movement, volcanism, and climate change affect organisms, influencing their evolution, and, in turn, affecting the way they affect the physical earth. This is a gigantic interaction, or feedback mechanism, that has been going on since life evolved on Earth. Paleontologists and geologists who ignore this interaction are likely to get the wrong answers as they try to reconstruct the past.

HOW PALEONTOLOGY WORKS

Some of Earth's ancient life has been preserved in rocks as fossils, and the study of these fossils is the science of paleontology. Paleontology deals with the interpretation of fossils as organisms, living, breeding, and dying in a real environment on a real, but past, Earth that we can no longer touch, smell, or see directly: we perceive a virtual Earth through our study of fossils and the rocks they are preserved in.

Most paleontologists don't study fossils for their intrinsic interest, although some of us do. Their greater value lies in what they tell us about ourselves and our background. We care about our future, which is a continuation of our past. One good reason for trying to reconstruct ancient life is to manage better the biology of our planet today, so we need to set up some kind of reasonable logic for interpreting life of the past.

Some basic problems of paleontology are like those of archaeology and history: how do we know we have found the right explanation for some past event? How do we know we are not just making up a story?

Anything we suggest about the biology of extinct organisms should make sense in terms of what we know about the biology of living organisms, unless there is very good evidence to the contrary. This rule applies throughout biology, from cell biochemistry to genetics, physiology, ecology, behavior, and evolution.

But suggestions are only suggestions until they are tested against evidence from fossils and rocks. Because fossils are found in rocks, we have access to environmental information about the habitat of the extinct organism: for example, the rock might show clear evidence that it was deposited under desert conditions or on a shallow-water reef. Fossils are therefore not isolated objects but parts of a larger puzzle. For example, it is difficult to interpret the biology of the first bird, *Archaeopteryx*, unless we consider environmental evidence from the Solnhofen Limestone in which it is preserved (Chapter 13).

An alert reader should be able to identify four levels of paleontological interpretation. First, there are *inevitable conclusions* for which there are no possible alternatives. For example, there's no doubt that extinct ichthyosaurs were swimming marine reptiles. At the next level, there are *likely interpretations*. There may be alternatives, but a large body of evidence supports one leading idea. For example, there is good evidence to suggest that ichthyosaurs gave birth to live young rather than laying eggs. Almost all paleontologists view this as the best hypothesis available and would be surprised if contrary evidence turned up.

Then there are *speculations*. They may be right, but there is not much real evidence one way or another. Paleontologists are allowed to accept speculations as tentative ideas to work with and to test carefully, but they should not be surprised or upset to find them wrong. For example, it seems reasonable to me that ichthyosaurs were warm-blooded, but it's a speculative idea because it's difficult to test. If new evidence showed that the idea was unlikely, I might be personally disappointed but I would not be distressed scientifically.

Finally, there are *guesses*. They may be biologically more plausible than others one might suggest, but for one reason or another they are completely untestable and must therefore be classified as nonscientific. For example, if I asked an artist to draw an ichthyosaur, I might suggest bold black-and-white color patterns, like those of living orcas, but another paleontologist might opt for more muted tones like those of living dolphins. Both ideas are reasonable, and are surely better than the green-with-pink-spots that one might find in a TV cartoon. But they are guesses, because there is no evidence at all.

You will find examples of all four kinds of interpretation in this book. Often it's a matter of opinion in which category to place different suggestions, and this problem has caused many controversies in paleobiology. Were dinosaurs warm-blooded? Some paleobiologists think this is an inevitable conclusion from the evidence, some think it's likely, some think it's only speculative, some think it's unlikely, and some think it is plain wrong. New evidence almost always helps to solve old questions but also poses new ones. Without bright ideas and constant attempts to test them against evidence, paleontology would not be as exciting as it is.

The fossil record gradually gets poorer as we go back in time, for two reasons. Biologically, there were fewer types of organisms in the past. Geologically, relatively few rocks (and fossils) have survived from older times, and those that have survived have often suffered heating, deformation, and other changes, all of which tend to destroy fossils. Earth's early life was certainly microscopic and soft-bodied, a very unpromising combination for fossilization. So direct evidence about the origin of life on Earth is very scanty.

THE ORIGIN OF LIFE

There is no good evidence of life, let alone intelligence or civilization, anywhere in the universe except on our planet, Earth. This fact of observation comes in the face of strenuous efforts by tabloid magazines, movie directors, and NASA publicists to persuade us otherwise. However, it is a fact as I write this in 2003, and we have to face up to its implications. This simple observation implies (but does not prove) that life evolved here on Earth. How difficult would that have been?

We can test the idea that life evolved here on Earth, from nonliving chemicals, by observation and experiment. Geologists and astronomers look for evidence from Earth, Moon, and other planets to reconstruct conditions in the early solar system. Chemists and biochemists determine how complex organic molecules could have formed in such environments. Geologists try to find out when life became established on Earth, and biologists design experiments to test whether these facts fit with the idea of the evolution of life from nonliving chemicals.

Complex organic molecules form in interstellar space, on comets and asteroids and interplanetary dust, and on the meteorites that hit Earth from time to time. These compounds probably form naturally in space, because gas clouds, dust particles, and cometary and meteorite surfaces are bathed in cosmic and stellar radiation. But life as we know it consists of cells, composed mostly of liquid water that is vital to life. It is impossible to imagine the formation of any kind of water-laden cell in outer space: that could only have happened on a planet that had oceans and therefore an atmosphere.

Planets have organic compounds delivered to them from space, from comets or meteorites, but it is unlikely that this process by itself leads to the evolution of life. Organic molecules must have been delivered to Mercury, Mars, Venus, and the Moon as well as to Earth, only to be destroyed by inhospitable conditions on those lifeless planets.

Experiments show that it is fairly simple to form large quantities of organic compounds in planetary atmospheres and on planetary surfaces, given the right conditions. Space-borne molecules may have added to the supply on a planetary surface, but they would never be the only source of organic molecules that led to the origin of life.

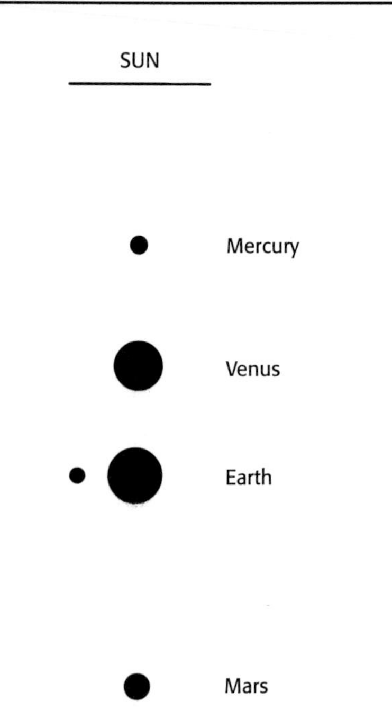

Figure 1.1 The Sun and the terrestrial or inner planets. Their relative sizes and their relative distances are correct, but their sizes are exaggerated by about five billion times compared with their distances.

The Inner Planets of the Solar System

Earth is one of the four terrestrial (rocky) planets in the inner part of our solar system (Figure 1.1). Venus and Earth are about the same size, and Mars and Mercury are significantly smaller. They all formed in the same way about 4570 million years ago (Ma), from dust and gas, and most likely, they were all largely complete as planets by 4500 Ma, though they were bombarded heavily for hundreds of millions of years as stray asteroids struck their surfaces. All of them were hot, and the heat energy released as the planets formed would have made them partly or totally molten. Earth in particular was struck by a huge Mars-sized body late in its formation, and that impact probably melted the entire Earth. We know from crater impacts and lunar samples that Earth and the Moon suffered a heavy late bombardment around 3900 Ma, and the same event probably affected all the other inner planets.

All the inner planets melted deeply enough to have hot surfaces that gave off gases to form atmospheres. But there the similarity ended, and each inner planet had its own later history. Even so, it is clear that an early planet is not a place where life could evolve.

Once a planet cools, conditions on its surface are largely controlled by its distance from the Sun and by any volcanic gases that erupt into its atmosphere. The geology of a planet therefore greatly affects the chances that life might evolve on it.

Liquid water is vital for life as we know it, so surface temperature is perhaps the single most important feature of a young planet. Surface temperature is mainly determined by distance from the Sun: too far, and water freezes to ice; too close, and water evaporates to form water vapor.

But distance from the Sun is not the only factor that affects surface temperature. A planet with an early atmosphere that contained gases such as methane, carbon dioxide, and water vapor would trap more of the Sun's radiation in the "greenhouse" effect, and would be warmer than an astronomer might predict just from distance.

In addition, distance from the Sun alone does not determine whether a planet has water, otherwise the Moon would have oceans like Earth's. The size of the planet is important, because gases escape into space from the weak gravitational field of a small planet. Gas molecules such as water vapor are lost faster from a small planet, and heavier gases are lost as well as light ones.

Gases may also be absorbed out of an atmosphere if they react chemically with surface rocks. They can be released again only by volcanic activity that melts those rocks. But a small planet cools faster than a large one, so any volcanic activity quickly stops as its interior freezes. After that, eruptions no longer return or add gases to the atmosphere. Therefore, a small planet quickly evolves to have a very thin atmosphere or no atmosphere at all, and no chance of gaining one.

Volcanic gases typically include large amounts of water vapor and CO_2 (Figure 1.2), and they are powerful greenhouse gases. Earth would have been frozen for most of its history without CO_2 and water vapor in its atmosphere. Together they add perhaps 33°C to Earth's average temperature.

With all these principles in mind, let's look at the prospects for life on the planets of our solar system. Both Mercury and the Moon had active volcanic eruptions early in their history, but they are small. They cooled quickly and are now solid throughout. Their atmospheric gases either escaped quickly to space from their weak gravitational fields or were blown off by major impacts. Today Mercury and the Moon are airless and lifeless.

Venus is larger than the Moon or Mercury, almost the same size as Earth. Volcanic rocks cover most of its surface. Like Earth, Venus has had a long and active geological history, with a continuing supply of volcanic gases for its atmosphere, and it has a strong gravitational field that can hold in most gases.

But Venus is closer to the Sun than Earth is, and the higher solar radiation hitting the planet was trapped so effectively by water vapor and CO_2 that water molecules may never have been able to condense out as liquid water. Instead, water remained as vapor in the atmosphere until it was dissociated, broken up into hydrogen (H_2), which was lost to space, and oxygen (O_2), which was taken up chemically by reacting with the hot surface rocks of the planet.

Today Venus has a dense, massive atmosphere made largely of CO_2. Volcanic gases react in the atmosphere to make tiny droplets of sulfuric acid (H_2SO_4), forming the clouds that hide the planetary surface. Water vapor has vanished completely. Although the sulfuric acid clouds reflect 80% of solar radiation, CO_2 traps the rest, so the surface temperature is about 450°C (850°F). We can be sure there is no life on Venus.

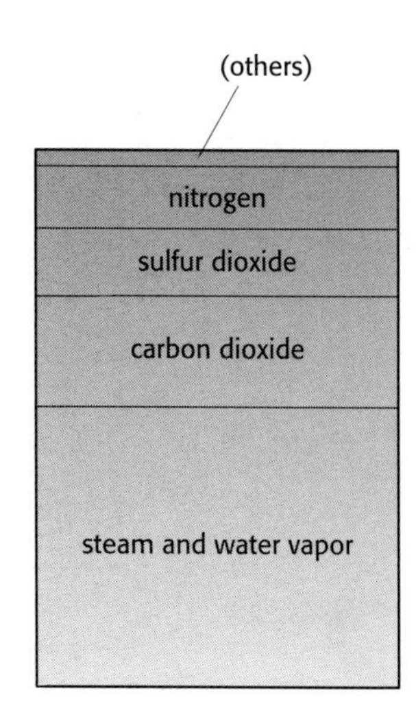

Figure 1.2 A planetary interior may be hot enough to melt rock. If molten rocks reach the planetary surface in volcanic eruptions, the gases they give off may help to form an atmosphere. The mixture of gases shown in this diagram was measured at Kilauea Volcano, in Hawaii; other volcanoes erupt different mixtures, but the basic ingredients are the same. Sulfur dioxide is caught up in raindrops, leaving water vapor, carbon dioxide, and nitrogen as the main constituents of the atmosphere.

Mars is much more interesting than Venus from a biological point of view. It is smaller than Earth, and farther from the Sun. But it is large enough to have held on to a thin atmosphere, mainly of CO_2. Mars today is cold, dry and windswept: dust storms sometimes cover half the planet.

No organic material can survive now on the surface of Mars. There is no liquid water, and the soil is highly oxidizing. But while Mars was still young, and was actively erupting volcanic gases from a hot interior, the planet may have had a thicker atmosphere with substantial amounts of water vapor. Cracks and crevices in the crust may still contain ice that could be set free as water, if large impacts heated the surface rocks deeply enough to melt it, or if climatic changes were to melt it briefly.

Mars had surface water in the distant past. Canyons, channels, and plains were shaped by huge floods, and other features look like ancient sandbars, islands, and shorelines. Ancient craters on Mars, especially in the lowland plains, have been eroded by gullies, and sheets of sediment lap around and inside them, sometimes reducing them to ghostly rims sticking out of the flat surface. However, the most recent summary of evidence indicates that Mars has always been cold and dry, with an occasional hot flash flood generated by impacts. The floods probably drained and dried very quickly, and there may never have been oceans.

Mars was too small to sustain geological activity for long. As the little planet cooled, its volcanic activity stopped (Figure 1.3). Its atmosphere and its water were blasted off by impacts, or lost by slow leakage to space, and by chemical reactions with the rocks and soil. The surface is now a dry frozen waste, and likely has been for three billion years. Even if floods were generated by a large meteorite impact, they could not last long enough to sustain life.

Did Mars once have life? In 1996, researchers reported they had found fossil bacteria in a meteorite that originated on Mars. (It was splashed into space by an asteroid impact, falling on Earth's Antarctic ice cap after spending thousands of years in space.) The researchers suggested that the bacteria were Martian. By now the evidence has been discounted: the objects are not bacteria and they are not evidence for life.

The asteroid belt lies outside the orbit of Mars. Some asteroids have had a complex geological history, but there is no question of life in the asteroid belt now. No planet or moon outside the orbit of Mars could trap enough solar radiation to form liquid water on its surface to provide the basis for life. Complex hydrocarbon compounds can accumulate and survive on asteroids, or in the atmospheres of the outer planets or on some of their satellites, but those bodies are frigid and lifeless.

Looking further afield, there is absolutely no evidence of life anywhere else in the Universe. The science-fiction writer Brian Aldiss thinks that the persistent search

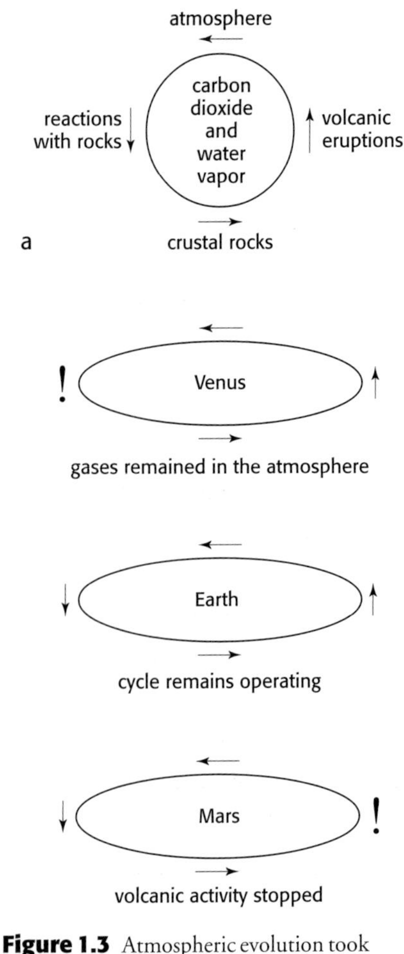

Figure 1.3 Atmospheric evolution took completely different courses on Venus, Earth, and Mars. On a model planet (a), water and carbon dioxide cycle around the rocks, the atmosphere, and the ocean. This still happens on Earth. Mars cooled so quickly that volcanic activity soon ceased, stopping the cycle with the gases frozen in the crust. Venus heated so much that carbon dioxide remained permanently in the atmosphere, stopping the cycle with a hot, dense atmosphere.

for such alien life reflects a deep-seated yearning for bogeymen deep in the human psyche.

So we return to Earth, the only known site of life. We can guess that impacts and eruptions released gases that formed a thick atmosphere around early Earth, consisting mainly of CO_2, with small amounts of nitrogen, water vapor, and sulfur gases (Figure 1.2). By about 4 billion years ago (4000 Ma or 4 Ga), but maybe as early as 4.4 Ga, Earth's surface was cool enough to have a solid crust, and liquid water that accumulated on it, forming oceans. Ocean water in turn helped to dissolve CO_2 out of the atmosphere and deposit it into carbonate rocks on the seafloor. This absorbed so much CO_2 that Earth did not develop runaway greenhouse heating as Venus did (Figure 1.4). Large shallow oceans probably covered most of Earth, with a few crater rims and volcanoes sticking out as islands.

Almost all geological evidence of these early times has been destroyed, especially by the catastrophic impacts around 3.9 Ga: the scenario of a cool watery Earth very early in its history is based on evidence from a few zircon crystals that survived as recycled grains in later rocks. But if there was early life on Earth, it would have been wiped out by the catastrophes at 3.9 Ga. The life forms that were our ancestors could not have evolved and survived until after the last sterilizing impact.

However, small late impacts may have encouraged the evolution of life. All comets and a few meteorites carry organic molecules, and comets in particular may have delivered some to Earth. But processes here on Earth also formed organic chemicals. Intense ultraviolet (UV) radiation from the young Sun acted on the atmosphere to form small amounts of very many gases. Most of these dissolved easily in water, and fell out as rain, making Earth's surface water rich in carbon compounds. The gases included ammonia (NH_3), methane (CH_4), carbon monoxide (CO), and ethane (CH_3), and formaldehyde (CH_2O) could have formed, at a rate of millions of tons a year. Nitrates built up in water as photochemical smog and nitric acid (HNO_3) from lightning strikes also rained out. But perhaps the most important chemical of all was cyanide (HCN). It would have formed easily in the upper atmosphere from solar radiation and meteorite impact, then dissolved in raindrops. Today it is broken down almost at once by oxygen, but early in Earth's history it built up at low concentrations in lakes and oceans. Cyanide is a basic building block for more complex organic molecules such as amino acids and nucleic acid bases. Life probably evolved in chemical conditions that would kill us instantly!

Life Exists in Cells

The simplest cell alive today is very complex: after all, its ancestors evolved through many billions of generations. We must try to strip away these complexities as we wonder what the first living cell might have looked like and how it worked.

A **living thing** has several properties: it has organized structure; the capacity to reproduce (replicate itself); stored information; behavior; and metabolism. Mineral crystals have the first two but not the last two.

First, a living thing **has a boundary** that separates it from the environment. As we shall see, a living thing operates its own chemical reactions, and if it did not have a boundary those reactions would be unable to work: they would be diluted by outside water or compromised by outside contaminants. So we refer to a living "**cell**" that has some sort of **cell membrane** or **cell wall** around it.

Second, there must be instructions that define the structure, the timing, and the ingredients for the chemical reactions needed to produce (and maintain) the organization of the cell. In all living cells today, that **information** is coded in nucleic acids (**RNA** or **DNA**), the instructions are carried round the cell by RNA, and proteins are involved in most of the reactions.

Third, a living thing can grow, and it can **replicate**: that is, it can make another structure just like itself. Both processes require complex chemistry which in turn requires a cell membrane. Growth and replication use materials that must be brought in from outside, through the cell wall.

Fourth, a living thing interacts with its environment in an active way: it has **behavior**. The simplest behavior is the activity involved with growth and reproduction: the chemical flow of substances in and out of the cell is an interaction with the outside world that can be turned on and off. The chemical flow will change the immediate environment, and, of course, the presence or absence of the desired chemicals will decide whether the cell turns on the flow. Temperature and other outside conditions will also affect the behavior of even the simplest cell.

Fifth, the chemical activity of the cell represents an energy flow: the energy flow is called **metabolism** in living things. The cell must operate reactions that synthesize molecules from simpler precursors, or break down complex molecules into simpler ones. If a cell grows or reproduces, it is building complex organic molecules, and those reactions need energy. The cell must obtain that energy from outside, in the form of radiation or "food" molecules that it can break down.

It's important to remember that these five attributes of a living cell are not five different things: they are all intertwined. They are all connected with the processes of gathering and processing energy and material into new chemical compounds (tissues), and continuing those processes into new cells. Any theory of the evolution of life, as opposed to its creation by a Divine Being, must include a period of time during which lifeless molecules developed the characters listed above and thereby became living. The phrase for this is **chemical evolution**. We have to be able to argue that every step in the process could reasonably have happened on early Earth (or somewhere else) in a natural, spontaneous way. It's easy to see that with the right starting compounds, a protocell could grow effectively. The critical turning point that defines life comes when "accurate" replication evolves.

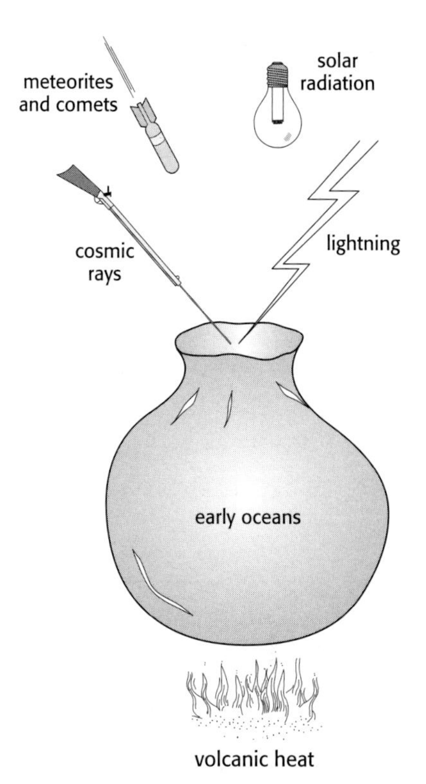

Figure 1.4 Some of the energy sources that would have been available to power chemical reactions on early Earth. (After Cowen, *History of Life*. © 1976 McGraw-Hill Book Company.)

Even with a time machine, we would find it very difficult to pick out the first living thing from the mass of growing organic blobs that must have surrounded it. But that cell survived and reproduced, however "accurately," and as time went by, cells that were more efficient remained alive and replicated, while those that were less efficient died or replicated more slowly. So at the same time that living things emerged, so did the processes of natural selection and extinction. Some lines of cells flourished, others became extinct. Of course, we do not see any descendants of those first cells living today with the same genetic and biochemical machinery their ancestors had: they have long had major upgrades of their original software.

That introduces one other concept into our discussion: **progress** or **improvement**. There is no question that the simplest living cells are more efficient than their distant ancestors. Arguments rage about the politically correct word to use to describe this. I used "improvement" in previous editions because many biologists have a phobia about the word "progress" (there are too many connotations associated with military and industrial technology, I suspect). But the reality of the fossil record is that here are many examples of improved performance that can be assessed mechanically. Thus modern horses run far more efficiently, living whales swim more efficiently, and living birds fly more efficiently than their ancestors did. I don't

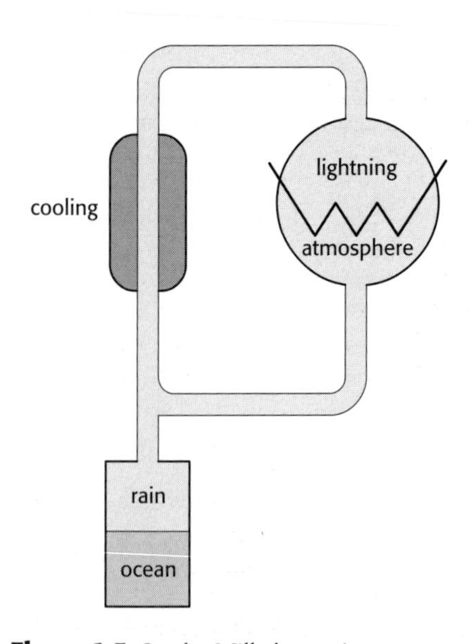

Figure 1.5 Stanley Miller's experiment was designed to simulate conditions on early Earth. An atmosphere of methane, ammonia, and hydrogen was subjected to lightning discharges, and the reaction products cooled, condensed, and rained out to collect in the ocean. The reaction products included amino acids.

see that one could doubt that similar trends have occurred in physiology, biochemistry, reproduction, and so on, though it would be more difficult to prove it. I can't think of a better word to describe this than "progress."

Now we turn to experiments that help us to see the steps by which life evolved from nonliving chemicals. In Chapter 2, we look at the rock record to try to find evidence of the earliest life on Earth: in particular, we look for evidence of structured fossils, and for any trace of their behavior, or chemical traces of the metabolic reactions they performed.

Making Organic Molecules

Certain conditions are necessary if life is to evolve from nonlife (Box 1.1). The first experiment to include some of them was published in 1953 by Stanley Miller, who was then a graduate student at the University of Chicago. He passed energy (electric sparks) through a mixture of hydrogen, ammonia, and methane in an attempt to simulate likely conditions on early Earth (Figure 1.5). Any products fell out into a protected flask. Among these products, which included cyanide and formaldehyde, were amino acids. This result was surprising at the time because amino acids are not simple compounds. Miller's experiment used a rather unlikely mixture of starting gases, but it encouraged many other experiments on the origin of life, and showed that such experiments were not only good science but were very exciting and rewarding.

Many experiments have shown that most of the amino acids found in living cells today could have formed naturally on early Earth, from a wide range of ingredients, over a wide range of conditions. They form readily from mixtures that include the gases of Earth's early atmosphere. The same amino acids that form most easily in laboratory experiments are also the most common in living things today. The only major condition is that amino acids do not form if oxygen is present.

Amino acids form easily on the surfaces of clay particles (Figure 1.6). Clay minerals are abundant in nature, have long linear crystal structure, and are very good at attracting and adsorbing organic substances: cat litter is clay and works on this principle.

Organic molecules occur in comets and meteorites. Again, the most common ones are also the most abundant in laboratory experiments. Sugars and nucleic acid bases are also found in meteorites, so all these compounds would have been supplied in quantity to early Earth. We do not know how much organic matter was supplied by natural synthesis on Earth and how much by comets and meteorites. Either way, the right materials were present on early Earth to encourage further reactions.

Linking sequences of amino acid molecules into chains to form protein-like molecules involves the loss of water, so scientists have tried evaporation experiments to

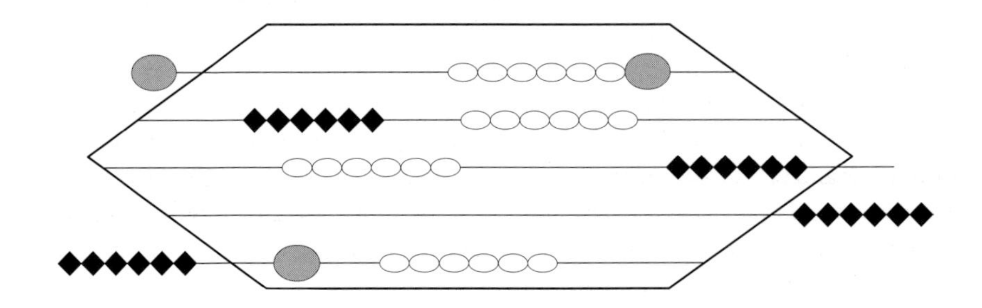

Figure 1.6 Clay minerals have long, straight cleavage planes. Linear organic molecules may line up along the cleavages, encouraging reactions that form long-chain organic molecules such as amino acids and nucleic acids.

BOX 1.1 Necessary Conditions for Chemical Evolution Toward Life

Energy. Energy is needed to form complex organic molecules in the absence of life, given the presence of simple precursors. Some laboratories have used electrical discharges to simulate lightning on early Earth; others use high-energy particles from a cyclotron in place of radioactivity from rocks and cosmic rays, heat for volcanic activity, shock waves or laser beams for meteorite impacts, or lamps for solar UV radiation (Figure 1.4). All these energy sources were present on the early Earth.

Protection. Continued energy input (especially heat) will destroy any complex organic molecules that form in reactions, so after they form they must quickly be protected from strong radiation. Laboratory experiments are often designed to allow organic molecules to drop into cold water away from the energy source. On early Earth, molecules may have been protected under shallow water, in sheltered tide pools or rock crevices, under rocks, ice, or particles of sediment.

Concentration. All chemical reactions run better at high concentration, but almost all reactions leading toward life give low yields in the laboratory. Life is water-based, yet too much water dilutes chemicals so that they react slowly. Some process must have concentrated chemicals on early Earth. Evaporation is the simplest one, but there are others such as absorption on to clay surfaces.

Catalysis. Catalytic converters in the exhaust systems of cars contain platinum as a catalyst that encourages the breakdown of pollutants. An organic substance that works as a catalyst is called an enzyme. All the reactions inside our cells and our bodies are aided by enzymes, which are necessary even in the simplest possible living cell. Suitable catalysts may have encouraged difficult reactions on early Earth, even at low energy levels and low concentrations. Later, the last stages leading toward life may have been aided by catalysts trapped on, or inside, membranes.

simulate the process under early Earth conditions. High temperatures help evaporation, but they also pose a problem, because organic molecules tend to break down if they are heated: the longer the molecule, the more vulnerable it is to damage. Here again, natural minerals provide an attractive alternative.

Nucleic acids (RNA and DNA) have structures made up of nucleic-acid bases, sugars, and phosphates. All four nucleic-acid bases can be made in reasonable laboratory experiments. Sugars form in experiments that simulate water flow from hot springs over clay beds. Naturally occurring phosphate minerals are associated with volcanic activity. Thus all the ingredients for nucleic acids were present on early Earth, and the cell fuel ATP (adenosine triphosphate) could also have formed easily.

Linking sugars, phosphates, and nucleic-acid bases to form fragments of nucleic acid called **nucleotides** is also a dehydration process, and the phosphates themselves can act as catalysts here. Long nucleotides form much more easily on phosphate or clay surfaces than they do in suspension in water (Figure 1.6).

Toward the First Living Cell

The basic organic molecules that make up cell membranes and cell contents may have been present in reasonable amounts in the oceans of early Earth. We must still explain how they evolved into a cell that could reproduce itself.

A membrane separates a cell from the environment. Many organic membranes are made of sheets of molecules called **lipids**. A lipid molecule has one end that attracts water and one end that repels water (Figure 1.7a). Lipids line up naturally with heads and tails always facing in opposite directions; a sheet of lipid molecules therefore repels water (Figure 1.7b). If the sheet of lipids happens to fold around to meet itself, it forms a waterproof membrane around whatever contents it has trapped. Such packets, called liposomes, form spontaneously in lipid mixtures.

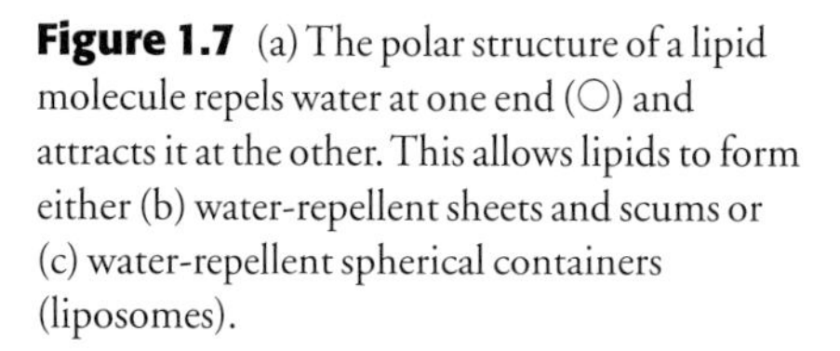

Figure 1.7 (a) The polar structure of a lipid molecule repels water at one end (○) and attracts it at the other. This allows lipids to form either (b) water-repellent sheets and scums or (c) water-repellent spherical containers (liposomes).

They are simple spheres with an outer membrane of lipids (Figure 1.7c). Whipping up an egg in the kitchen produces liposomes as the yolk is frothed around.

David Deamer (see Zimmer, 1995) discovered that liposomes can form from molecules that would have been present on early Earth: fatty acids, glycerol, and phosphates. Later, he found fatty acid molecules in meteorites, and made globules from them by drying them out and then rewetting them. When he added DNA to the original solution, it was sometimes trapped inside the liposomes as they formed, sometimes concentrated 100 times. Jack Szostak's research group (see Szostak et al., 2001) found that fatty acid globules form 100 times as fast as usual if clay is added to the mixtures.

This suggests that the formation of liposomes that had cell-like contents may not have been difficult on early Earth. They could have formed in great numbers as waves thrashed around lipid layers on water surfaces, or as lipid scums washed up on a muddy shore with clays in the water. These processes could have formed liposomes with greatly variable contents (some with amino acids, primitive forms of nucleic acid, and so on). The "best" ones would have operated chemical reactions much more efficiently than the "worst."

Chemicals react faster and more efficiently if they are concentrated. Four concentration mechanisms could have occurred naturally: evaporation; freezing; concentration in scums, droplets, or bubbles; and concentration on the surfaces of mineral grains.

"Naked Genes" in an RNA World

In living cells today, each protein is coded on long sequences of nucleic acid. The long sequences of DNA that specify these protein structures are themselves difficult to replicate, and replication requires many proteins to act as enzymes to catalyze the reactions. Protein synthesis and DNA replication are interwoven in cells today: they depend on one another, even though they use very different chemical pathways and probably evolved independently. How could these two processes have begun independently, then evolved to depend on each other?

The answer lies with the simpler nucleic acid, RNA. Some RNA sequences called **ribozymes** can act as enzymes and make more RNA even when no proteins are present. Other RNA sequences speed up the assembly of proteins. Perhaps the first living things were pieces of RNA, ribozymes caught up inside liposomes, which happened to have the right structure to act as enzymes that helped the RNA to replicate itself. In theory, RNA ribozymes on early Earth could have replicated themselves without proteins, slowly and inaccurately, and therefore could have been considered alive. The ribozymes are sometimes called "naked genes," but in reality they would have been inside liposomes, as described above.

Ribozymes that by chance had RNA that coded for protein enzymes would have replicated faster than other ribozymes. Increasingly successful ribozymes would very quickly have outcompeted all others to become the ancestors of all later life on Earth: I shall call them **protocells** from now on. The scenario that begins with ribosomes in an RNA world is currently the best hypothesis for the origin of life on Earth.

RNA is simpler than DNA, and likely evolved before DNA. But even simpler types of nucleic acid than RNA have been made in the laboratory, though they are not used now by living cells. They provide hints that the first living things may have evolved even before the RNA world. Molecules named PNA and TNA are now being used in experiments to see if they could have formed the basis for a living thing. PNA forms more easily in early Earth experiments than RNA does. In 2002, DNA was made in the laboratory from TNA.

Even the first genetic code might have been simpler than today's: experiments in 2002 showed that ribozymes can reproduce themselves even if they have only two nucleic acid bases, rather than the four found in RNA today.

WHERE DID LIFE EVOLVE?

Many different habitats have been seriously suggested as the environment in which life began (Box 1.2). However, some are less likely than others. **Soil surfaces** would not attract the quantity of organic material that would be available in water. **Interstellar space** and the **atmosphere** are too dry.

> **BOX 1.2 Possible Habitats for the Origin of Life**
>
> - Soils;
> - the upper atmosphere;
> - space;
> - lakes or lagoons;
> - glacial volcanic islands;
> - deep sea rifts.

Most theories of the origin of life suggest surface or shoreline habitats in lakes, lagoons, or oceans. But it's unlikely that life evolved in the sea. Complex organic molecules are vulnerable to damage from the sodium and chlorine in seawater. Most likely, then life evolved in **lakes**, or in **seashore lagoons** that were well supplied with river water. We have come to think of lagoons as tropical: the very name conjures up blue water and palm trees. Warm temperatures promote chemical reactions, and an early tropical island would most likely have been volcanic and therefore liable to have interesting minerals. But RNA bases are increasingly unstable as temperatures rise above 0°C: normal tropical water (25°C) may be about as warm as it could be for the origin of life.

So perhaps **cold volcanic islands** were the best environments favoring organic reactions on early Earth. In the laboratory, cyanide and formaldehyde reactions occur readily in half-frozen mixtures. Volcanic eruptions often generate lightning storms, so eruptions, lightning, fresh clays, and near-freezing temperatures (ice, snow, hailstones) could all have been present on the shore of a cold volcanic island.

Note that if this environment is the correct one, there had to have been land *and* sea when life evolved: fresh water can only occur on Earth if it is physically separated from the ocean.

Solar radiation or atmospheric phenomena are likely energy sources for the reactions leading toward life. But deep in the oceans lie the mid-ocean ridges, long underwater **volcanic rifts** where the seafloor is tearing apart and forming new oceanic crust. Enormous quantities of heat are released in the process, much of it through hot water vents on the floors of the rifts, and myriads of bacteria flourish in the hot water. Perhaps life began nowhere near the ocean surface, but deep below it, at volcanic vents.

Laboratory experiments have implied that amino acids and other important molecules can form in such conditions, even linking into short protein-like molecules, and currently the deep-sea hypothesis is popular. But if life evolved by way of naked genes, then it did not do so in hot springs. RNA and DNA are unstable at such high temperatures. Naked genes could not have existed (for long) in hot springs.

The deep-sea hypothesis, even though it looks unlikely (to me), has led to speculation that life might have evolved deep under the surface layers of other planets or satellites. (For example, Jupiter's moon Europa probably has liquid water under its icy crust.) The speculation helps to generate money for NASA's planetary probes. But the internal energy of such planets and moons is very low, and water-borne organic reactions are much less likely to work deep under the icy crust of Europa than in Earth's oceans. In any case, the under-ice oceans of icy moons are salty (that's how they were detected), so an origin of life is very unlikely in such environments.

New experiments are producing organic chemicals in conditions that simulate ices forming on dust grains in the freezing near-vacuum of space; in other words, on comets. Even so, the ices have to be thawed out to a water-based chemistry to react further. Processes in space may generate chemicals and deliver them to planets; but if a planet is hospitable, the right conditions for generating organic molecules and life already existed on that planet.

ENERGY SOURCES FOR THE FIRST LIFE

We have seen that living things use energy. Much of biology consists of studying metabolism and ecology: the ways in which living things acquire and use the energy they need to grow and reproduce.

The earliest cells likely evolved in a watery environment that contained large quantities of naturally formed organic molecules. So the first protocells had energy available to them in the form of ATP, amino acids, and other organic compounds that they could absorb from water. Those compounds had been accumulating for a long time, and they all have chemical energy stored in them, especially in the bonds between hydrogen and carbon atoms. If early protocells had the enzymes to break those bonds, the molecules would have provided plenty of fuel for cell growth and replication. But as protocells became more numerous and more effective in attracting and using organic molecules, there must have come a time when demand exceeded supply. As simple organic molecules became scarcer and scarcer, protocells encountered the world's first energy crisis. Paradoxically, this would have happened first in environments where protocells were most successful and abundant. Two very different reactions to a shortage of "food" can still be seen among living organisms nearly 4 billion years later.

Living organisms gain energy in two ways: **heterotrophy** and **autotrophy**. Heterotrophs obtain their metabolic energy by breaking down organic molecules they absorb from the environment: hummingbirds sip nectar and humans eat doughnuts. Heterotrophs do not pay the cost of building the organic molecules, they just have to operate the reactions that break them down: but they must live in an environment in which there are "food" molecules. Autotrophs do not need food molecules from outside: they make them inside the cell, typically paying the cost of building the molecules by absorbing *energy* from outside. Since they then break down the molecules again for growth and replication, they must operate in an environment that gives them outside energy.

Were the first cells heterotrophic or autotrophic? One can argue either case. Whichever is true, autotrophy and heterotrophy must have been exceedingly early developments. I argue that heterotrophy evolved first. Autotrophic cells must be able to break down the molecules they build, and they do so using the same universal biochemical reactions that heterotrophic cells use. It is easier to argue that existing heterotrophs added photosynthesis than to argue that the first cells evolved photosynthesis *and* heterotrophic breakdown reactions at the same time.

Heterotrophy and Fermentation

The first heterotrophic cells would naturally have used the simplest possible reactions to break down organic molecules. These are **fermentation** reactions, in which cells break down sugars such as **glucose**. Glucose is often called the universal cellular fuel for living organisms, and it was probably the most abundant sugar available on early Earth. [Humans use fermenting microorganisms to produce beer, cheese, vinegar, wine, tea, and yogurt, and fermenting organisms break down much of our sewage.]

As heterotrophs used up the easiest molecules to break down, there would have been intense competition among them to break down more difficult ones. One can imagine a huge advantage for cells that could break down a molecule that was not available to other cells. A whole set of fermentation reactions would quickly have developed. In turn that would have worsened the energy crisis, because heterotrophs would at first have been limited to the molecules that formed naturally in the atmosphere and ocean.

Autotrophy: Lithotrophy and Photosynthesis

Autotrophs generate their own energy, but in two completely different ways. They may extract chemical energy from inorganic molecules (**lithotrophy**), or gain energy by trapping radiation (**photosynthesis**).

Lithotrophy can occur when a microorganism rips an oxygen molecule off one inorganic compound and transfers it to another, making an energy profit in the process. That energy is then used to build organic food molecules. For example, microorganisms called **methanogens** gain energy from lithotrophy by breaking up carbon dioxide and transferring the oxygen to hydrogen, forming water and methane as by-products:

$$4H_2 + CO_2 \rightarrow CH_4 + 2H_2O + \text{energy}$$

Methanogens are as different from true bacteria as bacteria are from us, and are part of a special group of microorganisms, the **Archaea**. Since carbon dioxide and hydrogen would have been available in the early ocean, it is reasonable to suggest that this reaction could have been operated by very early cells. Indeed, based on their molecular genetics, Archaea were among the first living things on Earth.

If this ability evolved very early, it may have been the first time (but not the last) that living things modified Earth's chemistry and climate. By replacing the greenhouse gas carbon dioxide with the even more powerful greenhouse gas methane, the activity of methanogens might have warmed early Earth (Chapter 2).

Photosynthesis is simple in concept, but biochemically more complex than lithotrophy. Some molecules can absorb light and store it as energy in their structure. An early microorganism that happened to have such molecules could have captured light energy and used it to build up food molecules such as sugars.

Living things use **porphyrins** as the most important light-trapping molecules, and they could have formed from simpler substances on early Earth. **Chlorophylls** of various kinds are the porphyrins most widely used by living organisms to trap light: complex biochemical reactions have evolved to release and use that energy.

> **LIMERICK 1.1**
> Their bacterial plight was pathetic
> It's hard to be unsympathetic:
> Volcanic heat diminished,
> Organic soup finished,
> Their solution was photosynthetic.

The evolution of photosynthesis produced major ecological changes on Earth. Immediately, the energy trapped by chlorophyll was used to build more **biomass** (biological substance). Photosynthetic cells now had an energy store, a buffer against times of low food supply, that could be used as needed. It's easy to see how such cells could come to depend almost entirely on photosynthesis for energy: they did not have to compete so directly with heterotrophs. In many habitats, sunlight is a richer and more reliable energy resource than organic matter that must be sought and captured. Then, as photosynthesizers died and their cell contents were released into the environment, they inadvertently provided a new source of nutrition for heterotrophs. Photosynthesis dramatically increased the energy flow through biological systems on Earth, and for the first time considerable amounts of energy were being transferred from organism to organism, in Earth's first true ecosystem.

The earliest photosynthetic cells probably used hydrogen from H_2 or H_2S. For example, the reaction:

$$H_2S + CO_2 + \text{light} \rightarrow (CH_2O) + 2S$$

released sulfur into the environment as a by-product of photosynthesis. Later, some bacteria began to break up the strong hydrogen bonds of the water molecule. The step might first have been an act of desperation in a sulfur-poor environment, but the bacteria that successfully broke down H_2O rather than H_2S, like this:

$$H_2S + CO_2 + \text{light} \rightarrow (CH_2O) + 2O$$

immediately gained access to a much more plentiful resource. There was a penalty, however. The waste product of H_2S photosynthesis is sulfur (S), which is easily disposed of. The waste product of H_2O photosynthesis is an oxygen radical, monatomic oxygen (O), which is a deadly poison to a cell because it can break down vital organic molecules by oxidizing them. Even for humans, it is dangerous to breathe pure oxygen or ozone-polluted air for long periods.

Cells needed a natural antidote to this oxygen poison before they could operate the new photosynthesis consistently. **Cyanobacteria** were the organisms that made the first breakthrough to oxygen photosynthesis using water. They used a powerful antioxidant enzyme called superoxide dismutase to prevent the O from damaging them: essentially, the enzyme packaged up the O into less dangerous O_2 that was ejected out of the cell wall into the environment.

From then on, we can imagine early communities of bacteria made up of autotrophs and heterotrophs, evolving improved ways of gathering or making food molecules.

Photosynthesizers need nutrients such as phosphorus and nitrogen to build up their cells, as well as light and CO_2. In most habitats, the nutrient supply varies with the seasons, as winds and currents change during the year. Light, too, varies with the seasons, especially in high latitudes. Since light is required for photosynthesis, great seasonal fluctuations in the primary productivity of the natural world began with photosynthesis. Seasonal cycles still dominate our modern world, among wild creatures and in agriculture and fisheries.

We can now envisage a world with a considerable biological energy budget and large populations of microorganisms: Archaea, photosynthetic bacteria, and heterotrophic bacteria. So there is at least a chance that a paleontologist might find evidence of very early life as fossils in the rock record. In Chapter 2 we shall look at geology, rocks, and fossils, instead of relying on reasonable but speculative arguments about Earth's early history and life.

Further Reading

Planets

Buseck, P. R., et al. 2001. Magnetite morphology and life on Mars. *Proceedings of the National Academy of Sciences* 98: 13490–5. [Removes the last good evidence for life on Mars.]

Christensen, P. R. 2003. Formation of recent Martian gullies through melting of extensive water-rich snow deposits. *Nature* 422: 45–8. [But see Treiman, 2003.]

Chyba, C. F., and K. P. Hand. 2001. Life without photosynthesis. *Science* 292: 2026–7. [Under the ice of Europa, short version.]

Chyba, C. F., and C. B. Phillips. 2002. Europa as an abode of life. *Origins of Life and Evolution of the Biosphere* 32: 47–68. [Same, longer version.]

Gaidos, E. J., et al. 1999. Life in ice-covered oceans. *Science* 284: 1631–3. [Life under the ice of Europa is unlikely.]

Jacobson, S. B. 2003. How old is planet Earth? *Science* 400: 1513–14.

Kerr, R. A. 2003. Iceball Mars? *Science* 300: 234–6. [Latest conference on Martian climate, present and past.]

Malin, M. C., and K. S. Edgett. 2000. Sedimentary rocks of early Mars. *Science* 290: 1927–37.

Schilling, G. 1999. From a swirl of dust, a planet is born. *Science* 286: 66–8. [Current theories of planetary formation].

Segura, T. L., et al. 2002. Environmental effects of large impacts on Mars. *Science* 298: 1977–80. [Dry cold planet, hot flash floods.]

Treiman, A. H. 2003. Geologic settings of Martian gullies: implications for their origins. *Journal of Geophysical Research* 108: 8031–43. [The gullies are formed by dry avalanches, not water.]

Wood, J. A. 1999. Forging the planets: the origin of our Solar System. *Sky & Telescope* 97 (1): 36–48.

Origin of Life

Aldiss, B. W. 2001. Desperately seeking aliens. *Nature* 409: 1080–2. [People seem to want bogeymen.]

Bada, J. L., and A. Lazcano. 2002. Some like it hot, but not the first biomolecules. *Science* 296: 1982–3. [Cool environments.]

Bada, J. L., and A. Lazcano. 2003. Prebiotic soup — revisiting the Miller experiment. *Science* 300: 745–6.

Cohen, J., and I. Stewart. 2001. Where are the dolphins? *Nature* 409: 1119–22.

Cooper, G., et al. 2001. Carbonaceous meteorites as a source of sugar-related organic compounds for the early Earth. *Nature* 414: 879–83, and comment, 857–8.

Dobson, C. M., et al. 2000. Atmospheric aerosols as prebiotic chemical reactors. *Proceedings of the National Academy of Sciences* 97: 11864–8. [Aerosols, not as sites for forming life, but as chemical reactors.]

Dworkin, J. P., et al. 2001. Self-assembling amphiphilic molecules: synthesis in simulated interstellar/precometary ices. *Proceedings of the National Academy of Sciences* 98: 815–19.

Greenberg, J. M. 2000. The secrets of stardust. *Scientific American* 283 (6): 70–5.

Hanczyc, M. M., et al. 2003. Experimental models of primitive cellular compartments: encapsulation, growth, and division. *Science* 302: 618–622, and comment, 580–1. [New experiments by the Szostak group showing that clays catalyze liposome formation.]

Hazen, R. M. 2001. Life's rocky start. *Scientific American* 284 (4): 76–85. [Glitzy.]

Joyce, G. F. 2002. Booting up life. *Nature* 410: 278–9. [Computer models show that self-replicating might be easier than we had thought.]

Joyce, G. F. 2002. The antiquity of RNA-based evolution. *Nature* 418: 214–21. [RNA world.]

Kintisch, E. 2001. Is life that simple? *Discover* 22 (4): 66–71. [If this doesn't scare you, you're not paying attention!]

Levy, M., and S. L. Miller. 1998. The stability of the RNA bases: implications for the origin of life. *Proceedings of the National Academy of Sciences* 95: 7933–8. [RNA bases do not survive well much above 0°C.]

Lunine, J. I. 2001. The occurrence of Jovian planets and the habitability of planetary systems. *Proceedings of the National Academy of Sciences* 98: 809–14.

Miller, S. L. 1953. A production of amino acids under possible primitive earth conditions. *Science* 117: 528–9. [The paper that started it all: fifty years old and only two pages long!]

Miyakawa, S., et al. 2002. Prebiotic synthesis from CO atmospheres: implications for the origins of life. *Proceedings of the National Academy of Sciences* 99: 14628–31. [Adding carbon monoxide to Miller's experiments yields more biomolecules.]

Miyakawa, S., et al. 2002. The cold origin of life. *Origins of Life and Evolution of the Biosphere* 32: 195–218. [Argues strongly for formation of complex organic compounds in freezing conditions.]

Monastersky, R. 1998. The rise of life on Earth. *National Geographic* 193 (3): 54–81.

Nisbet, E. G., and N. H. Sleep. 2001. The habitat and nature of early life. *Nature* 409: 1083–91. [Outstanding summary.]

Pace, N. R. 2001. The universal nature of biochemistry. *Proceedings of the National Academy of Sciences* 98: 805–8. [Very readable review.]

Reader, J. S., and G. F. Joyce. 2002. A ribozyme composed of only two different nucleotides. *Nature* 420: 841–4.

Schšning, K.-U., et al. 2000. Chemical etiology of nucleic acid structure: the α-threofuranosyl-(3′2′) oligonucleotide system. *Science* 290, 1347–51, and comment, 1306–7. [TNA.]

Shock, E. L. 2002. Seeds of life? *Nature* 416: 380–1. [Amino acids can form in interstellar space. Do we care? (No.)]

Szostak, J. W., et al. 2001. Synthesizing life. *Nature* 409: 387–90. [Report on the state of the art.]

Valley, J. W., et al. 2002. A cool early Earth. *Geology* 30: 351–4.

Walter, N. G., and Engelke, D. R. 2002. Ribozymes: catalytic RNAs that cut things, make things, and do odd and useful jobs. *Biologist* 49: 199–203.

Ward, P. D., and D. Brownlee. 2000. *Rare Earth: Why Complex Life is Uncommon in the Universe.* New York: Copernicus Books.

Zimmer, C. 1995. First cell. *Discover* 16 (11): 71–8. [David Deamer's work on liposomes and the origin of cells.]

CHAPTER TWO

Earth's Earliest Life

When we move from astronomy and the laboratory to Earth itself to search for evidence about early life, we look for fossils. A **fossil** is the remnant of an organism preserved in the geological record. There are three kinds of fossils, **body fossils**, **trace fossils**, and **chemical fossils**. We are most familiar with body fossils, in which part or all of an organism is preserved. If an organism had body parts that were made of resistant materials (for example, shells, bones, or wood), it is much more likely than a "soft-bodied" creature to be preserved in the geological record. Such fossils may look more or less unchanged after death. Minerals may crystallize out of ground water to fill up large or small cracks, crevices, and cavities in the original substance, so body fossils may be denser and harder than they were in life. Sometimes the original shell or bone may be replaced by another mineral, making the fossil easier to recognize or easier to extract from the rock (Figure 2.1).

Obviously, the hard parts of an organism are more likely to be preserved than more fragile parts. But occasionally soft parts may leave an impression on soft sediment before they rot. Even more rarely, a complete organism may be encased in soft sediment that later hardens into a rock. Bees, ants, flies, and frogs have been preserved as fossils in **amber** (fossilized tree resin, Figure 2.2), and individual cells have been preserved in **chert**, a rock formed from silica gel that impregnated the cells and retained their shapes in three dimensions.

A **trace fossil** is not part of an organism at all, but it was made by an organism and therefore may tell us something about that creature. Trace fossils may be marks left by active organisms (footprints, trails, burrows [Figure 2.3]), or fecal masses, or even a spider's web (Figure 2.4). Trace fossils may give us insight into behavior that would not be available from a body fossil. (For example, although dinosaur skeletons suggest that some of them *could have* run, trace fossils of dinosaur footprints tell us that some of them certainly *did* run [Chapter 12].)

Chemical fossils are compounds produced by organisms and preserved in the rock record. They may be molecules that were originally part of the organism, or molecules that were produced in the metabolic processes the organism operated. They may provide information about the organisms that produced them. In the special cases where an organism selects one isotope of an atom over another, the chemical fossils of these isotopes can be used to give information too, as described in a section later in this chapter.

All kinds of agents may destroy or damage organisms beyond recognition before they can become fossils or while they are fossils. After death the soft parts of

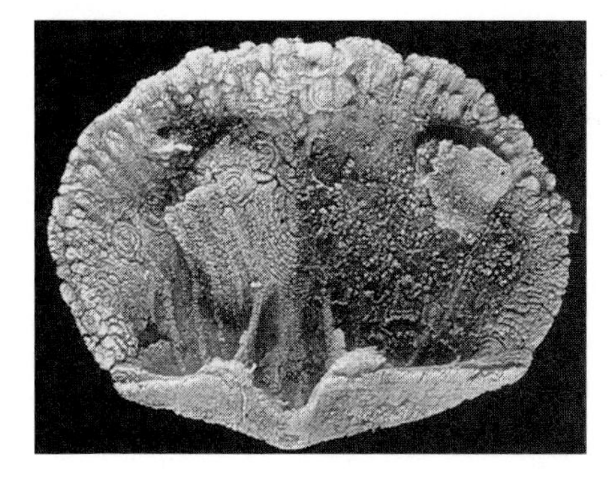

Figure 2.1 A brachiopod whose original calcite shell was replaced by silica. This made it fairly easy to dissolve the shell out of limestone for study.

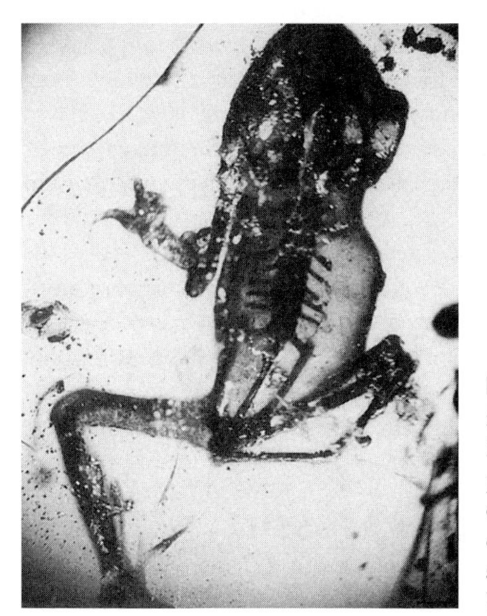

Figure 2.2 A frog preserved in amber. The frog is almost undamaged externally, though internally it has broken bones. It was probably seized by a predator (a hawk or an owl?) and carried to a tree where it was covered in resin and preserved in this spectacular fashion. (© George Poinar, University of California, Berkeley.)

Figure 2.3 Trace fossils like these were made in Nebraska during the Miocene by (large!) burrowing rodents, which occasionally left their bones at the bottom. The National Park Service provided the photograph and the ranger for scale.

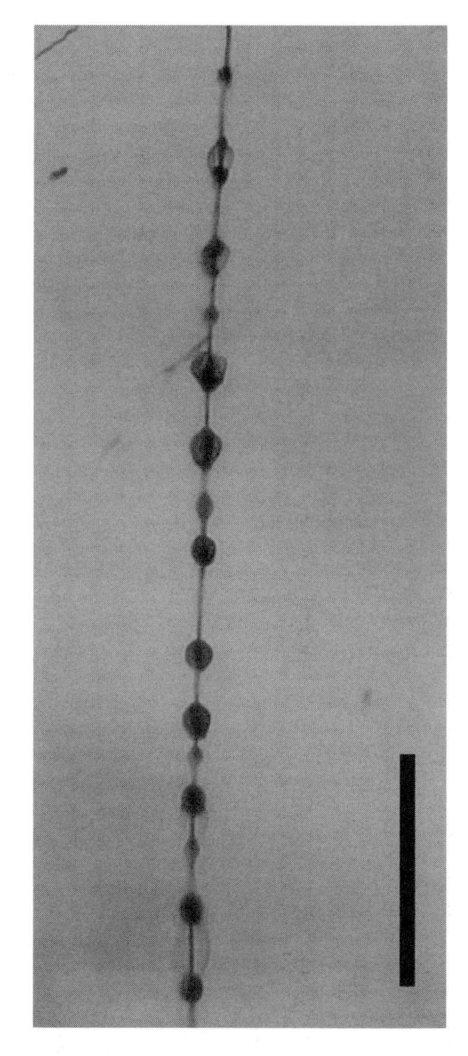

Figure 2.4 Spider silk that was preserved in amber for about 130 million years (from the Lower Cretaceous of Lebanon). The droplets of insect-trapping glue along the silk are like those produced today by araneoid spiders, so altogether this trace fossil tells us a lot about its maker. Scale bar 0.5 millimeters. This and other images were published in Zschokke (2003). (Courtesy of Samuel Zschokke, University of Basel. © Samuel Zschokke).

organisms may rot or be eaten. Any hard parts may be dissolved by water, or broken or crushed and scattered by scavengers or by storms, floods, wind, and frost. Remains must be buried to become part of a rock, but a fossil may be cracked or crushed as it is buried. After burial, groundwater seeping through the sediment may dissolve bones and shells. Earth movements may smear or crush the fossils beyond recognition or may heat them too much. Even if a fossil survives and is eventually exposed at Earth's surface, it is very unlikely to be found and collected before it is destroyed by weathering and erosion.

Even when they are studied carefully, fossils are a very biased sample of ancient life. Fossils are much more likely to be preserved on the seafloor than on land. Even on land, animals and plants living or dying by a river or lake are more likely to be preserved than those in mountains or deserts. Different parts of a single skeleton have different chances of being preserved. Animal teeth, for example, are much more common in the fossil record than are tail bones and toe bones. Teeth are usually the only part of sharks to be fossilized. Large fossils are usually tougher than small ones and are more easily seen in the rock. Spectacular fossils are much more likely to be collected than apparently ordinary ones. Even if a fossil is collected by a professional paleontologist and sent to an expert for examination, it may never be studied. All the major museums in the world have crates of fossils lying unopened in the basement or the attic.

When we look at museum display cases, it seems that we have a good idea of the history of life. But most of the creatures that were living at any time are not in a museum. They were microscopic or soft-bodied, or both, or they were rare or fragile and were not preserved, or they have not been discovered. We do have enough evidence to begin to put together a story. But that story is always changing as we discover new fossils and look more closely at the fossils we have found already.

HOW DO WE KNOW THE AGE OF A FOSSIL?

Fossils are found in rocks, and usually geologists try to establish the age of the containing rock or of a layer of rock that is not far under or over the fossil (and so might be close to it in age). The age of rocks is measured in two different ways, known as relative and absolute dating.

Age dating of rocks can only work if one identifies components of the rocks that change with time, or are in some way characteristic of the time at which the rocks formed. The same principles are used in dating archeological objects. Coins may bear a date in years (absolute dating), and one can be certain that a piece of jewelry containing a gold coin could not have been made before the date stamped on the coin. The age of waste dumps can be gauged by the type of container thrown into them: bottles with various shapes and tops, steel cans, aluminum cans, and so on. The age of old photographs, movies, or paintings can often be judged by the dress or hairstyle of the people or the cars shown in them.

Absolute geological ages can be determined because newly formed mineral crystals sometimes contain unstable, radioactive, atoms. Radioactive isotopes break down at a rate that no known physical or chemical agent can alter, and as they do so they may change into other elements. For example, potassium-40, ^{40}K, breaks down to form argon-40, ^{40}Ar. By measuring the amount of radioactive decay in a crystal, one can calculate the time since it was newly formed, just as one reads the date from a coin. The principle is simple, though the techniques are often laborious. For example, ^{40}K breaks down to form ^{40}Ar at a rate such that half of it has gone in about 1.3 billion years. If we measure the ^{40}Ar in a potassium feldspar crystal today, and find that half the original amount of ^{40}K has gone, then the age of the crystal is 1300 Ma or 1.3 Ga. (By convention, ages in millions of years are given in megayears [Ma], while time periods or intervals are expressed in millions of years [m.y.]. Ages in billions of years are gigayears [Ga].) Other dating methods use this same principle.

Absolute dating must be done carefully. Crystals may have been reheated or even recrystallized, resetting their radioactive clocks to zero well after the time the rock originally formed. Chemical alteration of the rock may have removed some of the newly produced element, also giving a date younger than the true age. Geologists are familiar with these problems, and go to immense trouble to find and use crystals that are pristine.

Most elements used for radioactive age dating are not used by living organisms to build shells or bones, so we usually cannot date fossils directly. Instead, we have to measure the age of a lava flow or volcanic ash layer as close to the fossil-bearing bed as possible (Figure 2.5), which does contain crystals we can use.

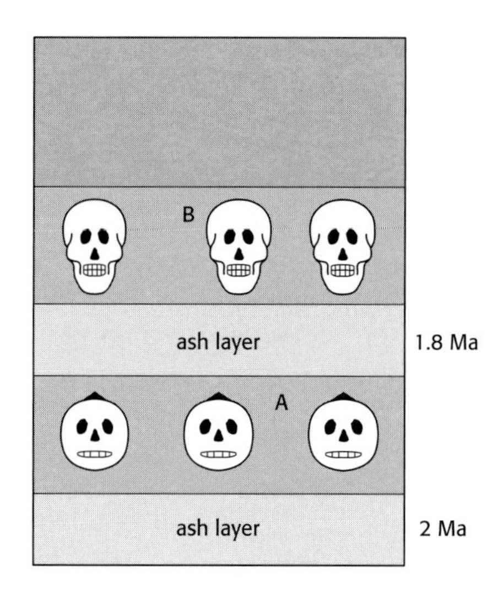

Figure 2.5 The skulls of these two fossil hominids do not contain any radioactive isotopes, but they lie close to two layers of volcanic ash that do. By using relative dating methods, one can say that hominid A is older than hominid B. Using absolute dating methods, the age of hominid A can be fixed closely between 1.8 Ma and 2.0 Ma because there are dated ash layers above and below it. All we know about the absolute age of hominid B is that it is younger than 1.8 Ma.

Paleobiologists more often deal with a **relative time scale**, in which one says "Fossil A is older than Fossil B" (Figure 2.5) without specifying the age in absolute years. This is much the same way that archaeologists date Egyptian artifacts. We know which Pharaoh followed which, even though we do not know the absolute ages for many of the earlier dynasties. So Egyptian history is scaled according to the reigns of particular Pharaohs, rather than recorded in absolute years. We can work this way with fossils, because it is a fact of observation that fossils preserved in the rock record at particular times are almost always different from those preserved at other times. The facts have been well established over the past two centuries by geologists working upward and downward in rock sequences in which sediments have been laid down in successive layers, each layer lying on and therefore being younger than the one underneath. With each study, a small part of geological time has had its sequence of fossils identified, much as archaeologists have identified successive events in each Pharaoh's reign.

With the occasional check from absolute methods, the geological record has been arranged into a standard sequence: the **geological time scale** (Figure 2.6). The time scale is divided into a hierarchy of units for easy reference, with the divisions between major units often corresponding to important changes in life on Earth. The names of the eons, eras, and periods are often unfamiliar and have bizarre historical roots. For example, the Permian Period was named after Sir Roderick Murchison and the Comte d'Archiac took a stagecoach tour of Russia in 1841, and discovered unfamiliar new rocks near the city of Perm. After a while, however, the names and their sequence become not a matter for laborious memorization but the key to a vivid set of images of ancient life.

LIFE ALTERS A PLANET

For too long, paleontologists thought of life as a set of passengers on a planet that had a certain geology, chemistry, and climate. Evolution took place within that life as creature interacted with creature, or a response to the physical environment. But we know now that biological processes dramatically affect the physical Earth, in a mutual interaction that has complex patterns. One cannot study any component of Earth's system on its own because the interplay is so important. This may make life difficult for Earth scientists, but we do our humble best.

For example, the fact that Earth's atmosphere has 21% oxygen reflects the continuous production of oxygen by photosynthesizers on land and in surface waters.

	Time divisions	Began (Ma)
Cenozoic	Quaternary	
	Holocene	0.01
	Pleistocene	1.8
	Tertiary	
	Pliocene	5
	Miocene	24
	Oligocene	34
	Eocene	55
	Paleocene	65
Mesozoic	Cretaceous	144
	Jurassic	206
	Triassic	250
Paleozoic	Permian	290
	Carboniferous	354
	Devonian	417
	Silurian	443
	Ordovician	490
	Cambrian	543
Precambrian	Proterozoic Eon	2500
	Archean Eon	3800
	(age of Earth)	(4550)

Figure 2.6 The relative time scale used by geologists, simplified from the standard scheme used by the Geological Society of America. I have not categorized the various time divisions into their formal Eon/Era/Period/Epoch ranks: for present purposes it is the sequence and the approximate absolute dates that are important. For details see .

Without life, oxygen cannot be present at more than a few parts per million. Chemically, 21% oxygen provides enough O_2 to form an ozone (O_3) layer in the high atmosphere, helping to shield the surface from UV radiation. Physically, 21% oxygen affects the chemistry of seawater (iron won't dissolve in it, for example), and it affects the reactions by which rocks break down on the surface and turn into sediment. (Our cars are built from iron ore that was oxidized by atmospheric oxygen 2 billion years or more ago.) The oxygen was extracted from CO_2, which reduced the concentration of that greenhouse gas in the atmosphere and ocean, and cooled the climate.

Methanogens turn CO_2 into methane, CH_4, which is a greenhouse gas that is much more effective than CO_2 in trapping solar radiation. So at times when methanogens were globally important autotrophs, their methane production may have warmed Earth.

As we follow the history of life, we shall see that major biological changes led to major environmental changes, which in turn led to further biological events, and so on. And in reverse, major physical changes led to biological changes, and so on. It is the dynamic interplay which is important. Life has such an important effect on a planet that NASA is working out strategies for detecting the presence of life on extrasolar planets by searching for its chemical signature. We could use the same strategy in trying to work out when life arose on Earth, and what form it took.

Isotope Evidence of Biological Processes

Most chemical elements have two or more isotopes, that is, atoms that behave as they should chemically, but have slightly different masses. Thus, most carbon atoms weigh 12 atomic mass units (the nucleus has 6 protons and 6 neutrons). But a few carbon atoms have an extra neutron, so they weigh 13 units, and are called carbon-13 or ^{13}C.

The extra mass does not affect the chemistry, but it has other effects. The heavier carbon moves a little slower than the lighter one. In the molecule CO_2, for example, molecules with ^{13}C move a little slower than molecules with ^{12}C. Photosynthesizers, in air or in water, take in CO_2 and break it up, building the carbon into their tissues. However, since they take in the lighter ^{12}C molecules more easily than ^{13}C, carbon that has gone through photosynthesis contains more ^{12}C: it has a ratio of ^{13}C to ^{12}C that differs from the ratio in the CO_2 it came from, skewed toward the lighter carbon isotope. The difference is called **isotope fractionation**, and can be measured in a mass spectrometer (at around $25 per sample). The isotope fractionation is expressed typically in parts per thousand (or "per mil"). Photosynthetic carbon in the ocean surface has an isotope fractionation, or ∂^{13}**C**, of about −20 per mil: the negative sign means that it contains more lighter carbon.

Note that this has nothing to do with radiocarbon dating, which deals with the radioactive carbon isotope ^{14}C. The isotopes used in the kind of work described here are nonradioactive or **stable isotopes**.

Different organisms operating different reactions may cause a different isotope fractionation. So methanogens, which split CO_2 and make methane, produce a $\partial^{13}C$ of about −60: in other words, methanogenic methane is characterized by very light carbon.

All this means that the photosynthetic activity on Earth, land and sea, tends to concentrate light carbon in living tissue. If that light carbon is then buried after death, photosynthetic production can leave a light-carbon "signature" in fossils and in sediments. Carbon isotopes can therefore be used as an indirect measure (or "proxy" indicator) of Earth's ecosystems in the past. Sudden changes in carbon isotopes do occur in the rock record, and these "isotope anomalies" may reflect dramatic, perhaps even catastrophic, changes in ancient ecosystems.

Nitrogen isotopes are used more in assessing ancient diets. If a heterotroph eats the tissue of another creature, it will digest, absorb, and perhaps lay down that nitrogen in organic components of bones or teeth. During that digestion, nitrogen isotopes ^{15}N and ^{14}N are fractionated by about +3 per mil. If that heterotroph is eaten by yet another, $\partial^{15}N$ increases by another +3 per mil. Nitrogen isotope signatures differ between terrestrial and ocean ecosystems. One can learn about the way of life of some animals by using N isotopes: for example, a species of Cretaceous fish was found to have annual variations in the N isotopes laid down in its ear bones. This was interpreted as showing the fish lived like a salmon, alternating between ocean and river water as the seasons changed. The diet of Neanderthals and other human fossils has been assessed in the same way, from N isotope signatures in teeth: to no-one's surprise, Neanderthals seem to have eaten a lot of meat.

The whole field of stable-isotope geology is exploding. Oxygen isotopes can carry information about ancient climate; sulfur isotopes carry evidence of the activity of sulfur bacteria that almost never preserve as cells; and new ways to interpret past environments and paleobiology are being perfected.

EARTH'S OLDEST ROCKS

The first one-third of Earth's history is called the **Archean**, a time when early Earth was very different from today's planet. There was little or no oxygen in the atmosphere. There was much less life in the seas and none on the land. Earth was young; its interior was hotter, and its internal energy was greater. We can guess that volcanic activity was much greater, but we have no idea whether it was more violent or just more continuous. Reconstructing conditions on early Earth is difficult and challenging.

Rocks older than 3.5 Ga (3500 Ma) are very rare. The oldest minerals on Earth are zircon crystals over 4 Ga, but they have been eroded out of their original rocks and deposited as fragments in younger rocks. These grains contain evidence that there were patches of continental crust on Earth before 4 Ga.

This is important because continental crust, dominated by granite, is only found on Earth. Its chemistry is very different from ocean crust, and includes important minerals that release phosphorus and potassium as they are broken down in weathering at the surface. Phosphorus in particular is vital for life (Chapter 1). Continental crust may be yet another unique feature of Earth that encouraged the evolution of life here.

The oldest known rocks on Earth are in West Greenland, and have been dated by several methods at about 3.85 Ga. Nearby, there are sedimentary rocks in a slightly younger set of rocks known as the Isua series (aged 3.7–3.8 Ga). The Isua rocks have been repeatedly folded, faulted, and reheated, but they can still tell us something about conditions on early Earth when they were formed. They were laid down in shallow water along a volcanic shoreline, because they include beach-rounded pebbles and weathering products from lava. Temperatures at the time may have been warm, but not extraordinary. In other words, conditions on Earth were hospitable to life at 3.7–3.8 Ga at the latest, including the fact that there was land as well as ocean.

The Isua rocks have carbon in them, and several groups of scientists have examined that carbon. Isotope fractionations from −22 per mil to −50 per mil have been measured, and on the face of it, this fractionation indicates life, photosynthetic or methanogenic. Others claim that the carbon isotope ratios can be explained by inorganic processes, without the implication of life. The question is still being debated. In younger rocks there would be no argument: the fractionation would be accepted as the well-known trace of photosynthesis, because the alternative hypothesis is more complex.

Figure 2.7 The oldest trace fossil yet found on Earth, a stromatolite from the Warrawoona rock sequence in Australia. It is associated with wave-affected sediments, and therefore it formed in very shallow water. By comparing this structure with those forming today in Shark Bay (Figure 2.8), it can be interpreted as having been formed by mats of bacteria around 3.5 Ga. (Courtesy of Stanley Awramik, University of California, Santa Barbara.)

EARTH'S OLDEST CELLS

Archean rocks are often rich in minerals, and Archean regions have been well explored geologically for economic reasons. One such district in northwestern Australia is called North Pole because it is so remote and inhospitable. It was originally explored for the mineral barite, but it has also yielded evidence for very early life.

The local rock sequence is called the **Warrawoona Series**, and its age is about 3.5 Ga. The rocks mainly consist of volcanic lavas erupted in shallow water, or nearby on shore, but there are sedimentary rocks too. The sediments contain storm-disturbed mudflakes, wave-washed sands, and minerals formed by evaporation in very shallow pools. The rocks have not been tilted, folded, or heated very much, and the environment can be reconstructed accurately. The rocks formed along shorelines that we can interpret clearly because we can match them to modern environments. There are three other components to the story from Warrawoona, and they all relate to evidence for life on Earth at that time.

Figure 2.8 Stromatolites are forming today in warm salty water in Shark Bay, Western Australia. The shovel gives the scale. Close study of these modern structures allows us to interpret the stromatolite from Warrawoona (Figure 2.7) as an early trace fossil formed by mats of cyanobacteria. (Courtesy of Paul Hoffman, Harvard University.)

Stromatolites

The Warrawoona rocks contain structures called **stromatolites**, low mounds or domes of finely laminated sediment (Figure 2.7). We know what stromatolites are because they are still forming today in a few places, for example, Shark Bay, Western Australia. Stromatolites are formed by mat-like masses of abundant microbes, and the microbes that form modern stromatolites usually include photosynthetic blue-green bacteria (cyanobacteria). Again, the simplest hypothesis is that cyanobacteria formed the Warrawoona stromatolites.

Stromatolites are forming today in Shark Bay in warm salty waters in long shallow inlets along a desert coastline (Figure 2.8). They form from the highest tide level down to subtidal levels, but the higher ones, closer to shore, have been better studied (sea snakes, not sharks, are the problem).

The cyanobacteria that grow and photosynthesize in Shark Bay so luxuriantly thrive in water that is too salty for grazing animals such as snails and sea urchins that would otherwise eat them. Like most bacteria, they secrete slime, and are also able to move a little in a gliding motion. Sediment thrown up in the waves may stick to the slime and cover up some of the bacteria. But they quickly slide and grow through the sediment back into the light, trapping sediment as they do so. As the cycle repeats itself, sediment is built up under the growing mats. Eventually the mats grow as high as the highest tide, but cannot grow higher without becoming too hot and dry. Some mats harden as the photosynthetic activity of the bacteria helps carbonate to precipitate from the seawater that saturates them, binding the sediment into a rock-like consistency that can resist wave action.

Some cyanobacterial mats are so dense that light may penetrate only 1 mm. The topmost layer of cyanobacteria absorbs about 95% of the blue and green light, but just underneath is a zone where light is dimmer but exposure to UV radiation and heat is also less. Green and purple bacteria grow here and also contribute to the growth of the mat. Under the second layer is a zone where light is too low for photosynthesis, and there heterotrophic bacteria absorb and process any dying and dead remains of the bacteria above them, typically by fermentation. Oxygen diffuses down from the surface, and sulfide diffuses upward from the zone below, creating an extraordinary zone where chemistry can change within minutes and within millimeters.

Night follows day, of course, and photosynthesis stops at night. The oxygen in the top layers of the stromatolite is quickly lost. Sulfide dominates the night-time hours, oxygen dominates the daylight hours, and all the bacteria must be able to adjust quickly to the daily change. The internal chemistry in stromatolites is as complex as the mix of bacteria: and there is no reason to suppose that ancient stromatolites were any different.

Figure 2.9 Stromatolites from the Fig Tree Group in southern Africa are nearly as old as the Warrawoona stromatolite shown in Figure 2.6. The coin is 2 cm across. (Photograph courtesy of Gary Byerly, Louisiana State University.)

Ancient Stromatolites

Stromatolites are trace fossils: they are formed by the action of living cells, even if those cells are hardly ever preserved in them as fossils. Because they are large, and because their distinctive structure makes them easy to recognize, stromatolites are the most conspicuous fossils for three billion years of Earth's history, from 3.5 Ga to the end of the Proterozoic at about 550 Ma. They are rare in Archean rocks, probably because there were few clear, shallow-water shelf environments suitable for stromatolite growth at the time. The few Archean landmasses were volcanically active, generating high rates of sedimentation that probably inhibited mat growth in many shoreline environments.

Stromatolites are small and rare in the Warrawoona rocks, but are much more numerous, complex, and varied in rocks of the Fig Tree Group in southern Africa, dated at about 3.4 Ga (Figure 2.9). Thus, bacterial life was well established by that time, even if stromatolites occurred only in local patches.

Solar UV radiation was intense in Archean time, with no oxygen (or ozone layer) in the atmosphere. The evolution of the stromatolitic way of life by cyanobacteria may have been a response to UV radiation. With light (and UV) penetrating only a little way into the slime and sediment of the mat, bacteria were able to live essentially at the food-rich water surface without damage from UV. Cyanobacteria were not just existing at Warrawoona: they were already modifying their microenvironment for survival and success. These stromatolites were not just the sites of simple microbial mats, but were complex miniature ecosystems teeming with life.

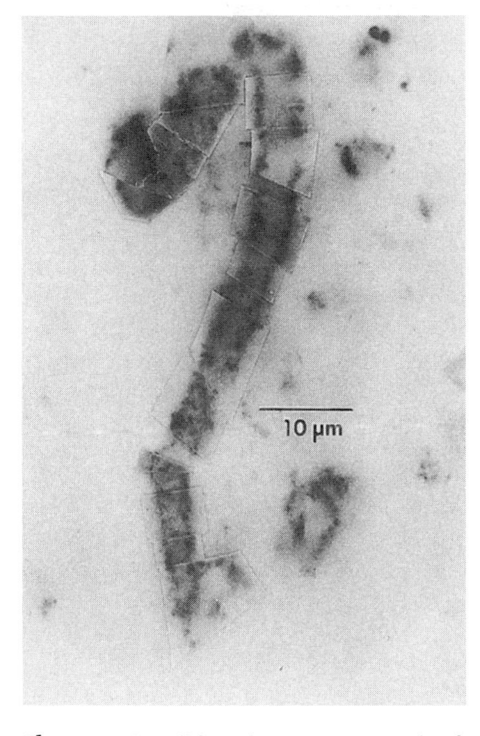

Figure 2.10 Cell from the Warrawoona rocks of Australia, 3.5 Ga. (Courtesy of J. William Schopf, University of California, Los Angeles.)

Fossil Cells in Warrawoona Chert?

Chert is a rock formed of microscopic silica particles (SiO_2). It does not form easily today because all kinds of organisms, including sponges, take silica out of seawater to make their skeletons. But silica-using organisms had not evolved in Archean times, so cherts are often abundant in Archean rocks. As chert forms from a gel-like goo on the seafloor, it may surround cells and impregnate them with silica, preserving them in exquisite detail as the silica hardens into chert. Once it hardens, chert is watertight, so percolating water does not easily dissolve or contaminate the fossil cells.

Fossil cells have been described from cherts in the Warrawoona rocks (Figure 2.10). At least some of them look very similar to cyanobacteria that produce oxygen

Figure 2.11 Stromatolites were more abundant in the Proterozoic than they had been in the Archean. This luxuriant array of stromatolites grew about 1.9 Ga on a shoreline where Great Slave Lake now lies on the Canadian Shield. (Courtesy of Paul Hoffman, Harvard University.)

today. Currently there is a vicious argument about the reality of these cells, so I should explain why it's not too important which way that argument is resolved.

If the Warrawoona "cells" are not real cells, it does not alter our picture of early life on Earth very much. The stromatolites are clear indicators of life, and carbon isotopes from Warrawoona rocks have a fractionation of around −30 per mil, indicating photosynthesis.

Were the Warrawoona stromatolites formed by cyanobacteria (implying oxygen production) or by other photosynthetic bacteria which may have produced, say sulfur? Clear cyanobacterial cells and stromatolites both occur in the Fig Tree Chert in South Africa, at 3.4 Ga (Figure 2.9). The claim for cyanobacteria at Warrawoona is important because cyanobacteria produce oxygen. But changing the date of oxygen photosynthesis from 3.5 Ga to 3.4 Ga really doesn't matter much (to a geologist)!

The Late Archean and the Early Proterozoic

The earliest stromatolites at Warrawoona and in the Fig Tree Chert lived around 3.4–3.5 Ga. By 3.1 Ga, there were two distinctly different styles of bacterial mat; by 2.9 Ga there is evidence of bacterial mats forming on soft silty seafloors, and by 2.8 Ga stromatolites are known from salt-lake environments as well as oceanic shorelines. So by Late Archaean times, bacterial life was abundant and varied in shoreline habitats.

There were important geological changes at the end of the Archean, which is defined as 2.5 Ga. Earth had cooled internally to some extent, and the crust was thicker and stronger. The thicker crust affected tectonic patterns: the way the crust moves, buckles, and cracks under stress. Continents became larger and more stable in Early Proterozoic times, with wide shallow continental shelves that favored the growth and preservation of stromatolites. Most Proterozoic carbonate rocks include stromatolites, some of them enormous in extent (Figure 2.11). Proterozoic stromatolites evolved into new and complex shapes as bacterial communities became richer and expanded into more environments.

EARTH'S EARLY ATMOSPHERE AND CLIMATE

Earth formed hot, heated by the energy in infalling meteorites, but we know from the rock record that it had cooled to "normal" by 3.7–3.8 Ga, when ocean waves were washing up on the beach at Isua. Yet our models of the physics of stars suggest that the Sun was cooler, perhaps as much as 30% cooler, during its early history. If so, Earth should have had a cool Archaean climate, perhaps largely frozen. This may have been good for the formation of organic molecules and of life, as we saw in Chapter 1, even if there were only a few places warm enough for the first living cells to flourish. However, the evolution of Archaea may have changed all that. Their production of the greenhouse gas methane could have had a positive feedback, warming Earth, promoting more biological activity, producing more methane, and so on, making the climate of Archean times warm and hospitable not just for Archaea but also for Bacteria. (Methane leaks from modern-day stromatolites, for example.)

Earth's early atmosphere had no oxygen: life could not have begun on Earth if it had. Therefore, Earth's early life lived in a world with little or no free oxygen, certainly less than 1%. Free oxygen destroys methane, but the warm climate of the Archean, maintained by a methane supply from methanogenic Archaea, was not jeopardized.

BANDED IRON FORMATIONS

At the beginning of the Proterozoic Era (around 2.5 Ga) we find increasingly massive accumulations of a peculiar rock type. **Banded iron formations (BIF)** are sedimentary rocks found mainly in sequences older than 1.8 Ga. They are alternations of iron oxide ore and chert, sometimes repeated millions of times in microscopic bands. No iron deposits like this are forming now, but we can make intelligent guesses about the conditions in which BIF were laid down.

The chemistry of seawater on an oxygen-poor Earth differed greatly from today's. Today there is practically no dissolved iron in the oceans. But iron dissolves readily in water that has no oxygen. Even today, in the oxygen-poor water on the floor of the Red Sea, iron is enriched 5000 times above normal levels. So Archean oceans must have contained a great deal of dissolved iron as well as silica.

Silica would have been depositing more or less continuously on an Archean seafloor to form chert beds, especially in areas that did not receive much silt and sand from the land. But the iron oxide can only have precipitated out of seawater in such massive amounts by a chemical reaction that included oxygen.

Therefore, to form one of the iron oxide layers in BIF, there must originally have been the usual Archean ocean without oxygen but with a lot of dissolved iron. Then a huge oxidation event must have dropped that iron on to the seafloor. Then the oxidation event stopped, the ocean once again became oxygen-free and loaded with dissolved iron, then the process began again.

So there were occasional or regular periods of iron ore formation against a background of regular chert formation. Between episodes of iron deposition, iron was replenished from erosion down rivers or from deep-sea volcanic vents.

The two main hypotheses about BIF, one inorganic and one organic, and both call for seasonal deposition of iron ore. Both may be true. In the leading inorganic model, UV radiation formed an iron compound in surface waters, which fell out to form a layer of iron ore on the seafloor. This reaction would have peaked during the summer, when sunlight was most intense. The inorganic model could explain why there are some BIF in the Isua rocks at 3.8 Ga.

However, even "inorganic" BIF may in fact have been deposited by bacteria. Bacteria can act to form tiny "seed" crystals of iron minerals, which then continue to grow inorganically. It would be extraordinarily difficult to detect that this was in fact begun by bacterial action. Laboratory cultures of bacteria can form iron that looks "inorganic," so it would be a brave person who would say that bacteria were not involved in the formation of BIF.

In the organic model, BIF were formed directly by photosynthesis. This may have worked in two ways: from cyanobacteria in stromatolites, or from purple bacteria floating in the water. Purple bacteria use photosynthesis to break up iron carbonate dissolved in the water, and produce oxidized iron as a byproduct. Both bacterial reactions would also have peaked during the intense light of summer. The same seasonal upwelling that brought nutrients to the bacteria also brought iron from deep water, leading to great bursts of iron ore deposition on the seafloor.

Today we probably see only a small fraction of the BIF that once formed on Archean seafloors, because most ocean crust is recycled back into Earth. But even the amounts remaining are staggering. BIF make up thousands of meters of rocks in some areas and contain by far the greatest deposits of iron ore on Earth. At least 640 billion tonnes of BIF were laid down between 2.5 and 2.0 Ga (that's an average of half a million tonnes of iron per year). [The metric tonne that is used internationally is 1000 kilograms, very close to an American ton.] The Hamersley Iron Province in Australia alone contains 20 billion tonnes of iron ore, with 55% iron content. At times, iron was dropping out in that basin at 30 million tonnes a year. Most modern steel industries are based on iron ores laid down by bacteria during that time. Cadillacs, Toyotas, and BMWs ultimately owe their existence to Precambrian photosynthetic bacteria.

THE OXYGEN REVOLUTION

BIF are deposited in bands that can be traced for hundreds of kilometers, so they were laid down uniformly, or at least continuously, over great areas. This suggests to me that photosynthetic bacteria were floating all across the ocean surface, not simply living in a thin zone of stromatolites in shallow seas around the edges of the small Archean continents. Even so, stromatolites were flourishing along those same shorelines, containing massive mats of cyanobacteria, all producing oxygen.

The toxic qualities of oxygen would first have affected the cyanobacteria inside which photosynthesis was taking place. As cyanobacteria evolved biochemical antidotes to oxygen poisoning (Chapter 1), they had the opportunity to control and then use oxygen in a new process, respiration (biological oxidation).

Fermentation leaves byproducts such as lactic acid that still have energy bonded within them. Using oxygen to break the byproducts all the way down to carbon dioxide and water, a cell can release up to 18 times more energy from a sugar molecule by respiration than it can by simple fermentation (Chapter 1).

Cyanobacteria can photosynthesize in light and respire in the dark. To do this, they must be able to store oxygen in a stable, nontoxic state for hours at a time. Most likely, they began to use oxygen in respiration very early: the energy advantages are astounding. The early success of cyanobacteria probably reflected their control over oxygen, which gave them an abundant and reliable energy supply in two different ways: by mastering water photosynthesis, and by breaking down food molecules in respiration rather than fermentation.

However, that same success poses a geological problem. If cyanobacteria had made such an important breakthrough, using a process that produces oxygen as a byproduct, why did the ocean and atmosphere not become oxygenated quickly? The rock record shows that stromatolites did not increase dramatically until the beginning of the Proterozoic Era at about 2.5 Ga, and BIF production does not reach a peak until then. Other evidence indicates that there was still no free oxygen in the atmosphere and ocean until perhaps 2.3 Ga. Furthermore, oxygen doesn't reach even close to modern levels until perhaps 500 Ma, after the Proterozoic. Why not?

There are four possible answers. The first is that Archaean stromatolites may have been formed by other bacteria, not cyanobacteria, and that oxygen-producing photosynthesis simply did not occur until perhaps 2.7 Ga. This is unlikely, given the evidence we have already described.

The second answer deals with the eventual fate of the organic tissue of the photosynthesizers. They split CO_2, use the carbon to build sugars, and release the O_2. In life, they build the sugars into tissue, or they oxidize the sugars for energy, combining the carbon back into CO_2. When they die, the carbon in their tissues may be recycled into other bacteria, and much of that eventually is oxidized back into CO_2. Most of the oxygen spends only a short time in the ocean and atmosphere before it is dragged out to oxidize back the same carbons that were involved in its release. The only way in which an oxygen molecule can be freed long-term by photosynthesis is if the carbon can be hidden away from ocean water and atmosphere: and that usually requires geological burial.

It is common knowledge that our major fuels, oil, gas, and coal, were produced as organic compounds from living things when buried. The formation of such sediments inevitably implies the release of released oxygen into ocean and atmosphere at the same time.

It is difficult to assess how much carbon was buried in ancient times, and therefore how much oxygen was freed. Oil companies generally do not expect to find major oil and gas fields in Proterozoic rocks, though a few small prospects do exist.

The third answer points to all the other ways that oxygen could be used up before it could be released into air and water. Oxygen reacts with sulfides and iron compounds to form sulfates and iron oxides (rust). Since the early oceans were saturated with such compounds, these reactions must have used up almost every spare oxygen molecule produced on early Earth. It may have taken hundreds of millions of years before large amounts of free oxygen began to accumulate in air and water. But did it take close to three billion years (from 3.5 Ga to 0.5 Ga)? That seems a long time!

The fourth answer is that photosynthesis (and oxygen production) was a lot slower than we might imagine. The reason is yet another geochemical process in which life and Earth have interacted over billions of years. Photosynthesis doesn't just need light, water, and carbon dioxide: the plants or bacteria that operate it need nutrients as well. In many environments, phosphorus is the limiting nutrient (many of our fertilizers contain phosphorus).

Phosphorus comes largely from continental crust, and Archaean continents were small. So phosphorus supplies may well have been limited, especially out in the vast surfaces of Archaean oceans. In addition, iron minerals absorb some phosphorus as they form, so BIF would have used up (and locked up) a lot of phosphorus as well as a lot of iron. Lack of phosphorus would then have slowed cyanobacterial growth, which would have slowed photosynthesis, which would have slowed BIF production. Renewed weathering and erosion would then have provided new iron and phosphorus supplies to ocean water, setting the stage for another BIF episode that removed phosphorus as well as iron. Altogether, this would have dramatically slowed the oxygenation of Earth, as well as helping to produce the huge cycles that characterized BIF production.

BIF production rose to a peak around 2.5 Ga, then fell off steadily until 1.8 Ga. As the rate of BIF formation slackened, the atmosphere and oceans began to accumulate permanent but very low amounts of free oxygen. BIF are rare after 2.0 Ga, as oxygen levels in the oceans reached a permanent level so high that seawater could no longer hold dissolved iron and BIF could no longer form, except in rare isolated basins.

Other geological evidence confirms the oxygenation of atmosphere and the oceans early in the Proterozoic. The uranium mineral uraninite cannot exist for long if it is exposed to oxygen, and it is not found in rocks younger than about 2.3 Ga. Sulfur isotopes in rocks suggest that sulfate levels rose, lowering methane production by Archaea, while the methane that was produced was quickly broken down by free oxygen. This change occurred between 2.4 and 2.1 Ga, perhaps with a very rapid change around 2.3 Ga.

In turn, the drop in methane levels in the atmosphere cooled Earth, and we see evidence of a very large ice age between 2.4 and 2.2 Ga.

On land, the presence of oxygen in air for the first time would have rusted any iron minerals exposed at the surface by weathering. Rivers would have run red as they flowed across Earth's surface before vegetation invaded the land. On land and in shallow seas, red beds, or sediments bearing original iron oxides, date from about 2.3 Ga.

Photosynthesis produces oxygen close to the ocean surface, because that is as far as usable light penetrates water. For the same reason (solar energy), the surface waters are warmer than deeper layers, so the water is less dense. So surface waters tend to stay at the surface, and the bulk of the ocean volume remains without oxygen. In today's oceans, the only places where surface waters can sink are where they become unusually dense: if they are very cold, if they are very salty, or both. (These areas are in polar seas—in the North Atlantic and around Antarctica—or in hot shallow tropical seas such as the Red Sea or Persian Gulf.)

Today, enough surface water sinks to carry oxygen to almost all the world ocean. But it would have been different in the Proterozoic. Clearly, the surface waters would have become oxygen-rich before the bulk of the ocean did. And the atmosphere, which can exchange gases with the surface waters, would also have become oxygen-bearing before the deep ocean did. We can imagine a Proterozoic world that had no free oxygen except in surface waters and the atmosphere. The deep water would still have been rich in dissolved iron and silica and sulfide, and inhabited by bacteria and methanogens, while the surface waters had photosynthesizers and oxygen-tolerant microbes. In such a world, occasional episodes of sinking surface water would have caused BIF to drop out from these deep waters, long after the surface oxygenation.

All these changes justify the term **oxygen revolution**: for the events around 2.3 Ga to 2.2 Ga. The revolution changed the chemistry of Earth's air, land, and water forever. One indirect result was vital for the further evolution of living things. Solar UV radiation acts on any free oxygen high in the atmosphere to produce ozone, which is O_3 rather than O_2. Even a very thin layer of ozone can block most UV radiation. Earth's surface, land and water, has been protected from damaging UV radiation ever since free oxygen entered the atmosphere. It then became possible for organisms to evolve that were longer-lived and more complex than bacteria or Archaea: the eukaryotes (Chapter 3).

Further Reading

Bjerrum, C. J., and D. E. Canfield. 2002. Ocean productivity before about 1.9 Gyr ago limited by phosphorus adsorption onto iron oxides. *Nature* 417: 159–62, and comment, 127–8.

Bosak, T., and D. T. Newman. 2003. Microbial nucleation of calcium carbonate in the Precambrian. *Geology* 31: 577–80. [The presence of bacteria encourages the precipitation of calcium carbonate and stromatolite formation at Precambrian chemistries, even if they are not cyanobacteria.]

Brocks, J.J., et al. 1999. Archaean molecular fossils and the early rise of eukaryotes. *Science* 285: 1033–6, and comment, 1025–6.

Dalton, R. 2002. Squaring up over ancient life. *Nature* 417: 782–4. [News summary of the vicious controversy over Warrawoona fossils. See also news story in *Science* 295, 1812–13. For the papers, see Brasier, M. D., et al. 2002. *Nature* 416: 76–81; Schopf, J. W., et al. 2002. *Nature* 416: 73–6.]

Des Marais, D. J. 2003. Biogeochemistry of hypersaline microbial mats illustrates the dynamics of modern microbial ecosystems and the early evolution of the biosphere. *Biological Bulletin* 204: 160–7. [Summary of stromatolite biology, ecology, and biochemistry.]

Greenberg, E. P. 2003. Bacterial communication: tiny teamwork. *Nature* 424: 134. [Biofilms.]

Hoehler, T. M., et al. 2003. The role of microbial mats in the production of reduced gases on the early Earth. *Nature* 412: 324–7.

Kasting, J. F. 2001. The rise of atmospheric oxygen. *Science* 293: 819–20. [Comment on an indigestible but important paper by Catling et al., *Science* 293: 839 ff.]

Kasting, J. F., and J. L. Siefert. 2001. The nitrogen fix. *Nature* 412: 26–7. [How little we know about the Archean atmosphere.]

Kerr, R. A. 1999. Early life thrived despite earthly travails. *Science* 284: 2111–13.

Lenton, T. M. 2003. The coupled evolution of life and atmospheric oxygen. In Rothschild, L. M., and A. M. Lister (eds), *Evolution on Planet Earth*. San Diego: Academic Press, 35–53.

Nisbet, E. G. 1987. *The Young Earth*. Boston: Unwin Hyman. [Dated now, but well written.]

Noffke, N., et al. 2003. Earth's earliest microbial mats in a siliciclastic environment (2.9 Mozaan Group, South Africa). *Geology* 31: 673–6. [Not the earliest microbial mats, but the first to form on a silty seafloor. Oxygen was probably present, which implies that cyanobacteria were too.]

Norris, R. D., and R. M. Corfield (conveners). 1998. *Isotope Paleobiology and Paleoecology. Paleontological Society Papers* 4.

Rasmussen, B., and R. Buick. 2000. Oily old ores: evidence for hydrothermal petroleum generation in an Archean volcanogenic massive sulfide deposit. *Geology* 28: 731–4.

Schopf, J. W. 1999. *Cradle of Life: The Discovery of Earth's Earliest Fossils*. Princeton: Princeton University Press.

Widdel, F., et al. 1993. Ferrous iron oxidation by anoxygenic phototrophic bacteria. *Nature* 362: 834–6, and comment, 790–1.

Wiechert, U. W. 2002. Earth's early atmosphere. *Science* 298: 2341–2. [Comment on papers by Farquhar et al., 2369–72, and Habicht et al., 2372–4.]

van Zullen, M. A., et al. 2002. Reassessing the evidence for the earliest traces of life. *Nature* 418: 627–30. [They think the carbon in Greenland rocks at 3.8 Ga was produced inorganically.]

Zschokke, S. 2003. Spider-web silk from the Early Cretaceous. *Nature* 424: 636–7.

CHAPTER THREE

Sex and Nuclei: Eukaryotes

Today **prokaryotes** (all living Archaea and Bacteria) differ fundamentally from **eukaryotes** (all other living organisms, Eukarya) (Figure 3.1, Box 3.1). Life began with prokaryote organisms, so eukaryotes must have evolved from some prokaryote ancestor. Who were these prokaryotic ancestors, and how and when did modern eukaryotes evolve from them?

Prokaryotes were and are successful in an incredible range of habitats, from stinking swamps to the hindgut of termites and from hot springs in the deep sea to the ice desert of Antarctica, and deep in rocks underground. They occur in numbers averaging 500 million per liter in surface ocean waters, 1 billion per liter in fresh water, and about 300 million on the skin of the average human. This diversity makes it difficult to search for plausible ancestors of eukaryotes among them!

SYMBIOSIS AND ENDOSYMBIOSIS

Symbiosis is a relationship in which two different organisms live together. Often they both derive a benefit from the arrangement. Examples range from the symbiosis between humans and cats to bizarre relationships such as that of acacia plants, which house and feed ant colonies that in turn protect the acacia against herbivorous animals and insects. The ultimate state of symbiosis is **endosymbiosis**, in which one organism lives inside its partner. Animals as varied as termites, sea turtles, and cattle can live on plant material because they contain bacteria somewhere in their digestive system with the enzymes to break down the cellulose that is unaffected by the host's digestive juices. Many tropical reef organisms have symbiotic partners in the form of photosynthesizing microorganisms. Living inside the tissues of corals or giant clams, these symbiotic partners have a safe place to live. In turn, the host receives a share of their photosynthetic production.

It is now clear that endosymbiosis was critical in the evolution of eukaryotes. The ancestors of the lineage Eukarya had a prokaryotic structure, and may have been as ancient a group of organisms as Bacteria and Archaea. We have found chemical fossils (molecules called sterols formed only by eukaryotic biochemistry) as far back as 2.7 Ga. No-one would claim that these were made by eukaryotes, but they could well be chemical fossils of a eukaryan ancestor.

At some point, a prokaryotic eukaryan made a dramatic evolutionary breakthrough: it obtained **organelles** in the form of other prokaryotes that came to live

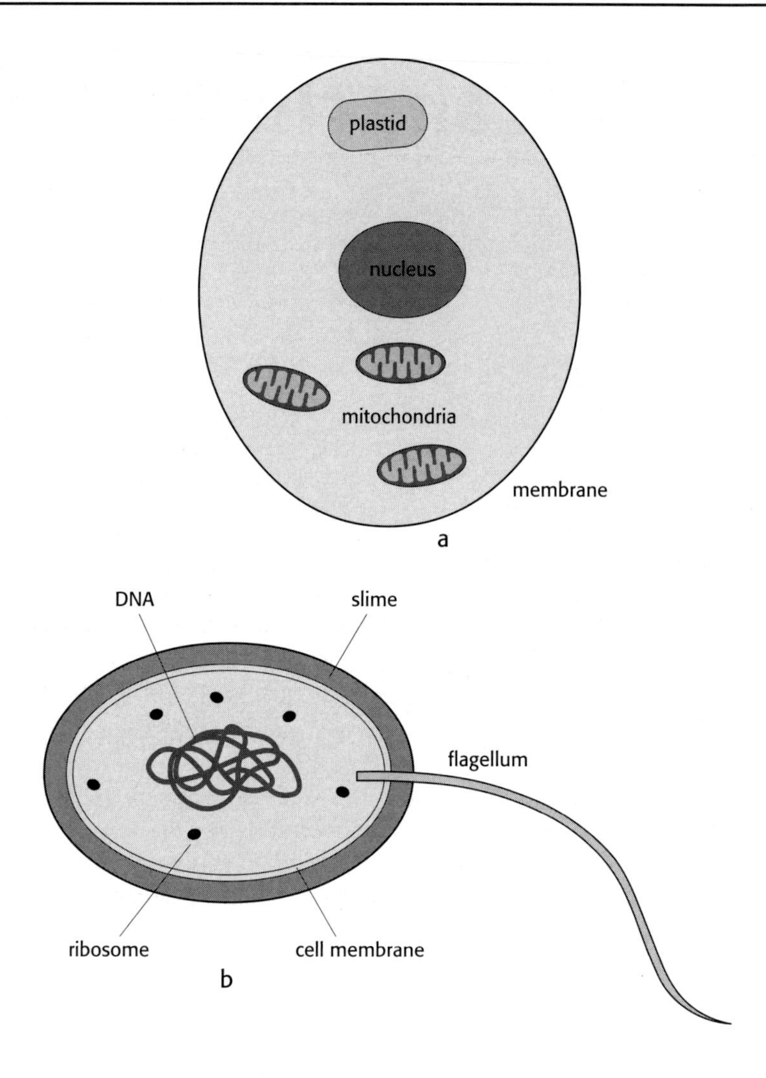

Figure 3.1 The structure of a eukaryote (a) and a prokaryote (b). The eukaryote is much bigger than the prokaryote.

inside it (as endosymbionts), and in doing so became a eukaryote rather than a prokaryote. **Mitochondria** and **plastids**, and perhaps even flagella, were once free-living bacteria, but they became so closely associated with a host prokaryote (ancestral eukaryan) that eventually they became, for practical purposes, part of it (Figures 3.1, 3.2). At least five major pieces of evidence show that organelles (and therefore eukaryotes) originated by endosymbiosis (Box 3.2).

Mitochondria and their Ancestors

How did a proto-mitochondrial bacterium get inside a eukaryan cell? Two current scenarios do not differ very much. They both assume that the ancestral eukaryan evolved a cell wall flexible enough to engulf other cells. For example, the eukaryan may have engulfed bacteria to digest them. However, if it engulfed a bacterium that could oxidize its waste products and release some of the calories back to the eukaryan, the eukaryan would benefit by not digesting the bacterium, but keeping it as a living internal guest (tenant, or slave, if you like). The engulfed bacteria came to reside permanently (and divide) inside the host as mitochondria.

Or perhaps eukaryan cells and bacteria began their relationship not as predators and prey, but as neighbors living side by side in very close contact, in *external*

BOX 3.1 Differences Between Prokaryotes and Eukaryotes

1 Eukaryotes have their DNA contained within a membrane, and under the microscope this package forms a distinct body, the **nucleus**. Prokaryotes have their DNA loose in the cell cytoplasm.
2 Prokaryotes have no internal subdivisions of the cell, but almost all eukaryotes have **organelles** as well as a nucleus. Organelles are subunits of the cell that are bounded by membranes and perform some specific cell function. **Plastids**, for example, are organelles that perform photosynthesis inside the cell, generating food molecules and releasing oxygen. **Mitochondria** contain the respiratory enzymes of the cell. Food molecules are first fermented in the cytoplasm, then passed to the mitochondria for respiration. Mitochondria generate ATP as they break food molecules down to water and CO_2, and they pass out energy and waste products to the rest of the cell. They also make steroids, which help to form cell membranes in eukaryotes and give them much more flexibility than prokaryote membranes.
3 Eukaryotes can perform **sexual reproduction**, in which the DNA of two cells is shuffled and redealt into new combinations.
4 Prokaryotes do not have flexible cell walls, so cannot expand to engulf other cells. The flexibility of eukaryotic cell membranes allows them to engulf large particles, to form cell vacuoles, and to move freely. Plant cells, armored by cellulose, are the only eukaryotes that have given up a flexible outer cell wall for most of their lives.
5 Eukaryotes have a well-organized system for duplicating their DNA exactly into two copies during cell division. This process, **mitosis**, is much more complex and precise than the simple splitting found in prokaryotes.
6 Eukaryotes are almost always much larger than prokaryotes. A eukaryote is typically ten times larger in diameter, but this means that it has about 1000 times the volume of a prokaryote.
7 Eukaryotes have perhaps a thousand times as much DNA as prokaryotes. They have multiple copies of their DNA, with much repetition of sequences. The DNA content of prokaryotes is small, and they have only one copy of it. There is little room to store the complex "IF . . . THEN . . ." commands in the genetic program that turn on one gene as opposed to another. Therefore, genetic regulation is not well developed in prokaryotes, which means that they cannot produce the differentiated cells that we and other eukaryotes can. Multicellular colonies of bacteria are all made up of the same cell type, repeated many times in a clone. Therefore, any species of bacterium is very good at one thing but cannot do others; its range of functions is narrow.

BOX 3.2 Evidence for Organelle/Eukaryote Symbiosis

1 The DNA in mitochondria and plastids is not like the DNA in the eukaryotic cell nucleus.
2 Mitochondria and plastids are separated from the rest of the eukaryotic cell by membranes; thus they are actually outside the cell. The cell itself makes the membrane, but inside it is a second membrane secreted by the organelle.
3 Plastids, mitochondria, and prokaryotes make proteins by similar biochemical pathways, which differ from those in the cytoplasm of eukaryotes.
4 Mitochondria and plastids are susceptible to streptomycin and tetracycline, and so are prokaryotes; eukaryotic cytoplasm is unaffected by these drugs.
5 Mitochondria and plastids can multiply only by dividing; they cannot be made by the cell cytoplasm. Organelles thus have their own independent reproductive mechanism. A cell that loses its mitochondria or plastids cannot make more.

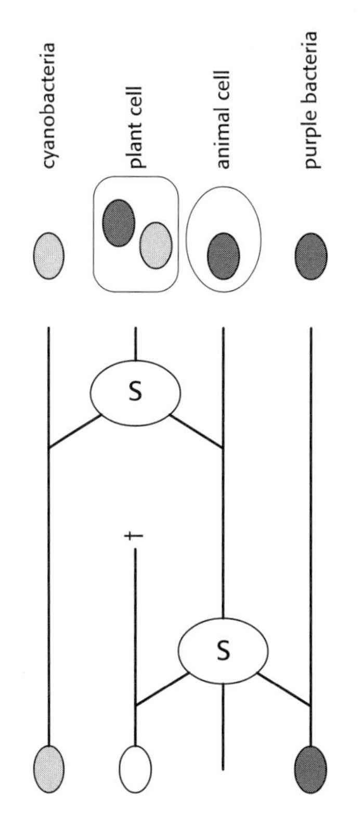

Figure 3.2 Plastids and mitochondria were once free-living cells until they became endosymbionts inside other cells. S symbols indicate symbiotic events.

symbiosis. Each produced substances that the other could use. Eventually, the eukaryan took the bacteria inside its tissues, converting them into internal symbionts (workers in company housing).

The host eukaryan and the bacteria eventually came to a stable relationship when they established mutual population control. The host cells grew, flourished, and divided, each daughter cell taking with it a population of bacteria. At some point the symbiont bacteria lost their cell walls and became mitochondria, and we can then call the host a true eukaryote.

LIMERICK 3.1
A eukaryote wanting to eat,
Saw bacteria as quite a nice treat.
Increasing closeness
Led to endosymbiosis,
And the modern plant cell was complete.
© Elizabeth Wenk, 1994

The critical point about mitochondria is their ability to oxidize molecules. Their host's energy budget was dramatically increased (it was a fermenter), and that energy budget is partly allocated to the mitochondria, so that both partners benefit. As early eukaryotes evolved to depend entirely on their mitochondria to provide them with ATP, the number of mitochondria had to be matched closely to the needs of the host. Therefore, the genes that controlled mitochondrial reproduction passed to the host's nucleus, leaving behind in the mitochondria (as far as we can tell) mainly the genes that control the oxidation they perform for the cell. This process cannot have been simple, and it must have been complicated by the efforts necessary to combat the poisonous effects of the oxygen needed for respiration.

Either of these scenarios could have produced **protists** (single-celled eukaryotes). Simple protists are amoeba-like animals, capable of moving and eating by engulfing other organisms. Food is fermented in the cytoplasm and oxidized in the mitochondria. The same process occurs in our cells today.

Plastids

Plants photosynthesize to accumulate energy. But plant photosynthesis is not performed in the cell cytoplasm but in organelles: in chloroplasts, or plastids. Some early protists acquired cyanobacteria as symbiotic partners which evolved into plastids (Figure 3.2), probably in the same way that the first eukaryotes had acquired mitochondria. The cyanobacteria benefited more from nutrients inside the host's tissues than they would as independent cells. In time, the protist came to rely so much on the photosynthesis of its partners that it gave up hunting and engulfing other cells, gave up locomotion, grew a strong cellulose cell wall for protection, settled in the light, and took on a way of life that we now associate with "plants."

This is a scenario for the evolution of the first eukaryotic photosynthesizers (the algae) (Figures 3.2, 3.3). Since that event, the plant-animal dichotomy has been one of the most important in the organic world. We place advanced plants and advanced animals in two different kingdoms; animals eat plants and one another.

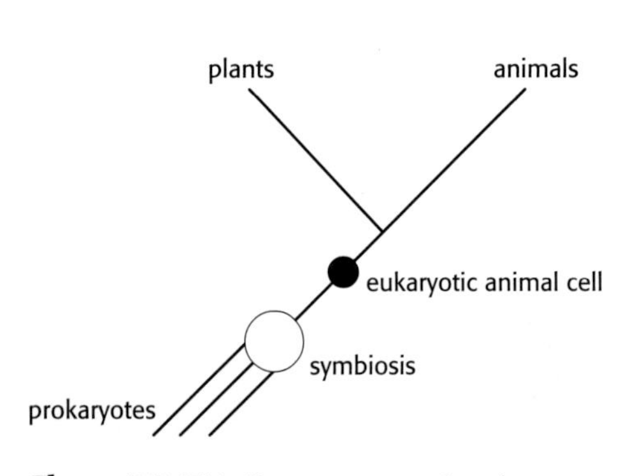

Figure 3.3 This diagram suggests that plants evolved from animal cells, as did Figure 3.2. The ancestors of both groups were single-celled, and the branch between them also occurred among protists.

EUKARYOTES IN THE FOSSIL RECORD

The endosymbiotic theory is convincing, but it is based on biological and molecular evidence. It's not easy to see how it can be tested in the fossil record, or how to identify the first eukaryotic fossils. Most fossil cells are small spherical objects with no distinguishing features. Although most eukaryotes are larger than prokaryotes, some bacteria are close to normal eukaryotic size. It is almost impossible to distinguish large rotting bacteria from small rotting eukaryotes. After death, the cell contents of prokaryotes can form blobs or dark spots that look like fossilized organelles or nuclei (Figure 3.4). Rotting colonies of cyanobacteria can look like multicellular eukaryotes, and filamentous bacteria can look like fungal hyphae. And finally, early eukaryotes were probably small and thin-walled, and therefore are most unlikely to be preserved as fossils.

We have to take the geological record and interpret it as best we can. Eukaryotes could not have evolved before oxygen became a permanent component of seawater. The major problem will be to identify them unambiguously as eukaryotes.

The best candidate for the oldest eukaryote is *Grypania spiralis*, a ribbon-like fossil 2 mm wide and over 10 cm long. It is abundant in rocks around 1400 Ma in China and Montana, but has been described also from banded iron formations in Michigan dated at 2100 Ma. *Grypania* looks very much like a seaweed blade: a eukaryotic alga. If it is, then algae evolved at or before 2100 Ma, and if Figure 3.3 is correct, simpler eukaryotes had evolved before that, even while major amounts of BIF were still being deposited. It looks as if eukaryotes evolved as soon as oxygen levels permitted them to.

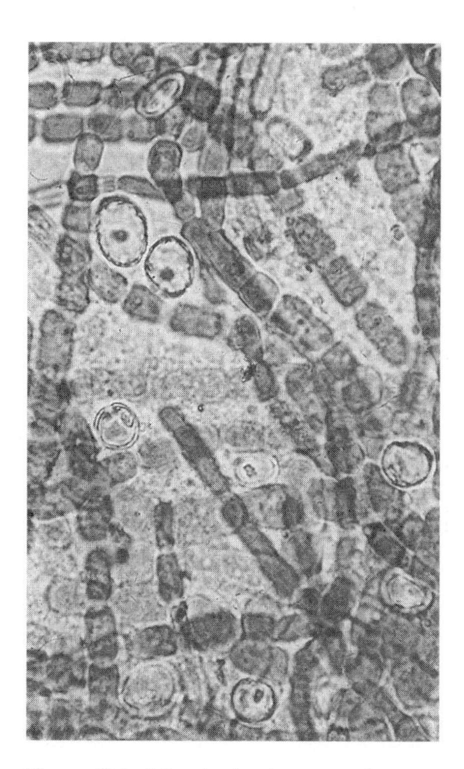

Figure 3.4 A few dead and rotting cells among these living cyanobacteria look as if they have organelles or nuclei inside them. The dead cells have split away from their chains and swollen to a spherical shape, and their rotting contents have formed dark patches and dots. So we cannot identify dots and spots in fossil cells as organelles or nuclei, even if they are. (Courtesy of Andrew Knoll, Harvard University.)

EVOLUTION AND NATURAL SELECTION

Living things grow by using the information coded into the nucleic acids they inherited from the parent cell that split in two in a **replication** event. The nucleic acids specify the right combination and sequence of proteins to be formed, so that the growing organism has a sound structure and biochemical system. But the growing organism lives in a fluctuating world, so conditions may or may not favor its growth and success. It may not survive to replicate, or it may have fewer successful replications than competing cells.

A **mutation** occurs if the nucleic acid is damaged. The mutation may cause a change in a protein that is coded in the nucleic acid sequence. The "wrong" protein may have a major effect on the organism if it is a very important one, or the effect of the change may be so minor as to be insignificant. Most often, but not always, individuals with major changes are unsuited to the environment, and they fail to replicate themselves and their changed DNA. However, small-scale mutations often allow their bearers to survive and reproduce, so variation in DNA and in the form and function of organisms can increase with time in a population. Mutations lead to the appearance of descendants that differ from their ancestors and are therefore likely to be better or worse suited to their environment. Such changes are likely to affect success in replication.

Prokaryotes reproduce dominantly by splitting (replication). But most prokaryotes are capable of coming together with other individuals and exchanging small pieces of DNA. This is **gene transfer**, and the process is called **conjugation**. Obviously, this process can make a significant difference to the gene content of a cell (its **genome**), for better or for worse. (Severe medical problems can occur if a bacterium gains a drug-resistant gene: it depends whether you are a bacterium or a hospital patient as to whether this is for better or for worse.) It's fair to ask how and why conjugation evolved: certainly the genomes of bacteria betray the fact that gene transfer has happened remarkably often in the past, even between distantly related organisms.

This question is much the same as another we could ask. **Sexual reproduction** in eukaryotes is basically very highly organized conjugation, in which the entire genomes of the sex cells of two individuals are entirely shuffled and recombined (in **recombination**). Sometimes an individual, prokaryotic or even eukaryotic may simply make **clones** (identical copies) of itself, in replication or **asexual**

reproduction. Usually, however, eukaryotes cooperate to produce a new individual with a mixture of nucleic acids from each parent, in sexual reproduction. While the new individual that grows as a result of sexual reproduction is genetically similar to both its parents, it is not identical to either. Sexual reproduction, and the recombination of genes that occurs during it, does not introduce any new genetic material into a population, but it does produce new, unique combinations of genes that are coded to produce a new, unique individual.

Therefore, mutation and sexual reproduction, in different ways, result in organisms that are subtly different from one another, and thus are likely to have different success in reproduction. The process by which a particular organism leaves more or fewer successful offspring than another is called **natural selection**. Many chance factors affect the survival of particular individuals in a population. But, in the long run, individuals that are well suited to a prevailing environment survive long enough to leave offspring, while individuals that are not so well suited die before they can reproduce, or leave fewer offspring than others. The offspring of comparatively successful individuals will tend to inherit the characters that made that parent successful, so that they in turn are likely to reproduce successfully if environmental conditions remain the same. Differential reproductive success, resulting directly from natural selection, is the link that connects organisms and their environment.

By now, natural selection has been operating on living things for billions of generations, acting at each generation to fine-tune the relationship between organisms and their environment.

THE EVOLUTION OF SEX

So, then, why conjugation, and why sexual reproduction? Each introduces large genetic changes into descendants, and such large changes, like large mutations, are often large mistakes that produce unsuccessful individuals.

Normally, every prokaryote is its own lineage, either dying, budding off, or replicating into daughter cells. Cell division is simple for prokaryotes. There are no mates to find, organelles to organize. Daughter cells are **clones**, with the same DNA as the parent cell, so they are already well adapted to the microenvironment. Prokaryotes gamble against a change in the environment: if a change occurs that kills an individual, the change will most likely wipe out all that individual's clones too. Prokaryotes have no way to affect the future of their genes, except occasional conjugation.

Most single-celled eukaryotes, and some complex ones, often reproduce by simple fission, cloning identical copies of themselves. An amoeba is perhaps the most familiar example, but corals, aphids, strawberries, and Bermuda grass can use this method too.

In most complex eukaryotes, however, the DNA of two individuals is shuffled and redealt to their offspring in sexual reproduction. Offspring are therefore similar but not identical to their parents: in fact, there is an impossibly low chance that any two sexually reproduced individuals are genetically identical, unless they developed from the same egg, as identical twins do.

The offspring of sexual reproduction resemble their parents in all major features but are unique in their combination of minor characters. Sexually reproducing species have built-in genetic variability that is often lacking in clones of bacteria. Individuals vary in the characters of their bodies, which often means that some individuals are slightly better fitted to the environment than others, so stand a better chance of reproducing. The particular sets of DNA in those individuals are thus differentially represented in future populations.

In organisms that reproduce by cloning, a favorable mutation can spread successfully only if it occurs in an individual that outdivides its competitors. The environment selects or rejects the whole DNA package of the individual, which either divides or dies. This is a one-shot chance, and many potentially successful mutations may be lost because they occur in an individual whose other characters are poorly adapted. However, a favorable mutation may allow one individual such success that it and its clones outcompete all the others, making the population uniform even though it may contain bad genes along with the good one. Uniform populations of yeast may be desirable to a baker or brewer, but in nature a uniform population may easily be wiped out by changes in the environment.

In contrast, a mutation in a sexually reproducing individual is shuffled into a different combination in each of its offspring. For example, a mutant oyster might find her mutation being tested in different combinations in each of her 100,000 eggs. Natural selection could then operate on 100,000 prototypes, not just one. Favorable combinations of genes can be passed on. A sexually reproducing population can evolve rapidly and smoothly in changing environments, and in favorable circumstances evolution can be greatly accelerated by sexual reproduction.

At the same time, sexual reproduction is conservative. Extreme mutations, good or bad, can be diluted out at each generation by recombination with normal genes. The genes may not disappear from the population, but may lurk as recessives, likely to reappear at unpredictable times as recombination shuffles them around.

Eukaryotes are so complex that only approximately similar individuals can shuffle their DNA together with any chance of producing viable offspring. Complex physical, chemical, and behavioral ("instinctive") mechanisms usually ensure that sex is attempted only by individuals that share much the same DNA. Such a set of organisms forms a **species**, which is defined as a set of individuals that are potentially or actually interbreeding. The composite total of genes that are found in a species is called the **gene pool**, the sum of all the genomes of the species.

Sexual reproduction has two great flaws. First, a sexual individual passes on only half of its DNA to any one offspring, with the other half coming from the partner. Therefore, to pass on all its genes, a sexual individual has to invest double the effort of an asexual individual. Second, the offspring of sexual parents are not identical. Sets of incompatible genes may be shuffled together into the DNA of an unfortunate individual, which may die early or fail to reproduce. At every generation, then, some reproductive wastage occurs. Yet so many species reproduce sexually that there must be very strong counterbalancing advantages of sex. Hundreds of pages of scientific papers and several books have been devoted to this question, but there are as yet no convincing answers.

Certainly sexual reproduction has advantages from the point of view of the species or population. It's good for a population to have variability to survive environmental crises; it's good for the future of the species to retain favorable mutations in the gene pool for comprehensive testing. But that's not how evolution works. Selection operates dominantly, perhaps exclusively, on individuals, and individuals do what's best for themselves and for their genes.

Because the problem of the origin of sex is not solved, let me try a suggestion of my own. The major advantage that I see in sexual reproduction is the ability to manipulate the fate of one's DNA. An individual is born with a certain set of DNA,

which cannot be altered. An individual that reproduces by cloning can only pass on what it carries, and it does that best if its environment (to which it is well adapted), does not change. But what if the physical, biological, or biological environment does change? Then a sexually reproducing organism may be superior, because it can modify the genes it passes on, by appropriate choice of a mate. Clearly, it cannot control the details, because recombination is a matter of shuffling and redealing DNA. But it has some choice over the DNA that is to be shuffled with its own, and this may, on average, produce offspring that survive better.

It is difficult to pin down the time when sexual reproduction began among eukaryotes. At best, we look for creatures in the fossil record whose descendants reproduce sexually today. For certainty, we must look to the appearance of multicellular animals, perhaps at 600 Ma. However, many protists reproduce sexually, and they are diverse in the fossil record by about 1000 Ma.

THE CLASSIFICATION OF EUKARYOTES

Eukaryotes occur in the natural world in ecological and evolutionary units called species: groups of individuals whose genetic material is drawn from the same gene pool but is almost always incompatible with that of another gene pool. Members of the same species can interbreed to produce viable offspring. They tend to share more physical, behavioral, and biochemical features (called characters) with one another than they do with members of other species. Defining and comparing such characters allows us to distinguish species of organisms. A species is not an arbitrary group of organisms, but a real, natural, unit.

Biologists use the Linnean system of naming species, after the Swedish biologist Carl Linné who invented it in the eighteenth century. A species is given a unique name (a specific name) by which we can refer to it unambiguously. Species that share a large number of characters are placed together into groups called **genera** (the singular is genus) and are given unique generic names.

Taxonomic names sometimes carry a meaning that may make then easier to remember. So Linné gave the specific name *noctua* to the little owl of Europe because it flies at night, and he gave it the generic name *Athene*, the Greek goddess of wisdom. (This owl is the symbol of the city of Athens, and appeared on its ancient coins.) However, formal taxonomic names do not have to carry a message, though a simple and appropriate name is easier to remember. One must be careful about names: *Puffinus puffinus* is not a puffin, but a shearwater, and *Pinguinus* is not a penguin but the extinct Great Auk! Thus, Linnean names are only a convenience, but a very valuable one. The bird that the British call the tawny owl, Germans the wood owl, and Swedes the cat owl, is *Strix aluco* among international scientists.

Genera may be grouped together into higher categories. For example, many species of owls are grouped together to form the Family Strigidae, or strigids, named after one of its genera, *Strix*. **Families** may be grouped into superfamilies, and after that into orders, classes, and phyla. Many other subdivisions can be coined for convenience (Figure 3.5).

kingdom
phylum
class
order
family
genus
species

super-
sub-
infra-

Figure 3.5 The Linnean hierarchy of taxa, or taxonomic units, as used by zoologists. Other categories such as "cohort" can be used, and the prefixes in the lower left-hand corner are used freely. Botanists and microbiologists use rather different categories, but the hierarchical principle remains the same.

A division or subdivision that is used to arrange organisms into groups is called a taxon (plural, taxa). Biologists who try to recognize, describe, name, define, and classify organisms are taxonomists or systematists, and the practice is called taxonomy or systematics or classification. Although slightly different ranks of categories are used for different kingdoms of organisms, the basic units of classification recognized by all biologists remain the species and genus.

Complex rules have grown up as more and more organisms have been described by taxonomists. Botanists and zoologists have their own rules for describing new species so that names are legally established.

Describing Evolution

We now recognize that members of a species together form a genetically based biological unit that is evolutionarily separate from the rest of the organic world. The recognition and naming of a new species, therefore, is a statement about evolution. It represents the hypothesis that the members of the species share the same gene pool, which is different from the gene pool of any other species because there has been evolutionary divergence over time.

Most taxonomists aim to form genera and higher categories that truly reflect evolution. In evolutionary classification, species are grouped into genera on the hypothesis that genera also share a unique set of characters that are not shared by other genera. In turn, genera are grouped into families, and so on, again giving evolutionary significance to Linnean terms.

Decisions about the course of evolution are not always obvious, so taxonomic decisions may have to be revised as new information becomes available. Species are moved around between genera and higher categories as taxonomists come to understand evolutionary history more accurately. The incomplete nature of the fossil record makes classification particularly difficult for paleontologists, and it often leads to uncertainties or arguments about classification.

As organisms evolve through time, their characters change, slowly or quickly, gradually or suddenly. As changes accumulate, species evolve characters that are changed from their original state. The new characters may be different enough that a biologist who could examine living specimens at the ends of the series would certainly regard them as separate species. But how can one draw a line between species in time? After all, descendants have always been genetically continuous with their ancestors. At some point, the first bird must have hatched out of a dinosaur egg. There is no discontinuity between ancestor and descendant species like that seen between contemporaneous species in the living world.

This is a special question facing paleontologists, and it makes the taxonomy of the fossil record rather difficult, and often viciously argued, as, for example, in the study of fossil hominids. The principle is simple, however. At one extreme one could say that all living organisms are the same species, because they all evolved, continuously, from a single ancestor that was the first living cell. But for convenience, and to reflect the reproductive gaps that exist between species at any given time in Earth's history, a paleontologist must draw lines somewhere between species (and genera, and families, and so on), knowing that the lines are artificial if they pretend to separate ancestors from descendants.

Fortunately or unfortunately, the fossil record is spotty enough, and the pace of evolutionary change is rapid enough, that truly intermediate fossils are very rarely found. The fossil record is therefore rather more easily divided into species and higher categories than one might expect.

CLADISTICS

This book uses a mild form of an approach to evolutionary taxonomy called

cladistics. The aim of cladistics is to trace the phylogeny (the evolutionary pathways) by which species appeared and to use that phylogeny to organize species into clades, that is, groups of species that all descended from one ancestral species. To use an analogy, a **clade** is a branch on a tree that represents all life: if all life on Earth is descended from the first living cell by a series of evolutionary branching events, life as a whole is one clade (the entire tree). But trees may branch many times, and branches may then branch, and so on, and clades of organisms also exist in a hierarchy of branches, with terminal branches representing single species. Every species belongs to a larger clade, which belongs to a larger clade, and so on. Every clade, no matter how large, began with a single branching event that produced the ancestor species of the clade.

Cladists try to identify groups of species that share a set of characters that evolved as new features in a common ancestor, and were then passed on to all descendant species. Such newly evolved characters mark a change from a primitive or original state to a novel or derived state. For example, all living mammals have fur, but no other living organisms do. Perhaps fur was inherited from a common ancestor of all living mammals that evolved a furry skin as a new derived character modifying a primitive one (a scaly skin). If that is true, then mammals are a clade. Examining the hypothesis, one finds other shared derived characters of living mammals that strengthen the argument: for example, all living mammals are warm blooded, and suckle their young.

Sometimes problems arise because similar derived characters are found in other species in another clade: those characters must have evolved more than once, in **parallel evolution**. For example, bats and birds both have wings, and in each group the wing is a derived character that has been modified from some other structure. But bats and birds share very few other derived characters, and the weight of evidence suggests that birds are a clade, bats are a clade, but [bats + birds] is not a clade.

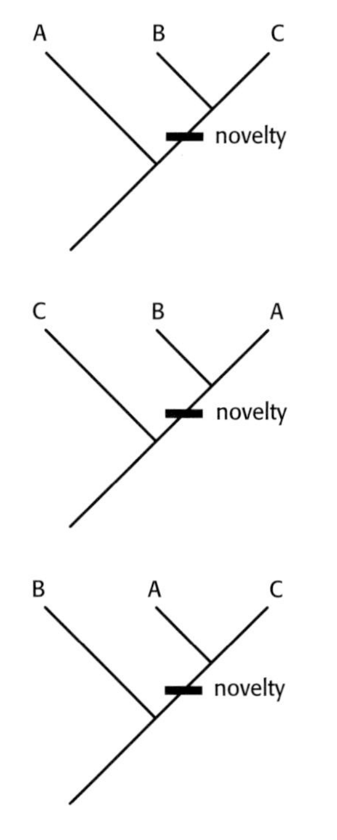

Figure 3.6 Three possible hypotheses (expressed as cladograms) could portray the evolutionary relationship of these three species. Perhaps one might feel unable to work out the relationship. In that case one would portray the three species as separating into three equal branches. Obviously, with more taxa under study, the number of possible hypotheses increases enormously.

Once a clade is established, we then search for characters that are novel in species within the clade. Subclades can be established based on the distribution of such derived characters within the group, until a single best hypothesis emerges about the total evolutionary history of the group of species. The hypothesis can be tested as further characters are examined or existing ones are reassessed, and as new species are discovered and fitted into the evolutionary framework.

As an example, three living species, A, B, and C, could be related along three possible evolutionary pathways (Figure 3.6). Which is correct? Which two of the three species are most closely linked? Two species may look very similar because they share similar characters, but if those are shared primitive characters that were also present in a common ancestor, they cannot tell us anything about evolution within the group, because they have not changed within that history. The useful character for solving the problem is the novelty, or derived character, which defines the group that has changed the most since the three species all shared the characters of their common ancestor (Figure 3.6).

Cladograms such as Figure 3.6 display the distribution of characters in a visual form, and the cladogram that requires the simplest and fewest evolutionary changes is assumed to represent best the phylogenetic history of the species. A cladogram therefore expresses a hypothesis about the phylogeny of a group. Two of the species are most closely linked, and form sister groups in a clade, while the third species becomes their sister group in a larger clade (Figure 3.6). Most often, the construction of a cladogram is aided by computer analysis of the characters. The computer can deal better with huge amounts of data than a normal human brain can, and the result is that a typical cladogram defines clades on multiple characters rather than those we once used (say, fur defines mammals).

Once the preferred cladogram is drawn to show the best hypothesis, one can make decisions about the best way to classify the species and to describe its evolutionary history. A cladogram in itself does neither of these things.

Names in Cladistics

One could introduce formal names for each clade on a cladogram. This would lead to a huge number of names, not all of which might be necessary for everyday discussion around the breakfast table. It conveys more information simply to print a cladogram and to use a minimum of hierarchical names.

The definition of names for clades is a continuing problem, just as it was in the days before cladistics came into general use. It is easy to see that a clade is defined by two things: the branch point, or **node**, or the branch just below that node (the **stem**) that define its first appearance, and also by the sum total of the members that make up the clade. These two definitions are, of course, identical.

Problems arise because we usually find ourselves in the situation that the later members of the clade are more derived (more evolved) than the founders. Typically, then, we have a picture of what a clade is, based on the more modern and more derived members that we are more familiar with. So the word "whales," to us, typically conjures up a picture of a blue whale or an orca. However, early members of that clade, the whale ancestors, had legs, and walked. The ultimate whale ancestor, that evolved the first whale-like characters, was related to artiodactyls (derived artiodactyls today are deer, cattle, pigs, hippos, and so on). So if we want to put names on the evolutionary history of whales, what do we do? What are whales (Cetacea)?

In this kind of situation there are two alternatives, typically favored by different groups of researchers who fill pages of criticism of one another's views, so that a lot of energy is diverted away from studying the creatures and into arguments about the philosophy of cladistics.

One can use a **crown-group definition**, a top-down view of whale evolution. Cetacea would be a clade defined by its living members, and it would include all whales down to the point or branch (node or stem) where one could find the common ancestor of all living whales. However, Cetacea once had close relatives that you and I would certainly call whales, but belonged to clades that are now extinct, rather than the clade that includes only living whales. What are these creatures, then? A crown-group definition excludes them from Cetacea, and we would have to find another name, or set of names, for them.

One can use a **stem-group definition**, or bottom-up definition. Then Cetacea would be defined as all those clades, living or fossil, that are more closely related to living whales than to living artiodactyls. This seems more reasonable (to me) in the sense that it includes everything you and I would identify as a whale, plus most of the walking ancestors of whales. The stem group then merges into a mess at the point where whale ancestors and artiodactyl ancestors are indistinguishable.

Notice that whatever system is chose, the difficult issues are always toward the root of the cladogram, where the fossil record tends to be poorer, and the creatures are more difficult to separate because they are very closely related.

Both definitions are logically defensible, which makes it hard to choose one

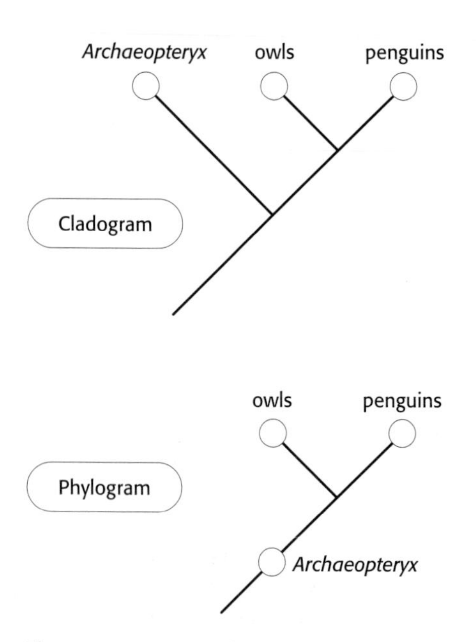

Figure 3.7 A cladogram and a phylogram of three groups of birds. In the cladogram, no group is shown as the ancestor of another, because a cladogram seeks only to show the degree of relationship between groups. Owls and penguins share derived characters that *Archaeopteryx* does not have. But suppose I wanted to make the additional hypothesis that *Archaeopteryx* was not only less derived than the other two, but was ancestral to them. To show that, I would draw a phylogram or phylogenetic tree of the same three groups to show the hypothesis that *Archaeopteryx* was the ancestor of owls and penguins. This time, *Archaeopteryx* is shown within the body of the tree, below the event that marks the evolutionary branching between owls and penguins.

LIMERICK 3.2
The human species has thrived,
In the short time since we arrived.
But the cladist affirms
That we are only worms,
Even though we are somewhat derived.

system over the other on logical grounds. Usually the decision is made on arbitrary grounds (which influential paleontologists work most on these animals, and what do they [mainly] agree they should do?). So Cetacea is normally defined on a stem-group basis, while Mammalia is not; Tetrapoda is under discussion. My personal preference is for a stem-group approach because it usually makes for a clearer understanding of the biology and ecology of a clade. Watch out for naming issues throughout the rest of the book, because they tend to make discussion of evolution more clumsy than it ought to be.

Remember, though, that the naming game does not affect the structure of the cladogram. Sometime in the distant future we will all agree on the cladogram of all life on Earth, but I expect we will still be arguing about names.

Phylograms

A cladogram is always drawn with all the species under study along one edge (Figure 3.7). No species in a cladogram is shown as evolving into another. Some cladists claim that one can never know true ancestor–descendant relationships, and in a strict sense this is correct because we don't have time machines. But sometimes a fossil is known that could well be an ancestor of a later fossil or of a living organism. At present, for example, it seems more reasonable (to me) to suggest that *Archaeopteryx* is the ancestor of owls and penguins than to suggest that these birds are descended from some ancestor that we haven't found yet.

Hypotheses like this are expressed on **phylograms** that may include time information: we may show a suggested ancestor within the diagram (Figure 3.7). Like cladograms, phylograms are not statements of fact but hypotheses, subject to continuous testing.

Counterintuitive patterns sometimes emerge in cladistics. We are all used to thinking about fishes, amphibians, reptiles, birds, and mammals as classes of vertebrates, in some way equal in rank to one another (Figure 3.8). But this is not a cladistic classification. Tetrapods are actually a clade within fishes, derived from them by acquiring some novel characters, including feet, and amphibians are a clade within tetrapods. Reptiles are in turn another derived subgroup of tetrapods, and birds are a clade of derived reptiles (derived dinosaurs, in fact).

There's nothing intimidating about this: it simply takes time to get used to it. The important thing is not to try to force the older taxonomic units into a cladistic framework, but to combine a simple and convenient classification with an explanatory cladogram or phylogram. I have tried to use cladograms and phylograms in this way.

If we classify all living reptiles into one group and draw a cladogram (Figure 3.8), we display the well-known fact that living reptiles and birds are more alike than either is to mammals. The cladogram also carries other information. It shows that warm blood, a derived character that living birds and mammals share, must have evolved independently at least twice, unless living reptiles have lost warm blood.

As we consider smaller subgroups of living and fossil reptiles, we find that the neat picture of reptile classification breaks down, and that we must revise our interpretations. Figure 3.9 shows that "living reptiles" is not a clade. We can define a clade called "reptiles," but we have to include birds in it. This means that turtles, birds, crocodiles, and lizards are all reptilian clades that have diverged from an ancestral, primitive reptile, although some are more derived than others in the sense that they have evolved more novel characters that their common ancestor did not

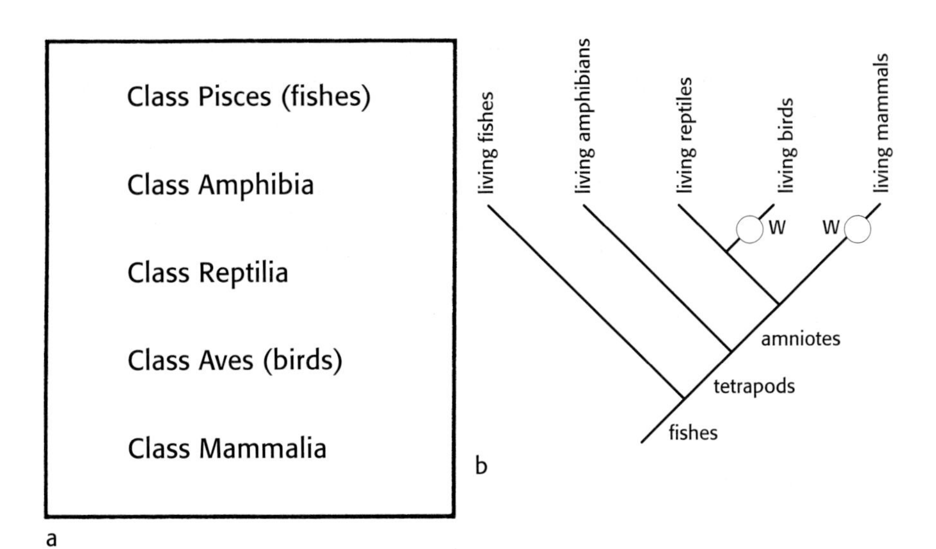

Figure 3.8 A conventional classification (a) of living vertebrates and a cladogram of the vertebrates. In the conventional classification, each class has equal rank to the others. In the cladogram (b), it is clear that all classes are not equivalent in rank: for example, living mammals are the sister group of [living birds + living reptiles]. Note that the novel character W, warm blood, has been independently derived in birds and mammals, according to the hypothesis expressed in this cladogram.

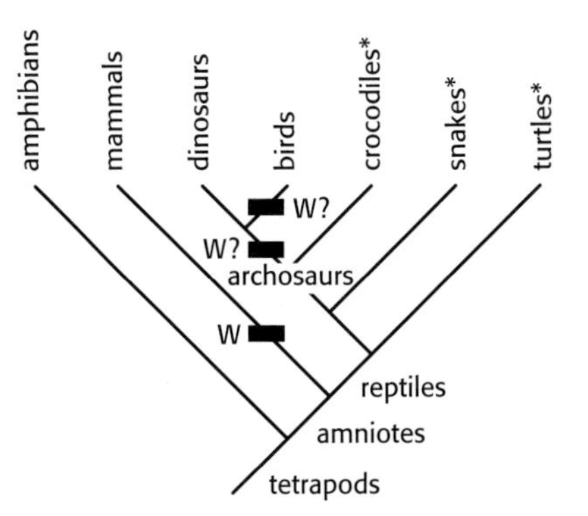

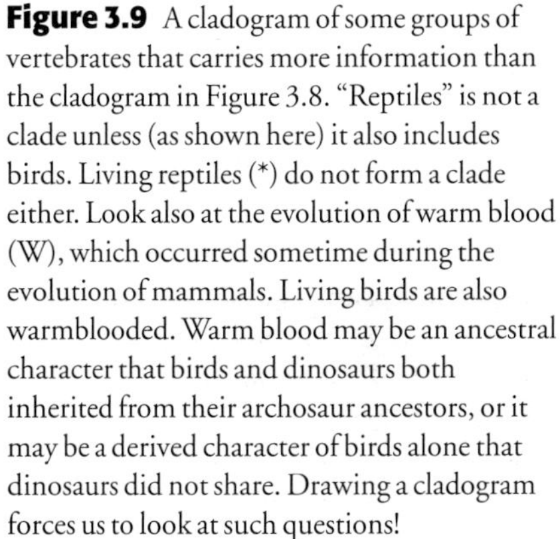
Figure 3.9 A cladogram of some groups of vertebrates that carries more information than the cladogram in Figure 3.8. "Reptiles" is not a clade unless (as shown here) it also includes birds. Living reptiles (*) do not form a clade either. Look also at the evolution of warm blood (W), which occurred sometime during the evolution of mammals. Living birds are also warmblooded. Warm blood may be an ancestral character that birds and dinosaurs both inherited from their archosaur ancestors, or it may be a derived character of birds alone that dinosaurs did not share. Drawing a cladogram forces us to look at such questions!

have. In the same way, humans are derived primates, derived mammals, and derived fishes, all at the same time. Once one is used to this unusual line of thinking, evolution becomes much more real, and we can see, for example, that humans, tapeworms, and the scums of Shark Bay are all prokaryotes, though some have accumulated more visible derived characters than others.

Further Reading

Brocks, J. J., et al. 1999. Archean molecular fossils and the early rise of eukaryotes. *Science* 285: 1033–6, and comment, 1025–6.

Gabaldón, T., and M. A. Huynen. 2003. Reconstruction of the proto-mitochondrial metabolism. *Science* 301: 609.

Han, T.-M., and B. Runnegar. 1992. Megascopic eukaryotic algae from the 2.1 billion-year-old Negaunee Iron Formation, Michigan. *Science* 257: 232–5; comment, v. 259, p. 835. [*Grypania.*]

Mojzsis, S. J. 2003. Probing early atmospheres. *Nature* 425: 249–251. [Comment on new research that suggests the Proterozoic atmosphere had high CO_2.]

Pavlov, A. A. 2003. Methane-rich Proterozoic atmosphere? *Geology* 31: 87–90.

Pilbeam, D. 2000. Hominoid systematics: the soft evidence. *Proceedings of the National Academy of Sciences* 97: 10684–6. [This is really about cladistic methods, rather than hominoid evolution.]

Roger, A. J., and J. D. Silberman. 2002. Mitochondria in hiding. *Nature* 418: 827–9.

Tenaillon, O., et al. 2000. Mutators and sex in bacteria: conflict between adaptive strategies. *Proceedings of the National Academy of Sciences* 97: 10465–70. [A difficult paper, but important. It suggests that sexual reproduction strongly selects against "strong mutators": genes that greatly increase mutation rates. In other words, sex tends to *slow down* the mutation rate. Recombination then plays a larger part in generating novelty.]

Whitman, W. B., et al. 1998. Prokaryotes: the unseen majority. *Proceedings of the National Academy of Sciences* 95: 6578–83.

CHAPTER FOUR

The Evolution of Metazoans

Eukaryotes come in two grades of organization: single-celled (protists) and multicellular (plants, animals, and fungi). The world today is full of **metazoans**, complex multicellular animals, and **metaphytes**, complex multicellular plants. How, why, and when did they evolve from protists?

PROTEROZOIC PROTISTS

A protist can carry chlorophyll (thus it can be an autotrophic, photosynthetic, "alga"), or it can eat other organisms (thus it can be a heterotrophic, "protozoan" "animal"), or it may do both.

Beginning about 1850 Ma, we find **acritarchs**, spherical microfossils with thick and complex organic walls. They were probably algae that grew thick organic coats in a resting stage of their life cycle, but spent most of their life floating (coat-free) in the **plankton**, the community of organisms that makes a living in the surface waters of oceans and lakes. Probably because of the different chemistry of surface and bottom ocean waters, Proterozoic protists seem to have been planktonic, while prokaryotes still lived in seafloor bacterial mats. Almost all eukaryotes, even single-celled ones, require some oxygen to run their mitochondria, so the Proterozoic seafloor would have been a very unfriendly place for eukaryotes.

Even at the surface, shortage of nutrients may have held back eukaryote evolution. It is only in the Late Proterozoic that we begin to see a good variety of planktonic protists. That may mean in turn that oxygen production did not increase much before, say, 800 Ma. Then increased oxygen levels would have gradually extended the oxygen-bearing zone deeper and deeper into the Proterozoic ocean, cutting down methane production in deep waters. In the end, this process would have cooled the global climate, and more and more of the seafloor would have become inhabited by protists as well as bacteria and Archaea. Not until the Late Proterozoic can we envisage seafloors crawling with successful populations of protists consuming the rich food supplies available in bacterial mats. Even more important, the Middle Proterozoic would not have been a good time for complex animals or plants to evolve.

EVOLVING METAZOANS FROM PROTISTS

A flagellate protist is a single cell with a lashing filament, a **flagellum** (plural, flagella), that moves it through the water (Figure 4.1a). Some species of flagellate protists, the **choanoflagellates**, form clones: they reproduce by budding off new individuals, which then stay together to form a compound animal or colony rooted to the seafloor (Figure 4.2a). The flagella beat to generate a systematic water current around and between the individuals, which filter bacteria from the water flow.

A **sponge** is the simplest multicellular variation on this theme. It contains many similar flagellated cells arranged so that they generate and direct water currents efficiently (Figures 4.1b; 4.2b; 4.3a). Sponges are more advanced than simple colonies of choanoflagellates because they also have specialized cells: some form a body wall, some digest and distribute the food they collect, and some construct a stiffening skeletal framework of organic or mineral protein that allows sponges to become large without collapsing into a heap of jelly (Figure 4.3b). Sponges are thus metazoans, not protists. Metazoans are not just multicellular, they have different kinds of cells that perform different functions.

Metazoans are most likely a clade, that is, they all descended from one kind of protist (Figure 4.4). All metazoans originally had one cilium or flagellum per cell, for example. Metazoans also share the same kind of early development. They form into folded balls of internal cells which are often free to move, and are covered by outer sheets of cells that form an external skin-like coating for the animal.

The first metazoans were small and soft-bodied, so we have no fossil record of them. But we can look at the tremendous variety of living animals and at the geologic record to try to reason out what the first metazoans might have looked like and what they might have done.

There are only three kinds of metazoans: **sponges** and their relatives; **cnidarians** and their relatives; and **bilaterians** (three-dimensional animals with distinct bilateral symmetry). All of them solved the problem of developing to greater size and complexity, but in different ways.

Sponges probably branched off first from the ancestral metazoan, and their immediate ancestor was probably like a colonial choanoflagellate (Figures 4.1, 4.2). Sponges evolved by extending the choanoflagellate way of life to large size and

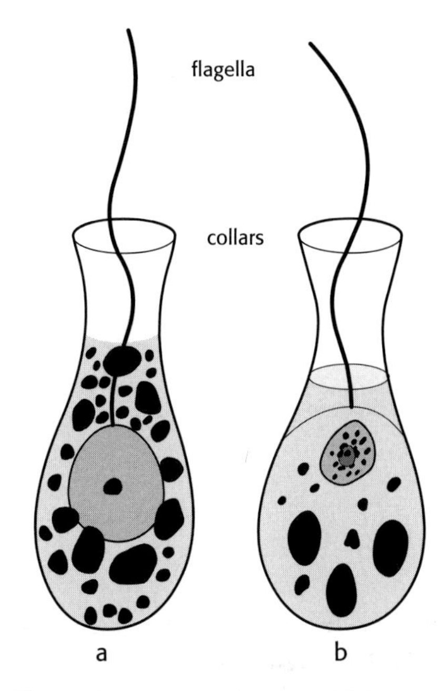

Figure 4.1 Protists and sponges. (a) A choanoflagellate, a protist that uses its flagellum to move water, thus propelling the animal through the water. Food is captured as water is pulled through the collar. (b) A "collar cell" or choanocyte from a sponge. It has almost exactly the same structure and function, except that the cell is anchored in the body of the sponge and does not move through the water. The flagellum beats to produce a water current that is entirely for feeding. (From Barnes et al., *The Invertebrates: A Synthesis.* 3rd edn © 2001 Blackwell Science.)

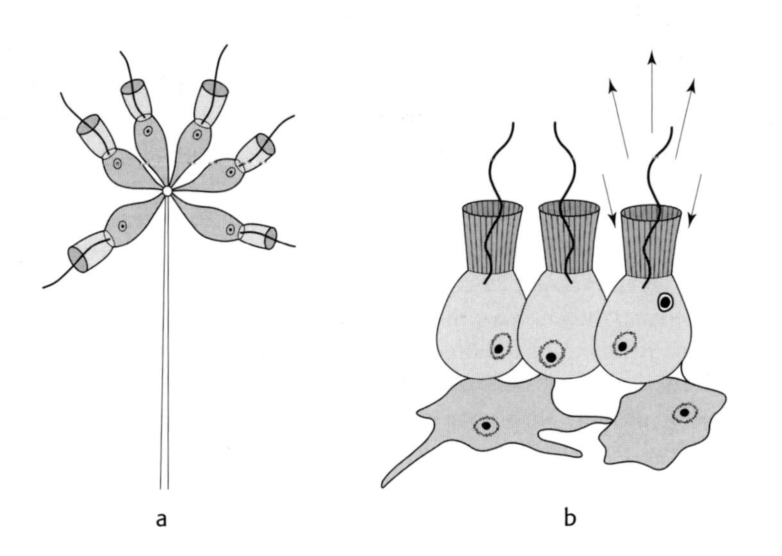

Figure 4.2 (a) A colony of choanoflagellates. New individuals that bud from a parent cell stay attached to one another and to the seafloor. The colony does not move, but the flagella of all individuals act together to generate a feeding current that draws water through the filters of individual cells. Most likely the first metazoan evolved through an organism that looked and functioned like this. (b) A group of choanocytes from a sponge. (From Barnes et al., *The Invertebrates: A Synthesis.* 3rd edn © 2001 Blackwell Science.)

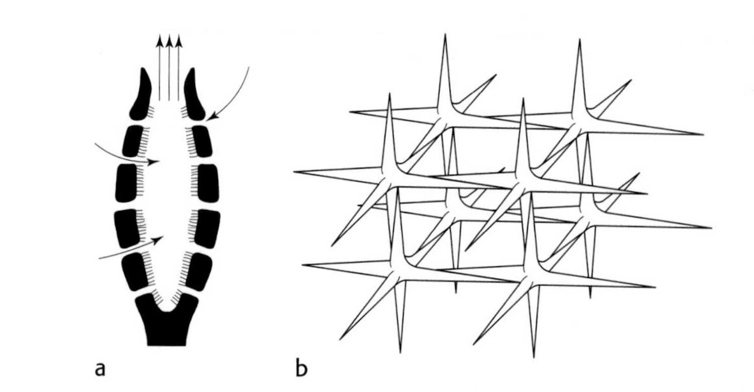

Figure 4.3 (a) The structure of a simple sponge, which is advanced over choanoflagellates in that the filters are more sophisticated and are enclosed inside a body wall. (After Barnes et al., *The Invertebrates: A Synthesis.* 3rd edn © 2001 Blackwell Science.) (b) Many sponges have supporting structures that help them retain their shape, even at large size. For example, some glass sponges have mineralized needles or spicules arranged into a skeletal framework like the steel girders in a skyscraper. (From Boardman et al. (eds), *Fossil Invertebrates.* © 1987 Blackwell Scientific.)

sophisticated packaging. They continued to pump water (and the oxygen and bacteria they capture from it) through their tissues, in internal filtering modules (Figure 4.3).

Cnidarians (or coelenterates), including sea anemones, jellyfish, and corals, are built mostly of **sheets** of cells, and they exploit the large surface area of the sheets in sophisticated ways to make a living. The cnidarian sheet of tissue has cells on each surface and a layer of jelly-like substance in the middle. The sheet is shaped into a bag-like form to define an outer and an inner surface (Figure 4.5a). A cnidarian thus contains a lot of seawater in a largely enclosed cavity lined by the inner surface of the sheet. The neck of the bag forms a mouth, which can be closed by muscles that act like a drawstring. A network of nerve cells runs through the tissue sheet to coordinate the actions of the animal.

In most cnidarians the outer surface of the sheet is simply a protective skin. The inner surface is mainly digestive, and absorbs food molecules from the water in the enclosed cavity. Because cnidarians are built only of thin sheets of tissue, they weigh very little, and can exist on small amounts of food. They can absorb all the oxygen they need from the water that surrounds them.

Cnidarians have **nematocysts** or stinging cells set into the outer skin surface (Figures 4.5b; 4.5c). The toxins of some cnidarians are powerful enough to kill fish, and people have died after being stung by swarms of jellyfish. Nematocysts are usually concentrated on the surfaces and the ends of tentacles, which form a ring around the mouth. They provide an effective defense for the cnidarian, but they are also powerful weapons for catching and killing prey, which the tentacles then push into the mouth for digestion in the cavity. The digestive cells lining the cavity leak powerful enzymes into the water. The tissues of the prey are broken down by the enzymes, and the cnidarian absorbs the food molecules through the cells of the inner lining of the cavity. A cnidarian can thus eat prey without a real mouth or a real gut.

Hardly any sponges can tackle food particles larger than a bacterium, though there are a few exceptions. Yet living cnidarians routinely trap, kill, and digest creatures that outweigh them many times by using their nematocysts. However, there is no guarantee that the first cnidarians had nematocysts: they may simply have absorbed dissolved organic nutrients from seawater.

The third and most complex metazoan group contains all the other metazoans, including vertebrates. These are the Bilateria or bilaterians, metazoans with a distinct bilateral symmetry that influences their biology enormously. Worms are the simplest bilaterians. Bilaterians consist basically of a double sheet of tissue that is folded around with the inner surfaces largely joining, to form a three-dimensional animal. In contrast to sponges and cnidarians, they have complex organ systems made from specialized cells, and those organ systems are built as the animal grows by special regulatory mechanisms coded in the genes.

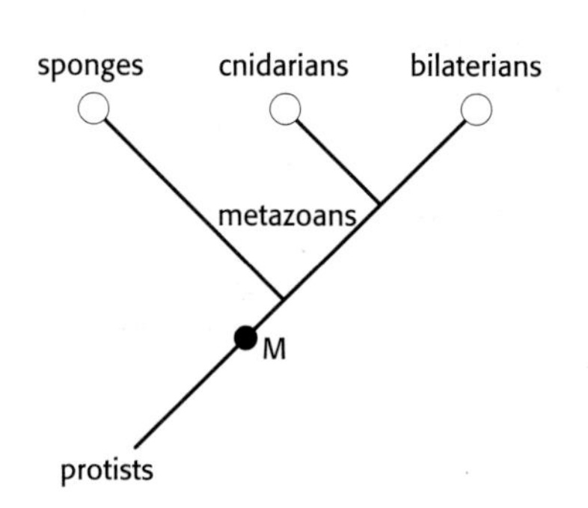

Figure 4.4 The three major groups of metazoans are a clade [Metazoa] that evolved from a protist by acquiring specialized multicellularity (M).

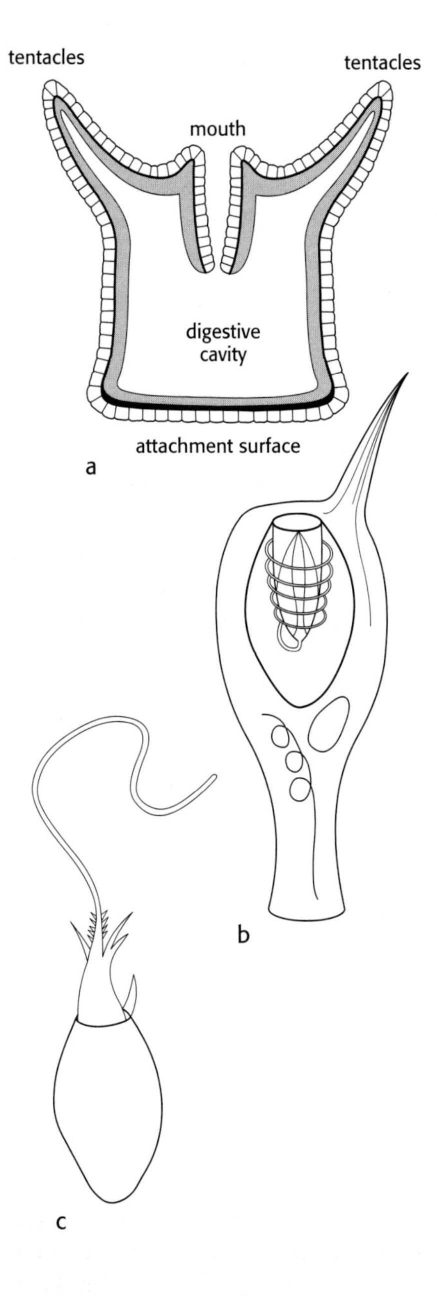

Figure 4.5 Important features of cnidaria. (a) Basic structure of a cnidarian. (From Boardman et al. (eds), *Fossil Invertebrates.* © 1987 Blackwell Scientific.) (b) Nematocysts are unique features of cnidarians, used in defense and feeding. This cell has not been triggered. (c) A discharged nematocyst, showing the range of the weapon. (Both after Barnes et al., *The Invertebrates: A Synthesis.* 3rd edn © 2001 Blackwell Science.)

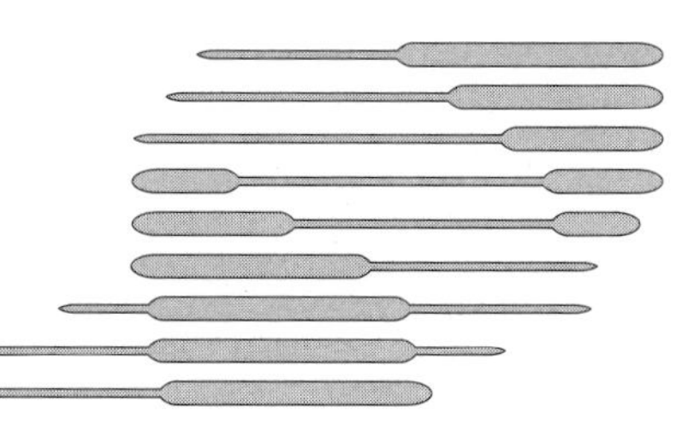

Figure 4.6 How bilaterians use the coelom as a hydraulic device for burrowing. This worm is burrowing from right to left. It extends the front end by squeezing fluid forward, then forms a bulb at the front to act as an anchor. Then it pulls the rest of the body up to the anchor, and begins the cycle once again. We use much the same logical sequence of actions, but a different mechanism, to winch a stranded truck out of mud. An advanced worm, partitioned into segments separated by valves, has an even more efficient burrowing action.

All sponges and most cnidarians are attached to the seafloor as adults, and depend on trapping food from the water. But many bilaterians were and are mostly free-living animals, making of them a living as mobile scavengers and predators. The bilateral symmetry is undoubtedly linked with mobility: any other shape would give an animal that could not move efficiently.

The first bilaterians would have been worm-like. Worms creep along the seafloor on their ventral (lower) surface, which may be different from the dorsal (upper) surface. They prefer to move in one direction, and a head at the (front) end contains major nerve centers associated with sensing the environment. A well-developed nervous system coordinates muscles so that a worm can react quickly and efficiently to external stimuli. The mobility of early bilaterians on the seafloor probably led to the differentiation of the body into anterior and posterior (head and tail) and into dorsal and ventral surfaces, as the various parts of the animal encountered different stimuli and had to be able to react to them.

LIMERICK 4.1

The paleontologist's view
Classes worms together with you
This is based on the claim
That instead of a plane
A worm's three-dimensional too.
© Elizabeth Wenk, 1994

The front end of bilaterians usually features the food intake, a mouth through which food is passed into and along a specialized one-way internal digestive tract instead of being digested in a simple seawater cavity. No sponge cell or cnidarian cell is very far away from a food-absorbing (digestive) cell, so these creatures have no specialized internal transport system. But the digestive system of bilaterians needs an oxygen supply, and the nutrients absorbed there have to be transported to the rest of the body. Bilaterians therefore have a circulation system, and the larger and more three-dimensional they are, the better the circulation system must be.

All but the simplest (?earliest) bilaterians have an internal fluid-filled cavity, the **coelom**, which may be highly modified in living forms. In humans, for example, the coelom is the sac containing all the internal organs. The coelom may have evolved as a useful hydraulic device. Liquid is incompressible, and a bilaterian with a coelom (a **coelomate**) can squeeze this internal reservoir by body muscles. Such squeezing pokes out the body wall at its weakest point, which is usually an end (Figure 4.6). Such a hydraulic extension of the body can be used as a power drill for burrowing into the sediment, to find food, or safety.

LIMERICK 4.2
"Complex" is a relative term
That is rarely applied to a worm.
But a worm that's coelomic
Can be rather comic:
It can burrow and wriggle and squirm.

The coelom could have provided another great advantage for bilaterians. Oxygen must reach all the cells in the body for respiration and metabolism. Single celled organisms can usually get all the oxygen they need because it simply diffuses through the cell wall into their tiny bodies. Sponges pump water throughout their bodies as they feed, and cnidarians and flatworms are at most two sheets of tissue thick. But larger animals with thicker tissues cannot supply all the oxygen they need by diffusion (Figure 4.7). Oxygen supply to the innermost tissues becomes a genuine problem with any increase in body thickness or complexity. If the animal evolved some exchange system so that its coelomic fluid was oxygenated, the coelom could then act as a large store of reserve oxygen. Eventually the animal could evolve pumps and branches and circuits connected with the coelom to form an efficient circulatory system.

Many advanced bilaterians have **segments**: their bodies are divided by septa that separate the coelom into separate chambers connected by valves. This arrangement is more efficient for burrowing than a simple, single coelomic cavity (Figure 4.6). The segmentation of many animals, including earthworms, may be derived from this invention on the Precambrian seafloor.

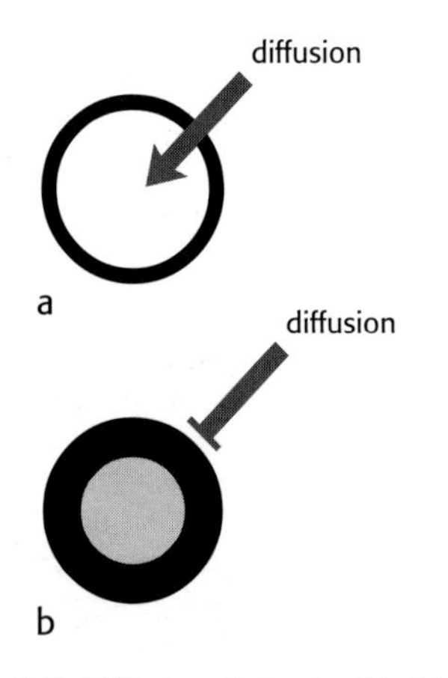

Figure 4.7 Diffusion. Animals with thin tissues (a) can rely on diffusion alone to supply the entire body with oxygen. But diffusion cannot supply oxygen to the interior of animals with thick tissues (b) special respiratory systems are needed so that the inner tissues are not starved of oxygen.

Respiration problems probably prevented early coelomates from burrowing for food in rich organic sediments, which are very low in oxygen. But a coelomate burrowing for protection might have evolved some special organs to obtain oxygen from the overlying seawater at one end while the main body remained safely below the surface. Many coelomates that live in shallow burrows have various kinds of tentacles, filaments, and gills that they extend into the water as respiratory organs (Figure 4.8). It is a very short step from here to the point where a coelomate collects food as well as oxygen from the water by filter feeding (Figure 4.8a), as in all bryozoans and brachiopods, in some molluscs, worms, and echinoderms, and in simple chordates.

Other coelomates have evolved an alternative solution. They burrow so actively that their body movements pump oxygenated water down the burrow over them. In these worms, respiration through the skin surface is sufficient for their oxygen needs as long as they also have an efficient internal circulatory system to distribute the oxygen (as in some worms and in many arthropods such as burrowing shrimp). Some of these animals have also evolved to collect food from the respiratory currents flowing down into their burrows (for example, Figure 4.8b), but most are still sediment scavengers and predators.

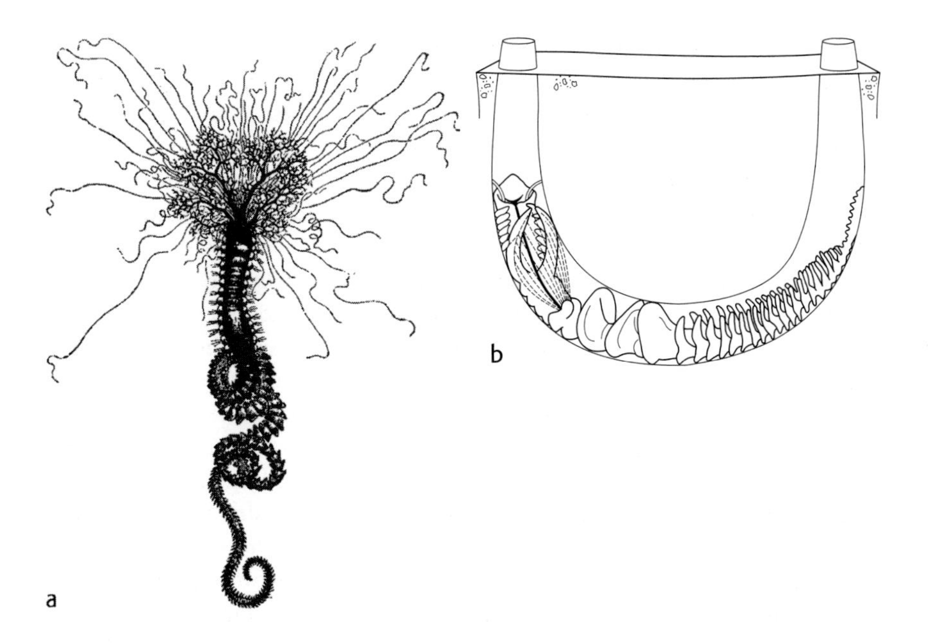

Figure 4.8 Respiration may be a problem as coelomate worms burrow into oxygen-poor sediments. Two solutions: the worm (a) evolved coelom-filled tentacles that project into the water for feeding and respiration. The worm (b) evolved body movements that pump water through its burrow, bringing food and oxygen. (From Barnes et al., *The Invertebrates: A Synthesis*. 3rd edn © 2001 Blackwell Science.)

EVOLUTION AND DEVELOPMENT IN METAZOANS

We can now read the entire genetic code of some organisms: all it takes is time and money. In addition, we have enough information that we can begin to recognize certain strings of DNA as **genes**, and understand what many of them do within the living organism. For example, the entire genome (genetic code) of the human parasite *Mycoplasma genitalium*, the smallest genome so far discovered, contains 580,000 molecules of DNA. Sorting, slicing, and dicing this genome, geneticists have concluded that *M. genitalium* has 480 genes that code for proteins, and 37 that code for RNA. We understand the emphasis on proteins, because they perform so many cell functions: building lipids for the cell membrane, transporting phosphate, breaking down glucose, and so on.

It is more complex to grow a viable multicellular animal than a protist. The genome must contain the information to build many kinds of cells rather than just one, and the information to grow them at the right time, to place them accurately in the body, and to develop the control mechanisms, sensory systems, transport systems, and whole-body biochemical reactions that operate in a multicellular organism.

The genetic programming that builds a metazoan from a single cell need not specify individual cells one by one. Like a well-written computer program, there can be tricks that promote efficiency. For example, one could program a computer to draw a flower, specifying the size, shape, and position of each petal. But the petals of any given flower are typically much alike, so one can use the same shape and size for each petal, and simply tell the computer to move the pen to the right place and draw the same petal each time (Figure 4.9).

Figure 4.9 One can construct a complex object by judicious placing of very simple units.

In the same way, metazoans have **structural genes** that build each piece of the animal, and **regulatory genes** that make sure the piece is built in the right place at the right time. For example, a set of regulatory genes could be used in combination with a set of "segment" genes to build all the segments along a growing worm. The same sort of regulatory genes could easily be used to build legs on, say, a millipede or a crab, by calling on a *leg* gene the appropriate number of times instead of a "segment" gene. By calling on slight modifications of the *leg* gene as growth developed, regulatory genes could build an animal whose legs were different along its length (as in insects), or build a vertebrate with different bones along the length of a backbone. For example, embryonic snakes have genetically programmed limb buds that show us where the legs were once placed on ancestral snakes. Today those buds never develop into legs because the regulatory genes do not send a growth instruction to them.

Developmental geneticists can identify regulatory genes by checking what goes wrong when a particular gene is damaged. As a result, we know that there are regulatory genes that control which way up an animal is formed, which is front and back, and how the animal varies along its length or around its edges. The most thrilling discovery is that much the same master genes occur throughout the metazoans. Such "**homeobox**" genes, or **Hox genes** for short, are so similar that they must have evolved from a common ancestor. They are sets of regulatory genes that sit close to one another in the DNA. Although they use much the same master "program," they can build an astonishing variety of metazoan bodies by calling on a variety of structural genes in a variety of patterns at different times and at different places in the body. In other words, one might well discover that two metazoan groups have Hox genes that are fairly similar, though the bodies they code for look very different.

What would one do in such a case? One would have to look at evidence of body morphology and evidence from genetics together, to suggest the simplest hypothesis that would connect these two metazoan groups with others. Remember that genetic analyses cannot be performed on extinct animals: perhaps evidence from fossils that are closer to the time when the metazoan groups actually originated would be useful, even vital! As with any set of data, discrepancies are going to occur and arguments will rage. In the end, however, evolution took one pathway, and everyone is trying to find what that pathway was. "**Evo-devo**," the study of evolutionary development, is giving us dramatic new evidence. Insights from the new techniques have helped to clarify fundamental aspects of metazoan evolution, in many cases overturning ideas that had been accepted for 100 years or more.

Hox Genes

Sponges have one set of Hox genes (and have simple structures), whereas mammals have 38 sets in four clusters, and goldfish have 48 in seven clusters. Hox genes control the growth of nerve nets, segments, and limbs throughout the metazoans, and their evolution and divergence must have accompanied the divergence in anatomy and physiology and ecology and behavior that we can interpret from the fossil record, and the divergence in genes and molecules that we can read from living animals. So Hox genes provide separate but complementary evidence to help us read the evolutionary history of the metazoans.

Protists don't need Hox genes to form a multicellular adult. But in early metazoans, Hox genes provided the genetic tool kit to guide the construction of viable complex animals. Hox genes control the lay-out of a sponge that gives efficiency of water currents passing through the body. In the simplest worms, Hox genes lay out the nerve nets that allow the worm to sense the environment all along the body. The earliest metazoans, wherever, whenever, and however they evolved, could quickly have radiated into a great variety of body shapes and structures, with natural selection acting equally quickly to weed out the shapes that were poor adaptations, and leaving a scrapbook of successful prototypes that proliferated.

THE VARIETY OF METAZOANS

When one animal group is radically different from another, and is also considered to be a clade, evolved from some single ancestral species (Chapter 3), it is a **phylum**, defined by its own particular body structure, ecology, and evolutionary history. Mollusca and Arthropoda are familiar phyla. They must once have had a common metazoan, bilaterian ancestor, but that ancestor wasn't a mollusc or an arthropod (by definition as well as common sense). There are arguments about the number of phyla among living metazoans, mostly because there is a bewildering variety of worm-like organisms, but most people would count about 30 phyla. Because only creatures with hard parts are easy to recognize as fossils, only nine or ten phyla are or have ever been important in paleontology (Box 4.1).

It is stunning to realize that all these phyla except Bryozoa are known from Early Cambrian rocks, between 540 Ma and 520 Ma, yet only two of them are known for certain in rocks older than Cambrian. That means two things, on the face of it: first, that there was an "explosive" burst of evolution at the beginning of the Cambrian period, and second, that we have no fossil record of the metazoan evolution that gave rise to the phyla that we recognize today.

Without fossil evidence of the metazoan radiation, then we are forced to look at evidence from their living descendants half a billion years later, and hope it tells some semblance of the truth! One would hope that the results would be compatible with the rich fossil record that begins with Cambrian rocks.

As we have seen, the two most simple metazoans are sponges and cnidarians. The bilaterians pose more of a problem: they are all complex and three-dimensional, they all have Hox genes controlling the placement of structures along their axis of symmetry. It is difficult to find compelling reasons for ranking them in terms of order of evolutionary branching, using standard arguments from anatomy or ecology. But molecular and genetic evidence has helped us attack the problem.

Advanced bilaterians form three major clusters of phyla, named Ecdysozoa, Lophotrochozoa, and Deuterostomia.

Ecdysozoa are animals that molt off their outer skins as they grow. Molting is characteristic of the Arthropoda, for example, and for many of them it is a major evolutionary burden. Crabs and lobsters must molt many times over the years of their (natural) lives, and each time they do so they are very vulnerable to predators and must spend a considerable time hiding while their new shell hardens. Some ecdysozoans have found a way to avoid this evolutionary constraint: insects do not molt as adults. However, they can only do this by having a very short adult life, with all their growth taking place in earlier life stages (as larvae). A short adult life is an extreme but successful way of avoiding a major evolutionary constraint! As far as the fossil record is concerned, Arthropoda are the dominant representatives of the Ecdysozoa.

Lophotrochozoa are animals with a cute fuzzy little floating larva and a way of life that originally involved filter-feeding from the water. Brachiopoda and Bryozoa are important fossil groups, worms have a poor fossil record, and the Mollusca are well known, well fossilized, well understood, and certainly more varied ecologically.

Deuterostomia also seem to have been originally filter-feeders with floating larvae, but their larvae are so different from those of lophotrochozoans that they cannot belong to the same clade. Deuterostomes include Echinodermata (sea-stars and relatives), and Chordata (including ourselves) and minor related groups.

Current speculation includes the suggestion of a fairly long period during which

BOX 4.1 The Major Phyla of Fossil Invertebrates

(† indicates an extinct group)

- Porifera or sponges (including † Archaeocyatha)
- Cnidaria
- Bryozoa
- Brachiopoda
- Mollusca
- Arthropoda
- Echinodermata
- Hemichordata (including † graptolites)
- Chordata (including vertebrates)

ancestral bilaterians diverged from the other metazoans, evolving maybe as many as seven sets of Hox genes as they did so. These first bilaterians may have looked like little flatworms, or perhaps like the planktonic larvae of flatworms. Then, perhaps shortly before the Cambrian, around 555 Ma, bilaterians become large enough to leave traces on seafloor sediment. During this time the three bilaterian groups diverged, but in ways that left no significant record of body fossils. Finally, the groups split into the phyla that did leave a rich fossil record, beginning at the base of the Cambrian.

But there is an ecological twist to the evolution of bilaterians. All simple metazoans build embryos as they begin cell division from the egg. And simple metazoans all build much the same sort of embryo, without using their Hox genes. They develop into free-living larvae that feed in the plankton. They are bilateral; they are made of perhaps 2000 cells with only a few cell types. At the end of the larval period, cells that have simply been riding along in the larva, without specific function, begin to divide and are organized, positioned, and differentiated under the direction of the Hox genes into a complex adult metazoan animal that usually bears no resemblance to its larva, and usually has a completely different habitat and ecology.

The discovery that this kind of development occurs in all simple metazoans has led to the suggestion that the earliest bilaterians were tiny planktonic animals that looked and functioned like the simple larval stages of many of their modern-day descendants. They would have floated and fed in the plankton. These adult metazoans would certainly have been small and soft-bodied, and unlikely to be fossilized, even as they diverged from one another. This scenario, if true, would correspond with the lack of a long Precambrian fossil record for metazoans, followed by the appearance of many different groups of large metazoans in the "Cambrian Explosion."

In this scenario, more complex development, involving a radical metamorphosis as the larva changed into an adult, and a much greater role for Hox genes, led to the "explosive" evolution of many different phyla in the Cambrian. In addition, their evolution of larger size led to their "invasion" of seafloor habitats, newly available with a greater oxygen supply in ocean water, and loaded with hitherto unexploited organic-rich sediment. It may not be an accident that so many Cambrian animals were "worms" and bottom-dwelling sediment feeders.

SNOWBALL OR SLUSHBALL EARTH

Recently, evidence has been building that Earth went through a series of dramatic cold periods late in the Proterozoic, from perhaps 750 Ma to 600 Ma. Many deposits from this period contain glacial debris, and many of them occur in regions that are reliably reconstructed near the Equator at the time. These deposits imply massive and widespread glaciation, much more extensive than any glaciations that have occurred since. The Late Proterozoic interval that includes the glaciations is coming to be known as the "Cryogenic" and one scenario that attempts to explain their wide distribution is usually known as "Snowball Earth." Furthermore, it has been suggested that there is a causal link between Snowball Earth and the metazoan radiation: that the dramatic physical events on Earth promoted the radiation.

The Snowball Earth model, as elaborated by Paul Hoffman and Dan Schrag, proposes that the surface of the ocean was frozen all the way to the Equator (except where volcanoes existed). Surface temperatures dropped to around −40°C. As the ice spread, photosynthesis was choked off, and most of the life in the oceans died off. In fact, the only surviving life would have been around seafloor hot vents, and (perhaps) in the surface ice. So much solar radiation would reflect back into space that you would think Earth would be locked permanently into a snowball state.

Volcanoes continued to erupt, however, putting carbon dioxide back into the atmosphere, until there was once again enough carbon dioxide to trap solar heat and melt the ice. However, calculations suggest it would have taken an enormous amount of carbon dioxide to break the grip of Snowball Earth. The ice did not melt until volcanoes had erupted around 350 times more carbon dioxide than there is in our present atmosphere.

The ice then melted quickly, but the enormous reservoir of carbon dioxide in the atmosphere rocketed the whole Earth directly into a "greenhouse" hot period, with temperatures averaging around 50°C (over 120°F). Tremendous (acid) rains then acted on the sterilized continents, the ocean was flooded with carbonate, and thick limestones were deposited very quickly on top of the glaciogenic rocks. Finally, weathering and photosynthesis brought down carbon dioxide levels, and the world recovered biologically. However, the geographic set-up that had begun the Snowball Earth cycle was still present, so the cycle then repeated itself, perhaps as many as four times.

Some people think that the scenario of Snowball Earth is more extreme than the evidence suggests, and that the linkage suggested between Snowball Earth and the metazoan radiation is highly unlikely. I would argue for a milder Slushball Earth that was nevertheless a significant influence on the history of life. Rather than squeezing metazoans almost to extinction, I argue that Slushball Earth provided an environment that was uniquely benign for plankton. In that planktonic paradise, eukaryotic life passed through an evolutionary filter that promoted the radiation of planktonic micrometazoans. When Slushball Earth ended, those micrometazoans evolved rapidly, perhaps explosively, into the larger and more ecologically diverse large metazoans that we see in the Ediacaran and Cambrian fossil record.

The glacial sediments include dropstones, which fell from floating icebergs into soft seafloor sediment. Such icebergs must have been floating in open water. The Slushball Earth concept is supported by climatic computer models that allow for a stable climate, with low-latitude continental ice-sheets and floating sea-ice over much of the world's oceans. These models project open tropical waters (between ~25°N and ~25°S, about 40% of today's open ocean surface) at cool to mild temperatures (up to 10°C at the Equator). The model atmosphere has only 2.5 times today's carbon dioxide, and the stability of the model implies that any small rise in those levels will easily revert conditions to "Earth normal," without the extreme greenhouse that the Snowball Earth idea calls for. In addition, methanogens in the anoxic waters under the ice could have produced and released enough methane to prevent an extreme Snowball Earth.

Most important in terms of evolution and paleontology, I suggest that Slushball Earth could have acted to nurture the origin and early evolution of metazoans, in the following way. Ocean water below the Slushball ice sheets would quickly have become anoxic all the way to the ice at the surface. Eukaryotes and metazoans (if any existed at the time) would have been wiped out under the ice and from the ocean floor, which would have contained an "Archaean biology" of anaerobic bacteria and Archaea.

Conditions in the open surface waters of the tropics would have been radically different. Here there would have been little or no seasonal fluctuation in climate. Active erosion by mountain glaciers on the equatorial continents would have provided a steady year-round supply of nutrients to ice shelves along the coastline and, via icebergs, to the surrounding waters. Iron enrichment from wind-blown dust would have been an important supplement to "normal" nutrient supply. Solar radiation in the tropical areas would have been uniform and intense, no matter what the surface temperature was. Within ocean waters, there would have been active vertical transport and corresponding upwelling of nutrient-rich water in the open equatorial areas, like the perennial upwelling in the Southern Convergence today, or along the Equator.

The nutrient-rich surface equatorial waters of Slushball Earth would have supported dramatic year-round productivity, uninterrupted by the seasonal darkness that cuts down today's Antarctic productivity in the winter months. Fall-out from the surface productivity would have driven the equatorial seafloor to anoxia, no matter how oxygen-rich the surface layers were. This is compatible with evidence that iron-rich sediments were deposited on the seafloor during the glacial periods.

The surface equatorial waters of Slushball Earth would have been a planktonic paradise, whose only analogy in today's world might be the upwelling area off the coast of Peru. In particular, the extraordinary, and permanent, productivity along the Slushball Equator would have provided a lot of oxygen in surface waters, and an ideal setting for the evolution of micrometazoans, small but effective metazoan predators on surface plankton.

Snowball Earth calls for catastrophe to trigger the metazoan revolution. That simply can't happen: major crises actually cause major extinctions. Repeatedly wiping the ocean free of oxygen is not the way to foster the evolution of metazoans. The eukaryotic survivors of a Snowball Earth would have been few indeed: on the other hand, eukaryotes would have flourished in the tropical waters of Slushball Earth. There would have been a virtual paradise for protists and for tiny metazoans specialized to feed on them.

As we saw, evidence from metazoan embryos suggests strongly that all the earliest metazoans were tiny, micrometazoans that would have looked like the larval stages of today's simple metazoans, and quite unlike the adult stages of their modern descendants. They would all have been planktonic, feeding on bacteria and protists. They would have reproduced at these very small sizes, perhaps with a great deal of cloning (as echinoderm larvae do today). This ecological reconstruction makes sense in terms of the oxygen levels in a "normal" Late Proterozoic ocean: most likely highest in the surface layers where photosynthesis occurred, and likely to have been rather low in the sediments of the ocean floor.

After the last Slushball episode, protists and micrometazoans would have been able to exploit the seafloors that had been accumulating organic sediment for the duration of the glaciation(s). In this setting, micrometazoans could have evolved adaptations for crawling and deposit-feeding, by evolving the Hox gene complexes that set up appropriate sensory and locomotory appendages in a metamorphosed, larger "adult" metazoan. Metazoans such as sponges could have adapted to seafloor life by specializing for bacterial capture. Cnidarians, perhaps already large planktonic feeders, may have evolved the sessile polyp configuration at this time. I can imagine these creatures rapidly evolving into metazoans that loved such conditions: sponges (filtering bacteria), cnidarians (which are essentially built of thin sheets of tissue that absorb organic molecules), and small worms (eating mud on the seafloor).

There is geochemical evidence that the carbon content of seafloor sediments dropped dramatically after the glaciations: I suggest that this is the effect of the colonization of the seafloor (for the first time) by tiny, then larger, metazoans.

In the end, the evolution of larger metazoans may have prevented the recurrence of the extreme glaciations of Snowball Earth or Slushball Earth. Surface productivity could no longer draw down carbon dioxide to critically low levels because primary producers were eaten back by new planktonic predators (small or larval metazoans). Stronger metazoan burrowers (arthropods and segmented worms) dug up buried carbon from seafloor muds and recycled it into carbon dioxide. And higher populations in nearshore waters intercepted nutrients before they reached the oceanic sea surface.

As micrometazoans evolved into metazoans and occupied a great number of ecological niches, we see the Vendian/Cambrian radiation. Geneticists have been arguing that metazoan roots are deep in the Precambrian, and paleontologists have been arguing that if so, there is no fossil evidence of them. This controversy is resolved by the Slushball Earth scenario outlined here. Simon Conway Morris may have mixed metaphors when he wrote "the motor of the Cambrian explosion was largely ecological," but it is clear that he is right. In the next chapter we will see how that worked out.

Further Reading

Adouette, A., et al. 2000. The new animal phylogeny: reliability and implications. *Proceedings of the National Academy of Sciences* 97: 4453–6. [Where are we in sorting out the cladogram of metazoan phyla? Guess why I haven't included one!]

Anbar, A. D., and A. H. Knoll. 2002. Proterozoic ocean chemistry and evolution: a bioinorganic bridge? *Science* 297: 1137–42. [Nitrate shortage may have held back eukaryote evolution.]

Barnes, R. S. K., et al. 2001. *The Invertebrates: A New Synthesis.* 3rd edn Oxford: Blackwell Science.

Baum, S. K., and T. J. Crowley. 2001. GCM response to Late Precambrian (~590 Ma) ice-covered continents. *Geophysical Research Letters* 28: 593–6. [Open tropical waters during Slushball Earth, based on climatic modeling.]

Bengtson, S. (ed.) 1994. *Early Life on Earth.* New York: Columbia University Press.

Bengtson, S., and Y. Zhao. 1997. Fossilized metazoan embryos from the earliest Cambrian. *Science* 277: 1645–8.

Condon, D. J., et al. 2002. Neoproterozoic glacial-rainout intervals: observations and implications. *Geology* 30: 35–8. [Open tropical waters during Slushball Earth, based on evidence from glacial dropstones.]

Conway Morris, S. 1998. Early metazoan evolution: reconciling paleontology and molecular biology. *American Zoologist* 38: 867–77. [Introductory background to later papers.]

Conway Morris, S. 2000a. Evolution: bringing molecules into the fold. *Cell* 100: 1–11. [Difficulties of melding molecular, morphological, and fossil evidence.]

Conway Morris, S. 2000b. The Cambrian "explosion": slow-fuse or megatonnage? *Proceedings of the National Academy of Sciences* 97: 4426–9. [It was ecological.]

Droser, M. L., et al. 2002. Trace fossils and substrates of the terminal Proterozoic Cambrian transition: implications for the record of early bilaterians and sediment mixing. *Proceedings of the National Academy of Sciences* 99: 12572–6. [Trace fossils mirror the body-fossil record; bilaterians seem to begin about 555 Ma; see Martin et al., 2000.]

Eaves, A. A., and A. R. Palmer. 2003. Widespread cloning in echinoderm larvae. *Nature* 425: 146. [Compare my model for the plankton of Slushball Earth.]

Erwin, D. H., et al. 1997. The origin of animal body plans. *American Scientist* 85: 126–37.

Grotzinger, J. P., et al. 1995. Biostratigraphic and geochronologic constraints on early animal evolution. *Science* 270: 598–604, and comment, 580–1.

Hoffman, P. F., et al. 1998. A Neoproterozoic snowball earth. *Science* 281: 1342–6.

Jensen, S., et al. 2000. Complex trace fossils from the terminal Proterozoic of Namibia. *Geology* 28: 143–6. [Pushes back the first occurrence of relatively powerful burrowers (worms) (but not very far).]

Knoll, A. H., and S. B. Carroll. 1999. Early animal evolution: emerging views from comparative biology and geology. *Science* 284: 2129–37. [Excellent review.]

Li, C.-W., et al. 1998. Precambrian sponges with cellular structures. *Science* 279: 879–82, and comment, 803–4.

Logan, G. A., et al. 1995. Terminal Proterozoic reorganization of biogeochemical cycles. *Nature* 376: 53–6, and comment, 16–17.

Lubick, N. 2002. Snowball fights. *Nature* 417: 12–13. [News report: status of the arguments in 2002.]

Martin, M. W., et al. 2000. Age of Neoproterozoic bilaterian body and trace fossils, White Sea, Russia: implications for metazoan evolution. Science 288: 841–5. [Also points to 555 Ma as the date for the appearance of significant bilaterians (see Droser et al., 2002.]

Peterson, K. J., and E. H. Davidson. 2000. Regulatory evolution and the origin of the bilaterians. *Proceedings of the National Academy of Sciences* 97: 4430–3.

Shen, Y., et al. 2003. Evidence for low sulphate and anoxia in a mid-Proterozoic marine basin. Nature 423: 632–5, and comment, 592–3.

Valentine, J. W. 2002. Prelude to the Cambrian explosion. *Annual Reviews of Earth & Planetary Science* 30: 285–306.

Waggoner, B. M. 1998. Interpreting the earliest metazoan fossils: what can we learn? *American Zoologist* 38: 975–82.

Walker, G. 2000. *Snowball Earth.* New York: Crown Books. [A good read if you like docudramas: see my critique .]

Wood, R. A., et al. 2002. Proterozoic modular biomineralized metazoan from the Nama group, Namibia. *Science* 296: 2383–6. [A sponge or a cnidarian of some sort.]

Wray, G. A., et al. 1996. Molecular evidence for deep Pre-Cambrian divergences among metazoan phyla. *Science* 274: 568–73, and comment, 525–6. [See Conway Morris, 1998.]

Wright, K. 1997. When life was odd. *Discover* 18 (3): 52–61. [Ediacaran fossils.]

Xiao, S., et al. 2000. Eumetazoan fossils in terminal Proterozoic phosphorites? *Proceedings of the National Academy of Sciences* 97: 13684–9. [Metazoan microfossils, probably cnidarians, in the Precambrian Doushantuo formation.]

CHAPTER FIVE

The Cambrian Explosion

EDIACARAN (VENDIAN) ANIMALS

When conditions returned to "normal" after the great glaciations, whether they were Snowball or Slushball, we see the earliest reasonable fossil record of animals in Late Proterozoic rocks, around 575 Ma. In south Australia the rocks that were laid down at this time still carry traces of soft-bodied animals (Figure 5.1). These animals make up the **Ediacaran fauna**, named after rocks found in Ediacara Gorge in the Flinders Ranges near Adelaide. (The **Vendian** fauna is named for fossils of the same type and the same age from northern Russia.)

Thousands of Ediacaran fossils have now been collected worldwide in dozens of different localities. Almost all the fossils occur between 575 and 543 Ma, with the highest abundance and diversity during the later period from 555 Ma to 543 Ma. After that, most Ediacaran animals seem to have become extinct. Most of them probably left no descendants; others gave rise to some of the Cambrian animals that followed.

There are a few Ediacaran sponges and bilaterians, but most Ediacaran fossils are cnidarians of some sort. Jellyfish (Figure 5.1) floated just like their living relatives, and became stranded on beaches in the same way (Figure 5.2). Colonies of sea pens were attached to the seafloor. Sea pens look like plants, but are cnidarians that capture and eat floating animals in the water.

Other Ediacaran fossils are worms that patrolled the seafloor. Some squirmed along in the surface sediment; others walked on the tufts of bristles located on their body segments. *Kimberella* (Figure 5.3) may be a flatworm or perhaps a lophotrochozoan evolving toward a slug-like early mollusc. Some Ediacaran fossils resist interpretation (Figures 5.4; 5.5).

Some Ediacaran animals fed on floating plankton, but others were mud eaters. There is no evidence of animals that grazed algae on the seafloor. (They may have been present but were not preserved.) Many Ediacaran animals are small, but some reached astonishing size for such early animals. Since Ediacaran animals were soft-bodied and apparently unprotected, there may have been no large carnivores on the seafloor.

Figure 5.1 *Mawsonites*, an Ediacaran animal from South Australia. It is similar in form to modern jellyfish, which suggests but does not prove that this animal too is a cnidarian. (Courtesy of Mary Wade of the Queensland Museum, Australia.)

Figure 5.2 Fossil of the future? This cnidarran is lying stranded on a California beach.

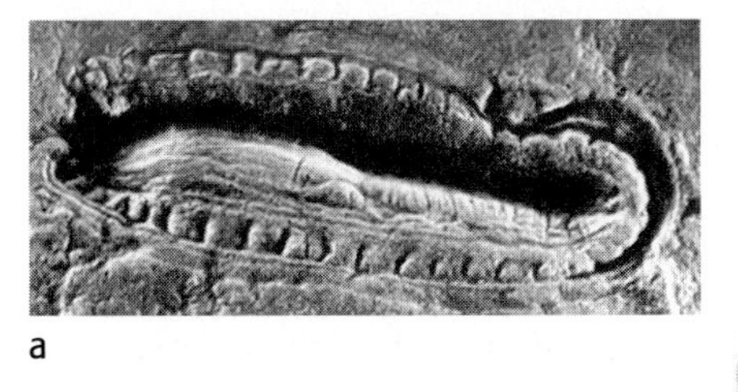

a

b

Figure 5.3 (a) *Kimberella*, perhaps a flatworm, perhaps a proto-mollusc. (Courtesy of Ben Waggoner) (b) Reconstruction, © Alexandra Cowen.

a

b

Figure 5.4 *Dickinsonia*, sometimes interpreted as a large, flat annelid worm, sometimes as a cnidarian. Rotate the page to different positions to try to decide which is more probable. ((a) Courtesy of the late Martin Glaessner, University of Adelaide; (b) Courtesy of Bruce Runnegar, University of California, Los Angeles.)

Figure 5.5 The enigmatic Ediacaran fossil *Tribrachidium*. (Courtesy of the late Martin Glaessner, University of Adelaide.)

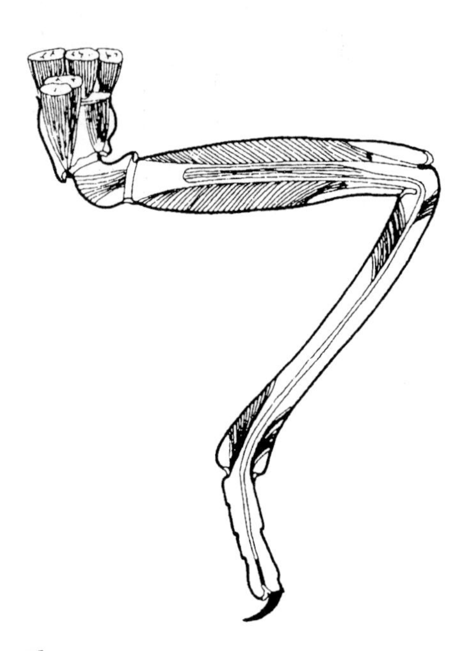

Figure 5.6 Arthropods have exoskeletons secreted from inside by the soft parts. The limbs must be operated from the inside by the muscles and ligaments, unlike the system evolved by vertebrates. (From Barnes et al., *The Invertebrates: A Synthesis*. 3rd edn © 2001 Blackwell Science.)

THE EVOLUTION OF SKELETONS

One of the most important events in the history of life was the evolution of mineralized hard parts in animals. Beginning rather suddenly, the fossil record contains skeletons: shells and other pieces of mineral that were formed biochemically by animals. Humans have one kind of skeleton, an internal skeleton or endoskeleton, where the mineralization is internal and the soft tissues lie outside. Most animals have the reverse arrangement, with a mineralized exoskeleton on the outside and soft tissues inside, as in most molluscs and in arthropods (Figures 5.6; 5.7). The shell, or test, of an echinoderm is technically internal but usually lies so close to the surface that it is external for all practical purposes. The hard parts laid down by corals are external, but underneath the body, so that the soft parts lie on top of the hard parts and seem comparatively unprotected by them. Sponge skeletons are simply networks of tiny spicules that form a largely internal framework. There is incredible variety in the type, function, arrangement, chemistry, and formation of animal skeletons; biomineralization is a whole science in itself.

With the evolution of hard parts, the fossil record became much richer, because hard parts resist the destructive agents that affect the soft parts of bodies. Almost as soon as geologists realized that fossils marked time periods in earth history, they also recognized that the quality of the fossil record depended on the style, structure, and composition of the hard parts of the organisms that were preserved (Chapter 2). For about a century, in fact, many geologists believed that there was no fossil record before hard parts evolved. The evolution of hard parts defines the beginning of a new eon in Earth history, the **Phanerozoic** Eon [the time of "visible life"], a new era, the **Paleozoic Era**, and the oldest subdivision of the Paleozoic, the **Cambrian Period**. In contrast, **Precambrian** time (the **Archean** and **Proterozoic** Eons) was first seen as a time of no life, and then as a time of soft-bodied, mainly bacterial life. Even today, the base of the Cambrian is defined at a time when major new fossils appear in the record.

Why did hard parts evolve in the first place, and why did they evolve when they did? What difference do hard parts make to the biology of an animal? (Do hard parts automatically imply that totally new groups of animals have evolved?)

Worms are soft-bodied. Sponges are sponges, whether they have tiny mineral spicules forming an internal skeleton, or a soft protein like that in bath sponges. But many metazoan groups have skeletons that are such an integral part of their body plans that they only exist as such when they have hard parts. Although there are molluscs without shells (slugs and squids, for example), it seems impossible to be a clam without a shell. Shells are so important to clams that if a clam evolved to be shell-less, its basic biology would be so changed that we would call it something else. An arthropod without a skeleton is basically a worm (unless it's a caterpillar). Thus any worm that evolved hard parts by definition evolved into some other major group: a new **phylum**.

The nine or ten phyla that appeared in the Early Cambrian had evolved hard parts that are very different from one another. Sponges have internal skeletons made of protein, or of calcite, or of fine silica needles. Molluscs and most brachiopods have an external shell made of calcium carbonate, but the two phyla use different minerals, and different crystal structure. Some brachiopods use calcium phosphate for their shells. Arthropods evolved chitin, but different groups of arthropods impregnated the chitin with calcium phosphate or with calcium carbonate. Echinoderms have an internal skeleton just under the skin, made up of small separate plates of calcium carbonate, each one a single calcite crystal.

Looking at the variety of minerals involved, it's clear that there wasn't some simple chemical change in the oceans (an increase in phosphate, for example) that triggered the invention of skeletons. Yet the evolutionary event was global, so it was probably triggered by some global biological or ecological factor. We need to know why these different creatures evolved different types of skeletons, and why they did it so rapidly that the event is often described as the "Cambrian explosion." We understand the three great groups of metazoans, but the phyla themselves branched too rapidly for us to reconstruct (yet). A symposium in 1998 heard several, radically different proposals for the metazoan radiation, and the organizers later reported that ten more years' research would help to begin to solve the question!

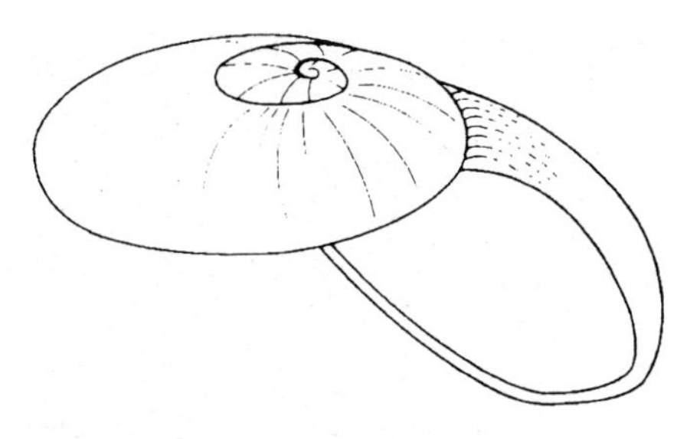

Figure 5.7 A tiny Cambrian animal, *Aldanella*, which may be a mollusc (though some scientists disagree). If it is a mollusc, *Aldanella* is the earliest known snail. The shell is only 1.5 mm across. (From Boardman et al. (eds), *Fossil Invertebrates.* © 1987 Blackwell Scientific.)

However, perhaps the Cambrian explosion was not so dramatic. The soft-bodied ancestors of Cambrian animals, already well evolved along different pathways, may suddenly have gained hard parts and large size, and "exploded" into the fossil record as they began to leave fossils on the seafloor. In doing so, they evolved the features that allow us to identify them anatomically and ecologically as the metazoan phyla that still survive. In this scenario, what evolved in the Cambrian explosion was ecological molluscs, ecological arthropods, and so on. Their separate ancestors may have had important differences in the DNA of important genes, but they had all been ecologically similar. This would make the Cambrian explosion an ecological event rather than a phylogenetic one. However, it is also possible that much of the evolution that formed the multiple metazoan phyla occurred just before and just into the Cambrian, making the Cambrian explosion genetic as well as ecological, and dramatic indeed. The truth is probably somewhere in between (as it often is).

For some purposes, it doesn't matter whether the Cambrian explosion genuinely represents the evolution of new phyla, or whether it represents the evolution of new characters within pre-existing but soft-bodied phyla. It is clear that the evolution of morphology that reflects new adaptation and novel ecology occurred very rapidly indeed at the beginning of the Cambrian; and that needs explanation. We'll review the evidence, and then look for explanation of the Cambrian explosion.

The Beginning of the Cambrian: Small Shelly Fossils

Technically, the base of the Cambrian, at about 543 Ma, is at a remote spot in a cliff face in Newfoundland, Canada, in rocks that have no fossils except a few worm tracks. But in Siberia, rocks of about the same age contain a whole suite of small shelly fossils, together with sponges of several different types. Most of the small shells are tiny cones and tubes that we don't understand properly, but at least some of them are complex animals, including the first molluscs (Figure 5.7). Soon archaeocyathid sponges (Figure 5.8) were forming reef patches.

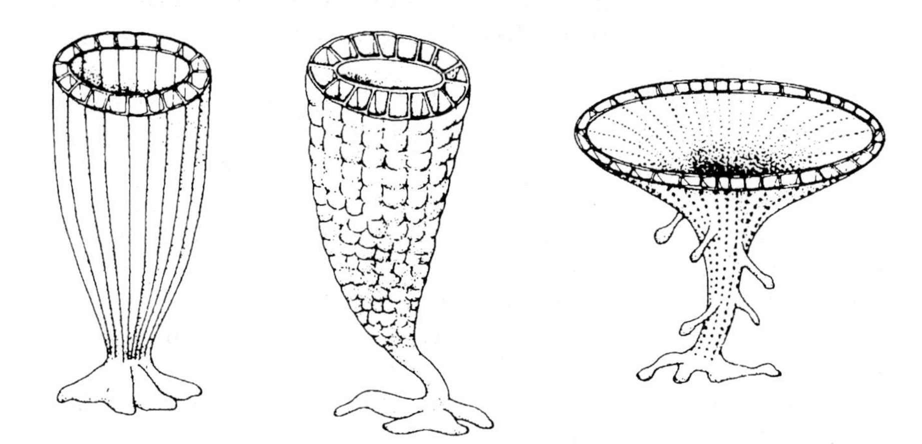

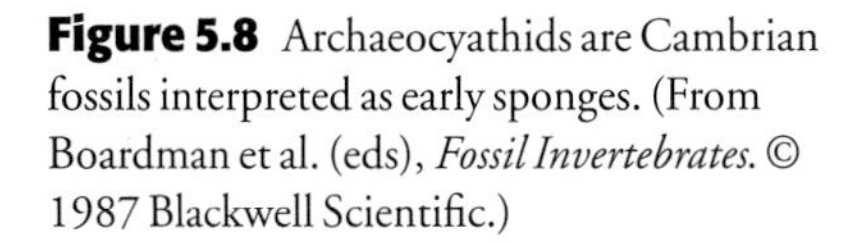

Figure 5.8 Archaeocyathids are Cambrian fossils interpreted as early sponges. (From Boardman et al. (eds), *Fossil Invertebrates.* © 1987 Blackwell Scientific.)

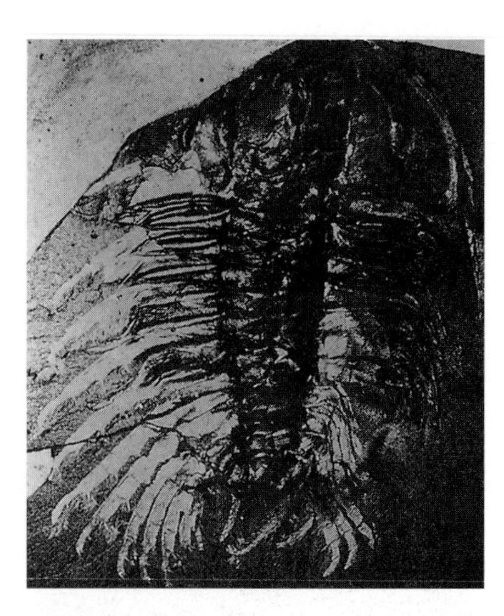

Figure 5.9 The Middle Cambrian trilobite *Olenoides*, from the Burgess Shale of British Columbia, Canada. This specimen is preserved with traces of its walking legs still visible under the thick, hard carapace that forms the usual trilobite fossil. (Courtesy of the Palaeontological Association.)

The same set of small shelly fossils is now known worldwide, and the world's fauna was completely revolutionized in the 20 m.y. that followed. The next stage of the Cambrian sees the appearance of more abundant and more complex creatures.

For perhaps 20 m.y., there were no animals larger than a few millimeters long except for archaeocyathid sponges. Then worldwide, in a few million years after 520 Ma, we see the appearance of a much larger variety of marine life. Dominant among these animals were trilobites, brachiopods, and echinoderms.

Larger Cambrian Animals

Trilobites are arthropods, complex creatures with thick jointed armor covering them from head to tail (Figures 5.6; 5.9). They had antennae and large eyes, they were mobile on the seafloor using long jointed legs, and they were something like crustaceans and horseshoe crabs in structure. They did not have the complex mouth parts of living crustaceans, so their diet may have been restricted to sediment or very small or soft prey. They burrowed actively, leaving traces of their activities in the sediment, and they are by far the most numerous fossils in Cambrian rocks. The number of fossils they left behind was increased by the fact that they molted their armor as they grew, like living crustaceans. Thus, a large adult trilobite could have contributed twenty or more suits of armor to the fossil record before its final death. Even allowing for this bias of the fossil record, it is clear that Cambrian seafloors were dominated by trilobites. Other large arthropods are also known from Early Cambrian rocks, although they are much less common.

Brachiopods are relatively abundant Cambrian fossils, creatures that had two shells protecting a small body and a large water-filled cavity where food was filtered from seawater pumped in and out of the shell (Figure 5.10). Cambrian brachiopods lived on the sediment surface or burrowed just under it.

These animals are large, and they are easily assigned to living phyla. For the first time, the seafloor would have looked reasonably familiar to a marine ecologist. Trilobites probably ate mud, and brachiopods gathered food from seawater. Yet some ecological puzzles remain. There are no obvious large predators among these earliest skeletonized Cambrian fossils, no obvious grazers unless trilobites ate algae, and no swimmers, only floating plankton.

Figure 5.10 Brachiopods. (a) A Cambrian brachiopod. (b) One of the better-known "lamp shell" forms found later in brachiopod history.

a

b

THE BURGESS FAUNA

I have so far discussed the Cambrian explosion as if it related entirely to the evolution of skeletons. While this is basically true in terms of fossil abundance, there was also dramatic evolution at the same time among animal groups with little or no skeleton. The abundance of trace fossils—tracks, trails, and burrows—increases at the beginning of the Cambrian, and soft-bodied animals appeared with some amazingly sophisticated body plans.

These soft-bodied animals are preserved in abundance in Early Cambrian rocks in south China (the Chengjiang Fauna), and in Middle Cambrian rocks in the Canadian Rockies (in the Burgess Shale). Similar fossils are now known from Cambrian rocks in several other places. I shall call them all the **Burgess Fauna**. (For many more illustrations see the web site.)

More than half the Burgess animals burrowed in or lived freely on the seafloor, and most of these were deposit feeders. Arthropods (such as *Marrella*, Figure 5.11) and worms dominate the Burgess Fauna. Only about 30% of the species were fixed to the seafloor or lived stationary lives on it, and these were probably filter-feeders, mainly sponges and worms. Thus, the dominance of most Cambrian fossil collections by bottom-dwelling, deposit-feeding arthropods is not a bias of the preservation of hard parts: it occurs among soft-bodied communities too. Trilobites (Figure 5.9) are fair representatives of Cambrian animals and Cambrian ecology.

The main delights of the Burgess Fauna are the unusual animals, which have provided fun and headaches for paleontologists. *Aysheaia* is a **lobopod**: it looks like a caterpillar, with thick soft legs (Figure 5.12). It has stubby little appendages near its head that may be slime glands for entangling prey. *Hallucigenia,* named because of its bizarre appearance, is a lobopod with spines.

There are predators in the Burgess fauna. Priapulid worms today live in shallow burrows and capture soft-bodied prey by plunging a hooked proboscis into them as they crawl by.

Anomalocarids are the most spectacular Cambrian predators. They are an extinct group of animals related to arthropods: pieces of Burgess animals suggest that they could have been 2 m long! *Opabinia* is a highly evolved anomalocarid, long and slim, with a vertical tail fin, so it probably swam about. It has five eyes and one large grasping claw on the front of its head (Figure 5.13).

Wiwaxia (Figure 5.14) is a flat creature that crept along the seafloor under a cover of tiny scales that were interspersed with tall strong spines. Halkieriids, best known from the Burgess fauna of Greenland, look like flattened worms, with perhaps 2000 spines forming a protective coating embedded into the dorsal surface. Yet two distinct subcircular mollusc-like shells are embedded in the upper surface close to each end. These creatures may be like *Wiwaxia*; if so, they too are worms—with armor.

The Burgess animals also include worm-like creatures that are identified as early chordates and vertebrates: in other words, the remote ancestors of ourselves and all other vertebrates (Chapter 7).

Altogether, the Burgess faunas give us a good idea of the sorts of exciting but extinct soft-bodied creatures that may always have lived alongside the trilobites but were hardly ever preserved.

Figure 5.11 *Marrella*, one of the commonest fossils in the Burgess Shale, is a very strange-looking arthropod. Soft parts are preserved: traces of long antennae are visible, legs can be seen under the strangely open carapace, and its guts were squeezed from the body and preserved as a dark stain on the rock surface. (Courtesy of the Palaeontological Association.)

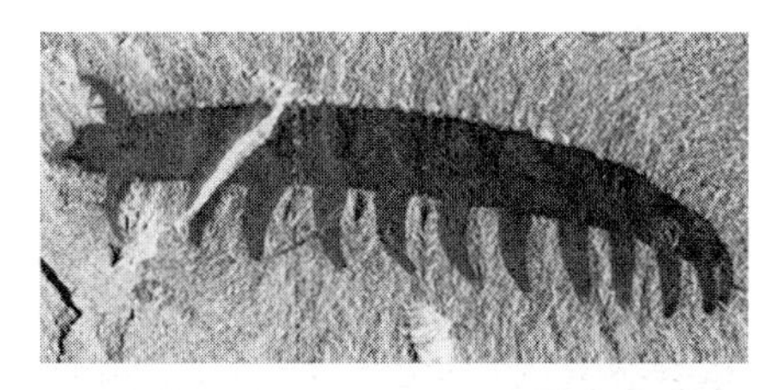

Figure 5.12 *Aysheaia* is a lobopod. (Courtesy of the National Museum of Natural History.)

SOLVING THE CAMBRIAN EXPLOSION

The Cambrian explosion remains a puzzle. The waves of evolutionary novelty that

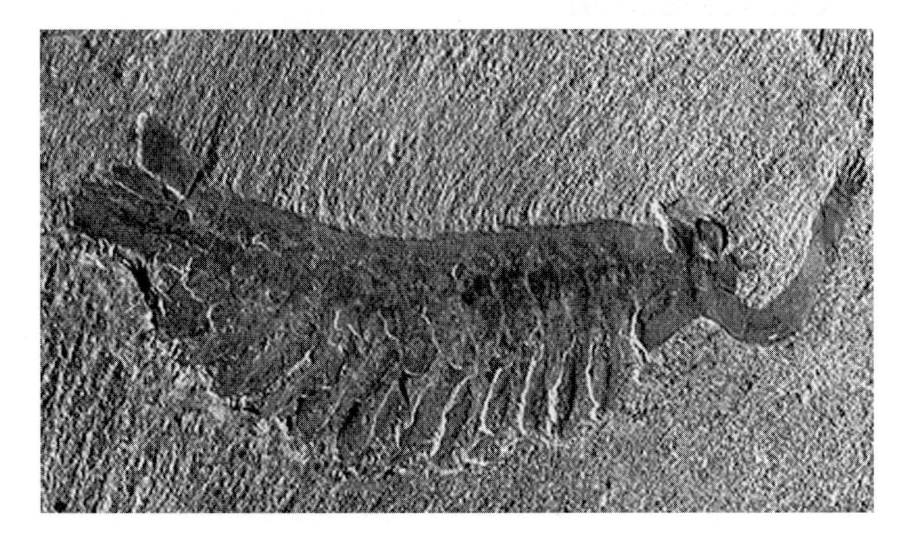

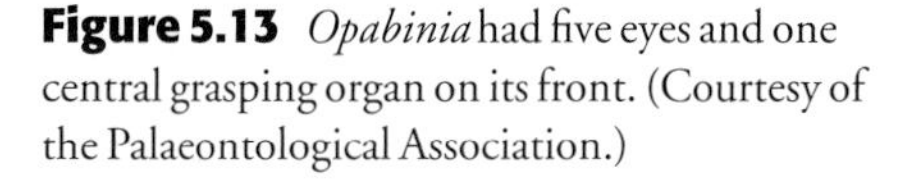

Figure 5.13 *Opabinia* had five eyes and one central grasping organ on its front. (Courtesy of the Palaeontological Association.)

appeared in the seas during the Early Cambrian have few parallels in the history of life. Many groups of fossils appeared quite suddenly in abundance, thanks to their evolution of skeletons, sometimes at comparatively large body size. This was not some ordinary event in the history of life.

Figure 5.14 *Wiwaxia* apparently crept along the seafloor (like a mollusc or a worm?) but was covered by an array of overlapping plates and spines. (Courtesy of the Palaeontological Association.)

Skeletons

A skeleton may support soft tissue, from the inside or from the outside, and simply allow an animal to grow larger. Therefore, sponges could grow larger and higher after they evolved supporting structures of protein or mineral, and they could reach further into the water to take advantage of currents and to gather food. Large size also protects animals from predators large and small. A large animal is less likely to be totally consumed, and in an animal like a sponge that has little organization, damage can eventually be repaired if even a part of the animal survives attack. As skeletons evolved, even for other reasons, they helped animals to survive because of their defensive value.

Early echinoderms had lightly plated skeletons just under their surfaces, and the most reasonable explanation of their first function is support, accompanied or followed by defense.

For other animals, skeletons provided a box that gave organs a controlled environment in which to work. Filters were less exposed to currents, so perhaps they would not clog so easily from silt and mud. A box-like skeleton would also have given an advantage against predation. Molluscs and brachiopods may have evolved skeletons for these reasons.

In yet other animals, hard parts may have performed more specific functions. We have already seen that worms tend to burrow head-first in sediment. But after penetrating the sediment they squirm through it. A worm that evolved a hardened head covering could use a different and perhaps better technique, shoveling sediment aside like a bulldozer. Richard Fortey suggested to me that the large headshield of trilobites was evolved and used in this fashion.

But arthropods, and especially trilobites, are strongly armored all over their dorsal surfaces, not just in the head region. Most likely their armor served for the attachment of strong muscles. Muscles pull and cannot push. Worms move by using internal hydraulic systems, as we have seen. On the other hand, walking demands that limbs push on the sediment, and that is very unrewarding if the other end of the leg is unbraced. Arthropods evolved a large, strong dorsal skeleton against which their jointed legs were firmly braced, allowing them to move much more efficiently than worms do.

Skeletons seem to have evolved for many different reasons, in many different chemistries, in many different animals, but why did they evolve in a very short geological time and in two abrupt waves? The only common factor is the dramatic invasion of new ways of life on the Cambrian seafloor that were impossible without support or sheltering of internal organs or muscular bracing.

Despite all the discussion of skeletons, the Burgess Fauna shows that dramatic evolution took place also in animals that did not have strong skeletons. However, many of these animals had outer coverings that were tough, but lightly mineralized: the Burgess arthropods are particularly good examples.

The common factor along successful groups of Cambrian animals is larger body size. This suggests that in some way the world had become hospitable to large animals, and in turn that tells us that the Cambrian event was driven by worldwide ecological factors. We do not yet know what they were. They could have been related to a change in food supply in the sea, which in turn depends on upwelling, which in turn depends on global climatic and geographic patterns. They could have been related to oxygen levels (large animals need more oxygen than small ones), but oxygen levels depend on productivity and burial of organic matter. We don't yet know enough about Cambrian geography, climate, and geochemistry to say anything sensible about these factors, but it's here that the answer probably lies and where future research should be focused.

Whatever the underlying global cause was, some specific mechanisms have been suggested to explain the Cambrian explosion.

Predation

The predation theory has two aspects. The first is a general ecological argument. The ecologist Robert Paine removed the top predator (a starfish) from rocky shore communities on the Washington coast and found that diversity dropped (see Stanley, 1973). In the absence of the starfish, mussels took over all available rocky surfaces and smothered all their competitors. Paine suggested that a major ecological principle was at work: effective predators maintain diversity in a community. If a prey species becomes dominant and numerous, the top predator eats it back, maintaining diversity by keeping space available for other species.

Steven Stanley used Paine's work to suggest that the evolution of predation triggered the Cambrian radiation (see Stanley, 1973). Stanley made an intellectual jump to suggest that predators can *cause* additional diversity in their prey. He argued that if predators first appeared in the Early Cambrian, they may have caused the increase in diversity at that time. Perhaps predators also encouraged the evolution of many different types of skeletonized animals.

Geerat Vermeij supported Stanley's idea, suggesting how new predators might indeed cause diversification among prey (at any time) (see Vermeij, 1989). In response to new predators, prey creatures might evolve large size, or hard coverings made from any available biochemical substance, or powerful toxins, or changes in life style or behavior (such as deeper burrowing), or any combination of these, all to become more predator-proof. And as the new predators in turn evolve more sophisticated ways of attacking prey, the responses and counter-responses might well add up to a significant burst of evolutionary change.

Are the characteristics of Early Cambrian fossils consistent with a predation theory? Predation as a way of life appeared long before the Cambrian. There are predatory protists, and many early micrometazoans were probably predators too. But this was on a tiny scale. The rules of the predator/prey game may have changed radically as large multicellular creatures evolved. Many Early Cambrian fossils have hard parts that look defensive. Some sharp little conical shells called sclerites may have been spines that were carried pointing outwards on the dorsal and lateral sides of animals, to fend off predators. There are armored and spined Early Cambrian animals, and some Early Cambrian trilobites have healed injuries that may indicate damage by a predator. Defensive structures made of hard parts could therefore have contributed to the increase in the number of fossils in Early Cambrian rocks.

Present evidence suggests that predation played an important part in generating the Cambrian event. It's difficult to be certain, because the only major predators we have discovered are the anomalocarids, and we have no evidence of what they ate: anything they could catch, probably! However, predation does not explain the timing of the Cambrian explosion: why not 100 m.y. earlier, or later? And predation alone cannot account for all the variety of skeletons that we see.

Oxygen Levels

Perhaps the Cambrian explosion is related to global oxygen levels. In one scenario, the evolution of large bodies and skeletons was made possible by high oxygen concentrations. Shells and thick tissues prevent the free diffusion of oxygen into a body (Chapter 4), so they could not have evolved unless there was a high enough oxygen level to push oxygen into the body through the few remaining areas of exposed tissue, through gills, for example. This also cannot be the whole story, because sponges and cnidarians could have evolved their skeletons (which do not inhibit respiration) in low oxygen conditions.

If the oxygen idea is true, however, it could explain much of the Cambrian explosion. Where did the increased oxygen come from? What would increase the amount of carbon buried in the seafloor?

Graham Logan and his colleagues pointed out that the evolution of complex metazoans (bilaterians) also involved the evolution of guts, and therefore of feces, usually in the form of compact fecal pellets. If carbon-rich fecal pellets are buried quickly, away from oxygen, then oxygen levels in the sea and in the atmosphere increase. In this view, then, the rise of metazoans large enough to produce reasonable quantities of fecal pellets was responsible for the rise in oxygen, which then permitted even more (and larger) metazoans to evolve, and so on, until it became advantageous for those metazoans to evolve skeletons. If my version of Slushball Earth is correct, the plankton paradise would have caused a good deal of carbon burial from the surface. This could have begun the oxygen rise that permitted larger and larger metazoans to evolve and then colonize the seafloor. New bilaterians would then produce fecal pellets in or on the seafloor, where they would be buried easily and quickly. (Remember that burrowing into the seafloor, that might have exhumed carbon to be oxidized, did not occur on any scale until the Cambrian began.)

Despite all this speculation, it is not clear that oxygen, or predation, or any other single factor was the reason for the nature, the scale, and especially the timing of the Cambrian explosion. One could argue (and people have) that the world is full of complex creatures, so complexity must have evolved sometime. Whenever it evolved, it was bound to cause a visible "burst" in the fossil record. Perhaps, then, there was no specific "trigger" for the Cambrian explosion. The first large animals evolved at that time, and it is hardly surprising that they spread rapidly and diversified into many body plans, with different groups evolving hard parts of different chemistry and structure. Remember that a newly evolved creature interacts with all of the other new creatures in its community, so the ecological ferment would have been intense.

After the dramatic events early in the Cambrian, the increase in numbers and diversity of fossils later in the period seems anticlimactic. Cambrian fossil collections are not very complex ecologically; they are dominated by trilobites, most of which lived on the seafloor and were deposit feeders. Filtering organisms are very much secondary, and although there are large carnivores, they are represented only by anomalocarids.

The Cambrian explosion is spectacular, but it is not unique; in my view the spectacular diversification of the diapsid reptiles, especially the archosaurs, in the Triassic is an analogous case (Chapter 11), as is the diversification of the mammals in the Paleocene (Chapter 17). These radiations stand out from "normal" evolutionary events just as "mass extinctions" stand out from the rest (Chapter 6). The Triassic on a real planet inhabited by real organisms, evolutionary rates are likely to vary in time and space, and evolutionary events are likely to vary in magnitude, duration, and frequency. We should not expect that ideal rules we might propose for an ideal planet would be followed by the natural world; instead, we have to find out from that natural world what the rules actually were.

Further Reading

Bengtson, S., and Y. Zhao. 1997. Fossilized metazoan embryos from the earliest Cambrian. *Science* 277: 1645–8.

Chen, J-Y., et al. 1991. The Chengjiang fauna: oldest soft-bodied fauna on Earth. *National Geographic Research & Exploration* 7: 8–19.

Conway Morris, S. 1997. *The Crucible of Creation: The Burgess Shale and the Rise of Animals.* Oxford: Oxford University Press. [Critical of Gould, 1989.]

Conway Morris, S. 1998. Early metazoan evolution: reconciling paleontology and molecular biology. *American Zoologist* 38: 867–77. [Introductory background to later papers.]

Conway Morris, S. 2000a. Evolution: bringing molecules into the fold. *Cell* 100: 1–11. [Difficulties of melding molecular, morphological, and fossil evidence.]

Conway Morris, S. 2000b. The Cambrian "explosion": slow-fuse or megatonnage? *Proceedings of the National Academy of Sciences* 97: 4426–9. [It was ecological.]

Droser, M. L., et al. 2002. Trace fossils and substrates of the terminal Proterozoic Cambrian transition: implications for the record of early bilaterians and sediment mixing. *Proceedings of the National Academy of Sciences* 99: 12572–6. [Trace fossils mirror the body-fossil record; bilaterians seem to begin about 555 Ma; see Martin et al., 2000.]

Gould, S. J. 1989. *Wonderful Life: The Burgess Shale and the Nature of History.* New York: W. W. Norton.

Hagadorn, J. W., et al. 2002. Stranded on a Late Cambrian shoreline: medusae from central Wisconsin. *Geology* 30: 147–50.

Jensen, S., et al. 2000. Complex trace fossils from the terminal Proterozoic of Namibia. *Geology* 28: 143–6. [Pushes back the first occurrence of relatively powerful burrowers (worms) (but not very far).]

Li, C.-W., et al. 1998. Precambrian sponges with cellular structures. *Science* 279: 879–82, and comment, 803–4.

Martin, M. W., et al. 2000. Age of Neoproterozoic bilaterian body and trace fossils, White Sea, Russia: implications for metazoan evolution. Science 288: 841–5. [Also points to 555 Ma as the date for the appearance of significant bilaterians (see Droser et al., 2002.]

Monastersky, R. 1993. Mysteries of the Orient. *Discover* 14 (4): 38–48. [The Chengjiang fauna.]

Nedin, C. 1999. *Anomalocaris* predation on nonmineralized and mineralized trilobites. *Geology* 27: 987–90.

Shu, D-G., et al. 1999. Lower Cambrian vertebrates from South China. *Nature* 400: 42–6, and comment in *Science* 286, 1064–5.

Shu, D-G., et al. 2001. An Early Cambrian tunicate from China. *Nature* 411: 472–3.

Stanley, S. M. 1973. An ecological theory for the sudden origin of multicellular life in the late Precambrian. *Proceedings of the National Academy of Sciences* 70: 1486–9.

Thomas, R. D. K., et al. 2000. Evolutionary exploitation of design options by the first animals with hard skeletons. *Science* 288: 1239–42.

Valentine, J. W. 2002. Prelude to the Cambrian explosion. *Annual Reviews of Earth & Planetary Science* 30: 285–306.

Vermeij, G. J. 1989. The origin of skeletons. *Palaios* 4: 585–9.

Waggoner, B. M. 1998. Interpreting the earliest metazoan fossils: what can we learn? *American Zoologist* 38: 975–82.

Wood, R. A., et al. 2002. Proterozoic modular biomineralized metazoan from the Nama group, Namibia. *Science* 296: 2383–6. [A sponge or a cnidarian of some sort.]

Xiao, S., et al. 2000. Eumetazoan fossils in terminal Proterozoic phosphorites? *Proceedings of the National Academy of Sciences* 97: 13684–9. [Metazoan microfossils, probably cnidarians, in the Precambrian Doushantuo formation.]

CHAPTER SIX

Changing Life in a Changing World

Life has evolved on a planet that has experienced changing geology, geography, and climate. Life did not evolve in random patterns, either; to understand the fossil record, we also have to look at the relationships between the physical and biological world on which it lives.

First, we need to know what the patterns of life through time have been. The fossil record has only been abundant since the beginning of the Cambrian, so we have the best chance of understanding the patterns of the Phanerozoic Eon: the Paleozoic, Mesozoic, and Cenozoic Eras. Furthermore, there has been abundant life in the sea for the entire Phanerozoic, but not on land; and fossils are more likely to be preserved in marine sediments (Chapter 2). The overall record of life is therefore best understood by studying the marine Phanerozoic record.

DIVERSITY PATTERNS IN THE FOSSIL RECORD

Jack Sepkoski spent over 20 years compiling published data on the fossil record of Earth through the Phanerozoic, concentrating most on marine fossils. At first he simply counted the number of families of marine fossils that had been defined by paleontologists from Ediacaran (Vendian) to Recent times; later he compiled genera. These data on global (marine) **diversity** show clear and reasonably simple trends (Figure 6.1). Few families of marine animals existed in Vendian times, but the beginning of the Cambrian saw a dramatic increase that followed a steep curve to a Late Cambrian level. A new, dramatic rise at the beginning of the Ordovician raised the total to a high level that remained comparatively stable through the rest of the Paleozoic. In the Late Permian there was a dramatic diversity drop in a very large extinction that marks the end of the Paleozoic Era. A steady rise that began in the Triassic has continued to the present, with a small and short-lived reversal (extinction) only at the end of the Cretaceous Period, which also marks the end of the Mesozoic Era.

This general pattern was known in 1860, when John Phillips defined the Paleozoic, Mesozoic, and Cenozoic Eras (Figure 6.2). The pattern is also familiar to any invertebrate paleontologist who has spent time rummaging broadly through the collections of a major museum. Sepkoski's contribution was to put the pattern in quantitative terms, and to lay out the data for anyone to analyze.

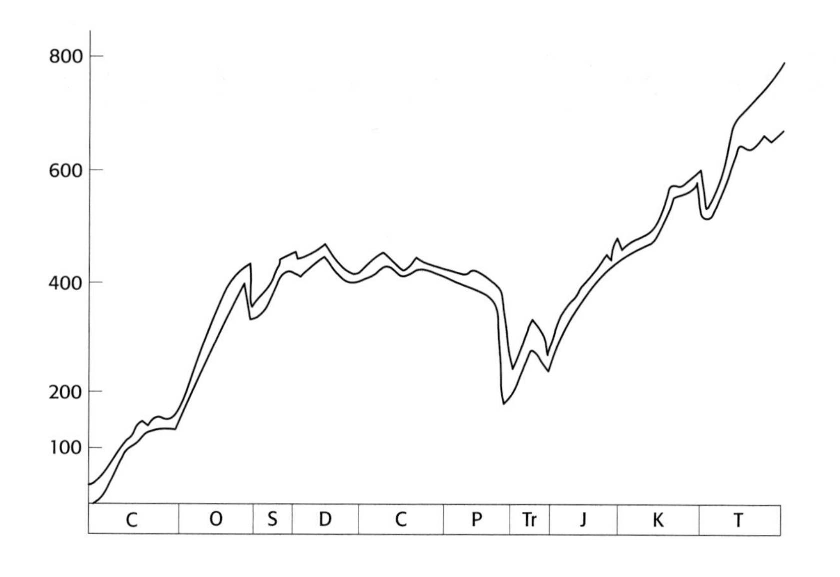

Figure 6.1 Jack Sepkoski's 1984 compilation of the diversity of families of marine organisms through time. The lower curve shows organisms with well-preserved hard parts; the upper curve shows all organisms. The similarity between the two curves shows that this bias of the fossil record is not very important, and the basic pattern has not changed with nearly 20 years' worth of new data. (Data from Sepkoski, 1981, 1984.)

It's easy to think of possible problems with Sepkoski's approach. For example, only some parts of the world have been thoroughly searched for fossils; some parts of the geological record have been searched more carefully than others; older rocks have been preferentially destroyed or covered over by normal processes such as erosion and deposition. Lengthy recent discussions have not shed a great deal of light on these biases.

I suspect that the trends that Sepkoski identified are real, even if the numbers attached to them may change with new research. Paleontologists have been searching the world for fossils for 200 years. The best-sampled fossil communities are shelly faunas that lived on shallow marine shelves, and our estimate of their diversity through time is likely to be a fair sample of the diversity of all life through time. Larger groups of animals are harder to miss than smaller groups, so we have probably discovered all the phyla of shallow marine animals with hard skeletons. Perhaps we have only found a few percent of the species in the fossil record, but we've probably discovered many of the families. In any case, if the search for fossils has been roughly random (and there's no reason to doubt it) the shallow marine fossil record as we now know it is a fair sample of the fossil record as a whole. So we can now ask what influenced the patterns that Sepkoski documented. The first question to ask is whether they have been affected by the changing geography of Earth, and if so, what were the causal connections?

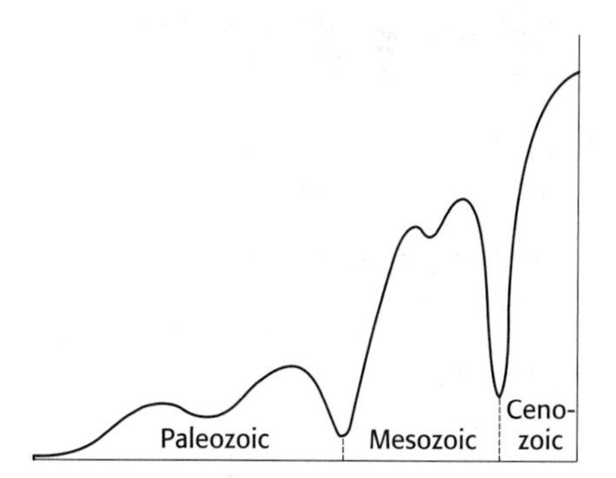

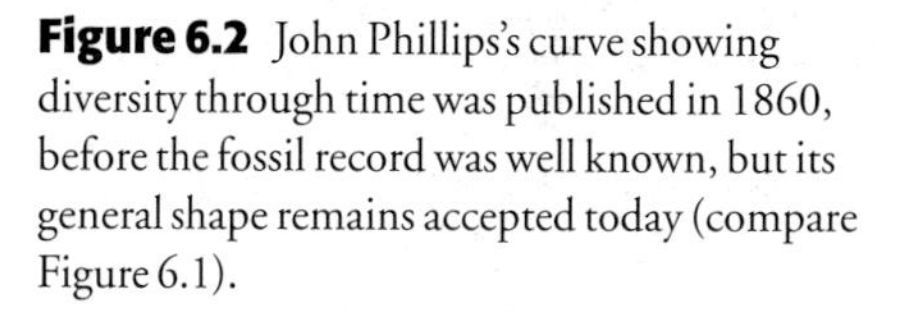

Figure 6.2 John Phillips's curve showing diversity through time was published in 1860, before the fossil record was well known, but its general shape remains accepted today (compare Figure 6.1).

GLOBAL TECTONICS AND GLOBAL DIVERSITY

We have known for over 30 years that Earth's crust is made up of great rigid plates that move about under the influence of the convection of Earth's hot interior. As they move, the plates affect one another along their edges, with results that alter the geography of Earth's surface in major ways. Two plates can separate to split continents apart, to form new oceans, or to enlarge existing oceans by forming new crust in giant rifts in the ocean floor. Two plates can slide past one another, forming long transform faults such as the San Andreas Fault of California. Plates can converge and collide, forming chains of volcanic islands and deep trenches in the ocean, volcanic mountain belts along coasts, or giant belts of folded mountains between continental masses. At times Earth has had widely separated continents; at other times

Figure 6.3 The paleogeography of the Permian. Two very large continents, Gondwana and Laurasia, combined to form the supercontinent Pangea, with the Tethys Sea lying between them along the Equator. This map depicts the Early Permian, about 280 Ma; Pangea is almost complete by this time. The great Permian extinction took place about 30 m.y. later.

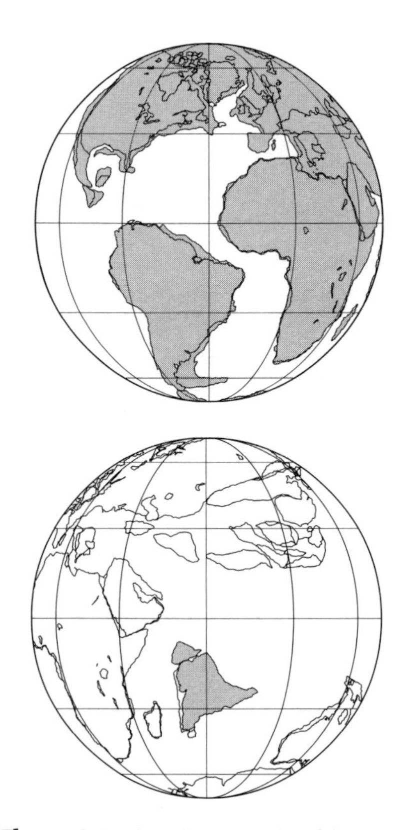

Figure 6.4 The paleogeography of the Cretaceous at about 80 Ma. Pangea is now fragmented, with the Atlantic open from New England southward. Africa and India have broken away from Antarctica and are drifting northward.

the continental crust has largely been gathered into just one or two "supercontinents." These movements and their physical consequences are studied in the branch of geology called **plate tectonics**.

James Valentine and Eldridge Moores suggested in 1970 that because plate tectonic movements affected geography, they could in turn affect food supply, climate, and the diversity of life. In other words, the tectonic history of Earth should have been a first-order influence on the diversity of the fossil record (See Valentine and Moores, 1972). Thirty years later, do we still see a correlation?

The brief answer is yes. A Late Proterozoic supercontinent, **Rodinia**, split progressively during the Cambrian and Ordovician, to form a number of small continents that were generally distributed in lower latitudes around Earth. The splitting events coincide with the great diversity rise in the Cambrian and Ordovician. There were several continental collisions from the Middle Paleozoic through the Permian, and larger land masses were formed. The great extinction at the end of the Permian coincides with the final merger of the continents into a giant global supercontinent, **Pangea** (Figure 6.3), composed of a large northern land mass, **Laurasia**, and a southern land mass, **Gondwana**.

The rise in diversity that began in the Triassic and continued into the Cenozoic coincides very well with the progressive break-up of Pangea. The breakup was under way by the Jurassic, and reached a climax in the Cretaceous (Figure 6.4). The continental fragments have continued to drift, and today the continents are perhaps as well separated as one could ever expect, even in a random world.

Thus the tectonic events that affected Earth over the past 550 million years are reflected in the diversity curve. What are the connecting factors?

Provinces

We all know that most creatures live only in a certain part of the world: for example, kiwis live only in New Zealand and sloths are South American. The pattern holds in detail too. Marine organisms typically occur in characteristic sets of species called **communities**, living together in certain types of habitat—rocky shore communities, mudflat communities, and so on. As one example, the northwest coast of North America, bathed by cool water, has a characteristic rocky shore community of plants and animals that looks much the same from British Columbia to Central California. In turn, the coastal communities of the world can be arranged into geographically separate **provinces**, with each province containing its own set of communities, such as the Oregonian and Californian Provinces of western North America (Figure 6.5).

Provinces are real phenomena, not artifacts of the human tendency to classify things. There are natural ecological breaks on Earth's surface, usually at places where geographic or climatic gradients are sharp, so that one may pass from one environmental regime to another in a short distance. A classic example is at Point Conception on the California coast. Here, the ocean circulation patterns cause a sharp gradient in water temperature. In human terms, Point Conception marks the northern limit of west coast beaches where one can surf without a wet suit, but it's easy to imagine that marine creatures would feel that difference too. The communities on each side of Point Conception are very different, so a provincial boundary is drawn here, with the Oregonian province grading very sharply into the Californian province (Figure 6.5).

As provinces are identified around the coasts of the world, it seems that the number of species in common between neighboring provinces is usually 20% or less.

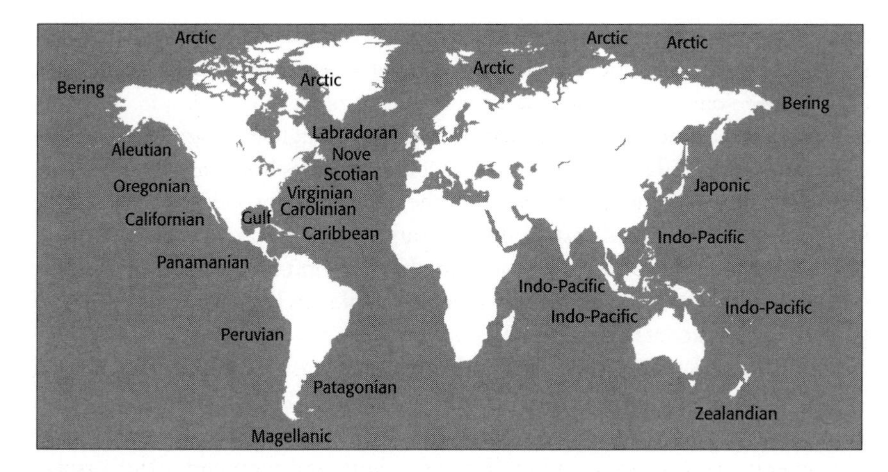

Figure 6.5 The marine biosphere can be divided into biological provinces. This figure does not show all 31 of them, but I have included all the provinces around the Americas, to stress the differences between east and west coasts and the strong latitudinal gradient that produces many provinces along north–south coastlines. In contrast, the Arctic province is very large, because organisms migrate easily along latitudes around the Arctic Ocean, and the Indo-Pacific province is enormous, because marine organisms migrate easily along east–west coastlines and island chains. The Zealandian province is small because it can occupy only a restricted area of shallow shelf. (Modified from Valentine, 1973.)

About 30 provinces have been defined along the world's coasts, mostly on the basis of molluscs, which are obvious, abundant, and easily identified members of coastal communities. Some provinces are very large because they inhabit long coastlines that lie in the same climatic belt (the Indo-Pacific, Antarctic, and Arctic Provinces); some are small, like the Zealandian Province, which includes only the communities around the coasts of New Zealand (Figure 6.5).

Each province contains its own communities and therefore carries unique sets of animals that fill various ecological niches. For example, the intertidal rocky shore community in New Zealand has its ecological equivalent in British Columbia, even though the families and genera of animals are quite different in the two communities.

The total diversity of the world's shallow marine fauna directly reflects the number of provinces, which in turn reflects climate and geography. But if tectonic movements were to change Earth's geography enough, they would also alter the number of provinces of organisms, and that in turn would increase or decrease world diversity. Other things being equal, a world with widely split continents would have greater lengths of shoreline, scattered around the world, giving many marine provinces and high diversity of life.

Poles and Tropics

The Equator has a fairly uniform climate, and the same applies to the broad tropical zone on Earth, which extends to 23.5° North and South. The sun is always strong, and the temperature variation between seasons is small. The general result, especially in the sea, is that food supply is stable, available at about the same level all year round. A species can specialize on one or two particular food sources and can rely on them always being available. As each species comes to depend on a narrow range of food sources, it adapts so well to harvesting them that it cannot easily switch to alternatives. Thus, a great variety of specialized species may evolve, competing only marginally with one another, at least for food. In the Serengeti plains of East Africa, for example, several species of African vultures are all scavengers on carcasses. But one species has a head and beak adapted for tearing through the tough hide of a fresh carcass, another is adapted only for eating the soft insides from an opened carcass, another is adept at cleaning bones, and another eats the scraps. Generally, each tropical species lives in a world of stable food resources, and the great diversity of tropical communities is a reliable reflection of this kind of complex ecosystem. In the sea, the tremendous diversity of life in and around coral reefs is a major contributor to the overall diversity of the tropics.

In high latitudes, however, food supplies may vary greatly from season to season and from year to year. Overall, food supply may be high. Tundra vegetation blooms in spectacular fashion in the spring. There are rich plankton blooms in polar waters during spring and summer, and millions of seabirds and thousands of whales migrate there to share in the abundant food that is produced. Antarctic waters teem with millions of tons of tiny crustaceans (krill) that eat plankton and in turn are fed on by fish, seabirds, penguins, whales, and seals. The Arctic tern migrates almost from pole to pole, timing its stay at each end of the world to coincide with abundant food supply. Yet for organisms that live all year in polar regions, spring abundance contrasts with winter famine. Plants will not grow in the winter darkness. Food variability is the problem.

Where food supplies vary, animals can not be specialists on only one food source; they must be versatile generalists. Generalists share some food sources, and probably compete more than specialists do. If so, fewer generalists than specialists can coexist on the same food resources. In seasonal or variable environments, where organisms must be generalists, diversity is lower. So there is a rather dramatic global diversity gradient, with high diversity at the equator and low diversity at the pole.

Tectonic movements can move continents around the globe. When there are many continents in the tropics, one might expect higher diversity than when many continents are in high latitudes.

Islands, Continents, and Supercontinents

Island groups tend to have milder climates — referred to as maritime or oceanic climates — compared with nearby continents, no matter whether they are tropical or at high latitudes. Thus the British Isles and Japan have milder climates than Siberia; the West Indies have milder climates than Mexico; and Indonesia has a milder climate than Indochina.

Large continental areas have especially severe climates for their latitudes. Asia, for example, is so large that extreme heat builds up in its interior in the northern summer, forming an intense low-pressure area. Eventually the low pressure draws in a giant inflow of air from the ocean, the summer **monsoon**, that brings a wet season to areas all along the south and east edges of the continent, from China to Pakistan (Figure 6.6a). In winter the interior of Asia becomes very cold, a high-pressure system is set up, and an outflow of air, the winter monsoon, brings very chilly weather to India, China, and Korea (Figure 6.6b). Land organisms respond to the great seasonality of the monsoon climate, and organisms in the shallow coastal waters are affected strongly too. As nutrient-poor water is blown in from the surface of the open ocean in the summer monsoon, food becomes scarce; as water is blown offshore in the winter monsoon, deeper water is sucked to the surface and brings nutrients and high food levels. As a result, the diversity of marine creatures along the coasts of India is far less than it is in the Philippines and Indonesia, which are far enough away from the Asian mainland that they feel the effects of the monsoons much less strongly. Reefs are scarce and poor in diversity along the Asian mainland coast; but they are rich and diverse in a great arc from the Philippines to the Australian Barrier Reef.

The effects of continental geography as opposed to oceanic geography thus have an important effect on global diversity, though their effects are still directly linked to the variation in food supply.

In an oceanic world, with continents small and widely separated, so that there are many provinces, each community in a province tends to have stable food supplies and high diversity. Therefore, the more the continents are fragmented into smaller units, the more oceanic the world's climate becomes, and the more diverse its total biota.

The other extreme occurs when all the world's continents are together in a supercontinent, such as Rodinia or Pangea: not only are there fewer provinces, but each province has low-diversity communities. Supercontinents had supermonsoons.

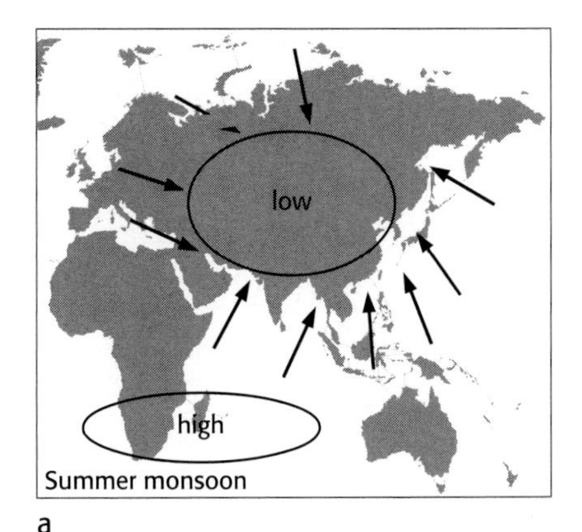

a

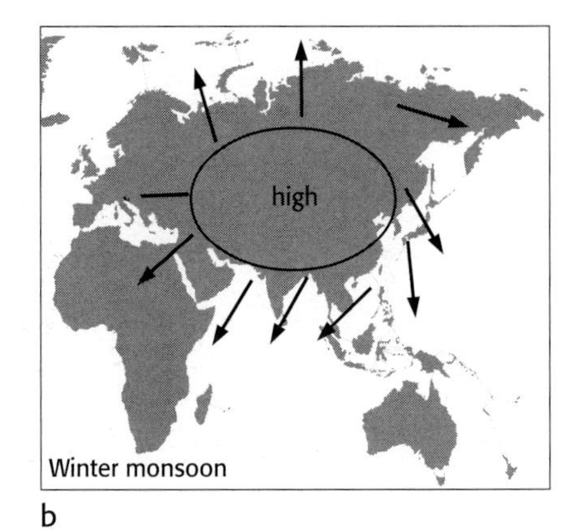

b

Figure 6.6 The monsoons of Southeast Asia. (a) In summer, heat builds up over the continent and generates low pressure that draws in moist air from the surrounding oceans. (b) In winter, high pressure over the continent generates cold winds that blow offshore.

Other Agents Also Affect Diversity

The overall pattern of diversity data through time does receive a first-order explanation from plate-tectonic effects. But that cannot be the whole story, for several reasons.

1 **Changing faunas through time.** If plate tectonics were the only control on diversity, much the same groups of animals should rise and fall with the changes in global geography. Instead, we see dramatic changes in different animal groups that succeed one another in time.
2 **Increase in global diversity.** The overall increase in global diversity from Ediacaran to Recent times is not predicted on plate tectonic grounds.
3 **Mass extinctions.** The major extinctions are much more dramatic than the major radiations. For example, the Permian extinction did not occur gradually over the 150 m.y. of the later Paleozoic, as the continents collided and assembled piece by piece. Most likely, the continental assembly set up the world for extinction, then an "extinction trigger" was pulled. There are too many sudden "mass extinctions" in the fossil record for a plate tectonic argument to be completely satisfactory. Even if plate tectonic factors set the world up for an extinction, we seem to need some separate theory to explain the extinctions themselves.

CHANGING FAUNAS THROUGH TIME

Three Great Faunas

Jack Sepkoski sorted his data on marine families through time to see if there were subsets of organisms that shared similar patterns of diversity. A computer analysis helped him to distinguish three great divisions of marine life through time, which accommodate about 90% of the data (Figure 6.7). Sepkoski called them the Cambrian Fauna, the Paleozoic Fauna, and the Modern Fauna. The faunas overlap in time, and the names are only for convenience. But they do reflect the fact that different sets of organisms have had very different histories.

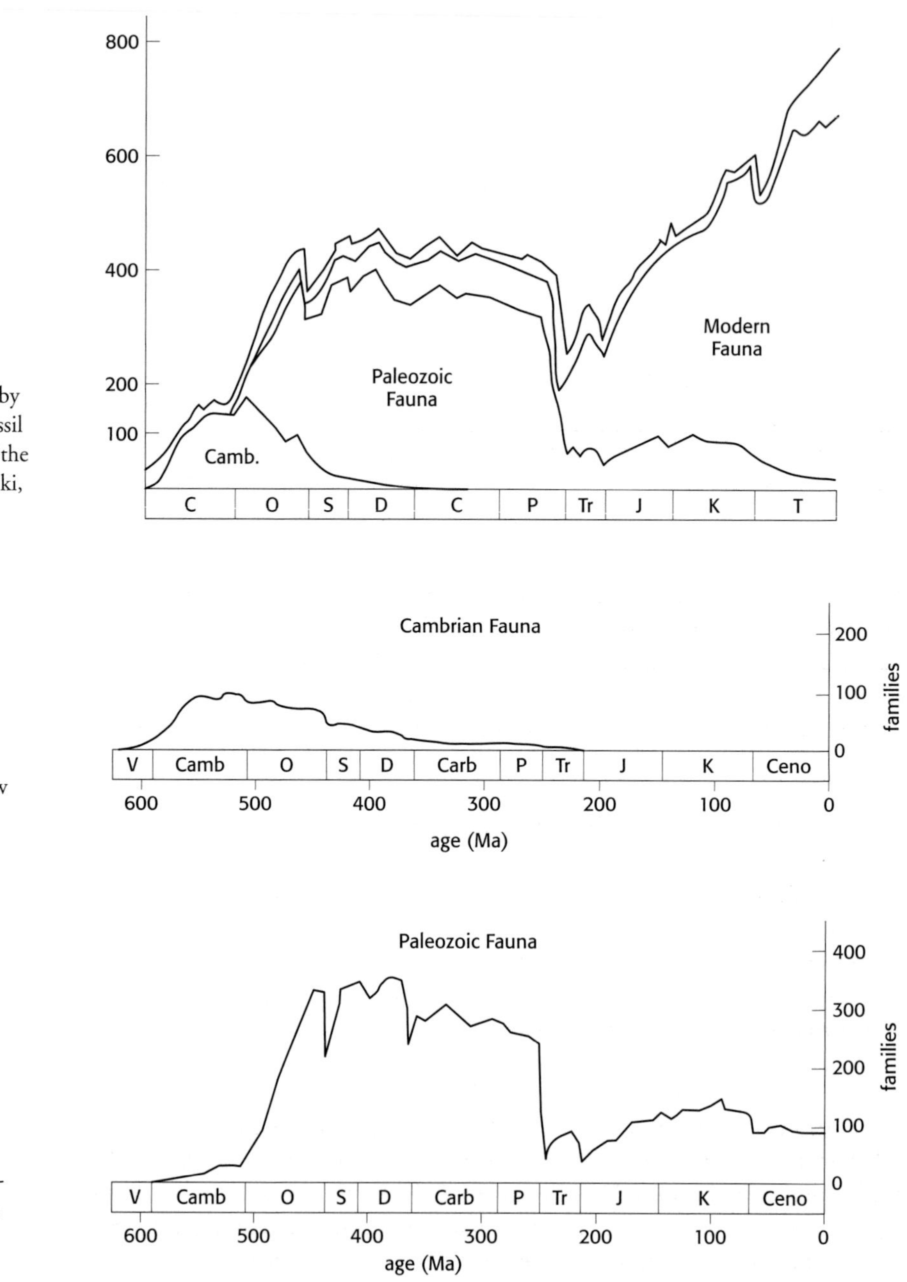

Figure 6.7 The three great faunas defined by Jack Sepkoski in his analysis of the marine fossil record at the family level. They are subsets of the data shown in Figure 6.1. (Data from Sepkoski, 1981, 1984.)

Figure 6.8 The Cambrian Fauna is dominated by trilobites and has generally low diversity (about 100 families). (Data from Sepkoski, 1981, 1984.)

Figure 6.9 The Paleozoic Fauna is dominated by suspension feeders and "lie-in-wait" predators and has higher diversity than the Cambrian Fauna (close to 400 families). (Data from Sepkoski, 1981, 1984.)

The histories of the three marine faunas are shown in Figures 6.8, 6.9, and 6.10. The **Cambrian Fauna** contains the groups of organisms, particularly trilobites, that were largely responsible for the Cambrian increase in diversity. But after a Late Cambrian diversity peak, the Cambrian Fauna declined in diversity in the Ordovician and afterward (Figure 6.8), even though other marine groups increased dramatically at that time (Figure 6.7).

In the same way, the success of the **Paleozoic Fauna** was almost entirely responsible for the great rise in diversity in the Ordovician, and slowly declined afterward (Figure 6.9). The Paleozoic Fauna suffered severely in the Late Permian extinction, and its recovery afterward was insignificant compared with the dramatic diversification of the **Modern Fauna** (Figure 6.10).

Figures 6.8, 6.9, and 6.10 show the composition of the three faunas. Their

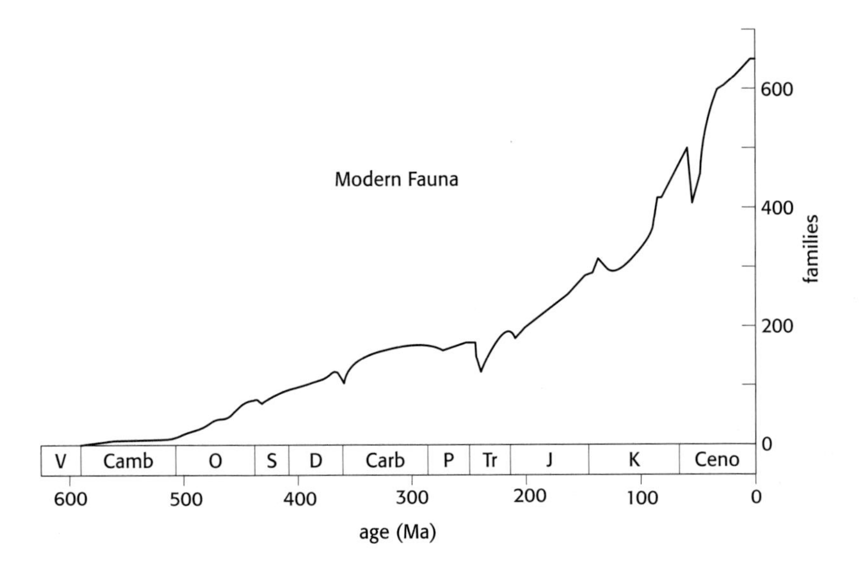

Figure 6.10 The Modern Fauna is dominated by molluscs and active predators and has even higher diversity than the Paleozoic Fauna (over 600 families). The Modern Fauna was largely unaffected by the Permian extinction. (Data from Sepkoski, 1981, 1984.)

definition is approximate, because Sepkoski's families are grouped at the level of classes or subphyla. (In hindsight, one could subdivide the marine animals into groups that would give even sharper divisions between the three Faunas: for example, one could separate Paleozoic corals from later ones.) There is no zoological affinity between the members of the three faunas, but they do have ecological meaning.

Explaining the Three Great Faunas

The diversity patterns imply that ecological opportunities in the world's oceans somehow changed through time to favor one particular ecological mixture and then to allow the diversification of others. Obviously there can be many different explanations of the facts, and I have room to discuss only a few suggestions. The diversity patterns have been known in outline for some time, so some of the explanations predate Sepkoski's analysis.

In the 1970s, Valentine (see Valentine, 1973) pointed to the different ways of life that are encouraged under different types of food supply. In the Cambrian, he argued, the continents were not widely separated, food supplies were variable, and the most favored way of life would have been deposit feeding: there is always some nutrition in seafloor mud. Thus Cambrian animals, wrote Valentine, are "plain, even grubby." The Burgess fossils may not be plain, but many of them were certainly mud grubbers. Even among soft-bodied animals, arthropods dominate Cambrian faunas in numbers and diversity, and most of them were deposit feeders.

The Paleozoic Fauna lived in more tightly defined communities, with a more complex trophic structure. The continents were more widely separated in the Ordovician, so one might expect much more reliable food supply in the plankton, which would have favored the addition of filter feeders to marine communities. One would also expect that a larger food supply in the form of stationary benthic filter feeders would have allowed slow-moving carnivores to become more diverse. Indeed, filter feeders reached higher in the water and fed at different levels, and there was more burrowing in the sediment. The overall trend was to add new ways of life, or **guilds**, to marine faunas. Altogether, Paleozoic animals seem to have subdivided their ways of life more finely through time.

If the Permian extinction was induced by the continental collisions that formed Pangea, one would predict that Paleozoic filter feeders and the predators that depended on them would have suffered a greater crisis than did other groups, because the food supply in the world's oceans would have become much more variable. In general, this prediction is correct: corals, brachiopods, cephalopods, bryozoans, and crinoids felt the Permian extinction most acutely. Again, predictably, the Permian communities that suffered the most in the Permian extinction were the reef faunas.

But it is more difficult to explain the rise of the Modern Fauna. Other things being equal, one would predict that as continents split again in the Mesozoic, Paleozoic-style predators and filter feeders would again have been favored. They had not become completely extinct, and could surely have been expected to recover. In fact, they did, but in a very subdued fashion. Most of the Mesozoic diversification was achieved by other groups that stand out in Sepkoski's analysis as the Modern Fauna. These new groups added more guilds, especially more infaunal, burrowing animals, and new predators.

Steven Stanley and Geerat Vermeij suggested predation as a major factor in the rise of the Modern Fauna, in what Vermeij called the Mesozoic Marine Revolution (see Vermeij, 1977). The new predators that appeared in the Middle Cretaceous seem to have been more effective than their predecessors at attacking animals on the seafloor. Modern gastropods evolved, capable of attacking shells with drilling radulae backed with acid secretions and poisons. Advanced shell-crushing crustaceans became abundant, as did bony fishes with effective shell-crushing teeth. Perhaps around this time the filter feeders of the Paleozoic Fauna, which were largely fixed to the open surface of the seafloor, became too vulnerable to predation. They were replaced by animals that can filter food from seawater that they pump down into their burrows. Burrowing bivalves with siphons and burrowing echinoids make up very important components of the Modern Fauna, together with effective, wide-roaming predators such as gastropods and fishes.

Another possible explanation is the bulldozer hypothesis, mostly the work of Charles Thayer (see Thayer, 1983). Thayer pointed out a major difference between the Modern Fauna and the Paleozoic Fauna: the relative abundance and diversity of strongly burrowing forms (especially worms, echinoids, crustaceans, and bivalves) in modern soft-sediment communities. Their continual churning of sediment means, among other things, that fixed filter feeders find it difficult to attach as larvae, continually run the risk of being overturned as adults, and are at least occasionally subjected to clouds of disturbed sediment that tend to clog their filters. Most filter feeders that are successful on soft sediments today are mobile forms, and fixed filter feeders are confined to the relatively restricted habitat of hard substrates. Most Paleozoic filter feeders were immobile, and they have not been able to compete successfully with the Modern Fauna even though the modern world is relatively oceanic and encourages filter feeders.

Thayer's work has been criticized on the grounds that there were powerful burrowing animals in the Paleozoic, but this criticism is not valid. Many elements of the Modern Fauna were present in the Paleozoic, but they were not dominant, so their bulldozing was limited. We need to explain that fact, of course, but the bulldozer hypothesis is not affected.

The predation hypothesis has run into similar criticisms. For example, occasional shell boring by gastropods has now been recorded as far back as the Devonian, but that does not affect the general validity of the hypothesis.

The predation effect and the bulldozer effect could both have operated. Predation and disturbance are both reasonable explanations of the failure of the Paleozoic Fauna to recover significantly in the Mesozoic, and both could help to explain the diversification of the Modern Fauna in the later Mesozoic.

INCREASE IN GLOBAL DIVERSITY

Why is there more diversity in the oceans today than there was during the diversity peak of the Paleozoic? It is not just the success of molluscs and crustaceans, because there were molluscs and crustaceans in the Paleozoic. There must have been some kind of overall change in world ecology that has favored greater diversity since the Late Mesozoic. Richard Bambach proposed the attractive name "the seafood hypothesis" (see Bambach, 1973) or an energy-related idea originally outlined by Geerat Vermeij. In this scenario, the additional energy pumped into marine ecosystems by runoff from the land, as it was covered first by advanced gymnosperms and then by angiosperm floras, supported more complex animals and ecosystems in very high diversity.

Vermeij has proposed an extended version of this line of reasoning. Most of the world's primary productivity (photosynthesis) occurs in the surface waters of the ocean, and it is likely that variation in ocean productivity through time has been just as significant as the additional runoff from the land. Vermeij argues that variations in the nutrient content of the ocean may have been very important. Nutrients could be high during periods of increased volcanic activity, for example, those that occur as ocean-floor spreading increases as continents separate. In Vermeij's view, this may be the connection between continental splitting and evolutionary radiation, and may be just as important as adding provinces or making environments more equable.

EXTINCTION AND MASS EXTINCTIONS

Extinction happens all the time. Martha, the last passenger pigeon left in the world, died of old age in the Cincinnati Zoo in 1914. This officially made her species, *Ectopistes migratorius*, extinct, though of course the species had been ecologically doomed when the last breeding birds were gone. Extinction occurs on all scales, from local to global, and it occurs at different rates at different times and in different regions. Some species have small populations that depend on a particularly narrow range of food, or habitat, and are vulnerable to even small scale ecological disturbance. So there must be a steady leakage of species, through extinction, out of the global biosphere. Occasionally, by bad luck, perhaps, one of these species will be the least of its family, and the loss of that family would become visible in a compilation like Sepkoski's.

Sooner or later, every species has some chance of becoming extinct: extinction is the expected fate of species, not a rarity. Of course, in a world with fairly steady diversity through time, existing species (and families) would become extinct about as often as new species (and families) evolved. We might expect that, other things being equal, global diversity might typically be fairly steady, or fluctuate gently up or down. This does seem to be the case for long stretches of the Phanerozoic. However, there have been times of extremely rapid extinction (Figure 6.1), and these events need explanations. Was some special extinction mechanism at work? If so, we have to try to identify it. Were large extinctions just extreme examples of normal (ordinary) extinction processes, or were they catastrophic events that were truly extraordinary?

Extinction events do vary in size. David Raup and Jack Sepkoski sifted through Sepkoski's data and identified extinction events large enough and sudden enough to be called **mass extinctions**. One can quibble about their methods, but the fact

remains that mass extinctions did happen and have been recognized in a nonquantitative way for decades. These seem to be the largest six:
• At the end of the Ordovician;
• at the end of the Frasnian stage of the Late Devonian (F–F);
• at the end of the Permian Period (Permo-Triassic or P–T);
• at the end of the Triassic period;
• at the end of the Cretaceous Period (Cretaceous-Tertiary or K–T);
• the current extinction, which of course does not show (yet) in the fossil record.

These six mass extinctions, by definition, occurred quickly: they are *events*. However, they need not all have had the same cause, and it is better to analyze each one on its own merits before trying to fit them all into a single pattern.

Explaining Mass Extinctions

Mass extinctions were global phenomena, so they have to be explained by global processes. The first that comes to mind is plate tectonics. However, tectonic changes are relatively slow in geological terms, while mass extinctions stand out because they are relatively sudden. If we are to use plate-tectonic explanations, we have to suggest that specific tectonic movements and processes combined to build a plausible trigger that suddenly fires a deadly "extinction bullet." It's not clear that there is any extinction bullet that is related to tectonics, so we must at least consider more extreme suggestions. Some plausible agents for global extinctions are:
• A failure of normal ocean circulation affects ocean chemistry enough to cause global changes in climate and atmosphere;
• a rapid change in sea level affects global ecology and climate;
• an enormous volcanic eruption affects global ecology and climate;
• an extra-terrestrial impact by an asteroid affects global ecology and climate.

Of these possible agents, enormous volcanic eruptions leave behind enormous masses of volcanic rocks, so are relatively easy to find in the rock record. Global changes in sea level will change the distribution of sediments laid down on Earth's surface: as long as the sea level change lasts long enough to leave behind this kind of evidence, we will be able to find it. But a failure of ocean circulation typically would be expected to leave behind only subtle chemical evidence, and it is likely to be short-lived, so evidence might be difficult to find and interpret. And an extra-terrestrial impact is by definition an instantaneous event that might leave behind only a thin layer of evidence. Unless we find a crater, or some unique piece of evidence that only an impact can produce, it may be very difficult to identify an impact, especially in more ancient rock. For the purposes of this chapter, it is important to remember that an asteroid impact large enough to cause a global impact should leave behind characteristic physical evidence. Three major indicators are:
1 A defined layer or spike of the element **iridium** (Ir), which occurs in greater abundance in meteorites than in Earth's crust;
2 **tektites** are tiny glass blobs (**spherules**), which are formed as a meteorite or asteroid splashes molten drops of rock at high speed into the atmosphere;
3 **shocked quartz** are quartz crystals with characteristic damage that can only be caused by intense shock waves.

At present, the leading hypotheses for the causes of the largest six extinctions are:

• End-Ordovician	Climate (an ice age)
• Late Devonian (F–F)	Oceanic crisis, or possibly an impact
• End-Permian (P–T)	Giant eruption **plus** an impact
• End-Triassic	Large eruption, possibly with impact

- End-Cretaceous (K–T) Giant eruption **plus** an impact
- The current extinction Human activity

I shall deal with the current extinction in Chapter 21. The end-Triassic mass extinction does not have enough clear data to discuss properly at this point (at least in my view). In this chapter I will discuss the others, but will leave the detailed story of the K–T event for Chapter 16.

The Ordovician Mass Extinction

The mass extinction at or near the end of the Ordovician seems to be closely linked with a major climatic change. A first pulse of extinction happened as a big ice age began, and the second occurred as it ended. Some paleontologists feel that as we collect fossils from more regions of the world, this "mass extinction" may turn out to have been a comparatively minor event.

The Late Devonian (F–F) Mass Extinction

A mass extinction took place, possibly in several separate events, at the boundary between the last two stages of the Devonian, the Frasnian and Famennian (the F–F boundary). There was a major worldwide extinction of coral reefs and their associated faunas, and many other groups of animals and plants were severely affected too. Evidence suggesting an asteroid impact has been reported from China and western Europe at or near the F–F boundary. However, there are also indications of climatic changes, and major changes in sea level and ocean chemistry, at the same time. Carbon isotope shifts indicate that global organic productivity was changing rapidly before the boundary.

George McGhee (see McGhee, 1996) favors an impact scenario. There are two cautions. First, the geological evidence requires that there were several closely spaced but medium sized impacts over perhaps two or three million years, rather than the one tremendous impact that seems to have occurred at the P–T and the K–T boundary; third, the evidence is incomplete in terms of our understanding of timing, of world geography at the time, and there are difficulties in going from evidence to interpretation. Kun Wang (see Wang et al., 1997) suggests that global ecosystems were already stressed when an impact occurred. There is no "magic marker" of impact phenomena at the extinction event, as there is at the P–T and the K–T boundary, and that makes the F–F boundary difficult to work with.

THE PERMO–TRIASSIC (P–T) EXTINCTION

The extinction at 250 Ma, at the end of the Permian, is the largest of all time: the "Mother of Mass Extinctions" according to Douglas Erwin. In 1993, Erwin felt obliged to suggest the "Murder on the Orient Express" hypothesis for the P–T extinction: that is, many factors, all acting together, led to the extinction. This is not a particularly "clean" hypothesis to accept or to test. We have ten more years' research now, by Erwin and others, and we can do better.

The P–T extinction was recognized and used by John Phillips 150 years ago to define the end of the Paleozoic Era and the beginning of the Mesozoic (Figure 6.2). An estimated 57% of all families and 95% of all species of marine animals became extinct. The Paleozoic Fauna was very hard hit, losing especially suspension feeders and carnivores, and almost all the reef dwellers. The P–T extinction is a major watershed in the history of life on Earth, especially for life in the ocean; the K–T extinction is small in comparison (Figure 6.1).

The P–T extinction was rapid, probably taking place in less than a million years. Although it was much more severe in the ocean, it affected terrestrial ecosystems too. A prolific swamp flora in the Southern Hemisphere had been producing enough organic debris to form coals in Australia, but the coal beds stop abruptly at the P–T boundary. No coal was laid down anywhere in the world for at least 6 m.y. afterward. A large change in carbon isotopes occurred across the P–T boundary, which signifies an important and global drop in photosynthesis.

The Permian extinction coincides with the largest known volcanic eruption in Earth history. In addition to plate tectonics, Earth also has **plume tectonics**. Occasionally, an event at the boundary between Earth's core and mantle sets a giant pulse of heat rising toward the surface as a plume. As it approaches the surface, the plume melts the crust to develop a flat head of basalt magma that can be 1000 km across and 100 km thick. Penetrating the crust, the plume generates enormous volcanic eruptions that pour hundreds of thousands of cubic kilometers of basalt—**flood basalts**— out over the surface. If a plume erupts through a continent, it blasts material into the atmosphere as well. After the head of the plume has erupted, the much narrower tail will continue to erupt for 100 m.y. or more, but now its effects are more local, affecting only 100 km or so of terrain as it forms a long-lasting hot spot of volcanic activity.

Plume events are rare: there have been only eight enormous plume eruptions in the last 250 m.y. The most recent is the Yellowstone plume: at about 17 Ma it burned through the crust to form enormous lava fields that are now known as the Columbia Plateau basalts of Oregon and Washington, best seen in the Columbia River gorge. North America drifted westward over this "hot spot," which continued to erupt to form the volcanic rocks of the Snake River plain in Idaho (Valley of the Moon and so on), and it now sits under Yellowstone National Park. The hot spot is in a quiet period now, with geyser activity rather than active eruption, but it produced enormous volcanic explosions about 500,000 years ago that blasted ash over most of the mountain states and into Canada.

A massive plume eruption took place exactly at the P–T boundary. A new plume burned through the crust in what is now western Siberia to form the "Siberian Traps," gigantic flood basalts that cover 4 million sq. km in area and are perhaps 2 to 3 million cu. km in volume. The eruptions coincided exactly with the P–T boundary, at 250 Ma, and lasted at full intensity for only about a million years: the largest known, most intense eruptions in the history of Earth.

There is a feeling, particularly among physical scientists, that if we can show that a physical catastrophe occurred at a boundary, we have an automatic explanation for an extinction. However, the connection has to be argued convincingly, not just assumed or asserted. Even so, the P–T extinction is the largest in Earth history, and so was the Siberian Traps eruption.

Volcanic Scenario for Extinction

In 1995, Paul Renne and his colleagues suggested a scenario based on the P–T plume eruption as a primary cause of the extinction. The plume rises toward the crust and erupts. The tremendous amount of sulfate aerosols would cool the climate enough to form ice-caps, rather quickly, and this in turn would cause a rather rapid drop in sea level along with global cooling, early in the eruptive sequence. In the rock record, we would expect to see changes in carbon and sulfur isotopes, and we do. Furthermore, as the plume erupts, the crust would be raised by the buoyant magma, perhaps enough to form a land footing for the large continental ice sheets that would grow in these high latitudes. Finally, as the eruption dies off, the crust would subside and the aerosols would disperse, making for a rapid end to the volcanically induced glaciation and another rapid change in climate. It is possible, but not calculated yet, that the volcanic gases that had built up during the eruption could have had a greenhouse effect for some time after the eruption ended, taking the earth from a volcanic glaciation to a volcanic hothouse.

Add to this scenario the "usual" effects of a giant eruption, such as acid rain, ozone depletion, a massive dose of carbon dioxide into the atmosphere, or any combination of the above, and the ingredients are in place for a mass extinction.

Though it is easy to imagine that a giant eruption might have caused a catastrophe at the P–T boundary, it is not certain that it would. We do not know how much dust, smoke, and aerosols would be produced, even though it is absolutely critical to calculations of their effects that we know those factors rather precisely. We do not know how far volcanic aerosols and stratospheric dust would be carried over Earth, or in detail what effects they would have. Dust and aerosols in the air can help absorb solar heat rather than reflect it, for example, and some models suggest that parts of Earth would warm, parts would cool, and parts would stay at about the same temperature.

The most persuasive scenarios of volcanic extinction are quickly summarized. Even a short-lived catastrophe among land plants and surface plankton at sea would drastically affect normal food chains. Large animals would have been vulnerable to food shortage, and their extinction after a catastrophe seems plausible. In the oceans, invertebrates living in shallow water would have suffered greatly from cold or frost, or perhaps from CO_2-induced heating. High-latitude faunas and floras in particular were already adapted to winter darkness, though perhaps not to extreme cold. Thus, tropical reef communities could have been devastated, but high-latitude communities could have survived much better.

These general patterns are observed at the P–T boundary, though high latitude floras were affected worse than one would have predicted. In 1996, Henk Visscher and his colleagues reported extreme abundances of fossil fungal cells in land sediments at the P–T boundary, and this has now been observed in South Africa as well. There are hints that the fungi-enriched "layer" is the record of a single, worldwide crisis, with the fungi breaking down massive amounts of vegetation that had been catastrophically killed (there were no termites yet). Such a fungal layer is unique in the geological record of the past 500 m.y. The best evidence we have suggests that there were major extinctions among gymnosperms, in Europe and among the coal-generating floras of the Southern Hemisphere. Early Triassic vegetation in Europe looks "weedy," that is, invasive of open habitats.

Caution About Eruptions

Most eruptions do not necessarily cause catastrophes. For example, the eruption of Krakatau in 1883 destroyed all life on the island and severely damaged ecosystems for hundreds of miles around. But those ecosystems completely recovered in 100 years, a geologically insignificant time. There's no biological trace of the much larger eruption of Toba, 75,000 years ago. No North American extinctions coincided with explosive eruptions from Long Valley caldera, California, from Crater Lake, Oregon, or from Yellowstone, all of which blew ash as far as Canada within the last million years. Other major plume eruptions are not linked with extinctions: examples include the Jurassic Karroo Basalts of South Africa and the Miocene Columbia Plateau Basalts.

However, one should beware of dismissing catastrophic explanations because small events do not trigger catastrophes. There may be a threshold effect: if the event is not big enough it will do nothing, but if it is big enough it will do everything. Only two eruptions in the last 500 m.y. are linked with mass extinctions, at the P–T boundary and the K–T boundary (Chapter 16). Is some other component required?

Alternatives to Eruption

Andrew Knoll and colleagues suggested in 1996 that the extinction was caused by a catastrophic overturn of an ocean supersaturated in carbon dioxide. This would result in tremendous, close to instantaneous, degassing that would roll a cloud of (dense) carbon dioxide over the ocean surface and low-lying coastal areas. An analog might be the 1986 catastrophic degassing of Lake Nyos, in the Cameroon, where hundreds of people were killed as tons of carbon dioxide bubbled out of a volcanic lake and cascaded down valleys nearby. The difference is that the proposed P–T disaster was global.

In this scenario, the carbon dioxide build-up results from global geography. All the continents were together in the supercontinent Pangea; the other two-thirds of the world was covered by the superocean, **Panthalassa**. Knoll and colleagues speculated that the deep ocean water in Panthalassa evolved into an anoxic mass loaded with dissolved carbon dioxide, methane, and hydrogen sulfide. At some point, the surface waters became dense enough to sink, triggering a catastrophe as the CO_2-saturated deep waters were brought up to the surface, degassing violently. The event would trigger a greenhouse heating and a major climatic warming.

It is not clear that a plume eruption could set off this kind of ocean overturning, though an asteroid impact could.

Geerat Vermeij and Dan Dorritie (see Knoll et al., 1986) suggested instead that methane release could have produced much the same results, without the need for such massive carbon dioxide build-up in Panthalassa. Today, methane hydrate (methane trapped in a gel-like form) builds up in sediments along continental shelves, and under the Arctic tundra. If there were methane hydrates in the Permian Arctic, the Siberian Trap eruptions could have triggered their release. (Other evidence for methane release at the P–T boundary was reported later.)

In 1998, Samuel Bowring and colleagues reported that the carbon isotope anomaly at the P–T boundary in south China was (geologically) rather short-lived: a "spike" only perhaps 165,000 years long. This suggests a major (catastrophic?) addition of nonphotosynthetic carbon to the ocean (read "methane"), rather than just a failure in the supply of organic carbon.

Greg Retallack and colleagues (see Retallack et al., 1999) found evidence in Australia of a prolonged greenhouse warming that began right at the P–T boundary. Several paleoclimatic indicators suggest the same story, which implies that the role of carbon dioxide, or methane, was the vital link between environmental disasters and extinctions. Greenhouse gases in the atmosphere could have been increased by volcanic eruptions, by oceanic turnover, and/or by methane hydrate release, and would have been accentuated and prolonged if plants were killed off globally. (World floras and oceanic plankton would have had to recover before carbon dioxide could be drawn down out of the atmosphere.)

Impact Scenario for Extinction

The whole discussion of the P–T extinction changed dramatically as I wrote this chapter. In late 2003, Asish Basu and colleagues reported convincing evidence for a major asteroid impact exactly at the P–T boundary. There had been hints of a P–T impact before, but the new evidence is conclusive. The team found pieces of a meteorite in rock beds right at the P–T boundary in Antarctica. The same bed contained shocked quartz but not iridium. Unusual metal fragments that are almost pure iron are found in the Antarctic beds as well as the P–T boundary beds in China and Japan. As we will see in Chapter 16, the same sort of evidence (though more abundant because it is younger and better preserved) is seen at the K–T boundary. Since the P–T evidence is basically global, the meteorite that produced it must have been a small asteroid, and the impact as it struck Earth would have been enormous.

We think we understand impacts and explosions rather well, after direct study of the Moon's surface, photographic surveys of cratered surfaces on planets and satellites, and our experience with nuclear blasts. We also know that asteroids strike Earth relatively frequently. Meteor Crater in Arizona, Manicouagan Crater in Canada, and scores of other impact craters can be seen from air photographs.

Some general predictions of the asteroid impact theory are clear and can be used as indirect tests of its plausibility. The impact of a large asteroid would blow a mass of vaporized rock and steam high above the atmosphere, forming an immense dust cloud that would slowly settle out through the atmosphere over a period of weeks, perhaps several months, perhaps several years. The blast and the cloud would spread material worldwide. The scenario has been discussed extensively because similar consequences—nuclear winter or at least nuclear fall—could result from a thermonuclear war. But realistic models are still not available, and at least some of the discussion is biased one way or another because the topic is so important politically.

Here is one possible impact scenario. Dust, smoke, and aerosols cut down the sun's rays for weeks or months, so that land plants and algal plankton in the ocean cannot photosynthesize. The dust also causes freezing air temperatures within days after the impact, and maintains them below freezing for weeks or even months. This may not be an unusual situation at a pole, and may not be a problem for an organism living deep in the ocean, but it is a catastrophe for organisms on continental land masses.

Later, once the dust and aerosols have settled out, the enormous amount of water vapor and CO_2 released into the atmosphere by the impact generates a greenhouse effect that elevates temperatures on Earth for a thousand years or more.

The most extreme impact scenario could be called the microwave summer because it contrasts so much with nuclear winter. It was put together by Jay Melosh and colleagues in 1990. In this scenario, some of the material produced in a very large asteroid impact is blasted upward at a velocity greater than Earth's escape velocity, although most of it eventually falls back into the atmosphere on ballistic trajectories after a travel time of about one hour.

One can calculate how much thermal radiation the mass of ballistic debris would have emitted as it reentered the atmosphere. Data on nuclear weapons suggest that the radiation pulse from infalling dust would have been 1000 times more than enough to ignite dry forests.

Ejecta radiation arrives spread over time, however, not in the single radiant pulse generated by an H-bomb. Even so, when we calculate this effect, the rates of worldwide radiation were somewhere between 30 and 100 times that of full sunshine, predominantly in the form of heat.

Of course, half of the radiation was directed upward into space, and some was absorbed by atmospheric water vapor and CO_2. Nevertheless, one-third reached Earth's surface. It would have taken most of the radiation to evaporate dense cloud, which would therefore largely have protected the surface beneath. Light cloud or no cloud would of course have given little or no shielding. Therefore, Melosh and colleagues estimate surface heating comparable with the heating in an oven set at "broil."

In general a surface temperature of 545°C is needed for wood to ignite spontaneously, and the radiation could not have produced this on a worldwide basis. But the volatile gases given off by hot wood will burst into flame after 20 minutes at 380°C, which is attained in the scenario. Even local variations in received radiation would have been sufficient to begin fires.

In perhaps the most bizarre of the "What if?" scenarios, if the tropical ocean surface were to reach 50°C, hypercanes (gigantic hurricanes) might have sucked up ice and dust and blown them into the stratosphere, blocking sunlight even more and destroying the ozone layer!

The Permo–Triassic Catastrophe

We know now that the largest extinction in the fossil record (and the major K–T extinction as well), both occurred at a time when an impact and a plume eruption coincided. This cannot be coincidence, and it shifts the arguments considerably. No reasonable person can doubt that the twin events must have caused these two major extinctions. But there are two important questions.

First, why the coincidence? Major plume eruptions are rare events, and so are major asteroid impacts. No-one would suggest that a plume eruption somehow attracts an asteroid impact, but could an asteroid impact set off a plume eruption, and if so, how? I would guess that a major asteroid impact could only coincide with a plume eruption if that eruption was somehow primed, ready to go off in any case: the impact simply acted as a trigger to bring the eruption forward. How far forward? A million years?

We know already that large earthquakes can cause astonishing effects on shallow liquid bodies thousands of miles away: water wells are particularly prone to disturbance, even if the earthquake is too far away to be felt. The great Alaska earthquake of 2002 wrecked boats in a marina in Louisiana, and California earthquakes have set off geysers in Yellowstone. Could an asteroid impact trigger a plume eruption that would have happened in any case in the next million years or so?

Second, just how do the twin insults operate ecologically to cause the extinctions? The extinctions are clearly not total, because so many organisms survived. But how (and where) did the survivors survive?

Fortunately, we now have two events to study. The Permo-Triassic extinction is further away in time than the K–T event, but it is much larger. We already have a huge mass of data ready to be discussed in terms of the new discovery, and the next few years' research should be very exciting. In particular, could a small mass extinction, at the end of the Triassic, have been a small replay of the P–T event?

EVOLUTIONARY RADIATIONS

New species appear all the time, just as species become extinct all the time. Occasionally we can look back into the record and see that a particular new species happened to be the first of a very successful group that we define as a family. The appearance of that species would thus be an event that would show up in a Sepkoski compilation of global diversity of families. As described above, a "normal" period in Earth history would have new families more or less balancing older ones that became extinct.

However, just as with extinctions, there are times when new families appeared much more often than old ones became extinct, so that we see a steep diversity rise. These events are called **evolutionary radiations**, and the name has meaning because one can often identify clades that entered a new way of life and evolved into several or even many families. Because of this, it is easier to understand the specifics of individual radiations than the specifics of individual extinctions. Radiations are likely to be evolutionary, whereas extinctions are likely to be disasters.

There is one general theme one can perceive about radiations. The radiation was a response to an **opportunity**. So what kind of opportunity would set off a radiation so large that it would show up in a compilation of global diversity? I can think of three:

1 **Mass Extinctions.** By their very nature, mass extinctions remove many organisms from the biosphere. If the mass extinction was a one-time massive physical disaster (plume eruption; asteroid impact), the physical world would probably recover quickly to "normal," yet have a biology that was missing major components. This situation provides a major opportunity for surviving organisms to evolve to fill those ecological gaps. The newcomers will not have the same anatomy, and will not reevolve the same characters as their extinct predecessors, so we are likely to see a wave of evolutionary novelty wash across the world.

Obvious examples include the radiation of the Modern Fauna after the P–T extinction; the radiation of land mammals after the extinction of most dinosaurs at the K–T extinction (Chapter 17); and the radiations of bats and whales after the extinction of most flying and swimming reptiles, again at the K–T extinction.

Such **recoveries** have been suggested as fruitful lines of research. They are, but they are not going to give us any fundamental principles we don't already know. Recoveries from mass extinctions are consequences of the extinctions, which provided the necessary opportunity.

One can say that mass extinctions remove the **incumbent effect**. This powerful image is easily understood by Americans, who live with a political system in which an elected representative to Congress (let's say) is very difficult to remove from office, once elected, even though the election process is entirely open and democratic. The reason is that the incumbent has name recognition, and has a lot of power and access to money, while any prospective challenger typically does not.

The incumbent effect works in biology too. Any species is well adjusted to its normal environment; it evolved in that environment, and its adaptations have been honed by natural selection for success there. Any invading species is likely to be less well fitted to that environment. As the incumbent effect pervades communities and provinces as well, ecosystems typically are stable over long time periods. Yet just as hurricanes can smash a local area of forest and allow weeds to flourish, or clean off a low-lying island that must be recolonized, so disasters such as mass extinctions can remove incumbents and allow survivors their place in the sun. Frightful as mass extinctions may be, in global terms they give surviving creatures an opportunity for major evolutionary innovation.

2 **Invading a New Habitat.** Evolution works by natural selection, which implies the continual testing of new mutations against the environment. Some organisms are always "pushing the envelope," and occasionally a lineage will evolve a body plan that allows it to invade a new habitat that may have been available for a long time,

but had been unexploited. If successful, that lineage may expand into a radiation as subclades explore the different ways of life that are possible in that new habitat. Obvious global examples include the first land plants and the first land animals (Chapters 8 and 9), and the first flying animals (Chapter 13).

This kind of opportunity probably exists at all scales. Land animals reaching a biologically "empty" isolated continent or island may radiate there: obvious examples are the marsupials of Australia and the mammals of South America during the Cenozoic (Chapter 18), not to mention the reptiles and birds of the Galápagos that influenced Darwin so much.

3 **New Biological Inventions.** Occasionally a lineage will evolve a body plan that allows it to do things that no organism has done before. If successful (if the timing and the ecology are just right), that lineage may expand into a radiation as subclades explore the different ways of exploiting that new invention. Obvious examples include the first eukaryotes (Chapter 3), the early metazoans (Chapters 4 and 5), and (again) the various groups that evolved the apparatus for flight (Chapter 13). Both dinosaurs (Chapter 12) and (eventually) mammals evolved warm blood and the erect limbs that allowed them a very active life style on land. Bats and whales evolved sonar, hominids invented the capacity to make tools . . . I could go on for pages, and do so in the chapters listed above.

Further Reading

Diversity through time

Adrain, J. M., and S. R. Westrop. 2000. An empirical assessment of taxic paleobiology. *Science* 289: 110–12. [Sepkoski's data base is filled with small errors, but they are random, so they do not affect the patterns that we see.]

Alroy, J., et al. 2001. Effects of sampling standardization on estimates of Phanerozoic marine diversification. *Proceedings of the National Academy of Sciences* 98: 6261–6; comment in *Science* 292: 1481. [There is significant sampling bias between different parts of the geological record.]

Bambach, R. K. 1993. Seafood through time: changes in biomass, energetics, and productivity in the marine ecosystem. *Paleobiology* 19: 372–97.

Courtillot, V., and Y. Gaudemer. 1996. Effects of mass extinction on biodiversity. *Science* 381: 146–8.

Jackson, J. C. B., and K. G. Johnson. 2001. Measuring past biodiversity. *Science* 293: 2401–4. [Let's collect more fossils, then we'll really see . . .].

Schiermeier, Q. 2003. Setting the record straight. *Nature* 424: 482–3. [The Paleobiology Data Base, meant to take Sepkoski's approach into the twenty-first century.]

Sepkoski, J. J. 1981. A factor analytic description of the Phanerozoic marine fossil record. *Paleobiology* 7: 36–53.

Sepkoski, J. J. 1984. A kinetic model of Phanerozoic taxonomic diversity. III. Post-Paleozoic families and mass extinctions. *Paleobiology* 10: 246–67.

Sepkoski, J. J. 1991. A model of onshore–offshore change in faunal diversity. *Paleobiology* 17: 58–77.

Sepkoski, J. J. 1993. Ten years in the library: new data confirm paleontological patterns. *Paleobiology* 19: 43–51.

Thayer, C. W. 1983. Sediment-mediated biological disturbance and the evolution of marine benthos. In Tevesz, M. J. S., and P. L. McCall (eds), *Biotic Interactions in Recent and Fossil Benthic Communities*, New York: Plenum, 479–625.

Valentine, J. W., and E. M. Moores. 1972. Global tectonics and the fossil record. *Journal of Geology* 80: 167–84.

Valentine, J. W. 1973. *Evolutionary Paleoecology of the Marine Biosphere.* Englewood Cliffs, N. J.; Prentice-Hall.

Vermeij, G. J. 1977. The Mesozoic marine revolution: evidence from snails, predators, and grazers. *Paleobiology* 3: 245–58.

Vermeij, G. J. 1987. *Evolution and Escalation.* Princeton: Princeton University Press. [Stresses the importance of predator/prey interactions in evolution. A fine set of essays, enormous list of references.]

Mass extinctions

Basu, A. R., et al. 2003. Chondritic meteorite fragments associated with the Permian-Triassic boundary in Antarctica. *Science* 302: 1388–92, and comment, 1314–16.

Benton, M. J. 1995. Diversification and extinction in the history of life. *Science* 268: 52–8. [No sign of periodic extinctions in a new data set.]

Bowring, S. A., et al. 1998. U/Pb zircon geochronology and tempo of the end Permian mass extinction. *Science* 280: 1039–45.

Byerly, G. R., et al. 2002. An Archaean impact layer from the Pilbara and Kaapvaal cratons. *Science* 297: 1325–7. [A giant global impact about 3470 Ma.]

Droser, M. L., et al. 2000. Decoupling of taxonomic and ecologic severity of Phanerozoic marine mass extinctions. *Geology* 28: 675–8.

Erwin, D. H. 1988. The end and the beginning: recoveries from mass extinctions. *Trends in Ecology & Evolution* 13: 344–9.

Erwin, D. H. 1993. *The Great Paleozoic Crisis: Life and Death in the Permian.* New York: Columbia University Press. [Erwin clearly labels thoughts and inferences so that we participate along with him as co-thinkers. More geology and paleontology than the title suggests. We badly need a new edition!]

Isozaki, Y. 1997. Permo-Triassic superanoxia and stratified superocean: records from lost deep sea. *Science* 276: 235–8.

Jin, Y. G., et al. 2000. Pattern of marine mass extinction near the Permian-Triassic Boundary in South China. *Science* 289: 432–6. [A very rapid event around 251.4 Ma.]

Knoll, A. H., et al. 1996. Comparative Earth history and Late Permian mass extinction. *Science* 273: 452–7, and comments, v. 274, pp. 1549–52, including Vermeij and Dorritie's suggestion of a methane crisis.

Kring, D. A. 2000. Impact events and their effect on the origin, evolution, and distribution of life. *GSA Today*. August 2000. [Snappy summary of impacts and their consequences, with good reference list.]

Looy, C. V., et al. 1999. The delayed resurgence of equatorial forests after the Permian-Triassic ecologic crisis. *Proceedings of the National Academy of Sciences* 96: 13857–2.

Looy, C. V., et al. 2001. Life in the end-Permian dead zone. *Proceedings of the National Academy of Sciences* 98: 7879–83.

McGhee, G. R. 1996. *The Late Devonian Mass Extinction: the Frasnian/Famennian Crisis*. New York: Columbia University Press. [McGhee covers all serious hypotheses. He favors multiple medium-sized impacts.]

Melosh, H. J., et al. 1990. Ignition of global wildfires at the Cretaceous/Tertiary boundary. *Nature* 343: 251–4.

Musashi, M., et al. 2001. Stable carbon isotope signature in mid-Panthalassa shallow water carbonates across the Permo-Triassic boundary: evidence for ^{13}C-depleted superocean. *Earth and Planetary Science Letters* 191: 9–20. [Evidence for huge influx of "light" carbon (read methane)].

Pfefferkorn, H. W. 1999. Recuperation from mass extinctions. *Proceedings of the National Academy of Sciences* 96: 13597–9. [Recovery was slow.]

Rampino, M. R., et al. 1988. Volcanic winters. *Annual Reviews of Earth and Planetary Science* 16: 73–99.

Raup, D. M. 1991. *Extinction. Bad Genes or Bad Luck?* New York: W. W. Norton.

Renne, P. R., et al. 1995. Synchrony and causal relations between Permian-Triassic boundary crises and Siberian flood volcanism. *Science* 269: 1413–16. [Volcanic scenario.]

Retallack, G. 1995. Permian-Triassic life crisis on land. *Science* 267: 77–80.

Retallack, G. J., et al. 1999. Postapocalyptic greenhouse paleoclimate revealed by earliest Triassic paleosols in the Sydney Basin, Australia. *Bulletin of the Geological Society of America* 111: 52–70.

Ryder, G., et al. (eds) 1996. *The Cretaceous-Tertiary Event and Other Catastrophes in Earth History*. Geological Society of America Special Paper 307. [The third major international conference on massive extinctions. Two previous conferences were published in volumes 190 and 247.]

Steiner, M. B., et al. 2003. Fungal abundance spike and the Permian–Triassic boundary in the Karoo Supergroup (South Africa). *Palaeogeography, Palaeoclimatology, Palaeoecology* 194: 405–14.

Stothers, R. B. 1984. The great Tambora eruption of 1815 and its aftermath. *Science* 234: 1191–8.

Suess, E., et al. 1999. Flammable ice. *Scientific American* 281: 76–83. [Introduction to methane hydrates].

Twitchett, R. J. 1999. Palaeoenvironments and faunal recovery after the end Permian mass extinction. *Palaeogeography, Palaeoclimatology, Palaeoecology* 154: 27–37. [Recovery was slow.]

Visscher, H., et al. 1996. The terminal Paleozoic fungal event: evidence of terrestrial ecosystem destabilization and collapse. *Proceedings of the National Academy of Sciences* 93: 2155–8.

Wang, K. et al. 1997. Carbon and sulfur isotope anomalies across the Frasnian Famennian extinction boundary, Alberta, Canada. *Geology* 24: 187–90. [An impact occurred when ecosystems were already stressed.]

Ward, P. D., et al. 2000. Altered river morphology in South Africa related to the Permian-Triassic extinction. *Science* 289: 1740–3, and comment, 1666–7.

Warme, J. E., and Sandberg, C. A. 1996. Alamo megabreccia: record of a Late Devonian impact in southern Nevada. *GSA Today* 6: 1–7.

CHAPTER SEVEN

The Early Vertebrates

Vertebrates dominate land, water, and air today in ways of life that combine mobility and large size (more than a few grams). Only arthropods (insects on land and crustaceans in the sea) come close to competing for these ecological niches. As vertebrates ourselves, we have a particular interest in the evolutionary history of our own species and our remote ancestors. It's hardly surprising that vertebrates should receive special treatment in this and almost every other book on the history of life.

It is easier for us to identify with vertebrates than with invertebrates. We can feel how ligaments, muscles, and bones work. We feed by using our jaws and teeth. We have sensory skin and good vision, and we sense vibrations in our ears. We walk, run, and swim. We have bodily sensations as we thermoregulate, and we understand by experience the bizarre system we have for getting oxygen and circulating it around the body. All vertebrates share some of these systems, and many vertebrates have them all. In contrast, most invertebrates have quite different body systems that are more difficult for us to identify with and to understand.

Our familiarity with vertebrate biology helps to make up for the rarity of vertebrate fossils. Vertebrates are rare even today in comparison with arthropods or molluscs, and vertebrate hard parts are held together only by skin, muscles, cartilage, and ligaments that rot easily after death. Even bones crumble and dissolve rather easily once they lose the organic matter that permeates them in life. Land vertebrates in particular live in a habitat that offers little chance of preservation. Bones are scattered and destroyed rather than buried and sheltered by sediment. Only the special interest in vertebrates shown by professional and amateur fossil collectors alike has compensated for the intrinsic poverty of vertebrate preservation. By now we have a very good idea of the major events in vertebrate history. (Explaining them is a different problem!!!)

VERTEBRATE ORIGINS

Vertebrates must have evolved from invertebrates, which are simpler in structure and have a longer fossil record. Vertebrates, of course, have a spine, a bony (or cartilaginous) column that contains a nerve canal and a **notochord**. The notochord is a specialized structure that looks like a stiff rod of dense tissue. It is a more fundamental character than the spine that surrounds it. It is a shared derived character that places vertebrates in the phylum **Chordata**, together with some soft-bodied creatures that have a notochord but do not have a head or a skeleton.

By using the stiffness of the notochord, a chordate without a spine can give its muscles a firm base to pull against, while retaining enough flexibility to allow a push against the water for efficient swimming. The notochord can store elastic energy that is released at the right moment to help swimming. I suspect that the evolution of the notochord, with this mechanism for energy storage and release, is the evolutionary novelty that promoted the success of soft-bodied chordates. It preceded by a long time the evolution of the skeleton of a typical vertebrate.

Urochordates and cephalochordates are two living groups of soft-bodied chordates that help to show us what a vertebrate ancestor might have looked like. Urochordates include **tunicates** (sea squirts), small box-like creatures that live as adults in colonies fixed to the seafloor. But their larvae swim actively, using the notochord and muscle fibers in a tail-like structure that is lost when they settle as adults (Figure 7.1). The tunicate *Ciona* has had its genome completely sequenced, and that genome looks much like that of vertebrates, but simpler.

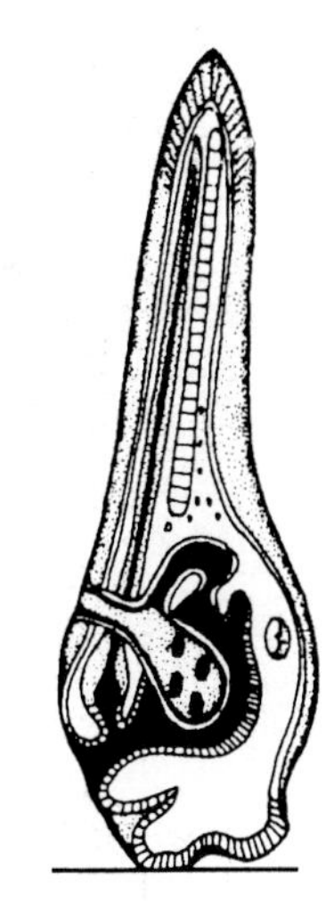

Figure 7.1 The larva of a urochordate has a notochord and swims freely. After a short time it attaches to the seafloor, as shown here, and metamorphoses into an adult that looks more like a sponge than a typical chordate. The adult is a sessile filter feeder. (From Barnes et al., *The Invertebrates: A Synthesis.* 3rd edn © 2001 Blackwell Science.)

Cephalochordates (Figure 7.2) are marine creatures that feed on small particles from seawater pumped through a special body chamber that is also used for respiration. The notochord runs along the dorsal axis and is surrounded by packs of body muscle arranged in V-shaped chevrons. Alternate contractions of the muscle packs flex the body from side to side in a wave-like pattern that allows it to swim. Sensory organs at the anterior end of the notochord mark the position of a primitive brain. In most of these characters, cephalochordates are much like fishes, even to the pattern of V-shaped muscle fibers that is so obvious when one dissects a fish carefully in a laboratory or a restaurant.

Branchiostoma, the amphioxus (Figure 7.2), is a typical cephalochordate. It lives and moves between sand grains and in open water, squirming and swimming in eel-like fashion with its muscle packs and notochord acting against one another.

These living animals are soft-bodied, so it once seemed very unlikely that we would ever have the fossil evidence to decide which group, if either, was the vertebrate ancestor. First, we are learning a great deal by studying the "evo-devo," the interplay between genes and development, in living tunicates, cephalochordates, and simple fishes. Perhaps more important (because it is evidence from the relevant time in Earth history rather than from living animals), we have a new treasure trove of soft-bodied fossils from the astounding Lower Cambrian Chengjiang fauna of China. They include a cephalochordate, a tunicate, and a generalized deuterostome animal. And there is a creature with a definite head, which is therefore a "craniate," which in turn makes it a basal fish (Figure 7.3). Accordingly, it has been named *Haikouichthys.* A lot of argument about interpreting these creatures does not hide the fact that very soon we should have a reasoned, documented, and detailed reconstruction of evolution toward the first vertebrates. Certainly we have a simple cladogram of their relationship (Figure 7.3).

I am using the everyday words "fish" and "fishes" to mean Craniata: those animals that are more closely related to living fishes than they are to Cephalochordata. In other words, I am using a stem-based concept of fish and fishes (Chapter 3).

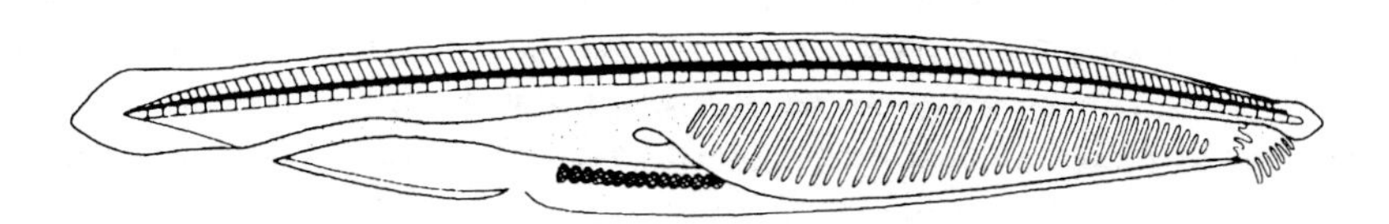

Figure 7.2 *Branchiostoma*, a living cephalochordate, the amphioxus. (After Barnes et al., *The Invertebrates: A Synthesis.* 3rd edn © 2001 Blackwell Science.)

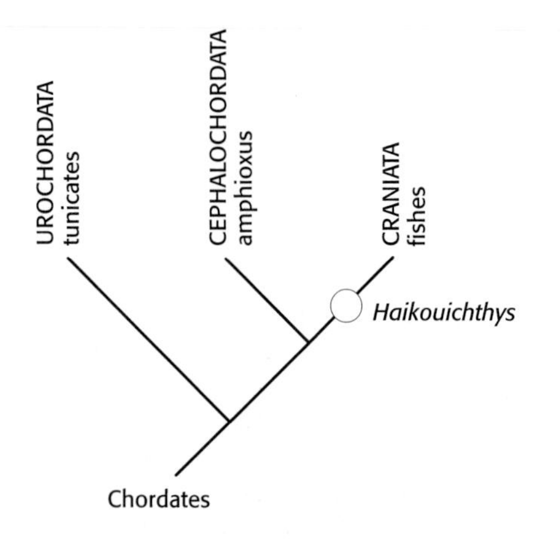

Figure 7.3 We now see in outline that vertebrates evolved from cephalochordates rather than tunicates. The earliest well-documented fish is *Haikouichthys* from the Lower Cambrian of China.

First Fishes

We now have more than 500 specimens of *Haikouichthys*. Although the soft parts are difficult to interpret, we can now say something about the first fishes from real fossils rather than theoretical speculations. First, a lobe at the front has dark areas interpreted as eyes, and perhaps nasal sacs were present too. The mouth and gills lie immediately under this "head" region. Some structures, probably cartilage, are wrapped around the notochord, and these are probably cartilaginous vertebrae. If so, then *Haikouichthys* is a genuine vertebrate.

The notochord probably evolved as a structure that aided in swimming. But the physics of hydrodynamics dictates that swimming efficiency increases with body length. As early chordates explored various ways of life, the more actively swimming species may perhaps have increased in body size. But there must come a body size for which efficient swimming requires more stiffness than a notochord can give, and some kind of cartilaginous or mineralized skeleton then becomes a cheap way of increasing efficiency. At the moment *Haikouichthys* is the first sign of this breakthrough in mechanical efficiency.

The earliest fishes with hard parts, from the Ordovician, did not have a bony internal skeleton. Instead, they had evolved mineralized bony plates that covered some or all of their bodies, adding stiffness and giving rise to the term **ostracoderm** ("plated skin") for them (Figure 7.4). In the same way, sharks today lack an internal bony skeleton but instead have a tough skin with strong fibers that stiffen the body considerably. The plates of ostracoderms would have provided protection too, from possible predators and from abrasion by sand and rock surfaces.

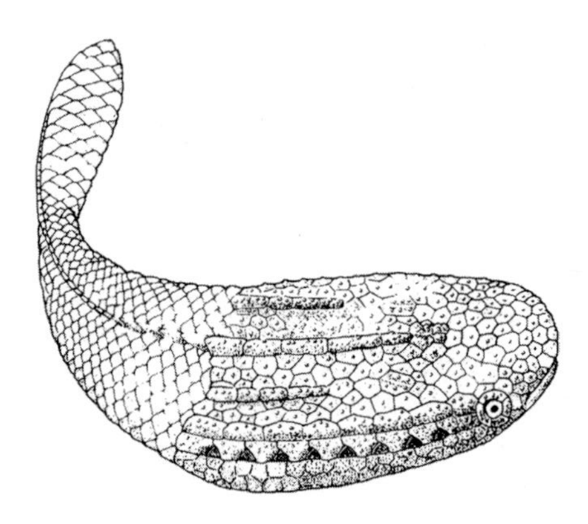

Figure 7.4 Reconstruction of an ostracoderm. *Astraspis*, from the Ordovician of Colorado, was about 13 cm (5 inches) long. (Courtesy of David K. Elliott, Northern Arizona University.)

Bone is dense, however, and the heavy plates made early fishes relatively clumsy, with slow acceleration. They probably swam slowly along the seafloor, inside heavy boxes.

Astraspis from the Ordovician of Colorado is one of the best-preserved early fishes (Figure 7.4). A headshield protected the anterior nerve center (which from this point can be called a brain), and also provided a stout nose cone for cutting through the water without flexing, and for probing into soft sediment. Behind the eyes were plates with openings to allow water to flow out past the gills. The tail was short, stubby, symmetrical, and small, and these fishes probably swam well but not fast.

None of the early fishes had jaws: they were all jawless (agnathan): they had a small simple mouth somewhere near the front and must have fed on small, easily digested objects.

The few jawless fishes living today (hagfishes and lampreys) have parasitic or otherwise strange ways of life, and they cannot tell us much about the ecology or evolution of their earliest ancestors. Two simplified cladograms show the relationships of the most primitive fishes (Figure 7.5). The living jawless fishes are the few survivors of a much broader range of early fishes.

Heterostracans

Heterostracans were the earliest abundant fishes (in the Silurian and Devonian). They had flattened headshields with eyes at the side, and they look well adapted for scooping food off the seafloor (Figure 7.6). Some had plates around the mouth that could have been extended out into a shoveling scoop. The rigid head and the stiff, heavy-plated body imply that propulsion came mainly from the tail in a simple

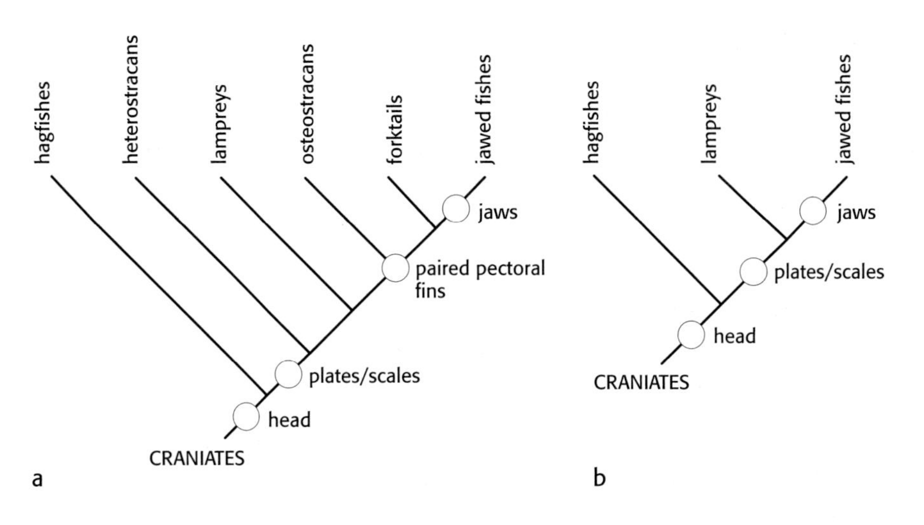

Figure 7.5 Two cladograms showing the relationships between groups of jawless fishes. The cladogram of living groups (b) gives limited information about their history. When extinct groups are included (a), the cladogram shows that they are only a remnant of the former diversity of jawless fishes. (Simplified from Janvier, 1996.)

swimming style, with none of the control surfaces provided by the complex fins of modern fishes. Nevertheless, the heterostracan way of life was successful, and their fossils are found all across the Northern Hemisphere.

Heterostracan fishes diverged quickly in Silurian times and their shapes evolved toward hydrodynamically more efficient shapes over time. Pteraspids (Figure 7.7) were most abundant heterostracans in Devonian times. They had beautifully streamlined armored headshields, with a sharp nose cone and a smooth curved shape that gave an upward motion to counteract the density of the armor. A spine projected backward over the lightly plated trunk, partly for protection and partly for hydrodynamic stability. Pteraspids had tails with the lower half longer than the upper; other things being equal, this too would have helped the fish to counteract the weight of the headshield. The mouth lay under the head, and a ventral plate covered the gills. Water was taken in through the mouth, and the exit passages were neatly tucked toward the back of the headshield, much like the exhausts of a twin-jet fighter aircraft (Figure 7.7). In some forms the headshield was very flattened, for gliding through water as a delta-wing aircraft glides through air.

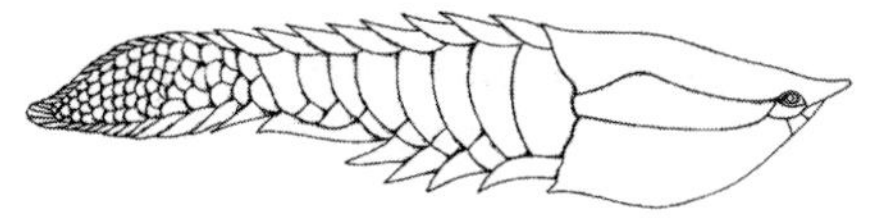

Figure 7.6 Many heterostracan fishes were adapted to scoop food from the sea floor, so the mouth lay on the under side of the head shield. This is *Anglaspis*. (After Kiaer.)

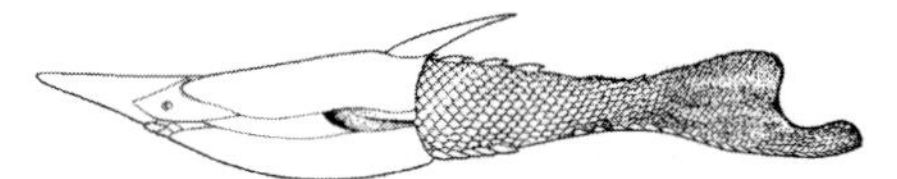

Figure 7.7 A reconstruction of *Pteraspis* shows the strong plated headshield and the flexible trunk and tail. (After White.)

Amphiaspids are another group of heterostracans that are best known from Siberia. Larisa Novitskaya found specimens from which she could reconstruct the gills (Figure 7.8). Pteraspid and amphiaspid gill systems look like the exhaust systems of 1930s racing cars, sharing their design for efficient passage of fluids.

But in all this successful evolution, heterostracans never evolved paired fins. Their swimming power came entirely from the trunk and tail, with perhaps a little help from the gill exhaust.

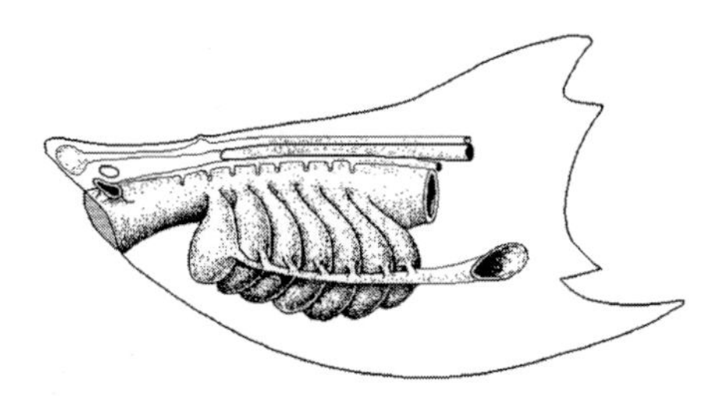

Figure 7.8 A diagram simplified from an internal reconstruction of amphiaspid gills by Larisa Novitskaya, showing their similarity to the exhaust system of an old supercharged multicylinder racing car of the 1930s.

Osteostracans

Other things being equal, any swimming creature would benefit by evolving powerful swimming and better maneuverability. We have seen this already among the heterostracans. An innovation came with the evolution of paired fins, leading to a Late Silurian radiation of new jawless fishes, the **osteostracans** (Figure 7.5). Osteostracans were like heterostracans in that they had a strongly plated headshield and a comparatively flexible body and tail that provided most of the propulsion.

The most important osteostracans, the **cephalaspids** (Figure 7.9), lived from Late Silurian to Late Devonian times. Their large solid headshields often had a large spine projecting forward and two spines or horns extending backward at each

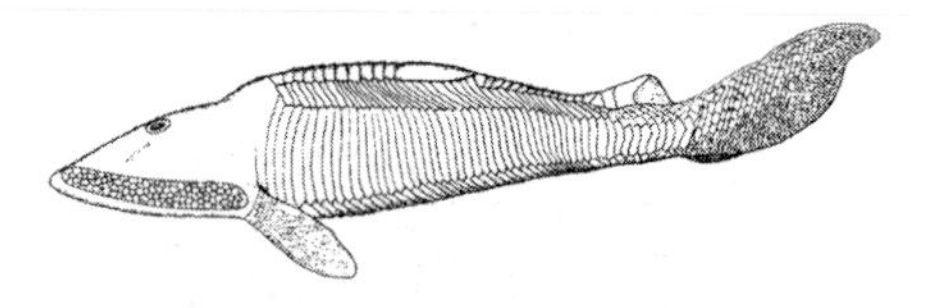

Figure 7.9 The cephalaspid *Hemicyclaspis* has paired fins and a large sensory area on each side of the solid headshield. (After Stensiö and Heintz)

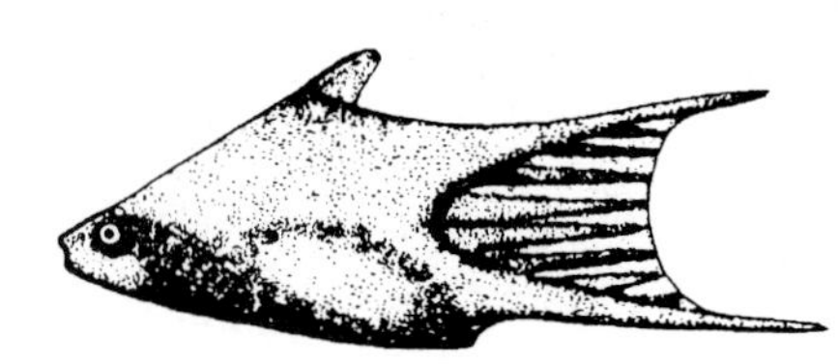

Figure 7.10 A generalized forktail from the Early Devonian of Canada. This diagram combines elements of more than one species of forktails. Note the bulging outline of the big stomach, unexpected in a jawless fish. (After Wilson and Caldwell, 1993.)

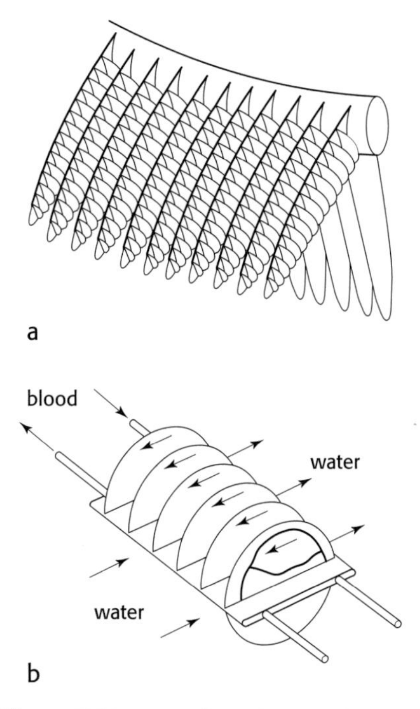

Figure 7.11 How gills work in most living fishes. Gills are arrays of thin plate-like structures set in rows supported on a strong axis (a). Oxygen-poor blood is pumped along a one-way system through each plate-like structure (b). Water is pumped the other way, exchanging gases with the blood by shedding oxygen and taking up carbon dioxide.

corner. Powerful paired fins were attached at the back corners of the headshield, just inside the protective spines. The body behind the headshield was laterally compressed, as in most living fishes, and small dorsal fins added stability. The cephalaspid tail was more versatile than the heterostracan tail, and it had some horizontal flaps that added new control surfaces.

Cephalaspids were bottom dwellers. The mouth was on the flat underside of the headshield. The eyes were small and close together on the top of the headshield. In addition, cephalaspids had large sensory areas on each side of the headshield, covered with very small plates (Figure 7.9). These organs may have served as pressure sensors in murky water, although they may have sensed electrical fields, as living sharks do.

Forktails

In 1998, a new group of jawless fishes was defined on the basis of some small but well-preserved fossils from northern Canada. They are the Furcacaudiformes, but I shall call them **forktails** because that is what the formal name means. Forktails were small and deep-bodied, unusual because mud is preserved inside them, filling a distinct stomach cavity (Figure 7.10). Filter-feeding fishes, which sort their food carefully before swallowing it, do not need large stomachs; mud-eating or carnivorous fishes do. Forktails appear to have been mud eaters, but the evolution of a distinct, specialized stomach may have been required before fishes could evolve jaws and eat large prey. In terms of jawless fish evolution, forktails sit nicely as the jawless sister group of jawed fishes (Figure 7.5).

THE EVOLUTION OF JAWS

By the Late Silurian, the jawless fishes filled many ecological roles. They were confined to eating small particles, such as plankton from the surface, sediment on the seafloor, or soft, easily swallowed food such as worms or jellyfish. But somewhere among them were fishes in the process of evolving jaws. The evolution of jaws was a major breakthrough for the feeding ecology of fishes; it apparently led to the decline of the surviving agnathans in competition with the newcomers.

Studies of anatomy and embryology suggest, but do not prove, that the bones that form the vertebrate jaw evolved originally from the gill arches of jawless fishes. In living fishes, water is taken in at the mouth and passes backward past the gills, where oxygen and CO_2 are exchanged with the blood system (Figure 7.11). Gills are soft, so they must be supported in the water current by thin strips of bone or cartilage called **gill arches**. The more water passing the gills, the more oxygen can be absorbed and the higher the energy the fish can generate. Living fishes usually have pumps of some kind to increase and regulate the flow of water passing the gills. Most fishes use a pumping action in which they increase and decrease the volume of the mouth cavity by flexing the jaws. Tuna swim so fast that they create a ramjet action that forces water past the gills, just as the airscoops of some jet fighters funnel air into the turbines.

If jaws evolved from a gill arch, the evolution of the jaw was probably connected originally with respiration rather than feeding. Water flow over the gills of jawless fishes may have been impeded by their small mouths and by a slow flow of water past the gills, so their swimming performance may have been limited by oxygen shortage. Perhaps a joint evolved in the forward gill arch so that it flexed to open the mouth wider, pumping more water backward over the gills (Figure 7.12). In the process, this gill arch was transformed into a true jaw. Opening up the jaws of a medium-sized bony fish such as a trout or salmon shows the similarity between the jaw and the internal gill arches.

Both food intake and oxygen exchange would have been increased by the evolution of the jaw, even if fishes continued to eat microscopic particles. I propose that jaws evolved in actively swimming fishes that were at the edge of their performance envelope (perhaps forktails).

Jawed fishes are **gnathostomes**, as opposed to the agnathans or jawless fishes. There is no general agreement on their classification and early evolution, mainly because they evolved and radiated so quickly around the Silurian–Devonian boundary that it's difficult to tell which groups are most closely related. A simplified conventional classification and tentative cladogram will help in discussing the various groups (Figure 7.13).

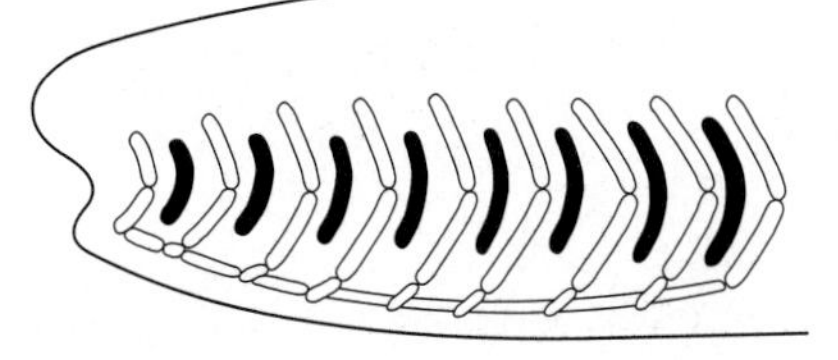

Figure 7.12 How gills probably worked in some early jawless fish. Water was taken in at the mouth, and passed out through several gill slits (black). Between each gill slit was a gill arch (white), a thin strip of bone that supported the gill. Possibly in an adaptation that increased water pumping, the first gill arch became hinged in the middle and evolved eventually into a feeding structure, the jaw.

Acanthodians

The earliest jawed fishes are small Silurian forms called acanthodians. They are lightly built, not well preserved, and not very well known (their placement in Figure 7.13 is controversial). I mention them only because they are among the earliest known fishes with jaws, and one fortunate find may upset our whole understanding of the earliest jawed fishes and the structure of Figure 7.13.

The question is important, because the evolution of jaws and a resulting extension of the potential food supply were keys to the tremendous ecological expansion and evolutionary success of jawed fishes. But teeth and jaws are only weapons: they must be applied to targets by a delivery system. The history of fishes since the Devonian has been largely one of increasing effectiveness in the mounting and hinging of the jaws, in the speed of strike, and in the hydrodynamics of propulsion and maneuverability.

Placoderms

Placoderms were dominant, worldwide fishes during Devonian times. Most placoderms had a well-developed headshield made of several plates, jointed to an armored girdle surrounding the front part of the trunk, making the fish very nose-

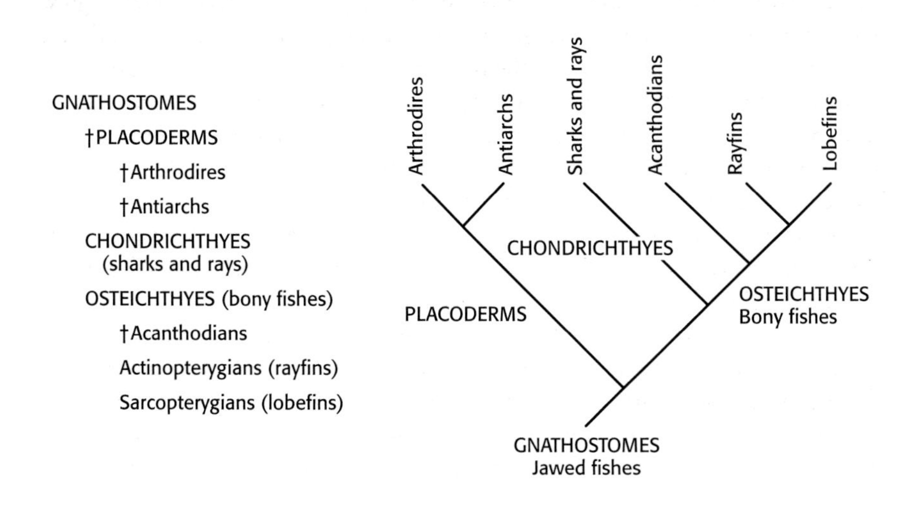

Figure 7.13 A conventional classification and a cladogram of the jawed fishes or gnathostomes.

Figure 7.14 The head of *Dunkleosteus*, a gigantic arthrodire from the Late Devonian. Its full body length was about 6 m (20 ft); the headshield alone was 2 m (6 ft) long. Tooth plates gave powerful cutting and slicing edges along the jaw line.

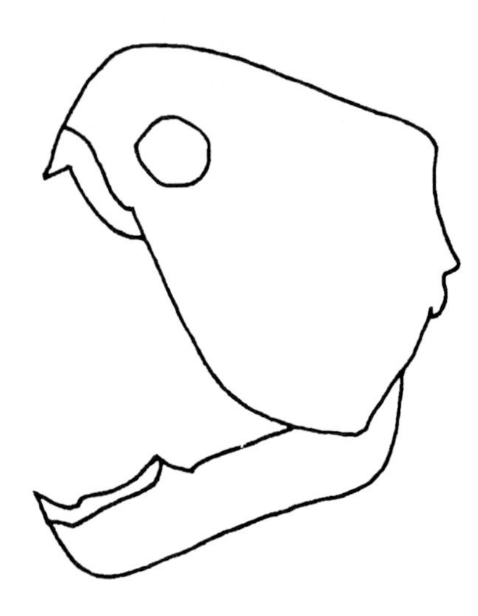

Figure 7.15 The head of *Dunkleosteus*, showing how its jaws gaped widely. The head could hinge upward on a special joint at the back of the skull, while the lower jaw dropped, opening up a very wide gape (compare with Figure 7.14).

heavy. The rest of the trunk was lightly scaled, and presumably the trunk and the long tail were very flexible. There were several pairs of fins, indicating good control over movement. But the body was usually flattened to some extent, and the eyes were usually small and set on the upper side of the headshield.

Placoderms may often have been powerful swimmers, but they cannot have been agile or capable of rapid maneuvers. The small eyes imply that they probably used other senses to a large extent, just as living sharks do. The jaws vary quite a lot, but some advanced placoderms had vicious sharp-edged tooth plates set into the jaw (Figure 7.14); these were large carnivores up to 6 m (20 ft) long. Others had large crushing tooth plates, perhaps for eating molluscs or arthropods.

Arthrodires were powerful, streamlined fishes, but their great weight of armor and their generally flattened body shapes may have limited their swimming performance. Large pectoral fins aided stability and provided lift for the heavy armored head. The joint between headshield and trunk armor was well developed. The head could be levered upward on that joint while the lower jaw dropped at the same time, opening a wide gape for effective hunting (Figures 7.14; 7.15), and possibly aiding water flow over the gills. Arthrodires were the dominant fishes of the Devonian, and included the most powerful predators of the time. Their distribution was worldwide, in both salt and fresh water. Giant arthrodires include *Tityosteus*, the largest known Early Devonian fish, with an length of about 2.5 m (8 ft). But *Tityosteus* is dwarfed by a Middle Devonian freshwater fish *Heterosteus* and by the Late Devonian *Dunkleosteus*, both of which grew to 6 m (20 ft) long (Figures 7.14; 7.15).

Antiarchs are much more specialized and difficult to understand, but these "grotesque little animals" (as one famous paleontologist called them) were successful worldwide, mostly in freshwater environments. They were small, with headshields up to 50 cm (20 in) long and a maximum known length just over a meter (3 ft). Their headshields were flattened against the bottom, with the eyes set close together high on the headshield. The mouth lay just under the snout. The body armor was long. Instead of pectoral fins, antiarchs had long, jointed appendages that look as if they were used for poling the fish along the bottom rather than swimming (Figure 7.16). Antiarchs had small mouths and probably ate mud, filling an ecological role that had been taken by earlier jawless fishes. It's clear that they were slow, rather clumsy swimmers.

Placoderms evolved at the beginning of the Devonian and were practically extinct by its close, even though they dominated the time between. It seems clear that eventually they were handicapped by their weight of armor, and they never seem to have conquered the problem of agile maneuvering in water. The cartilaginous fishes and the bony fishes, each in their own way, solved the problem and now dominate living fish faunas.

Cartilaginous Fishes (Sharks and Rays)

Sharks and rays, and all their ancestors we have been able to identify, have cartilaginous skeletons rather than bone. This distinction dates back to the Early Devonian, when this group of fishes was just one of the many early successful lines that had recently evolved jaws.

The fossil record of sharks and rays is poor, because they rarely preserve well as fossils. They have cartilage rather than bone, and a tough skin rather than heavy scales. They do have formidable teeth, which are often well preserved as fossils, but teeth alone give only a vague idea about the entire fish. Occasionally a rare find of a

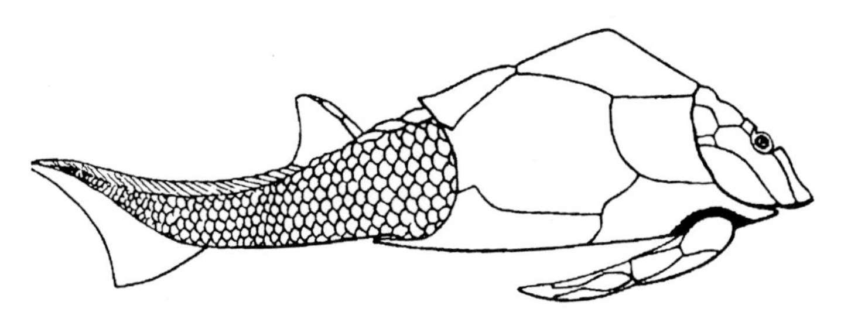

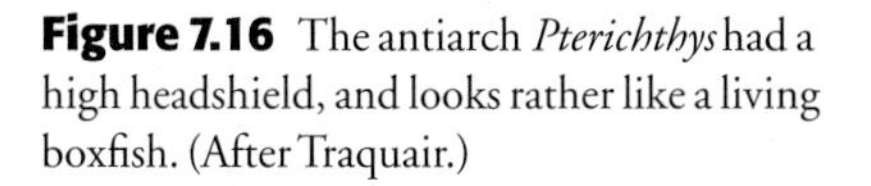

Figure 7.16 The antiarch *Pterichthys* had a high headshield, and looks rather like a living boxfish. (After Traquair.)

body outline allows us to see that sharks have not changed a great deal in overall body shape during their evolution (Figure 7.17).

Sharks have excellent vision and smell and an electrical sense, all of which combine to equip them well for hunting in all kinds of environments. They all have internal fertilization, and some have live birth. They are certainly not primitive. Sharks are simply a group of fishes that discovered a successful way of life several hundred million years ago.

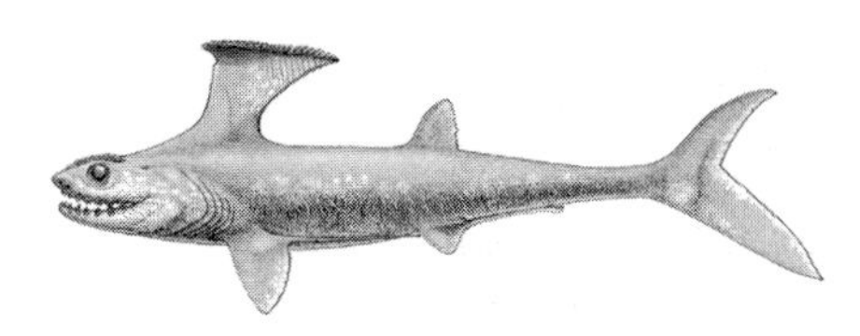

Figure 7.17 An exceptional find allowed this reconstruction of the fossil shark *Akmonistion*, from the Carboniferous. Its shape is uncannily like that of many living sharks, except for the prominent dorsal structure that is associated in shark biology today with mating. The teeth look like crushing teeth rather than slashing devices. (Courtesy of Michael Coates of the University of Chicago.)

BONY FISHES

The bony fishes, or Osteichthyes, evolved and radiated so fast in the Late Silurian and Early Devonian that we have very few fossils of the earliest forms. Almost as soon as we see them, they are divided into two major groups, the rayfins and the lobefins (Figure 7.18). The critical fish faunas involved in the very earliest radiation were recently discovered in south China, which was an isolated mini-continent at the time. Fossils from this region will probably give us more information in the next few years.

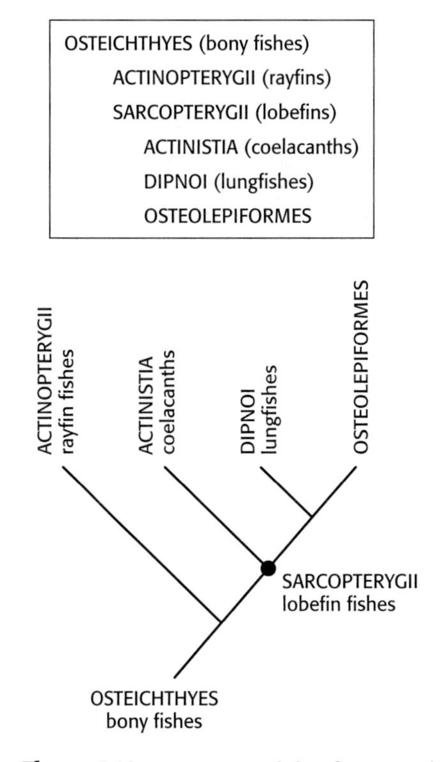

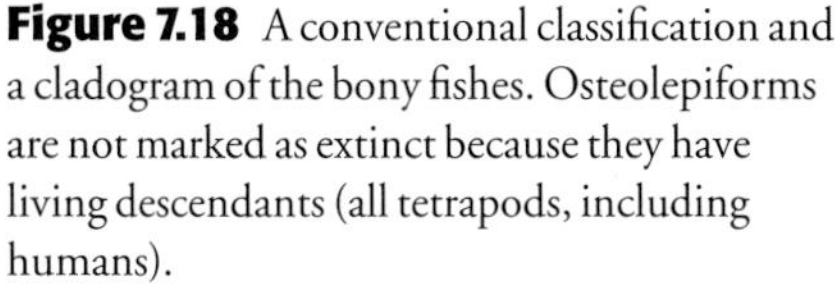

Figure 7.18 A conventional classification and a cladogram of the bony fishes. Osteolepiforms are not marked as extinct because they have living descendants (all tetrapods, including humans).

Actinopterygians (Rayfin Fishes)

Actinopterygians, or rayfins, have very thin fins that are simply webs of skin supported by numerous thin, radiating bones (called **rays**). Typically, rayfins are lightly built fishes that swim fast or maneuver very well. They have dominated marine and freshwater environments of the world since the decline of the placoderms at the end of the Devonian. It is tempting to suggest that their evolutionary success largely reflects their mastery of swimming and feeding in open water.

In general, the evolution of the rayfin fishes resulted in a lightening of the bony skeleton and the scaly armor, both of which improved locomotion. Increasing sophistication and variation in the shape and arrangement of the paired fins led to patterns that were optimum for specialized sprinters, cruisers, or artful dodgers. In the most advanced rayfins, swimming has come to depend more and more on the tail fin rather than on body flexing, while the other fins are modified as steering devices and/or defensive spines. Even flying fishes had evolved by Triassic times.

The jaws and skull of rayfins were gradually modified for lightness and efficiency. In particular, intricate systems of levers and pulleys allow advanced fishes to strike at prey more effectively by extending the jaws forward as they close. The same system also allows more efficient ways of browsing, grazing, picking, grinding, and nibbling, all encouraging the evolution of the tremendous variety of living fishes.

Sarcopterygians (Lobefin Fishes)

Sarcopterygians, the lobefin fishes, are distinguished as a separate group because they evolved several pairs of fins stronger than any found in a rayfin fish. Other differences in scale and skull structure confirm the separate evolution of these groups. They separated as early as the Silurian/Devonian boundary: fossils from southern China are basal lobefins, even though they are not far evolved from basal bony fishes. All the major lobefin groups originated in the Early Devonian, and at that time were closely similar. They diverged during the Devonian in shape, structure, and ecology, into three major clades: coelacanths, lungfishes, and a group named osteolepiforms which includes the ancestors of all tetrapods, including us (Figure 7.18).

The central part or lobe of a lobe fin is sturdy and contains a series of strong bones, while the edges have radiating rays as in ray fins (Figure 7.19). A lobe fin must beat more slowly than a ray fin of the same area because there is more mass to be accelerated and decelerated, but the resultant stroke is more powerful. Furthermore, and fundamental to later vertebrate history, a lobe fin that imparts a powerful stroke to the water has to have some kind of support at its base, just as an oar has to be stabilized in a rowlock. Therefore, lobefin fishes have internal systems of bones and muscles that help to tie one dorsal and two ventral pairs of lobe fins to the rest of the skeleton (Figure 7.19). These ventral linkages evolved to become the pectoral and pelvic girdles of land vertebrates, but of course that was not why they evolved: they evolved originally to allow early lobefin fishes to swim more effectively. All Devonian lobefins seem to have been effective swimmers and predators.

Lungfishes and coelacanths still survive, but only as rare and unusual fishes. Two species of coelacanth survive as two small populations, and one species of lungfish lives in each of the three southern continents: Australia, South America, and Africa. All these living lobefins have such an unusual biology and ecology that they must be interpreted with caution. They are much evolved from their Devonian ancestors in structure and in habits, so they may not be very good guides to the biology of those ancestors.

Coelacanths

A living coelacanth, *Latimeria* (Figure 7.20), was unexpectedly discovered in 1938. Coelacanths had been known as fossils for decades, but it was thought that they had died out after the Cretaceous. We know now that they are rare, probably endangered, and live in cold or deep water off the east coast of South Africa, and in at least one Indonesian locality. Living coelacanths are lazy swimmers, and do not have lungs. The females bear live young, as many as 26 at a time, which develop internally from very yolky eggs.

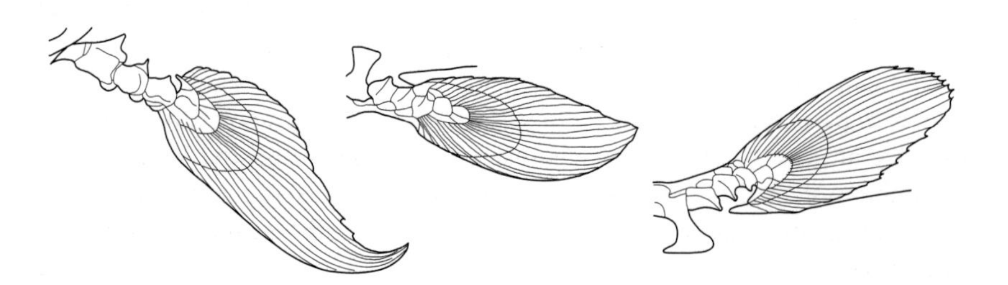

Figure 7.19 From left to right: the anterior ventral, the posterior ventral, and the posterior dorsal lobe fins of the coelacanth *Latimeria*. The anterior (or pectoral) fin is significantly larger than the other two, but the posterior dorsal fin is just as large as its ventral counterpart and has just as strong an internal bony skeleton. (After Millot and Anthony.)

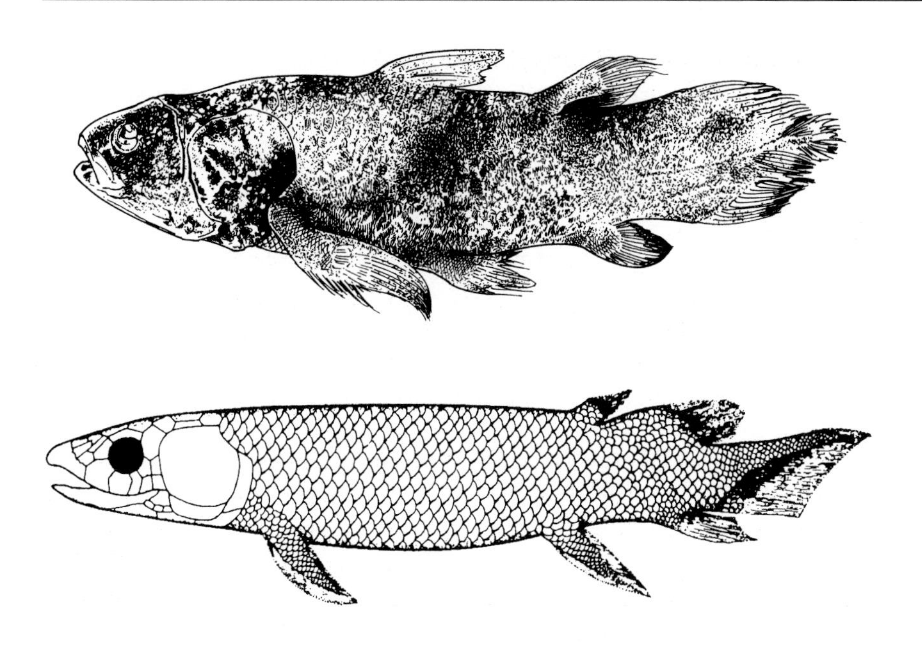

Figure 7.20 The coelacanth, *Latimeria chalumnae*. This is the South African species, not the newly discovered Indonesian one. (Drawing by Bob Giuliani. © Dover Publications Inc., New York. Reproduced by permission.)

Figure 7.21 Devonian lungfishes such as *Dipterus*, shown here, were more active than their living descendants. (After Traquair.)

Lungfishes

Living lungfishes are medium-sized, long-bodied fishes found in seasonal freshwater lakes and rivers in tropical areas. They seem best designed for rather slow swimming. Living lungfishes, or Dipnoi, can breathe air, allowing them to survive periods of drought or low oxygen in seasonal lakes and rivers in tropical climates. Lungfishes probably survive today because they can tolerate environments that would kill most other fishes. The African lungfish can even tolerate a dry season in which its river dries up. It digs a burrow, seals itself inside, and estivates (turns its body metabolism to a very low level) until the rainy season sends water down the river and into the burrow, reviving it.

Lungfishes have evolved considerably to their present anatomy, biology, and ecology. The first lungfishes were marine fishes, and look as if they were much more active swimmers than their living descendants (Figure 7.21). Living lungfishes are descended from a clade of Devonian ancestors that evolved the ability to live in fresh water, where they evolved changes in teeth and jaws that mark a shift in feeding from other fishes to molluscs and crustaceans. (Living forms have flattened teeth shaped like plates, for crushing their prey.) Early lungfishes also evolved a way of dealing with drought that has not changed much either, it seems. Carboniferous lungfishes have been found fossilized in burrows 300 m.y. old.

Osteolepiforms

Osteolepiforms are the sister group of lungfishes (Figure 7.18). The old name for them, rhipidistians, does not meet modern standards of cladistic precision. Osteolepiforms include the ancestors of the land-going vertebrates, and are discussed in more detail in Chapter 8.

While placoderms were the dominant fishes of the Devonian, at least in size, the osteolepiforms and the other lobefins were most successful in shallow waters around coasts and in inland waters, but were hardly dominant. After the Devonian the rayfin fishes came to be the most successful group, with their combination of lightness and maneuverability, while the lobefins were gradually confined to unusual habits and habitats. Perhaps in the process of being squeezed, ecologically speaking, Late Devonian osteolepiforms evolved adaptations that allowed them to expand in an unexpected direction — toward life in air.

Further Reading

Erdmann, M. V., et al. 1998. Indonesian 'king of sea' discovered. *Nature* 395: 335. [New coelacanth population.]

Forey, P., and P. Janvier. 1993. Agnathans and the origin of jawed vertebrates. *Nature* 361: 129–34.

Fricke, H., and J. Frahm. 1992. Evidence for lecithotrophic viviparity in the living coelacanth. *Naturwissenschaften* 79: 476–9

Gee, H. 1996. *Before the Backbone: Views on the Origins of Vertebrates.* London: Chapman and Hall.

Gee, H. 2002. Return of a little squirt. *Nature* 420: 755–6. [Sequencing the genome of a tunicate.]

Janvier, P. 1996. *Early Vertebrates.* Oxford: Clarendon Press.

Koob, T. J., and J. H. Long. 2000. The vertebrate body axis: evolution and mechanical function. *American Zoologist* 40: 1–18. [Evolution of notochord and backbone.]

Long, J. A. 2001. On the relationships of *Psarolepis* and the onychodontiform fishes. *Journal of Vertebrate Paleontology* 21: 815–20. [The wonderful *Onychodus*.]

Ruben, J. R. 1989. Activity physiology and evolution of the vertebrate skeleton. *American Zoologist* 29: 195–203.

Sanderson, S. L., et al. 2001. Crossflow filtration in suspension-feeding fishes. *Nature* 412: 439–41, and comment, 387–88. [Could apply to jawed fishes like the acanthodians.]

Shu, D-G., et al. 1996. A *Pikaia*-like chordate from the lower Cambrian of China. *Nature* 384: 157–8. [*Cathaymyrus*.]

Shu, D-G., et al. 2001. An Early Cambrian tunicate from China. *Nature* 411: 472–3. [*Cheungkongella*.]

Shu, D-G., et al. 2003a. Head and backbone of the Early Cambrian vertebrate *Haikouichthys*. *Nature* 421: 526–9.

Shu, D-G., et al. 2003b. A new species of yunnanozoan with implications for deuterostome evolution. *Science* 299: 1380–4. [A basal deuterostome, *Haikouella*.]

Stokes, M. D., and N. D. Holland. 1998. The lancelet. *American Scientist* 86: 552–60.

Thomson, K. S. 1991. *Living Fossil: The Story of the Coelacanth.* New York: W. W. Norton. [Not much paleontology, but a lot of science, and a good read.]

Thomson, K. S. 1993. The origin of the tetrapods. *American Journal of Science* 293A: 33–62. [For discussion of lungfishes and coelacanths.]

Thomson, K. S. 1999. The coelacanth: Act Three. *American Scientist* 87: 213–15.

Wilson, M. V. H., and M. W. Caldwell. 1993. New Silurian and Devonian fork-tailed "thelodonts" are jawless vertebrates with stomachs and deep bodies. *Nature* 361: 442–4.

Wilson, M. V. H., and M. W. Caldwell. 1998. The Furcacaudiformes: a new order of jawless vertebrates with thelodont scales, based on articulated Silurian and Devonian fossils from Northern Canada. *Journal of Vertebrate Paleontology* 18: 10–29. [Longer account of forktails.]

Zhu, M., et al. 2001. A primitive sarcopterygian fish with an eyestalk. *Nature* 410: 81–4. [New Devonian fish *Achoania*: important find that will help us to understand the early radiation of bony fishes.]

Zimmer, C. 2000. In search of vertebrate origins: beyond brain and bone. *Science* 287: 1576–9. [Broad perspective.]

CHAPTER EIGHT

Leaving the Water

Plants, invertebrates, and finally vertebrates evolved to live on land in the Middle Paleozoic. There were major problems in doing so (Box 8.1), related not so much to the land surface as to exposure to air. Many marine animals and plants spend their lives crawling on the seafloor, burrowing in it, or attached to it. As a physical substrate, the land surface is not very different. But land organisms are no longer bathed in water. There are predictable consequences for the evolutionary transitions involved, many of them based on the laws of physics and chemistry.

Organisms weigh more in air without the buoyant effect of water, so support is more of a problem. Air may be very humid, but it is never continuously saturated, so organisms living in air must find a way to resist desiccation. Tiny organisms are particularly sensitive to drying out in air, because they have relatively large surface areas but cannot hold large reserves of fluid. Therefore reproductive stages and young stages of plants and animals are very sensitive to drying. Temperature extremes are much greater in air than they are in water, exposing plants and animals to heat and cold. Oxygen and carbon dioxide behave differently as gases than they do when dissolved in water, so respiration and gas exchange systems must change in air. The refractive index of light is lower in air than in water, and sound transmission differs too, so vision and hearing must be modified in land animals.

There are also ecological consequences. Seawater carries dissolved nutrients, but air does not, so some organisms, especially small animals and plants, have a food supply problem in air. It's unlikely that the same food sources would be available to an animal that crossed such an important ecological barrier, so invasion of the land would often be associated with a change in feeding style.

All the major adaptations for life in air had to be evolved first in the water, as adaptations for life in water. Only then would it have been possible for organisms to emerge into air for long periods. We must reconstruct a reasonable sequence of events during the transition, then test our ideas against evidence from fossil and living organisms.

BOX 8.1 Problems in Adjusting to Life in Air

For Plants and Animals
- No buoyancy, so support needed;
- Danger of drying out;
- Extremes of temperature;
- Gases behave differently;
- No nutrients in air.

For Animals Only
- Refraction of light changes;
- Hearing must be modified.

THE ORIGIN OF LAND PLANTS

We have no idea when plants first colonized land surfaces. Plants must have emerged gradually into air and onto land from water, and the first "land" plants must have been largely aquatic, living in swamps or marshes.

Almost all the major characters of land plants are solutions to the problems associated with life in air. Land plants grow against gravity, so they have evolved structural or hydrostatic pressure supports (hard cuticles or wood) to help them stay upright. They cannot afford evaporation from moist surfaces, so they have evolved some kind of waterproofing. Roots gather water and nutrients from soil and act as props and anchors. Internal transport systems distribute water, nutrients, and the products of photosynthesis around the plant. Even so, all these adaptations for adult plants are useless unless the reproductive cycle is also adapted to air. Cross-fertilization and dispersal require special adaptations in air. All these adaptations must have evolved in a rational and gradual sequence. But because the first stages would have been soft-bodied water plants, the fossil record of the transition is difficult to find.

The following scenario for the evolution of land plants is modified from suggestions by John Raven. Water-dwelling plants, probably green algae, were already multicellular. Green algae grow rapidly in shallow water, bathed in light and nutrients. One might think that cells in a large alga are comparatively independent of one another: in the water, each cell has access to light, water, nutrients, and a sink for waste products. But the fastest growing points of algal fronds need more energy than the photosynthesis of the cells there can supply, so some green algae have evolved a transport system between adjoining cells to move food quickly around the plant. They presumably do this because they can then grow more rapidly.

Raven's scenario begins with green algae living in habitats that were subject to temporary drying. The algae might already have evolved to disperse spores more effectively by releasing them into wind instead of water. Spores, even in algae, are reasonably watertight and could easily have been adapted for release into air from special sporangia (spore containers) growing high enough to extend out of the water on the uppermost tips of otherwise aquatic plants. As plant tissue extended into air, photosynthesis increased because light levels in air are higher than they are in water, especially at each end of the day, and are free from interference by muddy water. Furthermore, CO_2 is more easily extracted from air than it is from water.

As plants grew out into air, some tissues were no longer bathed in the water that had provided nutrients and a sink for waste products. Internal fluid transport systems between cells became specialized and extended. Photosynthesis was concentrated in the upper part of the plant that was exposed to more light. Photosynthesis fixes CO_2, so there had to be continual intake of CO_2 from the air. However, plant cells are saturated with water, but air is usually not, so the same surfaces that take in CO_2 automatically lose water. Sunlight heats the plant, encouraging evaporation. The water loss had to be made up by transporting water up the stem from the roots to the photosynthesizing cells.

Water is transported much more effectively as liquid than as vapor. Early land plants evolved a simple piping system called a **conducting strand** of cells to carry water upward. The conducting strand, found in living mosses, is powerful enough to prevent water loss in small, low plants, if soil water is abundant. But mosses quickly dry up if soil water is in short supply.

Early land plants began to evolve a **cuticle** (a waxy layer) over much of their exposed upper surfaces. The cuticle helps the plant through alternating wet and dry conditions. In wet times it acts as a waterproof coating. It prevents a film of water from standing on the plant that could cut off CO_2 intake. In dry times it seals the plant surface from losing water by evaporation. A cuticle may also have added a little strength to the stem of early plants, and its wax probably helped to protect the plant from UV radiation and from chewing arthropods.

But the cuticle also cut down and then eliminated, from the top of the plant downward, the ability to absorb water-borne nutrients over the general plant surface. Nutrients were taken up more and more in lower parts of the plant, eventually taking place on specialized absorbing surfaces at the base (**roots**) which probably evolved from the runners that these plants often used to reproduce asexually. As roots grew larger and stronger, they helped to anchor and then to support the plant.

The cuticle sealed off CO_2 uptake over the general plant surface. As cuticle evolved, plants also evolved pores called **stomata** where CO_2 uptake could be concentrated. If it is too hot or too dry, stomata can be closed off by **guard cells** to control water loss. As CO_2 uptake was localized, plants evolved an **intercellular gas transport system** that led from the stomata into the spaces between cells, improving CO_2 flow to the photosynthesizing cells. The same system was also used to solve an increasingly important problem. As roots enlarged, more and more plant tissue was growing in dark areas where photosynthesis was impossible; yet those tissues needed food and oxygen. Soils are low or lacking in O_2, especially when they are waterlogged. The intercellular gas transport system feeds O_2 from the air down through the plant to the roots, sometimes through impressively large hollow spaces.

Later plants evolved **xylem**, an improved piping system for better upward flow of water from the roots round the plant. Xylem is made of elongated dead cells arranged end to end to form long pipes up and down the stem. Evaporation at the top of the plant essentially sucks water upward through the empty dead cells. Even a narrow xylem can transport water much faster than can normal plant tissue, and, once begun, the evolution of xylem was probably a rapid process that immediately gave plants greater tolerance of dry air. The xylem also carries important dissolved substances as it flows upward, and it is the primary intake and transport system for plant nutrients, particularly for phosphate. Plants that have xylem are called **vascular plants**.

Xylem cells are dead, so xylem transport is passive, driven entirely by the suction—or negative pressure—of evaporation from the upper part of the plant. The forces generated can be very large, so the long narrow walls of xylem cells may tend to collapse inward. Xylem cell walls came to be strengthened by a structural molecule, **lignin**. Once lignin had evolved, it was used later to strengthen the roots and stem as plants grew taller and heavier. Later still it also became a deterrent to animals trying to pierce and chew on plant tissues.

As plants became increasingly polarized, with nutrient and water being taken up at the roots and photosynthesis taking place in the upper parts, the xylem and gas transport systems improved, but neither of them was able to transport liquid downward. This problem was solved by the evolution of another transport system called **phloem** from the cell-to-cell transport system of green algae. Phloem cells transport dissolved substances round the plant in a process that is still not properly understood. Phloem carries photosynthate from photosynthesizing cells to growing points such as reproductive organs and shoots, and to tissues such as roots that cannot make their own food.

Throughout the process, the advantage that encouraged plants to extend into air in spite of the difficulties involved was the tremendous increase in available light. Marine plants are restricted to the narrow zone along the shore where light has to penetrate sediment-laden, wave-churned water. Growth above water increases light availability. Furthermore, competition for available light tended to encourage even more growth of plant tissues above the water surface, and more effective adaptations to life in air (Figure 8.1). Once plants could grow above the layer of still air

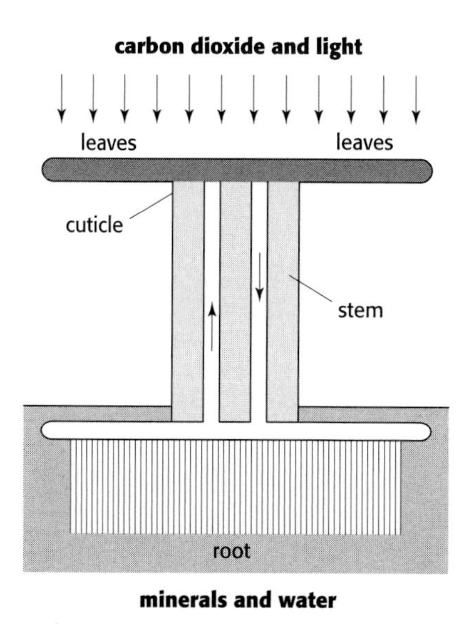

Figure 8.1 The basic land plant.

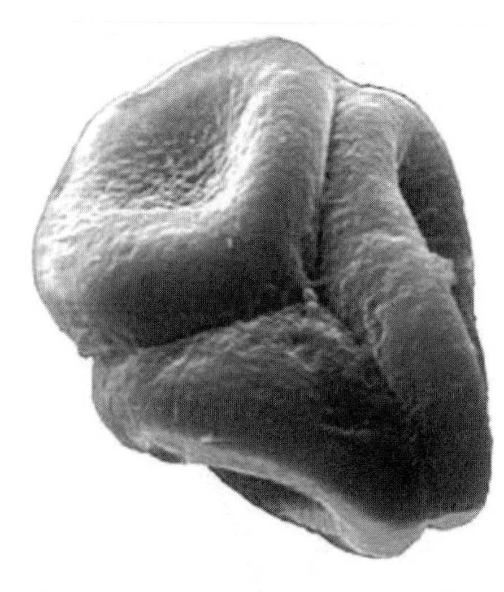

Figure 8.2 Some of the earliest spores of land plants, from the Ordovician of Libya. Research reported in Wellman et al., 2003. (Courtesy C. Wellman of Sheffield University, UK.)

near the water surface, spores could be released into breezes. Greater plant height and the evolution of spore containers (**sporangia**) on the tips of branches were both adaptations for effective dispersal.

THE EARLIEST LAND FLORAS

Pre-Devonian plants all occur fossilized in marine or coastal sediments. But plants can be swept downstream in floods and deposited far from their habitats. Mats of floating vegetation can be found today off the mouth of the Amazon, sometimes complete with roots, soil, and sediment. So the fossil record doesn't necessarily imply that pre-Devonian plants lived in the sea. (In fact, Raven suggests that the osmotic systems of land plants imply that they originally evolved in freshwater marshes.)

The earliest spores that belonged to land plants come from Middle Ordovician rocks (Figure 8.2). They look like the spores of living liverworts, and there are fragments of their parent plants preserved with them. Later spores from Silurian rocks look as if they came from more advanced plants (though we cannot tell which ones). Ordovician and Early Silurian land plants were probably tiny and weakly constructed, and difficult to preserve as fossils. Molecular evidence suggests that all land plants today are descended from liverworts (or plants much like liverworts), but molecular evidence cannot say anything about *extinct* groups of plants that might have been the pioneers of plant life in air.

Late Silurian and Early Devonian Plants

Well-preserved land plants are not found in rocks older than Late Silurian. Though it is Devonian in age, *Aglaophyton* (Figure 8.3) probably has a grade of structure that evolved in Silurian times. It grew to a height of less than 20 cm (about 8 in). Although it had most of the adaptations needed in land plants (cuticle, stomata, and intercellular gas spaces), it did not have xylem, only a simple conducting strand.

The Late Silurian fossils almost certainly include vascular plants. *Cooksonia* (Figure 8.4) was only a few centimeters high and had a simple structure of thin, evenly branching stems with sporangia at the tips, and no leaves. But it also had central structures that were probably xylem rather than simple conducting strands. Later species of *Cooksonia* from the earliest Devonian have definite strands of xylem preserved, and cuticles with stomata, so they probably had intercellular gas spaces and were better adapted for life in air.

Early Devonian land plants were dramatically more diverse. They grew up to a meter high, although they were slender (1 cm diameter). For support they must have grown either in standing water or in dense clusters, aided by the fact that they reproduced largely by budding systems of rhizoids for asexual, clonal reproduction, as strawberries do today (Figure 8.3). This style of reproduction not only gave mutual support to individual stems, but, by "turfing in," a cloned mass of plants could help to eliminate competitive species. Plants like this could have grown and reproduced very quickly, a way of escaping the consequences of relatively poor adaptations for living in air.

Rhynia (Figure 8.5), like *Cooksonia*: is a genuine vascular plant. *Rhynia* gave its name to the rhyniophytes, an extinct group of early vascular plants. Meanwhile, other major groups of vascular plants also evolved in the Early Devonian. Lycopods

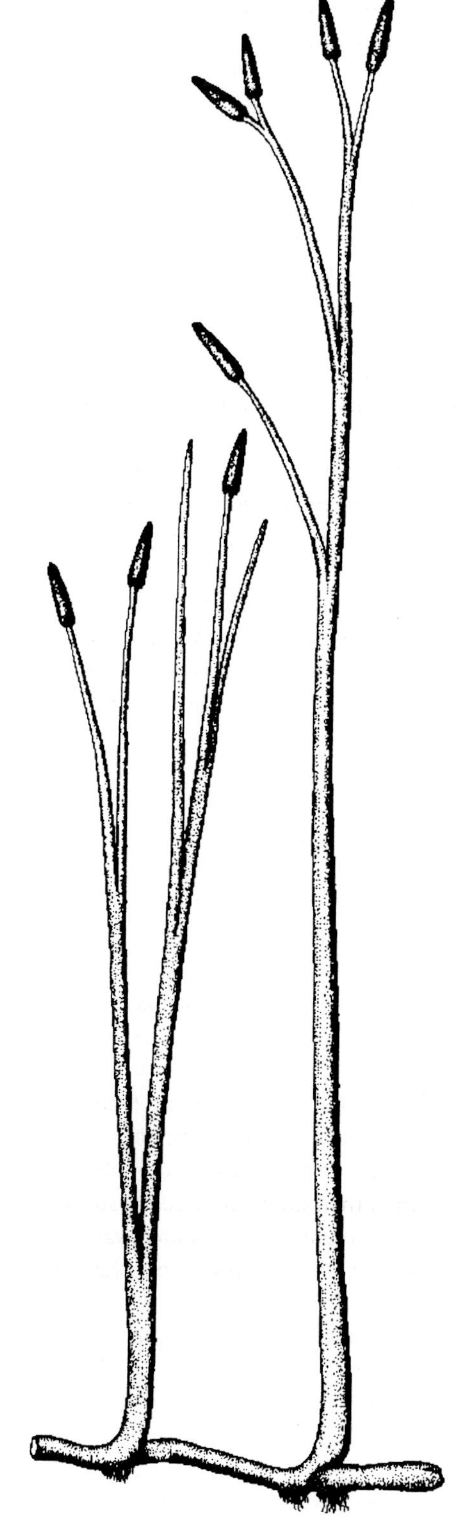

Figure 8.3 *Aglaophyton*, an early land plant from Devonian rocks in Scotland, was nonvascular. It had a simple conducting strand for transporting water up the stem. It budded new plants from a rhizome at the base. (After Kidston and Lang.)

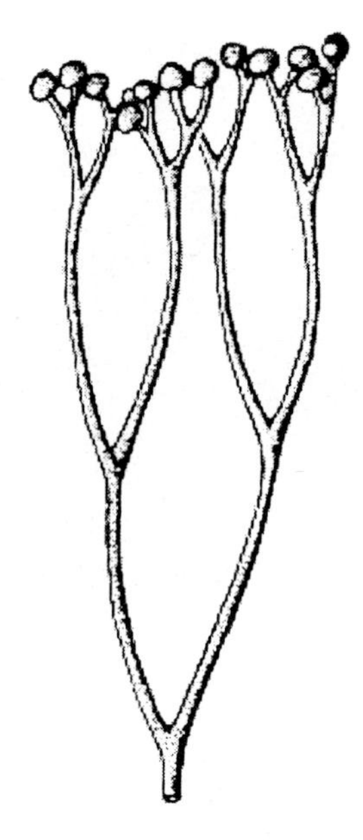

Figure 8.4 *Cooksonia*, an early vascular land plant. (After Edwards.)

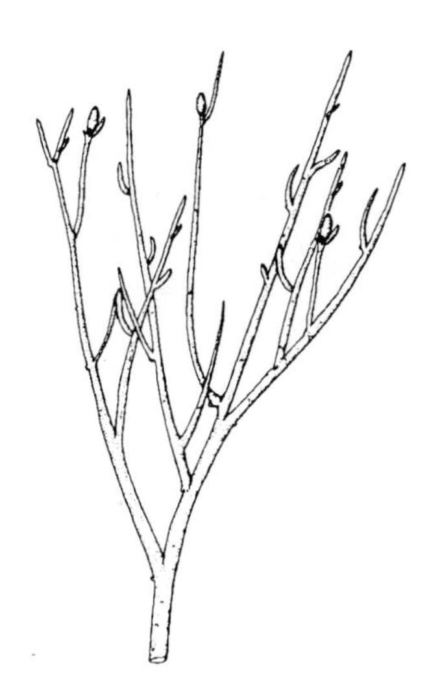

Figure 8.5 *Rhynia*, a Middle Devonian vascular plant with well-developed xylem, from the Rhynie Chert of Scotland. *Rhynia* gives its name to a major group of early land plants, the rhyniophytes. (After Edwards.)

are the ancestors of living club-mosses, but they formed trees and forests in the coal swamps of the Carboniferous period. We still cannot identify the rapid evolutionary changes that gave all the other main groups of living vascular plants.

Figure 8.6 The Devonian plant *Psilophyton* represents a new grade of structure in land plants, advanced over rhyniophytes in its strong construction. (After Hopping.)

Later Devonian Plants

New structural advances are seen in Later Devonian and Early Carboniferous plants. The successive floras all lived in lowland floodplains, which have a good fossil record. It looks as if we are seeing waves of ecological and evolutionary replacement on all levels, from individual plants to world floras, as structural innovations allowed each plant group to outcompete its predecessor.

For example, *Rhynia* had only 1% of its stem cross-section made of xylem, but *Psilophyton* had 10%, and the whole plant was more strongly constructed (Figure 8.6). So psilophytes could grow taller (up to 2 m high) and photosynthesize more efficiently than rhyniophytes. Other improvements in reproduction and light gathering, through the evolution of leaves rather like those of living ferns and through more complex branching, also aided plant efficiency.

Later plants evolved even more xylem, so the structure became even stronger. Secondary xylem—**wood**—gave great strength and allowed higher growth, producing the first trees (Figure 8.7). Late Devonian trees also evolved true **roots**, more complex branching patterns, and larger, flatter leaves. Reproductive systems were improved still further: most Late Devonian plants had two different kinds of spores, female megaspores and male microspores.

Figure 8.7 The first tree *Archaeopteris* shown silhouetted against a Late Devonian sunrise. *Archaeopteris* may very well have been the plant that carried the first known seed, *Archaeosperma*. (After Beck.)

Seed-bearing plants evolved in the Late Devonian. A Late Devonian seed-like structure called *Archaeosperma* looks as if it belongs to a tree very much like *Archaeopteris* (Figures 8.7; 8.8). This was a great advance: all previous plants had needed a film of water in which sperm could swim to fertilize the ovum, but seed plants can reproduce away from water.

By the end of the Devonian, all the major innovations of land plants except flowers and fruit had evolved. Forests of seed-bearing trees and lycopods had appeared, with understories of ferns and smaller plants.

The dominant process in Devonian plant evolution seems to have been selection based on simple efficiency—in size and stability, photosynthesis, internal transport, and reproductive systems. Plant groups replaced one another as innovations appeared. Perhaps the most interesting part of this story of early plants is the rate at which innovations appeared. There is no obvious reason why the process should not have gone faster, or slower. The innovations we have discussed should have given immediate success whenever they appeared. But it took the length of the Devonian (about 50 m.y.) for rhyniophyte-type plants to evolve to seed-plants. Even "obvious" innovations may take time to evolve and accumulate.

Devonian Plant Ecology

The best-known Devonian plants grew in swampy environments near the equator, where they were very likely to be preserved. The fossil record may not be biased, however; early land plants probably did live in low-lying, damp, tropical regions, where there was little seasonal fluctuation in temperature, light, or humidity.

By Middle Devonian times there were many fern-like plants with well developed leaves. Fossil tree trunks from the Middle Devonian of New York suggest plants over 10 m (30 ft) high, with woody tissue covered by bark. Once plants reached these heights, shading of one species by another would have led to fairly complex plant communities.

The evolution of the seed seems to have been the foundation for competitive success in the Early Carboniferous. Seed plants invaded drier habitats, and seed dispersal by wind (rather than water) became important; we have winged seeds from Late Devonian rocks. Seed dispersal allowed some plants to specialize as invaders, avoiding the increasingly dense and competitive habitats in wetlands and along rivers.

The increasing success of land plants, especially their growth to the size of trees, must have produced ever-larger amounts of rotting plant material in swamps, rivers, and lakes, leading to very low O_2 levels in any slow-moving tropical water (O_2 is used up in decay processes). At the same time, the increasing photosynthesis by land plants drew down atmospheric CO_2 and increased atmospheric oxygen.

All this probably helped to encourage air breathing among contemporary freshwater arthropods and fishes, and it led to better preservation of any fossil material deposited in anoxic swamp water. Coals are known from Devonian rocks, but truly massive coal beds formed for the first time in Earth history in the Carboniferous Period, which was named for them.

Figure 8.8 A frond from the top of an *Archaeopteris* tree.

COMPARING PLANT AND ANIMAL EVOLUTION

Whether one counts spores or plant macrofossils, there is a striking increase in land plant diversity from Silurian to Middle Devonian time, when a diversity plateau was reached that extended into the Carboniferous. A second increase in Carboniferous land plant diversity was followed by a long period of stability. A third, Late Mesozoic expansion in land plants and animals raised diversity to current levels.

The pattern looks rather like the pattern of Sepkoski's three major faunas in the oceans. But the radiations among land plants and marine animals did not occur at the same times, so they were not directly linked. Extinctions among plants are different from those among animals, which suggests that plants and animals may respond to quite different extinction agents. Specifically, Andrew Knoll suggested three major factors:

- Plants are more vulnerable to extinction by competition;
- plants are more vulnerable to climatic change;
- plants are less vulnerable to mass mortality events.

These differences reflect basic plant biology. All plants do much the same thing. They are all at the same primary trophic level, so they cannot partition up niches as easily as animals can. A new arrival in a flora may be competitively much more dangerous than a new arrival in a fauna.

For example, CO_2 uptake must be accompanied by the loss of water vapor, since the plant is open to gas exchange. Many plant adaptations are responses to the problem of water conservation. Because it is so basic a part of their biology, an innovation here could provide a new plant group with an overwhelming advantage. Other plant systems such as light-gathering are equally likely to be improved by innovation.

Plant distributions are sensitive to climate. If climate changes, plants must adapt, migrate, or become extinct. In extreme circumstances, there may be no available refuges. Thus, the tree species of north-west Europe were trapped early in the ice ages between the advancing Scandinavian glaciers to the north and the Alpine glaciers to the south, and were wiped out. In contrast, similar species in North America were able to move their range south along the Appalachians, then north again as the ice retreated.

However, plants are well adapted to deal with temporary stress, even if it is catastrophic to animals. Plants readily shed unwanted organs such as leaves or even branches in order to survive storms and extreme weather. Many weeds die, to overwinter as seeds or bulbs. Even when plants are removed, by fire or drought, the soil is always rich with seeds, so that mass mortality of full-grown plants does not mean the end of the population. Plants are the dominant biomass in communities recovering after volcanic eruptions, tropical storms, or other catastrophes have devastated an area. In many land communities, the removal of dominant trees by storm, fire, or human agency is followed by the rapid growth of species that are specialized to colonize disturbed areas.

THE FIRST LAND ANIMALS

As plants extended their habitats into swamps and onto riverbanks and floodplains, they would have provided a food base for animal life evolving from life in water to life in air. The marine animals best preadapted to life on land were arthropods. They already had an almost waterproof cover and were very strong for their size, moving on sturdy walking legs. The incentive to move out into air might have been the availability of organic debris washed ashore on beaches, or perhaps the debris left on land by the first land plants. (Foraging crabs are obvious members of many beach communities.) Plant debris, whether it's on a beach or on a forest floor, tends to be damp; it provides protection from solar radiation and is comparatively nutritious. Thus it is not surprising that the earliest land animals were arthropods that ate organic debris, and other arthropods that ate them.

Different arthropods probably moved into air by different routes. The easiest transition would seem to have been by way of estuaries, deltas, and mudflats, where food is abundant and salinity gradients are gentle.

The earliest land arthropods are known from Late Silurian trace fossils of their footprints, rather than the animals themselves. Very small arthropods have been found at several Early Devonian localities. Most of them (mites and springtails) were eating living or dead plant material, and in turn were probably eaten by larger carnivorous arthropods such as early spiders. Larger arthropods are usually found in tiny fragments, but it is clear that some were large by any standard. We have pieces of a scorpion that was probably about 9 cm (over 3 in) long, and a very large millipede-like creature, *Eoarthropleura*, which probably lived in plant litter and ate it. At 15–20 cm (6–8 in) long, this was the largest terrestrial animal of the Early Devonian.

This early terrestrial ecosystem did not include any vertebrates as permanent residents, but no doubt the entire food chain, including fishes in the rivers, lakes, and lagoons, benefited from the increased energy flow provided by plants and their photosynthesis.

Osteolepiforms

The invasion of the air by plants, and by invertebrates that exploited them for food and shelter, led to a large increase in organic nutrients in and around shorelines. In the Devonian we see the first signs that fishes were beginning to exploit the newly enriched habitats near the shore and near the surface. But we must not imagine that vertebrates adapted quickly to life in air, or that they readily left the water.

As we saw in Chapter 7, lobefin fishes evolved into different ways of life by the end of the Devonian. Lungfishes came to specialize in crushing their prey, small clams and crustaceans. If we can judge by the last surviving species of coelacanth, this group came to hunt in the water by stealth, followed by a quick dash. The osteolepiforms seem to have been the Late Devonian lobefins best adapted to hunting fishes in shallow waters along sea coasts and into brackish shoreline lagoons and freshwater lakes and rivers. They look more active than coelacanths and were probably fast-sprinting ambush predators.

Coelacanths use their powerful tails for the final surge after prey, and the paired lobe fins act mainly to adjust attitude and speed as they stalk. The ventral fins beat rather like wings, up and down in the water. The beats are synchronized, with the right pectoral fin beating at the same time as the left pelvic fin, and vice versa. It's not clear why they do this, but it probably has to do with hydrodynamic eddying: it may be difficult to "fly" underwater with two sets of wings beating in pairs. (Dragonfly wings beat in air in the same pattern as coelacanth fins.) The fascinating aspect of the coelacanth pattern of ventral fin beats is that it is exactly the same one tetrapods use to walk on land. So a pattern of fin or limb movement that is used in deep water today by coelacanths for swimming may have been shared by their Devonian relatives the osteolepiforms, who also passed it on to tetrapods because it turned out (by chance) to be a useful pattern for squirming and walking across a muddy substrate.

Osteolepiforms had long, powerful, streamlined bodies with strong lobe fins and tail (Figure 8.9), adapted for strong swimming. They had long snouts, especially the larger ones. Perhaps as a result, osteolepiforms evolved a skull joint that allowed them to raise the upper jaw as well as, or instead of, lowering the lower jaw as they gaped to take prey. This could have had two important effects, both related to life in shallow water. First, the snout movements would have changed mouth volume, perhaps allowing extra water to be pumped over the gills without moving the lower jaw. Second, osteolepiforms could have caught prey in shallow water by raising the snout without dropping the lower jaw. Crocodiles do exactly the same thing as they take prey in shallow water. Some osteolepiforms may have been able to chase prey right up to or even beyond the water's edge. Their powerful ventral lobe fins, set low on the body, may have allowed them to drive after prey on, over, or through shallow mud banks, thus making rapid trips over surfaces that could be called "land."

The main sprinting propulsion in osteolepiforms came from the tail. The lobe fins were set on the dorsal side of the body as well as ventrally (Figure 8.9). In deep water, osteolepiforms could attack prey from any angle. But in shallow water the ventral fins took on additional importance. The pectoral ventral fins could be used against the bottom as supports, strengthening the posture of the anterior trunk and acting as props in chasing; the pelvic ventral fins acted to grip and push on the substrate so that maximum effort could be expended against it, adding to the thrust. In some advanced osteolepiforms, the lobe fins evolved toward limbs, not as an adaptation for walking but to become a more efficient fish.

Osteolepiforms are usually pictured as living exclusively in freshwater lakes and rivers, but that is not necessarily true of the early ones. All Devonian lobefin groups were initially marine fishes. Osteolepiforms radiated into fresh water mostly in

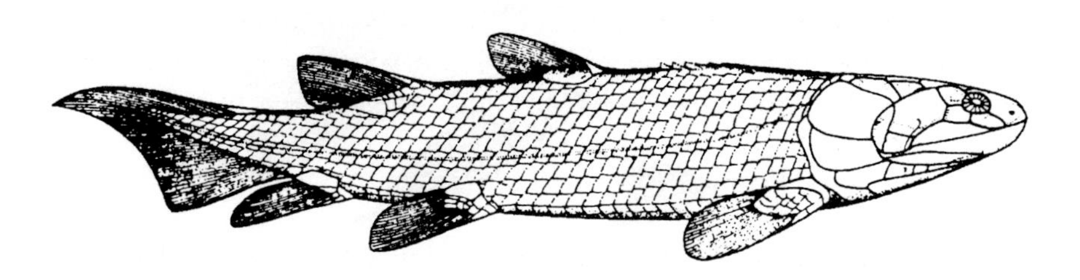

Figure 8.9 A typical osteolepiform, with powerful low-set lobe fins. In fact, this is *Osteolepis*, which gave its name to the group. (After Jarvik.)

the Middle and Late Devonian, at the same time that land plants were radiating abundantly.

We have a picture, then, of a varied group of osteolepiforms, all hunters but some adapted to shallow-water habitats. All of them were fishes; none was adapted to be active out of water for any length of time.

AIR BREATHING

Some primitive rayfin fishes that survive today breathe air, with lungs. As we have seen, living lobefins, lungfishes, do the same. But one group of Late Devonian osteolepiforms typified by *Eusthenopteron* (Figure 8.10) had evolved the nostrils that are now used in all tetrapods for air breathing, and the internal air passage, the choana, that allows air breathing through the nostrils. Therefore these osteolepiforms were breathing air, they had lungs, and are closer to tetrapod ancestry than any other lobefins. Air breathing may in fact have evolved in the common ancestor of rayfins and lobefins, which would have been an early bony fish living in the sea. Why would marine fishes evolve the ability to breathe air?

The answer I used for many years involved low oxygen levels in water, but Colleen Farmer has suggested a better idea (see Farmer, 1999), and I have used her work to compose much of this section.

Respiration in animals has built-in universal features. Animals take in oxygen to burn their food in respiration, and they produce CO_2 as a waste product. Carbon dioxide is toxic because it dissolves easily in water to form carbonic acid. Animals can tolerate only a small buildup of CO_2 before passing it out of the system. (For example, it is high CO_2 in our lungs that makes us want to breathe out, not shortage of oxygen.)

Gases are exchanged with the environment, whether it is water or air, as body fluids are passed very close to the body surface. For example, blood flows close under the lung surface in our own breathing. As long as the environment has higher O_2 and lower CO_2 than the body, diffusion acts to pass O_2 in and CO_2 out. The rate depends on several factors: the surface area and the thickness of tissue through which the gases must diffuse; the rate at which the external and internal fluids pass across the surface; and the concentrations of gases in the internal fluid and in the external medium. In normal fishes, CO_2 and O_2 diffuse in opposite directions across the gill surface.

Respiration in Early Fishes

The earliest fishes of the Early Cambrian probably had a respiration system like that of living cephalochordates (Chapter 7). Water is pumped into a basket-like

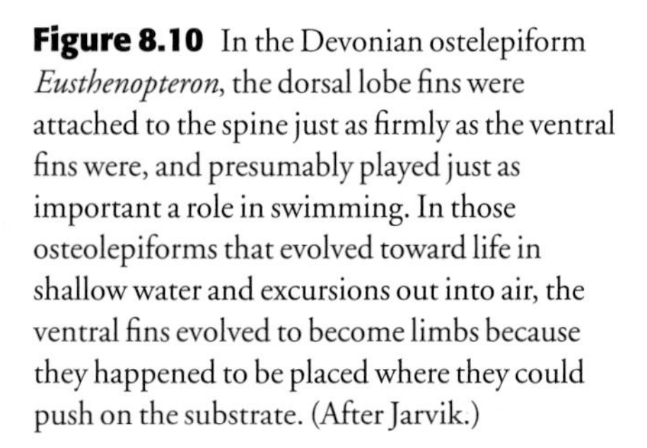

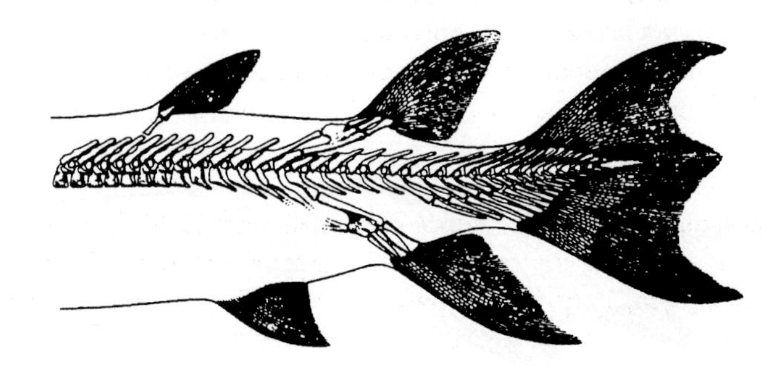

Figure 8.10 In the Devonian ostelepiform *Eusthenopteron*, the dorsal lobe fins were attached to the spine just as firmly as the ventral fins were, and presumably played just as important a role in swimming. In those osteolepiforms that evolved toward life in shallow water and excursions out into air, the ventral fins evolved to become limbs because they happened to be placed where they could push on the substrate. (After Jarvik.)

structure, and food particles and oxygen are taken out of it. Oxygen is carried in a blood system pumped by a heart, and travels through the body tissues, delivering oxygen, until it reaches the heart.

This system was inherited by later jawless fishes. They may have evolved sophisticated gills (Figure 7.8), but their system had a basic flaw: blood arrived at the heart depleted in oxygen, because it had served the body first. The more active the fish, the more likely it was to suffer heart failure! Though this is true of all animals with hearts, it is not an ideal piece of engineering. (But you inherit what your ancestors give you.)

Relatively slow-moving, largely bottom-dwelling jawless fishes flourished well for many millions of years with this system. (When you think about it, the success of early fishes in Cambrian seas confirms that global oxygen levels were reasonably high, at least in shallow water.)

With the evolution of jaws and the Siluro-Devonian radiation of jawed fishes, more lineages must have become more active foragers and predators, and there must have been strong selective pressures to modify the ancient system to cope with the extra oxygen demands of a more active life.

Even today, there are often natural "fish kills" in which massive mortality occurs among fishes. Often these episodes are related to a lowering of oxygen in the environment: for example, in shallow pools, rivers, or lagoons that heat up too much. The immediate culprit is the environmental insult, of course, but the crisis is worsened because the fishes were using a gill system for breathing that could not handle the oxygen shortage.

Air breathing provided a different option for respiration, so we should compare the process of respiration in air and in water.

Oxygen Intake

It may be easier and cheaper to extract oxygen from air rather than water. Water is hundreds of times denser and more viscous than air, and even at best it contains less oxygen. Many gill-breathing animals have to pump external water across their gill surfaces at ten times the rate they pump their internal blood. Gills have to be designed to resist the leakage of dissolved body salts, and the tissues across which oxygen is exchanged cannot be as thin as they can in air, so gas exchange is rarely anywhere near 100% efficient.

Because oxygen diffuses 100,000 times more quickly in air than in water, oxygen-poor air is rare. But oxygen-poor water does occur quite often, especially in tropical regions, wherever warm freshwater or saltwater lakes, ponds, or lagoons are partly or completely isolated, especially in a hot season. Warm, rotting debris can quickly use up oxygen, especially if there is little or no natural water flow. Even if the effect is only seasonal, it may still be critical for fishes and other organisms living in the water. The water is stagnant, hot, and full of rotting debris, often teeming with bacteria that may also release toxic substances.

Why would fishes want to live in oxygen-poor water, where gill breathing is difficult? The food supply may be rich for fishes that can tolerate it, and there are situations in which fishes might benefit from swimming into areas of warm, often oxygen-poor water near the surface.

Many carnivorous fishes today are bottom feeders, hunting for small prey that live on or in the surface of the sediment. In warm latitudes, the bottom waters are often much cooler than the surface waters, which are heated by the sun. Digestion can be very slow in cold-blooded animals, especially if they live in cold environments. It may be the critical factor holding back growth and development. In such cases, increasing the digestive rate by swimming into the warm surface water can produce faster growth, earlier maturity, and more successful reproduction.

But what happens to a fish that swims into surface water because it is warm, only to find that it is also oxygen-poor? Even if surface waters are generally low in oxygen, there is always a thin surface layer of water, about a millimeter thick, that gains oxygen from the air by diffusion. Many living fishes in tropical environments come to this surface layer to bathe their gills in the surface oxygen layer. They can breathe, but they have to solve other problems too. If they break the surface, their bodies extend out into the air, losing some buoyancy.

Some living fishes in this situation bite off bubbles of air and hold them in their mouths for positive buoyancy, to remain at the surface without active swimming. Some living species of gobies breathe this way. Once they have an air bubble, they can extract oxygen from it in the back of their mouths much more efficiently than at the gills. When the oxygen level in the mouth bubble falls, reducing its size and its buoyant effect, the fish must then get rid of the bubble and bite off another. Rhythmic air breathing might evolve as a result of this action, as fishes begin to get rid of CO_2 from their mouth bubbles while they are still losing it at gill surfaces as well.

Oxygen intake in the mouth enriches the blood supply there, and a fish can store oxygen in an air bubble. An air bubble that takes up only 5% of body volume can increase oxygen storage by 10 times compared with a fish without a bubble. Therefore, bubble breathing doesn't mean that a fish is tied to the surface; it can make extended dives to the bottom. This is true today of all air breathers with low metabolic rates, including crocodiles and turtles: many water-dwelling insects and spiders use air bubbles too.

LIMERICK 8.1
An oxygen-low situation
Is bad for gill respiration
For some fishes, no trouble
They bite off a bubble:
Inspirational improvisation!

Once air breathing has evolved, it can dominate respiration, even among fishes that never leave the water. This is true for the African lungfish and for the South American electric eel, which lives in the warm and often stagnant waters of the Amazon.

Because of the advantages of air breathing, many early fishes probably did so. Increasingly, it looks as if it was a very early development among bony fishes, rayfins as well as lobefins. Most living rayfins lost the ability, turning the lung into the swimbladder that helps to maintain buoyancy in the water. Colleen Farmer suggests that most rayfin lineages reverted to gill respiration (in Mesozoic times) as a result of predation by aerial hunters: first pterosaurs, and then seabirds. Remember that nineteenth-century whalers relied on spotting whales "spouting" at the surface to locate and kill them. Submarines face the same problem: a submarine is most vulnerable at or on the surface.

Shedding Carbon Dioxide

It is usually easy to get rid of CO_2 in normal seawater. CO_2 dissolves and diffuses readily in it. So gills would be retained for this purpose, even if oxygen was mostly supplied by air breathing. Osteolepiforms had gills as well as nostrils, of course, so they did have alternative ways of gaining oxygen. If they lived in shallow-water habitats close to shore (and the fossil record suggests that this was always true), and if they were exposed to air for short periods, they would have run (or slithered) into a CO_2 problem. In air, gills become inefficient at shedding CO_2, because their filaments stick together. It's difficult to lose CO_2 by diffusion into the air in the mouth, because carbonic acid builds up rapidly as CO_2 and water react. But if gills don't work, another method must be found.

There are two possible alternatives: to lose some CO_2 through a wet skin surface rather than the gills (living amphibians lose most of their CO_2 this way), or to breathe rapidly and rhythmically, getting rid of the air bubble in the mouth quickly, before it gets too acidic. Many people think that osteolepiforms and early tetrapods used skin surfaces for shedding CO_2, but I doubt it. Carbon dioxide loss at the skin is efficient only in small animals with a large area of skin relative to a small body volume, but osteolepiforms and early tetrapods were large and scaly. (A tough, relatively impermeable skin would have been very helpful in avoiding loss of water and dissolved salts during exposure to air.) Therefore early tetrapods probably lost CO_2 by rapid breathing.

It is probably not an accident, then, that some osteolepiforms evolved into land-going vertebrates and the other lobefins became extinct or nearly did so. Only the lungfishes hang on as "living fossil" shallow-water air breathers, in very specialized ways of life. The dominant shallow-water air breathers today are amphibians, reptiles, and mammals.

LIMBS AND FEET: WHY BECOME TETRAPOD?

Why would osteolepiforms, as fast-swimming predators in the water, have evolved lobe fins that increasingly came to look and operate like the tetrapod limb? How would an osteolepiform have benefited from an ability to push on a resistant substrate, rather than using a swimming stroke in water? Why take excursions out into air, rather than simply breathe air at the water surface? In other words, why would an osteolepiform become a tetrapod? Evolutionarily, it can have resulted only from a chain of events that produced an improved osteolepiform.

The old story about this transition was that an ability to withstand air exposure helped an osteolepiform find another pool of water if the one it inhabited dried up in a drought. This idea is probably wrong. In the Florida Everglades, animals around drying waterholes stay with the little supply there is, rather than striking off into parched country in the hope of finding more: it's simply a better bet for survival.

Basking?

The evolution of strong, low-slung lobe fins on osteolepiforms probably helped them to hunt small prey in shallow water by poling their bodies through and over mudbanks. The fins became powerful enough to support the weight of the fish, at least briefly, while it gasped and thrashed its way along. The brief exposures to air would not have been long enough to pose much danger of drying out, but they would have preadapted osteolepiforms for longer periods of exposure.

If some osteolepiforms evolved the habit of sunning themselves on mudbanks to warm up their bodies, their digestion would have been faster than in the water; other things being equal, they would have grown faster, matured earlier, and reproduced more successfully than their competitors did. Basking behavior would have been effective even if the fish exposed only its back at first, supported mainly by its own buoyancy. But such effectiveness would have encouraged longer and more complete exposure. Some fishes, and many living amphibians and reptiles (including alligators and crocodiles), bask while they digest (Figure 8.11).

As a basking, air-breathing osteolepiform became more exposed, more of its weight would have rested on the ground, threatening to suffocate it by preventing

Figure 8.11 Basking is important for many animals, including crocodiles, which are perhaps the closest living ecological analogues to early tetrapods. (From Lydekker.)

the thorax from moving in respiration. The pectoral fins in particular would have become stronger, to take more and more of the body weight during basking. Part of the shoulder girdle originally evolved to brace the gill region, and part to link with the pectoral fins. So the pectoral fins of osteolepiforms were still strongly linked with the skull and backbone, retaining the neckless appearance of all fishes.

Basking behavior may have made a more competitive fish, but we would still have recognized it as an osteolepiform. What other factors might have encouraged its evolution into a completely new kind of creature in a completely new environment?

Reproduction?

The most vulnerable parts of the life cycle of a fish are its early days as an egg and hatchling. If some osteolepiforms could make very short journeys—even a meter or so to begin with—over land, or over very shallow water, they would have been able to find small, warm pools, lagoons, ponds, and sheltered backwaters nearby to spawn in. There would have been fewer predators in these side pools than in open water, and eggs and young would have survived better there. In much the same way, and for the same reasons, salmon struggle to swim far upstream to spawn, and many freshwater fishes swim into seasonally flooded areas to breed.

Isolated warm ponds would also have been ideal breeding grounds for small invertebrates such as crustaceans and insects, which would have formed a rich food supply for the young osteolepiforms. Then, reaching a size at which they could handle larger prey and that would give them some protection against being eaten themselves, the young osteolepiforms could make their way back to the main stream and take on their adult way of life as predators on fishes. Among young crocodiles, the greatest cause of death (apart from human hunting) is being eaten by an adult crocodile. Crocodiles provide intensive parental care while their young are small. Iguanas tend to separate juvenile and adult habitats. Osteolepiforms perhaps solved the same problem in a different way, by arranging for their young to spend time away from other adults. Note that this advantage would still have operated if osteolepiforms, like coelacanths, had live birth.

Nitrogen Loss

Gills do more than exchange oxygen and CO_2 in fishes: they are also involved in ridding the body of nitrogenous waste. Although body wastes such as ammonia and/or its secondary product urea can be lost through kidneys and intestines, the fact is that most fishes lose most of their nitrogen through the gills. Even lungfishes, which breathe in oxygen through their lungs, retain gills through which they lose CO_2 and nitrogen. Gills have a great deal of water flowing past them, so they can lose these wastes effectively without building up a concentrated store of them in and near the gill tissues. In contrast, excreting nitrogen via the kidneys means building organs that not only can but must have high ammonia concentrations, released in relatively concentrated fluids.

Jawed fishes typically converted ammonia to urea, and this process was inherited in the lobefin fishes that evolved into tetrapods. This was an advantages in situations where the fish might become water-stressed. Urea is less toxic than ammonia, and can be tolerated at higher concentrations, for example, during times of exposure in air. Lungfishes lose waste as ammonia when they are in water, but they convert the ammonia to urea and store it while they are exposed to air, only getting rid of it when they are once again under water. In fact, no living fishes excrete nitrogen into air, by any means.

Typical amphibian tadpoles have gills and use them for excreting ammonia. When they metamorphose into adults, they lose their gills. At that point they lose carbon dioxide through the lungs and skin, and lose nitrogen (as urea) through the kidneys. Urea is more concentrated than ammonia, so causes less water loss as it is eliminated.

On the face of it, one of the great breakthroughs toward life in air was the evolution of nitrogen excretion into air. So why would the first tetrapods or their ancestors have acquired this ability? Christine Janis and Colleen Farmer, who laid out the "kidney story" I have summarized here, found it "difficult to conceive of an [appropriate] adaptive scenario . . . in the aquatic environment."

I cannot resist pointing out that such an ability would have been strongly adaptive for lobefins that were first basking, and then making short excursions into air (possibly for spawning), in the scenario that I have argued here. Short terrestrial "phases" may have been very important in the life history of many adult lobefins that were otherwise aquatic, because they included the all-important reproductive component in their lives.

From Osteolepiform to Tetrapod

Only one line of osteolepiforms took the evolutionary path toward tetrapods, while the others remained as normal fishes in rivers, lakes, and shallow coastal waters. Osteolepiforms such as *Eusthenopteron* (Figure 8.10) have long been recognized as the most likely tetrapod ancestors. The skull bones, the pattern of bones in the lobe fins, and the general size, shape, and geographic distribution of these fishes are close to those of the earliest tetrapods. Osteolepiforms have lobe fins with bony elements corresponding to a 1–2–several–many pattern. Our limbs do the same: our arms have humerus; radius + ulna; wrist bones (carpals); hand bones; and our legs have femur; tibia + fibula; ankle bones (tarsals); foot bones. We and all other tetrapods share the same pattern, inherited from osteolepiforms.

Osteolepiform locomotion in shallow water and on shallow mudbanks would have been improved by stronger fins, especially stronger fin edges. Land locomotion consisted at first of the same undulatory twisting that salamanders still have, with the fins acting simply as passive pivots (Figure 8.12). The fins gradually

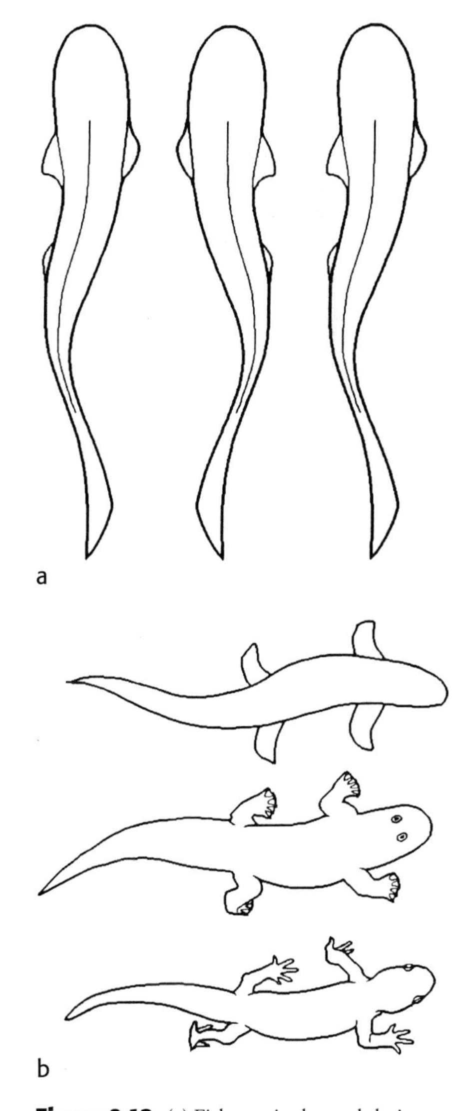

Figure 8.12 (a) Fishes swim by undulating the body significantly, while the head swings less. (b) The same basic body movements are used by an osteolepiform fish swimming or squirming over a mud bank, by an early tetrapod crawling, and by a salamander walking on dry land. No sudden or large shifts in locomotory mechanism were required for the transition, even though the fish has fins and the salamander has feet.

exerted stronger traction on the substrate, which may have encouraged the multiple rays in the fins to become fewer and stronger until toed feet evolved. In the process, the pectoral fins came to support the thorax, while the pelvic fins came to be better suited to push the body forward. The pelvic fin evolved a hinge joint at a "knee" and a rotational joint at an "ankle," a pattern that persisted into tetrapods. This difference was inherited by all later vertebrates: elbows flex backward, knees flex forward.

As the pectoral and pelvic girdles evolved better linkage with the fins, the fins evolved gradually to become clearly defined limbs. The hindlimbs quickly became as powerful as the forelimbs and evolved strong links with the backbone through the pelvis.

Other changes also took place as osteolepiforms evolved into tetrapods. A leathery skin evolved to resist water loss, fins were strengthened into limbs for locomotion and support, and senses improved for an air medium. Ecologically, tetrapods and osteolepiforms divided up the habitat as they diverged. Aberrant osteolepiforms (evolving tetrapods) spent more and more time at and near the water's edge, sunning and basking, while normal osteolepiforms remained creatures of open water.

The first tetrapod was the first animal to evolve feet rather than fins (note that this is a stem definition of a tetrapod). It is becoming clear that toes—digits, to be anatomically precise—are a new structure, added on to bones that were present in osteolepiform fins. Feet, when they evolve, are separated from bones further up the limb by a joint at what we can now call an ankle or wrist.

Since feet, toes, and ankle (= wrist) make up a complex of bony structures, they would have evolved gradually. Obviously, if we had a lot of fossils we could see that happen. We have only scraps of fossils of very early tetrapods, so people argue about which scrap can be defined as the first tetrapod.

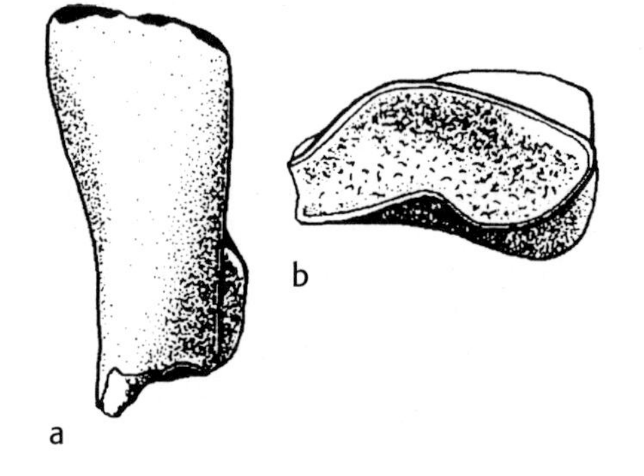

Figure 8.13 *Elginerpeton*, from the Late Devonian of Scotland. The tibia (a) is well enough preserved to show from the ankle joint (b) that this animal could walk. (After Ahlberg.)

THE FIRST TETRAPODS

Elginerpeton, from Late Devonian rocks in Scotland (perhaps 368 Ma) has a tibia with a well-preserved joint surface at its lower end, evidence that it had an "ankle" (Figure 8.13). Its jaw is also tetrapod-like rather than fish-like. *Elginerpeton* was large: the jaw alone is 40 cm (a foot or more) long, and the whole animal must have been several times that. There is some hesitation in calling *Elginerpeton* a tetrapod because we still do not have its feet, but on balance the evidence indicates that it is.

Slightly later, we find bits and pieces of tetrapods from localities widely spread across the tropics of the latest Devonian. *Ichthyostega* and *Acanthostega* are the best known, from almost complete skeletons from Greenland; and *Tulerpeton* from Russia is a pile of bones that may have come from more than one animal. No other Devonian tetrapods are complete enough to say much about their biology (or in fact their evolutionary placement). The best way of dealing with that is to call them "stem tetrapods." Stem tetrapods lived within a 2-m.y. time period close to the Devonian/Carboniferous boundary (363 Ma). Small arthropods and plants were not a suitable food supply for these early tetrapods, which were all large (more than a meter long). These animals ate fishes in the water. After that, there are interesting variations.

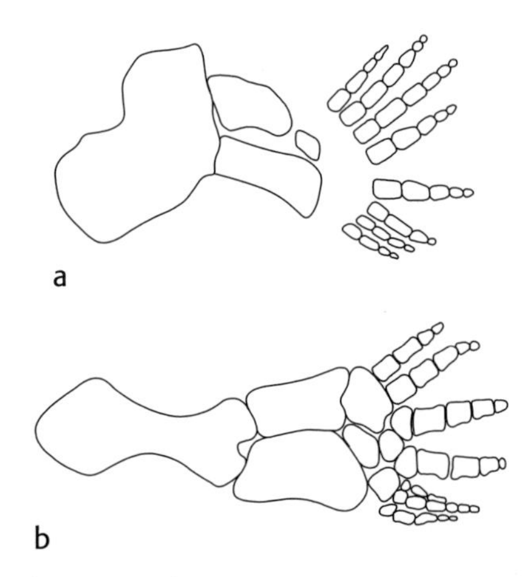

Figure 8.14 Early feet and toes, from two tetrapods from the Late Devonian of Greenland. (a) The front limb of *Acanthostega*. Although the limb clearly has (eight) toes, it looks more like a functional flipper rather than a walking foot. (b) The hind limb of *Ichthyostega*. There are seven toes (well, six and a half). (After Clack and Coates and Thomson.)

The stem tetrapods had many digits. The number varied, but it was not five: *Acanthostega* had eight toes and *Ichthyostega* had seven (Figure 8.14); *Tulerpeton* had six. Though the number of examples is limited, the loss of toes seems to be linked to the relative use of the foot in pushing on the bottom. *Tulerpeton* could have walked

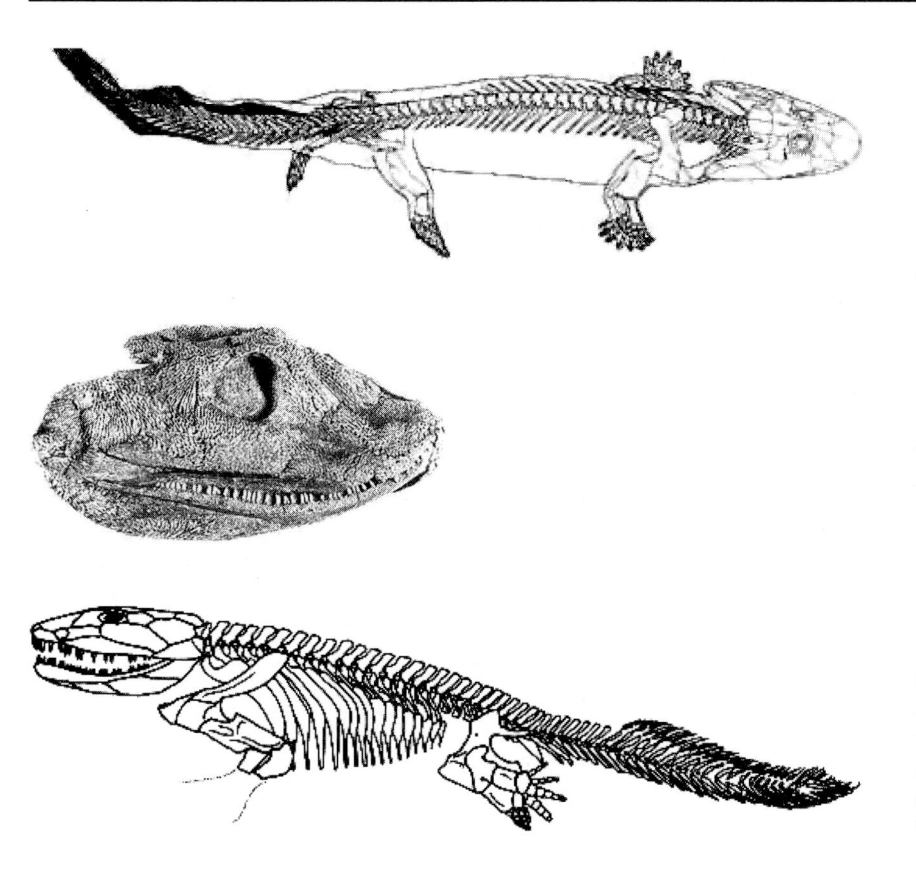

Figure 8.15 *Acanthostega.* A new reconstruction reflects the discovery that the limbs of *Acanthostega* are adapted to work well in the water rather than on land. Nevertheless the structure is tetrapod. Based on Coates, 1996. (Reconstruction courtesy of Jenny Clack of the Zoology Museum, Cambridge University; © Jenny Clack.)

Figure 8.16 *Acanthostega.* This is the (fish-eating) skull of the individual Copenhagen Museum, MGUH 1300, otherwise known as Grace. (Courtesy of Jenny Clack of the Zoology Museum, Cambridge University; © Jenny Clack.)

Figure 8.17 *Ichthyostega.* A new reconstruction reflects the discovery by Coates and Clack (1995) that the hind limbs (the only feet preserved) are not very good walking devices. (Reconstruction courtesy of Jenny Clack of the Zoology Museum, Cambridge University; © Jenny Clack.)

quite well on land, *Acanthostega* was much more adapted to life in water, and *Ichthyostega* was somewhere in between.

Acanthostega (Figures 8.15; 8.16) seems to have had the most fish-like biology of the stem tetrapods. It still had functional gills, for example, which means it could breathe underwater as well as in air. Its forelimbs were rather weak, its ribs did not curve round to support its weight well, and its eight-toed lower limbs were still somewhat flipper-like. It may have been best adapted to eating fish in weed-choked shallows, and it may not have been able to support its weight for long out of water.

Ichthyostega had a massive skeleton but was otherwise very much like Late Devonian osteolepiforms in spine, limb, tooth, jaw, palate, and skull structure (Figure 8.17), and probably in diet and locomotion. Like osteolepiforms, it had a tail fin, but unlike them it had a strong rib cage and limbs and feet rather than lobe fins. *Ichthyostega* solved the problem of supporting the chest for breathing on shore by having a massive set of ribs attached to the backbone (compare with those of *Acanthostega*, Figure 8.15). These ribs (I would argue) were adaptations for excursions into air.

As an adult, *Ichthyostega* was probably much like a living crocodile in ecology. New specimens show that its hind feet are not well designed for walking, and so far its front feet are unknown (though they were placed on very strong bones). Jenny Clack compares its potential for land movement with that of the living elephant seal (though at smaller size). Elephant seals basically haul themselves around on the beach, using the strong front feet for propulsion and the hind feet as props and skids. Nevertheless, unwary tourists have been damaged by the sudden and rapid charge of a big male elephant seal driven to rage by twittering intruders into its "territory."

The aquatic hunting of *Ichthyostega* was aided by a unique ear structure. A large air-filled pocket in the skull probably amplified any underwater sound reaching it, then transmitted the signals through a long thin stapes bone to the inner ear. No other tetrapod has anything quite like it. However, all the stem tetrapods could have come out into air and on to land for reasons other than food, such as the digestive and reproductive advantages I have mentioned, and the new information about *Ichthyostega* is certainly compatible with that. After all, elephant seals leave the water for display, fighting for dominance, mating, and breeding.

Further Reading

Bateman, R. M., et al. 1998. Early evolution of land plants: phylogeny, physiology, and ecology of the primary terrestrial radiation. *Annual Reviews of Ecology and Systematics* 29: 263–92. [Long review of early land plants.]

Beerling, D. J., et al. 2001. Evolution of leaf-form in land plants linked to atmospheric CO_2 decline in the Late Palaeozoic era. *Nature* 410: 352–4, and comment, 309–10.

Carroll, R. L. 2001. The origin and early radiation of terrestrial vertebrates. *Journal of Paleontology* 75: 1202–13. [Nice review.]

Clack, J. A. 2002. *Gaining Ground: The Origin and Evolution of Tetrapods.* Bloomington: University of Indiana Press. [This is not a light-weight book in size or level, but it is very well written, and gives a complete picture of research up to 2002, including references to previous papers by her and Michael Coates.]

Clack, J. A., et al. 2003. A uniquely specialized ear in a very early tetrapod. *Nature* 425: 65–9. [The wonderful inner ear of *Ichthyostega.*]

Cleal, C. J., and B. A. Thomas. 1999. *Plant Fossils: The History of Land Vegetation.* Woodbridge, England: Boydell Press. [Very nicely written, with beautiful photographs. Covers Paleozoic plants very well.]

Coates, M. I. 1996. The Devonian tetrapod *Acanthostega gunnari* Jarvik: postcranial anatomy, basal tetrapod interrelationships and patterns of skeletal evolution. *Transactions of the Royal Society of Edinburgh, Earth Sciences* 87: 363–421. [Monograph on *Acanthostega.*]

Coates, M. I., and J. A. Clack. 1995. Romer's gap: tetrapod origins and terrestriality. *Bulletin du Museum National d'Historié Naturelle, Paris* 17: 373–88.

Daeschler, E. B., and N. Shubin. 1998. Fish with fingers? *Nature* 391: 133.

DiMichele, W. A., and R. W. Hook (rapporteurs) 1992. Paleozoic terrestrial ecosystems. Chapter 5 in A. K. Behrensmeyer et al., *Terrestrial Ecosystems Through Time.* Chicago: University of Chicago Press.

Doyle, J. A. 1998. Phylogeny of vascular plants. *Annual Reviews of Ecology and Systematics* 29: 567–99. [Overall survey of all land plants.]

Farmer, C. G. 1999. Evolution of the vertebrate cardio-pulmonary system. *Annual Reviews of Physiology* 61: 573–92.

Graham, L. E., et al. 2000. The origin of plants: body plan changes contributing to a major evolutionary radiation. *Proceedings of the National Academy of Sciences* 97: 4535–40.

Heatwole, H., and R. L. Carroll (eds) 2000. *Anphibian Biology.* Vol. 4 *Palaeontology.* Chipping Norton, N.S.W., Australia: Surrey Beatty & Sons. [The first seven chapters are especially relevant to early tetrapods.]

Heckman, D. S., et al. 2001. Molecular evidence for the early colonization of land by fungi and plants. *Science* 293: 1129–33.

Janvier, P. 1996. *Early Vertebrates.* Oxford: Clarendon Press.

Janis, C. M., and C. Farmer. 1999. Proposed habitats of early tetrapods: gills, kidneys, and the water–land transition. *Zoological Journal of the Linnean Society* 126: 117–26.

Keeley, J. E., et al. 1984. *Stylites,* a vascular land plant without stomata absorbs CO_2 via its roots. *Nature* 310: 694–5, and comment, 633.

Knoll, A. H. 1984. Patterns of extinction in the fossil record of vascular plants. In M. H. Nitecki (ed.). *Extinctions.* Chicago: University of Chicago Press, 21–68.

Niklas, K. J., et al. 1985. Patterns in vascular land plant diversification: an analysis at the species level. In J. W. Valentine (ed.). *Phanerozoic Diversity Patterns.* Princeton: Princeton University Press, 97–128.

Raven, J. A. 1984. Physiological correlates of the morphology of early vascular plants. *Botanical Journal of the Linnean Society* 88: 105–26.

Shear, W. A. 1991. The early development of terrestrial ecosystems. *Nature* 351: 283–9.

Shear, W. A., et al. 1996. Fossils of large terrestrial arthropods from the Lower Devonian of Canada. *Nature* 384: 555–7.

Shubin, N., et al. 1997. Fossils, genes and the evolution of animal limbs. *Nature* 388: 638–48.

Wellman, C., et al. 2003. Fragments of the earliest land plants. *Nature* 425: 282–5, and comment, 248–9.

Wurtsbaugh, W. A., and D. Neverman. 1988. Post-feeding thermotaxis and daily vertical migration in a larval fish. *Nature* 241: 846–8.

Zhu, M., et al. 2002. First Devonian tetrapod from Asia. *Nature* 420: 760–1.

Zimmer, C. 1995. Coming onto the land. *Discover* 16 (6): 118–27. [Feature article on Jenny Clack and *Acanthostega.*]

CHAPTER NINE

Tetrapods and Amniotes

Once the first tetrapods evolved from osteolepiform fishes, they radiated quickly into a great variety of sizes, shapes, and ways of life. Although this is the most poorly known part of the terrestrial vertebrate record, it is also one of the most exciting areas in which research is constantly turning up new fossils. In this chapter I will give a progress report on the story as we see it now, and some of the problems that these early land animals faced and solved.

EARLY TETRAPODS

The earliest tetrapods, *Acanthostega* and *Ichthyostega* (Chapter 8) spent their adult life in water, with (in my scenario) only occasional journeys into air for basking and spawning. One could describe this way of life as amphibious, but that does not make the animals amphibian in formal terms. The early tetrapods have left living descendants that are divided sharply into amphibians (Amphibia) and amniotes (Amniota, which are living reptiles, birds, and mammals); many groups of early tetrapods became extinct, and are difficult to classify even when we have good skeletons preserved. Amphibia and Amniota are defined in a crown-group manner, so we are looking backward for the earliest ancestor of each group. That leaves a mass of early tetrapods that are neither Amphibia or Amniota: these are stem tetrapods (Figure 9.1).

Ecologically, early tetrapods were the first large animals to exploit the environment in and around the water's edge. Their variety reflects different adaptations to different habitats and different ways of life. Some were dominantly terrestrial, some aquatic, and some genuinely amphibian. Naturally, there were variations even within each group.

Living amphibians are all small-bodied and soft-skinned, and in these respects are quite unlike early tetrapods. They are the newts and salamanders, the frogs and toads, and the caecilians, which are burrowing legless amphibians. Living amphibians are usually classed together as "smooth amphibians" or Lissamphibia, though in a crown-group definition this is the same as saying Amphibia. This clade is very much derived and probably did not evolve until Late Permian or even Triassic times. The biology of living amphibians is fascinating, but may not be any guide at all to the origin, the paleobiology, or the classification of the early tetrapods of the Late Devonian and Carboniferous. In exactly the same way, living amniotes have been

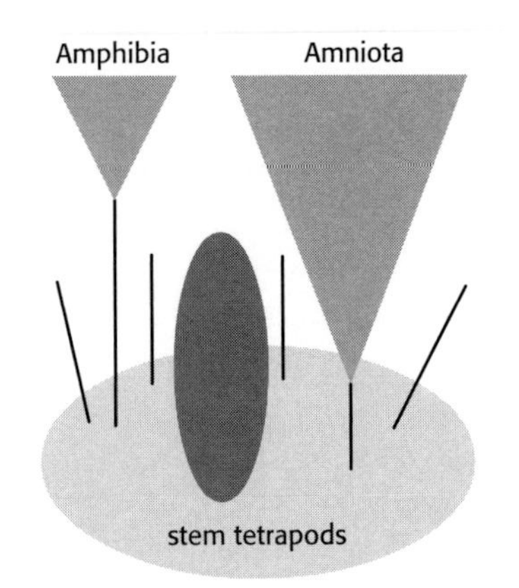

Figure 9.1 Diagram to show the relationship between Amphibia, Amniota, and "stem tetrapods." Stem tetrapods include forms for which we have no strong hypotheses of relationship. Some clear clades can be defined, but they left no survivors. The two clades that did, Amniota and Amphibia, are crown-based clades. We have a very clear picture of their living members, but, as often happens in paleontology, it is difficult to distinguish their earliest members among the background of stem tetrapods. The late-surviving stem tetrapod clade of temnospondyls stands out; a temnospondyl may have been the ancestor of Amphibia, but that is a very weak hypothesis and is not shown here as a formal relationship.

evolving for 300 m.y. or so, and are equally poor indicators of the ways of life of their distant ancestors. Altogether, evolutionary patterns among early tetrapods are difficult to work out.

Several tetrapod groups were evolving in parallel in the Late Devonian, some like *Acanthostega* toward a more aquatic life than *Ichthyostega*, some like *Tulerpeton* toward a more terrestrial life (Chapter 8). One possible evolutionary scheme is shown in Figure 9.2. But it seems that the picture changes each time a new fossil is studied carefully, or an old one is cleaned. Our ideas are likely to change a lot in the next few years.

Early tetrapods radiated quickly into many lineages. Two of them were the ancestors of Amphibia and Amniota, so still survive, and the others are more closely or more distantly related to these two clades. The fossil record is biased, because it favors large animals over small, and it favors preservation in water rather than on land. We simply do the best we can with it.

Pederpes is a meter-long tetrapod, first reported in 2002 from Early Carboniferous rocks of Scotland (age around 350 Ma). Its feet have only five toes, unlike earlier tetrapods. It still looked and probably behaved like a small crocodile, spending most of its time in the water. It is one of only a handful of tetrapods from Early Carboniferous rocks, and is the first tetrapod with feet that are genuinely adapted for walking on land.

A really good collection has come from rocks dating to about 335 Ma at East Kirkton in Scotland, where there was a complex tropical delta environment at the time, including shallow pools fed by hot springs. No fishes were found in the same levels as the tetrapods, possibly because the pools were too hot for them at those times. The tetrapods are likely members of an early community of animals that lived in the rivers and swamps near the pools, walked by them and sometimes fell into them or were washed into them. These animals included scorpions and millipedes, the earliest known harvestman, and, of course, several tetrapod groups. These included two groups of larger tetrapods, the anthracosaurs and the temnospondyls, but other tetrapods were smaller in body size and varied in ecology.

It looks as if the temnospondyls led (eventually) to living amphibians and the anthracosaurs led (eventually) to living amniotes (Figure 9.1), so they have received

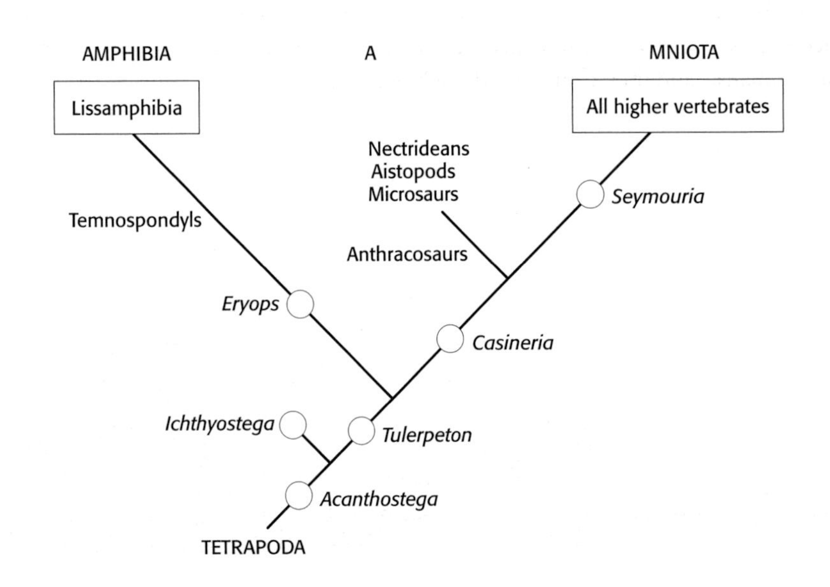

Figure 9.2 One possible phylogram for living tetrapods, and their relationship to early tetrapods. There are several competing alternatives, none of them strong. This is a phylogram, not a cladogram, so I have placed some of the tetrapods mentioned in the text in the body of the diagram.

most interest. Only a brave person would actually call them stem amphibians or stem amniotes, however.

To show how difficult the East Kirkton fossils can be, consider the new fossil *Eucritta melanolimnetes*, "the creature from the black lagoon," which has a skull rather like an anthracosaur and a body rather like a temnospondyl. It looks rather like a salamander, but of course it is not one. *Eucritta* adds complexity to the puzzle, rather than simplifying it.

Ancestors of Living Amphibians? Temnospondyls

Temnospondyls are the largest and most diverse group of Carboniferous tetrapods: 40 families and 160 genera have been described altogether. There are over 30 skeletons of temnospondyls in the East Kirkton collections.

Temnospondyls were large, with teeth like those of osteolepiforms and *Ichthyostega*. The most common temnospondyl at East Kirkton, *Balanerpeton*, is about 50 cm (20 in) long. It has heavy bones, and its beautifully preserved feet look strong and well adapted for walking. Temnospondyls probably had a biology much like that of crocodiles, and they were fisheaters as adults (see *Eryops*, Figure 9.3). The jaw was designed to slam shut on prey, and the skull was very strong.

More terrestrially adapted temnospondyls had very massive, strong skeletons capable of supporting them on land, even when they were rather small (Figure 9.4). Some temnospondyls even became more terrestrial during their lifetimes. For example, young *Trematops* had a jaw designed for eating small, soft food items, but adults had a carnivorous jaw and a lightly built skeleton capable of rapid movement. As adults they were probably land-going predators.

As part of their adaptation to life in air, temnospondyls had an ear structure that suggests they could hear airborne sound. The stapes bone is strong and seems to have conducted sound to an amplifying membrane that sat in a special notch in the skull. Other early tetrapods, including early amniotes, did not evolve such an advanced system.

Temnospondyls survived into the Cretaceous in isolated areas of Gondwana. Some Triassic forms were giant marine animals, and they ranged as far south as Antarctica (living amphibians can live in cold water as long as they spawn in spring or summer). A Late Paleozoic temnospondyl was probably the ancestor of living amphibians, though we lack a detailed fossil record of the transition. If the connection is eventually confirmed, then we *would* call temnospondyls "stem amphibians," and their clade would not be extinct.

Figure 9.3 Some temnospondyls were capable of land locomotion even though they spent most of their time in water. This heavily built animal, *Eryops*, was much like a crocodile in size and probably in ecology. The limbs were sprawling, the head was massive, and an adult must have been very clumsy on land. (Negative 35632. Photograph by A. E. Anderson. Courtesy of the Department of Library Services, American Museum of Natural History.)

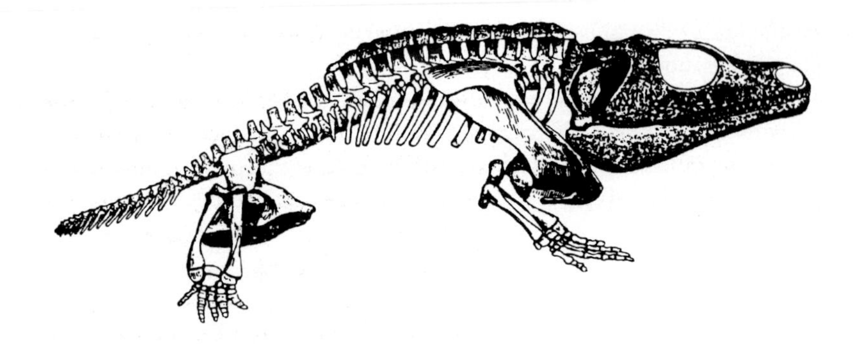

Figure 9.4 The terrestrially adapted Permian temnospondyl *Cacops* was small, with a body length of only 40 cm (16 in), but massively built. The strength of the front limbs and shoulder girdle is associated with the heavy skull. (After Williston.)

Figure 9.5 An aistopod, which perhaps lived like a little water snake. A complete specimen is about 75 cm long (30 in). (After Fritsch.)

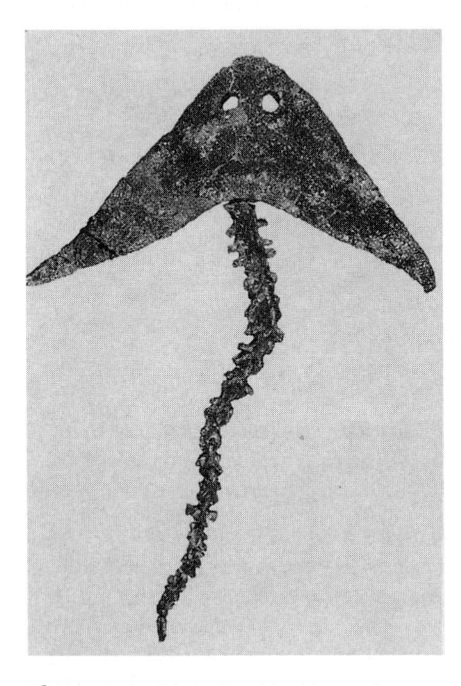

Figure 9.6 The skull and backbone of a horned nectridean, *Diplocaulus*. (Negative 35679. Photograph by Thomson. Courtesy of the Library Services Department, American Museum of Natural History.)

Small but Interesting Groups of Early Tetrapods

We do not yet know how to classify some of the small, mainly aquatic early tetrapods. They may be closer to amniotes than to amphibians, as shown in Figure 9.2, but that is a weak hypothesis.

Microsaurs were small, with weakly calcified skeletons. Their remains are usually fragmentary, and they include many juvenile forms. **Aistopods** were small, slim tetrapods that had lost almost all trace of their limbs. They probably lived rather like little water snakes (Figure 9.5). They do not preserve well and their fossils are quite rare, so they have not been properly studied.

Nectrideans are better preserved and understood. They had a short body and a long, laterally flattened tail that made up two-thirds of their total length and was probably used for swimming. The vertebrae were linked in a way that allowed extremely flexible bending, and nectrideans probably swam like eels.

Horned nectrideans are fascinating. They had flat, short-snouted skulls with the upper back corners extended backward on each side. In early forms the extensions were quite small, but later they evolved to look like the swept-back wings of a jet fighter (Figure 9.6).

Anthracosaurs

Anthracosaurs are the other numerous and diverse group of early tetrapods. *Tulerpeton* may be the earliest one (Chapter 8), though we need more complete specimens to confirm this. There are certainly two anthracosaurs in the East Kirkton fauna. Most anthracosaurs were adapted for life primarily in water, as long-snouted and long-bodied predators, presumably crocodile-like fisheaters, with jaws designed for slamming shut on prey. Their limbs were not very sturdy, but they may have been very good at squirming among dense vegetation in and around shallow waters. They must have swum in an eel-like fashion. It's unlikely that they had the speed and power to compete with fishes in open water, even though some were quite large, up to 4 m (13 ft) long.

A few anthracosaurs were smaller, slender animals, adapted to terrestrial life. These tetrapods seem to be the closest relatives of early reptiles (Figure 9.1), even though their ears remained adapted for low-frequency water-borne sound. *Seymouria* (Figure 9.7) and *Diadectes* are well-known members of this large clade.

Though the consensus is that amniotes evolved from somewhere near or in anthracosaurs, no one hypothesis is strong. The problem is that amniotes began small, whereas anthracosaurs were large. I suspect that a major shift in habitat and ecology was involved here, so convincing evidence is going to be difficult to find.

Figure 9.7 *Seymouria* was one of the few anthracosaurs well adapted for terrestrial life. The scale is 10 cm, so these animals, from the Permian of Germany, were about about 60 cm long (2 ft). (Courtesy of David Berman of the Carnegie Museum, Pittsburgh, and the Society of Vertebrate Paleontology.)

AMNIOTES AND AMNIOTA

The word **amniote** means a tetrapod that forms eggs inside a membrane. The word **Amniota** is a formal name for the clade of tetrapods that includes all living amniotes (mammals, reptiles, and birds) and their common ancestor. This is an uneasy combination of terms. The common ancestor of living amniotes may or may not have been the first tetrapod to lay an amniotic egg (how would we know?). But if we are to use a crown-based method of classification, we have to accept its awkward aspects along with its power.

The Amniotic Egg

Living amphibians differ from living amniotes in several characters of the skeleton that can be recognized in fossils, and in other characters that affect the soft parts and cannot be recognized in fossils. The major soft-part character of living amniotes is that they have eggs surrounded by a membrane, rather than the little jelly-covered eggs of fishes and amphibians. This fundamental difference in biology needs special attention because it was so important in the evolution of tetrapods into entirely terrestrial habitats.

How did the amniotic egg evolve, and who evolved it? (Perhaps the earliest amniotes laid amphibian-style eggs, and the amniotic egg evolved later in the lineage. Perhaps some early tetrapods laid amniotic eggs before the common ancestor of living amniotes appeared. This is not an easy argument to grasp at first, but only someone uneasy about the process of evolution would have any logical problem with it.)

Amphibians have successfully solved most of the problems associated with exposure to air. But their reproductive system was and is linked to water, and it remains very fishlike. Almost all amphibians spawn in water and lay a great number of small eggs that hatch quickly into swimming larvae. The eggs do not need any complex protection against drying, because if the environment dries, the larvae are doomed as well as the eggs. Thus, selection has acted to encourage the efficient choice of suitable sites for laying eggs, rather than devices to protect eggs. Both fishes and amphibians may migrate long distances for spawning, and favored sites are often disputed vigorously.

Living reptiles have a different system. Their juveniles hatch into air as competent terrestrial animals, often miniature adults. Yet the stages of embryological development are strikingly similar to those of amphibians. The difference is that reptiles develop for a longer time inside the egg, which in turn means that the egg must be larger and must provide more food and other life-support systems. Reptiles typically lay far fewer eggs than amphibians of comparable body size, so they have evolved more complex adaptations to ensure greater chances of survival for each individual egg.

A reptile (amniotic) egg is enclosed in a tough membrane covered by an outer shell made of leathery or calcareous material. The membrane and shell layers allow gas exchange with the environment (water vapor, CO_2, and oxygen) for the metabolism of the growing embryo, but they also resist water loss. Reptiles lay eggs on land, so eggs are not supported against gravity by water. Instead, the shell gives the egg strength, protects it, holds it in a shape that will allow the embryo room to grow freely, and buffers it against temperature change and desiccation. Birds' eggs are much like reptile eggs, usually with harder shells, and in most mammals the whole egg (without a shell) is nurtured internally so that the embryo emerges from the amnion at the time it emerges from the mother ("live birth").

Inside the amniotic egg, the embryo is nourished by a large **yolk**, and special internal sacs act as gas-exchange and waste-disposal modules. The most fundamental innovation, however, is the evolution of another internal fluid-filled sac, the **amnion**, in which the embryo floats. Amniotic fluid has roughly the same composition as seawater. The amniotic egg acts toward the embryo like a spacecraft nurturing an astronaut in an alien environment: it has food storage, fuel supply, gas exchangers, and sanitary disposal systems (Figure 9.8).

Because the embryo inside an amniotic egg is encased in membranes, and often inside a shell, the female's eggs must be fertilized before they are finally packaged. Internal fertilization must have evolved among with the amniotic egg.

The evolution of the amniotic egg broke the final reproductive link with water and allowed tetrapods to take up truly terrestrial ways of life. Its evolution demanded changes in behavior patterns and in soft-part anatomy and physiology. The transitional forms either evolved into or were outcompeted by more advanced animals, so they are now extinct and unavailable for direct study.

The amniotic egg was probably evolved by an early tetrapod that looked like a little reptile. I present here a reasonable scenario for the evolution of amniotes. It may be supported or contradicted by future evidence, or replaced by a simpler or more elegant story.

With the increasing ability of tetrapods of all kinds to make forays onto land, their breeding grounds became much less secure. Sites that had once been safe refuges for young animals gradually became more susceptible to raiders. The same evolutionary pressures that I suggested in Chapter 8 for the origin of tetrapods from fishes now drove some tetrapods to seek still safer refuges for breeding and the development of their young. In so doing they evolved into the first amniotes, and were preserved in environments quite different from those of their ancestors.

Forays farther from water became more practicable as the evolution of flourishing plant life provided a myriad of damp hiding places in and around Carboniferous swamps. Small tetrapods could have found small, sheltered, hidden places that were damp enough to foster egg development but not obvious enough to attract predators. They may have had behavior patterns like those of many living amphibians, particularly tree frogs and tree-dwelling salamanders.

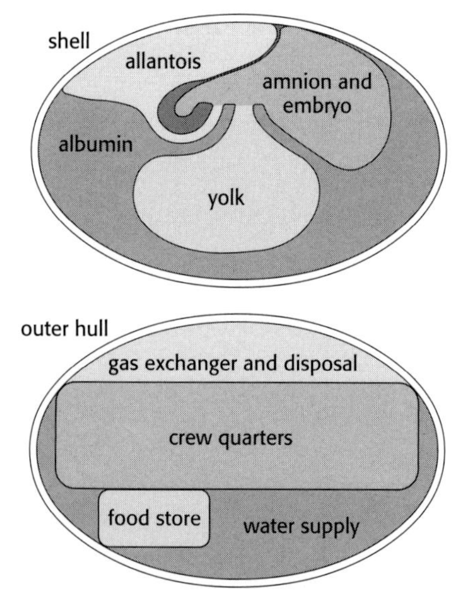

Figure 9.8 The amniotic egg is in many ways analogous to a spacecraft.

The major problem in laying unshelled eggs away from larger water bodies is drying, at spawning time and during development. A crude sort of egg membrane would have been a partial solution to the developmental problem. Further refinement of the system is then fairly easy to imagine. Internal fertilization was probably a preliminary solution to the problem of desiccation while spawning. (Some living amphibians have independently evolved a crude kind of internal fertilization.)

Most frogs and toads undergo a complex development after hatching. A drastic metamorphosis from tadpole to adult involves not only a major anatomical reorganization but a major change in life style. The problems associated with this kind of amphibian reproduction can be solved, sometimes in spectacular ways. Tree frog eggs often hatch into tadpoles in places where there is little water. Some frogs carry their tadpoles one by one to little pools in bromeliad plants; some carry them in pouches on their bodies, where the young develop into miniature adults; and in some Australian frogs the females swallow their eggs after they are fertilized and hatch and develop the young internally in the digestive tract (they don't feed while they incubate!).

Early tetrapods may have had a much more direct development, hatching as miniature adults. A few frogs today lay large eggs, 10 mm across, which hatch into miniature adults. These eggs show no sign of evolving toward amniotic eggs, but they show that some living amphibians can lay large eggs that then develop without a complex metamorphosis. Presumably, as the amniotic egg evolved, the reproductive problems faced today by living frogs were avoided by simply allowing the embryo to develop longer and longer inside the egg. Longer development could, of course, have evolved gradually along with increased size and complexity of the egg.

Evolution toward laying eggs in damp air rather than water may have been encouraged by the better oxygen supply available from air. An egg laid in water, especially a small, warm body of water, may be exposed to anoxic conditions, especially in hot tropical habitats of the Earky Carboniferous. As long as the egg does not dry out, it may have a better oxygen supply in damp air than in water.

This story provides a unifying theme that links the evolution from fish to amniote through the early tetrapods. Throughout, evolutionary change is linked with successful reproduction. As a by-product, successful animals are encouraged to enter new habitats. As they do so, they evolve ways of exploiting those habitats, and new ways of life become not only possible but encouraged. Simple themes that explain many facts are always satisfying: but they are only stories until they are tested against evidence.

Why Were the First Amniotes Small?

The osteolepiforms and the large early tetrapods had a long, heavy skull, with a jaw designed for slamming shut on larger prey, which was then swallowed whole or in large pieces. There was little chewing, and the jaw muscles generated little pressure along the tooth line.

However, some Carboniferous tetrapods, and all early amniotes, were small. All early amniotes were about the same size as living lizards, and much like them in body proportions, posture, and jaw mechanics (Figures 9.9; 9.10). They were probably like them in ecology too. They had a notably small skull with a short jaw well suited to hold, chew, and crush small, wriggling prey, and to shift the grip for repeated bites. The small head was set on a neck joint that allowed very swift three-dimensional motion, whereas in many early tetrapods the long, heavy skull moved

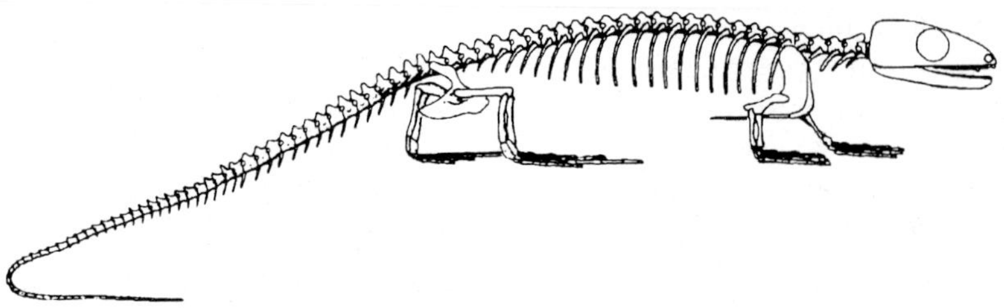

Figure 9.9 *Hylonomus*, an early amniote from the Late Carboniferous tree-stumps of Nova Scotia.

Figure 9.10 Life reconstruction of *Hylonomus* as an analog of a living lizard. (From Lydekker.)

mainly up and down and the neck must have seemed very stiff (this adaptation is effective for swimmers). Why small size?

Animals that spent longer time on land did not do so simply because they were seeking places to breed, but because there were potential food supplies there. Carboniferous forests were rich in worms, insects, and grubs. Young or small animals would have been best suited for foraging after this kind of food. Worms, insects, and grubs are small, though highly nutritious; they are easy to seize, process, and swallow; and they can be found among cracks and crevices in a maze of plant growth in a complex three-dimensional forest setting. Most of these potential prey items are slow-moving, and a successful predator need not have been quick and agile at first. But small and light-bodied animals could have quickly evolved greater agility as their repertoire of prey extended to the expanding number of large insects in Carboniferous forests.

Small body size may also have been favored by a thermoregulatory effect. Animals encounter greater temperature extremes in air than they do in water, and small animals can shelter more easily from chilling or overheating among vegetation, in cracks and crevices, or in hollow tree trunks, than can animals the size of *Ichthyostega*. Small bodies are also quicker and easier to heat by basking in the sun. Again, this suggests that terrestrial and/or arboreal excursions would have most benefited juvenile or small animals.

A reasonable scenario of amniote evolution, mostly due to Robert Carroll, is that they evolved on a forest floor covered with rotting material, leaf litter, fallen branches, and tree stumps, ideal places for prey to hide and amniotes to search (Figure 9.11, for example).

The scenario sounds reasonable enough to people used to temperate forests. But tropical forest floors are clean. The shade of the forest canopy is so thick and continuous that no vegetation grows at ground level, except along river banks where water barriers break the continuity of the canopy, or where storms (or people) have carved an open track of fallen trees. Fallen branches, leaves, bodies, and other pieces of organic debris are broken down and recycled so quickly on the ground by fungi and insects that vertebrates find it hard to make a living there. In contrast, the canopy and the river banks teem with small vertebrates: reptiles, amphibians, mammals, and birds in the canopy, and fishes in the water.

The best candidates for first amniotes are found in the Late Carboniferous. They were living in forests of Early Pennsylvanian (Late Carboniferous) age. Many tree stumps and tree trunks have been fossilized upright in life position in the rocks of Nova Scotia (Figure 9.12), and the amniotes have been found preserved inside some of the hollow stumps. *Hylonomus* (Figure 9.9) is one example. This may not be a freak of preservation. The amniotes may have lived inside hollow tree trunks, as little insectivorous mammals do today in tropical rain forests, or perhaps they sheltered in the hollow stumps or were washed into them in floods.

Whichever suggestion one prefers, I would argue that amniotes were living in the canopy forest in the Late Carboniferous. The Late Carboniferous coal forests

Figure 9.11 Reconstruction of a Carboniferous swamp. Trees were tall but often had shallow roots and weak structure, so they were frequently felled by storm and flood. A rich fauna of insects, spiders, and other arthropods lived in this ecosystem, and I think it is likely that small early reptiles lived as much in the tree tops as under them. This scene has much growth on and near the forest floor because it is near a natural open space — the water body in the middle distance. (Negative 333983. Courtesy of the Department of Library Services, American Museum of Natural History.)

included giant lycopod trees 30 m (100 ft) high (Figures 9.11), and dense tree ferns up to 10 m high formed an understory (Figure 9.13). The lycopod trees were fragile and shallow-rooted, and they may have been hollow in life, as many tropical trees are today. After storms or old age felled them (Figure 9.11), their hollow stumps sometimes remained standing (Figure 9.12).

Vertical climbing is easy with a small body size, so small Carboniferous vertebrates could have been tree dwellers, as many salamanders are today. Trees offer damp places in which to lay eggs, and rich insect life high in the canopy forest provides abundant food. Even today, salamanders (and spiders) are the top carnivores in parts of the Central and South American canopy forest. The rich fossil record of Late Carboniferous insects, scorpions, spiders, and amniotes may portray the ecosystem of the canopy rather than the forest floor.

Figure 9.12 In 1868, the Canadian geologist William J. Dawson described fossils of early reptiles preserved inside hollow tree stumps still standing erect in a fossil forest at Joggins, Nova Scotia. Dawson gave a vivid and sensible word picture of this ancient environment, and this engraving illustrated his account.

CARBONIFEROUS LAND ECOLOGY

Little is known yet about the land ecology of Early Carboniferous times; the East Kirkton fossils are the best-known tetrapod fauna from this time, and they are preserved in an unusual setting. All the Devonian and Carboniferous tetrapods so far discovered lived close to the equator.

The evidence is much better when we turn to the Late Carboniferous or Pennsylvanian. Late Carboniferous coalfields have been intensely studied for economic reasons, yielding a lot of information that gives us a good picture of the flora and global paleoecology of the time. Swamp forests in tropical lowlands were dominated by lycopods, and the vegetation and organic debris that were deposited in oxygen-poor water formed thick accumulations of peat, now compressed and preserved as giant coal beds stretching from the American Midwest to the Black Sea.

By the time all this carbon was buried, there may have been high levels of oxygen in the seas and atmosphere of the Carboniferous. The evolution of flight in insects, a very fuel-intensive activity, may have been made possible by a richly oxygenated atmosphere, but at the moment both the data and the inference are very speculative.

There were no herbivores among early land arthropods, possibly because of

Figure 9.13 Tree ferns such as *Psaronius* formed a dense understory several meters high in Carboniferous swamp forests, under the 30-m canopy forest. (After Morgan.)

lignin (Chapter 8). This universal substance in vascular plants is formed through biochemical pathways that include toxic substances which are often stored in cell walls and dead plant tissue. From the Silurian to the Late Carboniferous, lignin and its associated biochemistry probably made vascular plants invulnerable to potential herbivores.

But eventually, of course, both invertebrates and vertebrates made the breakthroughs that allowed direct herbivory. Bacteria and fungi can break down the toxins in dead plants, and it's possible that symbiosis (Chapter 3) with one or both allowed some animals to eat living plant material for internal enzyme-assisted digestion. Also, early land plants evolved larger sporangia and seeds (Chapter 8) that were very nutritious and low in toxins, and therefore more liable to attack by arthropods. Insects quickly evolved the anatomy to feed on the reproductive tissues of plants. Seeds may have evolved not only for better waterproofing of the embryo but also to deter insect predation.

The first insect is Devonian, but the dominant fact of early insect evolution is the explosive radiation of winged insects in the Late Carboniferous, about 325 Ma. Some had mouthparts for tearing open primitive cones, and their guts were sometimes fossilized with masses of spores inside. Others had piercing and sucking mouthparts for obtaining plant juices. Overall, it seems that leaf eating was rare among early insects; instead, they ate plant reproductive parts, sucked plant juices, or ate other insects. Gigantic dragonflies were flying predators on smaller arthropods; Late Paleozoic dragonflies were the largest flying insects ever to evolve, with wingspans up to 60 cm (Figure 9.14).

Explosive evolution had occurred among land-going invertebrates by the Late Carboniferous, much of it linked with the evolution of herbivory among insects: 137 genera of terrestrial arthropods are recorded from the Mazon Creek beds of Illinois, including 99 insects and 21 spiders, with millipedes present also. Most of the living groups of spiders had evolved by the Late Carboniferous, with only the sophisticated orb-web spiders missing. Centipedes were important predators.

Millipedes are important forest recyclers today, feeding on decaying plant material. They include flattened forms that squirm into cracks in dead wood and literally split their way in, reaching new food and making space for shelter and brood chambers at the same time. Carboniferous millipedes reached half a meter in length, and a giant relative, *Arthropleura*, reached 2.3 m (7 ft) long, and 50 cm

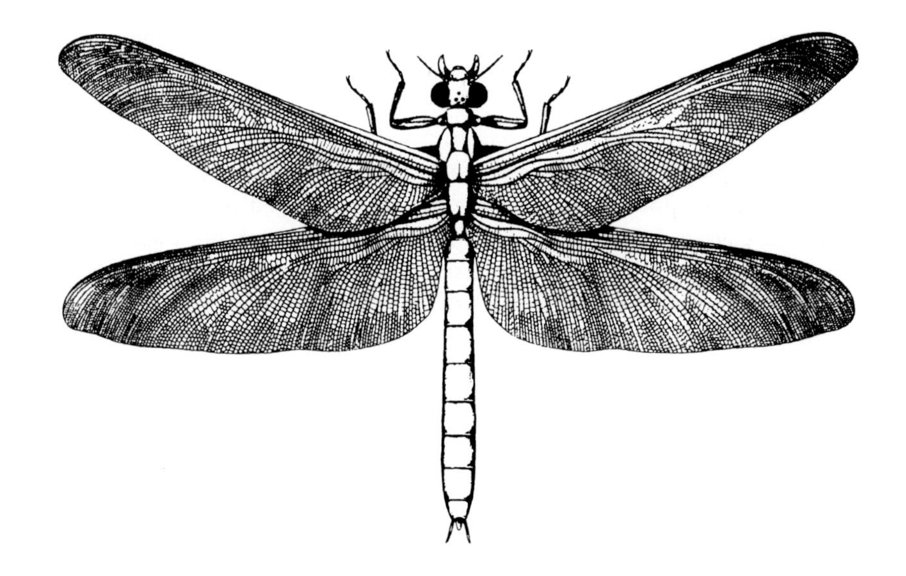

Figure 9.14 A giant dragonfly from the Late Carboniferous. This one had a wingspan of 60 cm (2 ft). (Reconstruction after Handlirsch.)

(18 in) across. The gut contents of *Arthropleura* suggest that it ate the woody central portion of tree ferns.

Most early vertebrates, however, were carnivorous. Fishes, small tetrapods, and giant dragonflies all ate insects, and in turn were eaten by larger carnivores. For land vertebrates, then, Carboniferous swamp plants provided shelter and cover, but not food: herbivory by vertebrates evolved late, as we shall see in the next chapter.

Further Reading

Beerling, D. J., and R. A. Berner. 2000. Impact of a Permo-Carboniferous high O_2 event on the terrestrial carbon cycle. *Proceedings of the National Academy of Sciences* 97:12428–32.

Bickford, D. 2002. Male parenting of New Guinea froglets. *Nature* 418: 601–2.

Carroll, R. L. 2001. The origin and early radiation of terrestrial vertebrates. *Journal of Paleontology* 75: 1202–13. [Nice review.]

Clack, J. A. 1998. A new Early Carboniferous tetrapod with a mélange of crown group characters. *Nature* 394: 66–9. [*Eucritta.*]

Clack, J. A. 2002. An early tetrapod from 'Romer's Gap'. *Nature* 418: 72–6, and comment, 35–6. [*Pederpes.*]

Clack, J. A. 2002. *Gaining Ground: The Origin and Evolution of Tetrapods.* Bloomington: University of Indiana Press. [This is not a light-weight book in size or level, but it is very well written, and gives a complete picture of research up to 2002.]

Duellman, W. E., and L. Trueb. 1986. *Biology of Amphibians.* New York: McGraw-Hill.

Heatwole, H., and R. L. Carroll. (eds). 2000. *Amphibian Biology.* Vol. 4. *Palaeontology.* Chipping Norton, Australia: Surrey Beatty and Sons.

Labandeira, C. C. 1998. Early history of arthropod and vascular plant associations. *Annual Reviews of Earth & Planetary Sciences* 26: 329–77. [Concentrates most on Late Paleozoic associations, but reviews the whole story lightly.]

Monastersky, R. 1999. Out of the swamps: how early vertebrates established a foothold—with all 10 toes—on land. *Science News*, May 22, 1999. [*Casineria.*]

Paton, R. L., et al. 1999. An amniote-like skeleton from the Early Carboniferous of Scotland. *Nature* 398: 508–13. [*Casineria.*]

Perry, D. 1986. *Life Above the Jungle Floor.* New York: Simon and Schuster. [An account of the fauna inside hollow tree trunks reads like a visit to the Pennsylvanian, except for the bats. The section on coal forests is good, the sections on dinosaurs are not so sound.]

Ruta, M., et al. 2003. Early tetrapod relationships revisited. *Biological Reviews* 78: 251–345. [Latest analysis; huge reference list.]

Shear, W. A. 1991. The early development of terrestrial ecosystems. *Nature* 351: 283–9.

Smithson, T. R. 1989. The earliest known reptile. *Nature* 342: 676–8. [*Westlothiana.* It is not a reptile and may not even be an amniote.]

CHAPTER TEN

Early Amniotes and Thermoregulation

The radiation of amniotes was probably encouraged by ecological opportunities away from water bodies. But away from water, microenvironments have lower humidity, more exposure to solar radiation and to colder nights, less vegetation and shelter, and greater temperature fluctuations. Some degree of temperature control or thermoregulation is needed to live in such habitats, and the varied responses of reptiles to environmental and physiological challenges are major themes in their evolutionary history.

Tetrapods emerged onto land and the first amniotes evolved in warm, humid, tropical regions along the southern shores of the great northern continent Euramerica. Life away from such swamps and forests demands adaptations for dealing with seasons, where temperature, rainfall, and food supply vary much more and are less predictable than in the tropics. In many ways, such challenges to early land vertebrates were simply extensions of the problems involved in leaving the water. In this chapter we shall follow the early history of amniotes and discuss the adaptations that allowed them their great terrestrial success.

THE AMNIOTE RADIATION

Amniotes came to be dominant large animals in all terrestrial environments in Permian times. The radiation probably began in Euramerica, because hardly any land vertebrates are known from Siberia, from East Asia, or from the whole of Gondwana before Middle Permian times. Seas and mountain ranges may have blocked land migration; alternatively, problems of thermoregulation may have confined land vertebrates to the tropics of Euramerica until the Middle Permian. The invasion of other continents and/or climates was accompanied by a spectacular evolution of varied body types. Since amniotes rather than amphibians radiated so successfully, perhaps it was their solution to thermoregulatory problems that allowed them to invade regions in higher latitudes.

Three major groups of amniotes had diverged by the Late Carboniferous and Early Permian (Figure 10.1). The earliest amniotes had **anapsid** skulls (they had no openings behind the eye) (Figure 10.2a). This character was inherited from fishes and early tetrapods. The two major amniote clades have derived or advanced skull types, in which there are one or two large openings behind the eye socket. **Synapsids** (with one skull opening behind the eye socket) (Figure 10.2c), diverged first, from an anapsid amniote ancestor that we have not yet identified. Synapsids dominated Late Paleozoic land faunas. They include the Late Paleozoic pelycosaurs and their descendants, the therapsids and mammals. Synapsids never evolved the water-saving capacity to excrete uric acid rather than urea, a character that all other surviving amniote groups share (Figure 10.1).

Diapsids are amniotes with two skull openings behind the eye socket (Figure 10.2b). They include the dominant land-going groups of the Mesozoic (including dinosaurs and pterosaurs) and all living amniotes except mammals (reptiles and birds). Turtles have no skull openings, so are technically anapsid. But their ancestors were most likely diapsids that lost the two skull openings. Turtles evolved in Late Triassic times from diapsid ancestors we have not yet identified.

The earliest well-known diapsid is *Petrolacosaurus* (Figure 10.2), which looked like and probably lived like a lizard (but then the earliest amniotes did too: see Figure 9.10). Compared with later diapsids, *Petrolacosaurus* had a heavy earbone, the stapes, that could not conduct airborne sound. As in most early tetrapods, the massive stapes probably transmitted ground vibrations through the limb bones to the skull.

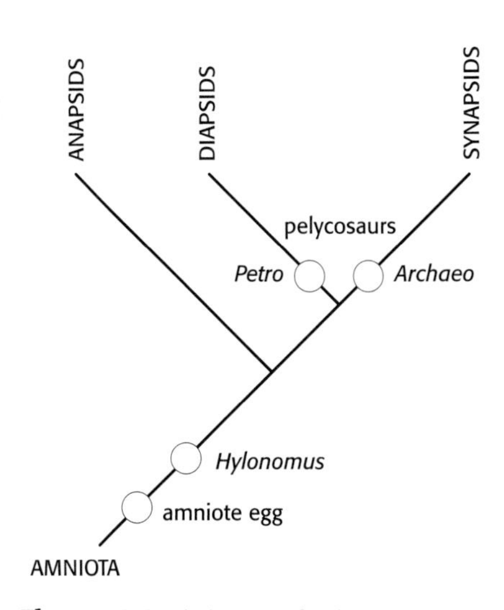

Figure 10.1 Phylogram of early amniotes. Two lineages diverged quickly from the basal anapsid amniotes. The earliest well-described amniote is *Hylonomus. Petro* is *Petrolacosaurus*, the earliest well-known diapsid; *Archaeo* is *Archaeothyris*, the earliest synapsid.

PELYCOSAURS

Although the first diapsid evolved in the Late Carboniferous, the major radiation of diapsids took place much later, in the Triassic (Chapter 11). The dominant Late Carboniferous and Permian reptiles were synapsids, including five of the six other amniotes found with *Petrolacosaurus*.

Early synapsids are classed together as **pelycosaurs**, the most famous of which are the sail-backed Permian forms such as *Dimetrodon* (Figure 10.2c). They were already the most important group of fully terrestrial tetrapods. Over 50% of Late Carboniferous amniotes were pelycosaurs, and over 70% of Early Permian amniotes. After that they appear to decline, but only because one clade of them had evolved into the dominant therapsids of the Late Permian. Despite their variety, early pelycosaurs are rare. They may have first evolved and lived in habitats that were drier than those favored by earlier tetrapods. Later they became much more abundant as fossils, and widespread geographically (Figure 10.3).

Archaeothyris, found in the fossil tree trunks of Nova Scotia, was one of the earliest pelycosaurs. It was small, perhaps 50 cm (20 in) long including the tail, and it was lizard-like in general appearance. But it had the characteristic skull structure of

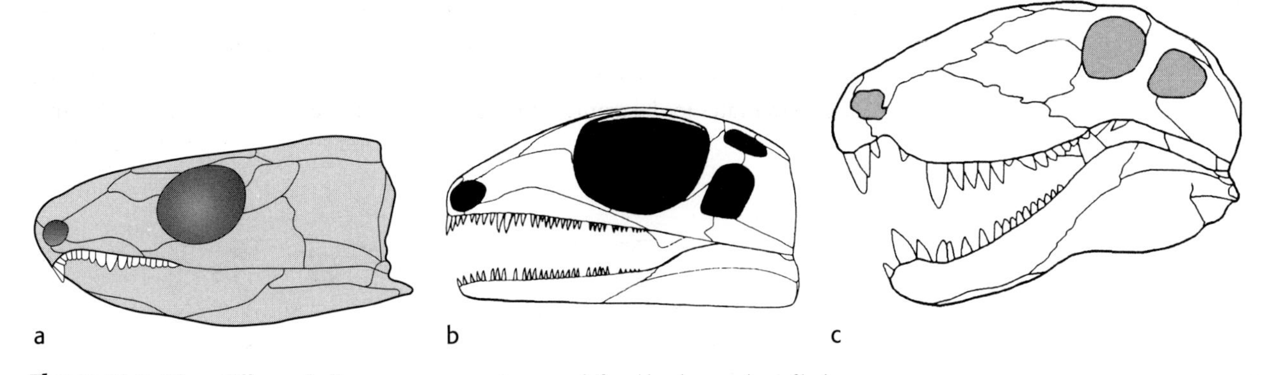

Figure 10.2 Three different skull types among amniotes are defined by the number of holes in the skull behind the eye socket. (a) Anapsid, represented by *Captorhinus*. (b) Diapsid, represented by *Petrolacosaurus*. (c) Synapsid, represented by *Dimetrodon*.

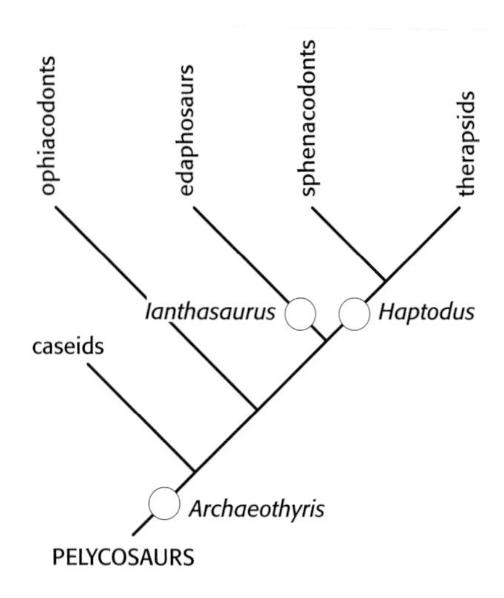

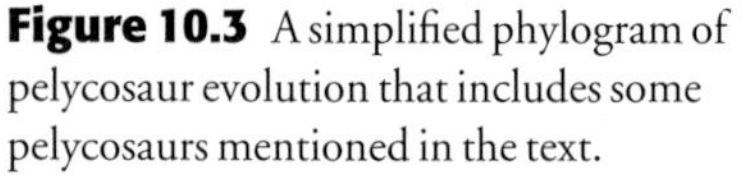
Figure 10.3 A simplified phylogram of pelycosaur evolution that includes some pelycosaurs mentioned in the text.

synapsids, with a single opening behind the eye socket instead of the double opening of diapsids. The edges of this opening gave secure attachment for powerful jaw muscles.

PELYCOSAUR BIOLOGY AND ECOLOGY

Locomotion

Pelycosaurs are well enough known that we can reconstruct how they walked. The massive front part of the body was supported by a heavy, sprawling forelimb. The lighter hind limb had a greater range of movement, although it was also a sprawling limb. There was no well-defined ankle joint, and the toes were long and splayed out sideways as the animal walked. Thus, the feet provided no forward thrust but simply supported the limbs on the ground. The forelimbs were entirely passive supports that prevented the animal from falling on its face, while the hind limbs provided all the forward thrust in walking with powerful muscles that rotated the femur in the hip joint.

Think of two children playing "wheelbarrow." The propulsion and steering are both from the rear, and the wheelbarrow is stable only as long as the leading child stays stiff. Spinal flexibility is important to many swimming animals, particularly those that actively pursue fishes. But in pelycosaurs, a strong stiff backbone prevented the body from collapsing in the middle under its own weight and allowed thrust from the hind limb to be converted directly into forward motion. Therefore, most pelycosaurs were predominantly terrestrial animals. If they swam, they were slow swimmers that hunted by stealth rather than speed. Only one pelycosaur (*Varanosaurus*) had a really flexible spine, and it may have been almost entirely aquatic.

Carnivorous Pelycosaurs

Early pelycosaurs were all carnivorous: they all have the pointed teeth and long jaws of predators. Two groups remained completely predatory. **Ophiacodonts** included the first small pelycosaur, *Archaeothyris*, from Nova Scotia, but later forms became quite large. *Ophiacodon* itself was 3 m (10 ft) long and probably weighed over 200 kg (450 pounds). Many ophiacodonts have long-snouted jaws (Figure 10.4), with many teeth set in a narrow skull. The hind limbs tended to be longer than the forelimbs.

Ophiacodonts may have hunted fishes in streams and lakes of the swamps and deltas of the Late Carboniferous and Permian, although they were perfectly capable of walking on drier, higher ground, and like crocodiles, their prey may well have

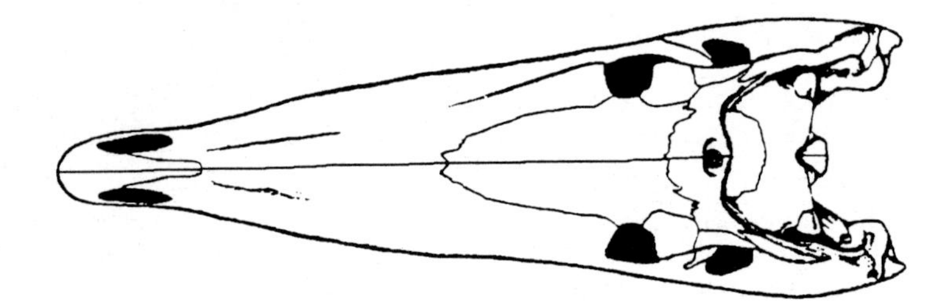

Figure 10.4 The skull of the crocodile-sized pelycosaur *Ophiacodon*, seen from above, is long and pointed, as in many fish-eating animals. (After Romer and Price.)

included terrestrial animals coming down to the water to drink. Their general lack of spinal flexibility (except in *Varanosaurus*) may suggest that they were slow swimmers, possibly eating more tetrapods than fishes.

Sphenacodonts were specialized carnivores on land. Many of their skull features betray the presence of very strong jaw muscles, and the teeth were very powerful. They were unlike typical early amniote teeth in that they varied in shape and size and included long stabbing teeth that look like the canines of mammals. The sphenacodont body was narrow but deep, and the legs were comparatively long. Both of these characters suggest that sphenacodonts were reasonably mobile on land.

The earliest sphenacodont was *Haptodus* (Figures 10.3; 10.5). It was a little less than a meter long and fairly lightly built. Similar forms existed throughout the Permian, but later sphenacodonts were much larger. The group is best known from spectacular fossils of *Dimetrodon* (Figure 10.6). *Dimetrodon* had vertebrae extended into spines projecting far above the backbone. (I'll discuss these structures later.)

Evolution within carnivorous pelycosaurs reflected their prey capture. The jaws slammed shut around the hinge, with no sideways or front-to-back motion for chewing. With this structure, a long jaw made it easier to take hold of prey, but the force exerted far from the hinge was not very great. Small prey could perhaps be killed outright by slamming the jaw on them.

In ophiacodonts, which may have hunted in water for fish, the difficult part of feeding would have been seizing the prey; their teeth were subequal in length in a long, narrow jaw. Most fisheaters swallow their prey whole.

In sphenacodonts, which were terrestrial carnivores, the head was bigger and stronger. Long, stabbing teeth were set in the front of the jaw (Figure 10.2c). Struggling prey could be held between the tongue and some strong teeth set into the

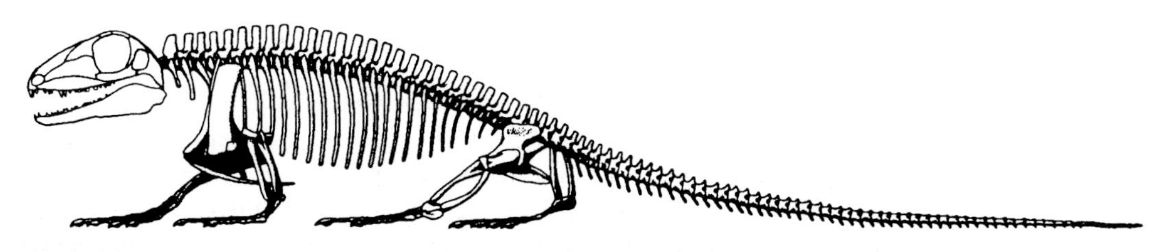

Figure 10.5 *Haptodus*, the earliest known sphenacodont, a little under a meter long (3 ft).

Figure 10.6 The skeleton of the most famous Permian synapsid, *Dimetrodon*, a carnivorous pelycosaur with extended vertebrae that formed a "sail" on its back. (Negative 315862. Photo by Julius Kirschner. Courtesy of the Department of Library Services, American Museum of Natural History.)

palate, and could be subdued by powerful crushing bites from the teeth at the back of the jaw. Robert Carroll suggested that the success of pelycosaurs in the Carboniferous and Permian, compared with diapsids, was due to their massive jaw muscles, which were strong enough to hold the jaws steady against the struggles of large prey. The carnivorous pelycosaurs thus could become large predators, not simply small insectivores.

VEGETARIAN PELYCOSAURS

Carroll's suggestion cannot be the whole story, because there were also vegetarian pelycosaurs. Caseids and edaphosaurs were the first abundant large terrestrial animals, and were among the first terrestrial herbivores. Caseids and edaphosaurs had similar body styles, presumably because they were similar ecologically. They had about the same range of body size as the carnivorous sphenacodonts, but they had smaller, shorter heads that gave more crushing pressure at the teeth. There were no long canines, and the teeth were short, blunt, and heavy. In addition, smoothing of the bones at the jaw joint allowed the lower jaw to move backwards and forwards slightly, grinding the food between upper and lower teeth. Caseids ground their food between tongue and palatal teeth, while edaphosaurs had additional tooth plates in their lower jaw that they used to grind food against palatal teeth. Vegetation is low-calorie food compared with meat, so herbivores need a large gut to contain a lot of food. As one would expect, the bodies of all these vegetarians pelycosaurs were wide to accommodate a large gut. The limb bones were short but heavy.

Caseids were more numerous than edaphosaurs. They included *Cotylorhynchus*, which was over 3 m long (10 ft), and weighed over 300 kg (650 pounds) (Figure 10.7). Caseids had small heads for their size, which perhaps implies that they did not chew very much, and perhaps had powerful digestive enzymes or gut bacteria to help break down plant cellulose.

Edaphosaurs are best known from *Edaphosaurus* itself (Figure 10.8), which had vertebrae extended into spines rather like *Dimetrodon*. The earliest edaphosaur *Ianthasaurus*, however, was not a vegetarian but a small carnivore found at Garnett. It had a small sail on its back, sharp pointed predatory teeth, and probably ate insects. We find vegetarian pelycosaurs only in the Early Permian, and they must have evolved herbivory as a new ecological way of life. We shall find the intermediate edaphosaurs some day, and perhaps they will show us how herbivory evolved in these early synapsids. Until that time we can only make some guesses.

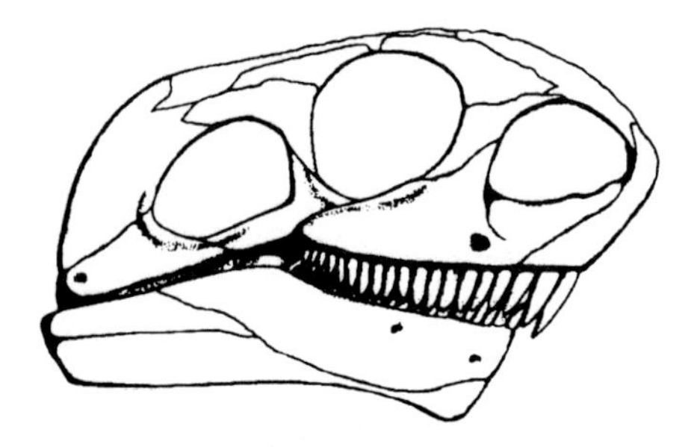

Figure 10.7 A typical caseid, *Cotylorhynchus*, a vegetarian pelycosaur. The teeth are sharper than one might expect, suggesting that food was macerated in the gut or gizzard, not chewed in the mouth. In keeping with this fact, the skull is small for the body size, only about 20 cm (8 in) long. The nostrils are very large, and one could speculate about the reasons for that. (After Romer.)

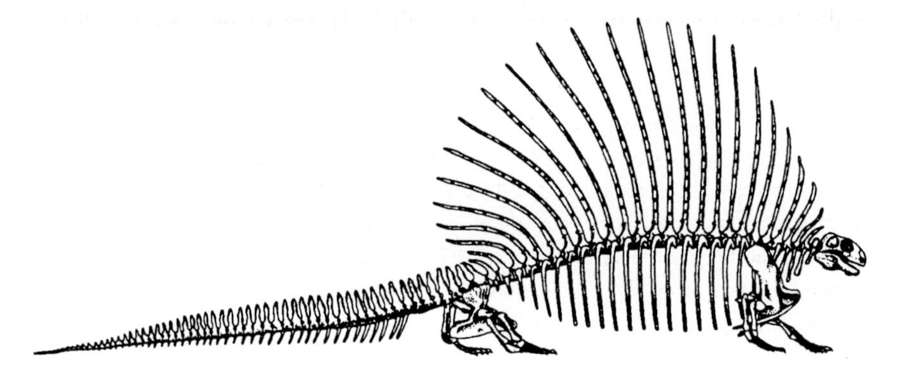

Figure 10.8 *Edaphosaurus*, a vegetarian pelycosaur that independently evolved a thermoregulatory sail comparable with that of *Dimetrodon*. Total length of body and tail was about 3.5 meters (11 ft). (Skeletal outline after Romer.)

HOW DOES HERBIVORY EVOLVE IN TETRAPODS?

Most plant material is difficult to digest. Vertebrates can break down cellulose only if they chew it well and have some way of enlisting fermenting bacteria as symbionts (Chapter 3) to aid digestion. Living vegetarians do this: cattle and many other grazers have bacteria in a stomach compartment called the rumen (so they are called ruminants). Horses and rabbits have gut bacteria lower in the digestive tract. Any vertebrate that begins to eat comparatively low-protein plant material must process large volumes of it, and so must have a rather large food intake at a rather large body size. Some plant material is high in protein or sugar, especially the reproductive parts, but only a small animal can selectively feed on plant parts.

In other words, there are only two possible evolutionary pathways toward herbivory. One of them begins with animals that are small, active, and selective in their food gathering, so high-calorie foods such as juices, nectar, pollen, fruits, or seeds can be collected from the plant. Examples today are small mammals, hummingbirds, and insects. If an animal enlists gut bacteria as symbionts, however, the diet can contain more and more cellulose and a larger vegetarian can evolve, as in many mammal groups, including leaf-eating monkeys and gorillas. Large birds can also be herbivores: the extinct moas of New Zealand are good examples.

The other pathway begins at rather large body size with rapid and rather indiscriminate feeding, possibly omnivory, so that a large volume of low-calorie food can be processed. Bear-like mammals are examples of a group in which some members have evolved away from a carnivorous way of life toward omnivory and then to a completely vegetarian diet, as in pandas. Later, as more efficient chewing and digestive systems are evolved, large vegetarians can survive at midsize.

Because vegetarianism depends so much on body size, diets must change with growth. Most living reptiles and amphibians change their diet as they grow. Food requirements and opportunities change as they reach greater size and can catch a different set of prey. Among living reptiles, small and young iguanas are carnivorous or omnivorous, while large iguanas are largely vegetarian but take meat occasionally. Living amphibians today are almost all small and carnivorous. I suggest that herbivory evolved among Late Paleozoic tetrapods only after they evolved to large body size.

The giant Carboniferous coalfields (Chapter 9) contain rock sequences in which many beds consist almost entirely of carbon formed from plant debris such as leaves, trunks, roots, spores, and pollen, plus half-rotted and unrecognizable fragments. The coalfields of the Northern Hemisphere, particularly those in the Late Carboniferous (Pennsylvanian), formed the energy basis for the Industrial Revolution in western Europe and North America and still provide large quantities of fuel for many industrial countries (Figure 10.9). Carboniferous coalfields have been studied so intensively that we can reconstruct their plant communities very well; we can tell, for example, that plants had spread away from the rivers and lakes into so-called uplands—probably not very high above sea level but with distinctly drier air and soil than the lowland swamps.

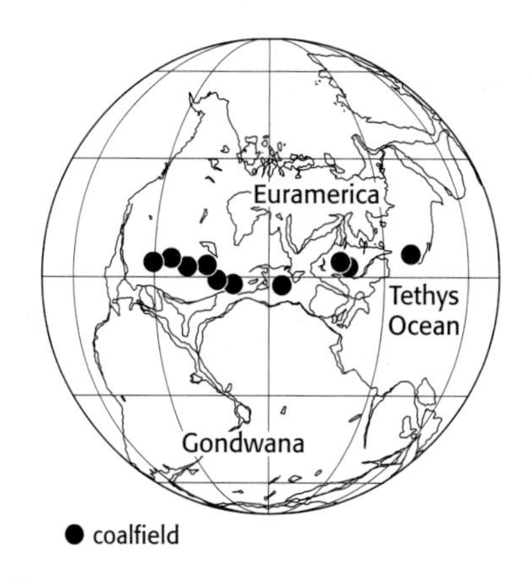

Figure 10.9 The paleogeography of the Carboniferous coalfields, which formed in swampy tropical forests. Each black dot represents a major coalfield that has survived to be exploited today.

The rich floras provided a food base for insects at first, but large terrestrial herbivores appeared in the latest Carboniferous and Early Permian of Euramerica. The advanced anthracosaur *Diadectes*, for example, ranged up to as much as 4 m (13 ft) long as an adult, and synapsids of this size were also common. Large herbivores appeared at the same time as a major change in plants, when upland plants replaced the coal swamp forests.

Why were tetrapods relatively slow to evolve herbivory? First, because the wet tropical forest in which they evolved is a poor habitat for ground dwelling herbivores. As in today's tropical forests, most leaves were in the canopy, and leaf litter was broken down quickly by fungi and arthropods. The first vertebrate herbivores could not have found much green material on the floor of the coal forests (Figure 9.11). Vertebrates could not have evolved herbivory until they could survive well on the forest margin, away from the watery habitats most likely to be preserved.

Second, any large-bodied vegetarian eats large volumes of low-calorie plant material and needs gut bacteria to help digest the cellulose. Gut bacteria work well only in a fairly narrow range of temperature, so an additional requirement for the first successful large-bodied vegetarians was some kind of thermoregulation. Before we can attack this question, we should look at the processes of thermoregulation in living vertebrates.

THERMOREGULATION IN LIVING REPTILES

Body functions are run by enzymes, which are sensitive to temperature. Other things being equal, enzymes work best at some optimum temperature; any other body temperature implies a loss of efficiency—in digestion, in locomotion, in reaction time, and so on. Birds and mammals have a sharp peak of efficiency that drops off radically with a small rise or fall in body temperature. Reptiles are called cold-blooded, but in fact they take on the temperature of their surroundings, so can be warm or cold. Their bodies can function over quite a range of internal temperature, but they also have an optimal temperature, and reptiles try to control it at that level by **behavioral thermoregulation**.

Generally, reptiles try to maintain their body temperatures at the highest level that is consistent with safety and cost. Although it takes energy to stay warm, the higher activity levels that are possible at higher temperatures give greater hunting or foraging efficiency, greater food intake, faster digestion, and faster growth: remember the section on basking in Chapter 8. As long as the climate is warm and food supply is abundant enough to fuel a reptile, thermoregulation that produces or maintains warm body temperature gives a net gain in reproductive rate and so is selectively advantageous. The same principles should apply to all cold-blooded vertebrates.

Body size is a vital factor in thermoregulation. Small bodies have a low mass with a relatively large surface area. Small reptiles bask in the sun, sit in the shade, hide in burrows or in leaf litter, or exercise violently (often with push-ups) to change their body temperatures. Their small mass allows them to respond quickly to temperature changes by behavioral means, giving them sensitive control over their body processes. Large reptiles have a natural resistance to temperature change because of their mass: it takes a long time to heat them up or cool them down (just as it takes a long time to boil a full kettle of water).

Behavioral thermoregulation is more energy-consuming and much less responsive for larger reptiles than for smaller ones. So large reptiles today live in naturally mild tropical climates with even temperatures day and night and season to season (like the large monitor lizards of Indonesia, Australia, and Africa), or they live near or in water, which buffers any changes in air temperature (like crocodiles and alligators, which even then are never found far outside the tropics). There are no large lizards at high latitudes.

Thermoregulation in Pelycosaurs

The spectacular pelycosaurs *Dimetrodon* and *Edaphosaurus* were not closely related, but they were both large (over 3 m long). They both independently evolved very long spines on some of their vertebrae, forming a row of long vertical spines along the backs of these creatures (Figures 10.6; 10.8). In life, the bones were covered with tissue to form a huge vertical sail. Most people think that the sail was used for thermoregulation.

Here is the simplest version of the story. *Dimetrodon* and *Edaphosaurus* were too large to hide from temperature fluctuations (in a crevice or tree-stump or burrow, for example). Each probably used its sail to bask in the early morning and the late afternoon, turning its body so that the sail intercepted the sunshine. By pumping blood through the sail, it could collect solar heat and transfer it quickly and efficiently to the central body mass (solar panels work this way to heat water). Once warm and active, the pelycosaur would face no further problem unless it overheated. It could shed heat from the sail by the reverse process, turning the sail end-on to the sun. At night, heat would be conserved inside the body by shutting off the blood supply to the sail (Figures 10.6; 10.8; 10.10).

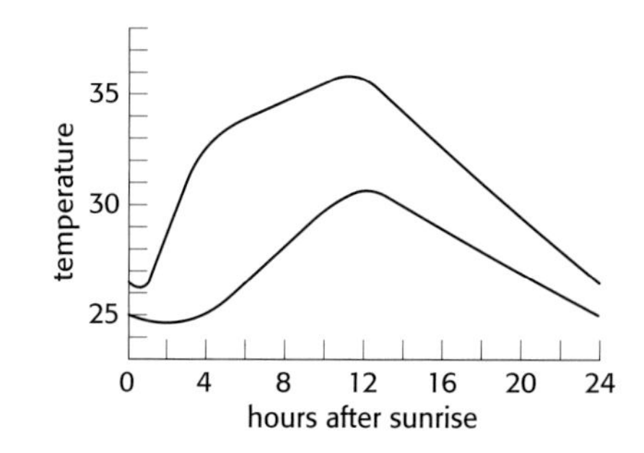

Figure 10.10 Calculations of the thermoregulatory capacity of a large *Dimetrodon* with (upper curve) and without (lower curve) its sail help to confirm why the sail evolved. It allowed the animal to warm up quickly in the morning and to reach a high body temperature close to ours. I mistrust this particular model when it suggests that *Dimetrodon* would cool faster at night with a sail than without, and that its body temperature would fluctuate more with a sail than without. (Data from Haack, 1987.) It is astonishing that this question has not been revisited since Haack's publication.

The sail, as an add-on piece of solar equipment, allowed rapid and sensitive control over body temperature. Enzyme systems could have been fine-tuned to work at high biochemical efficiency within narrow temperature limits, and the animal could have foraged even in environments where air temperatures fluctuated widely. The activity levels, locomotion, and digestive systems of *Dimetrodon* and *Edaphosaurus* would all have improved. Smaller reptiles that lived alongside them would have been able to heat up quickly in the morning, simply because they were smaller, and it would have been important for the large animals to be equally active at that time—*Dimetrodon* for effective hunting and *Edaphosaurus* for effective escape or defense.

This solar scenario is usually cast in terms of an arms race between the two giants, with *Dimetrodon* warming up in the morning in order to chase *Edaphosaurus*, and *Edaphosaurus* warming up to escape. But that isn't necessarily the whole story. *Edaphosaurus* would have been an easy prey for a group of midsized predators if it were paralyzed by cold, and *Dimetrodon* no doubt took other prey. Furthermore, *Edaphosaurus* would have needed thermoregulation if it had gut bacteria to help it digest plant material.

Some pelycosaurs did not have a sail at all, and the small pelycosaur *Ianthasaurus* had only a small sail. Young *Dimetrodon* and *Edaphosaurus* had small versions too. The area of the sail was clearly related to body size, which makes sense only if pelycosaurs were behavioral thermoregulators like living lizards. The heat inertia of a large body means that it warms and cools slowly, but more energy is needed for a high activity level. Birds and mammals burn large amounts of food in a built-in, high, internal metabolism that allows them to be consistently "warmblooded," but for a behavioral thermoregulator, energy must come from outside. So some large pelycosaurs evolved sails, add-on solar technology for effective thermoregulation, making them super-synapsids perhaps, but not birds or mammals.

In 1996, S.C. Bennett made two very astute observations which complicate the simple story. First, *Edaphosaurus* has knobs on the bony spines on its sail. By testing a model in a wind tunnel, Bennett found that the knobs would have generated eddies in breezes blowing past the sail. This would have no effect on solar collection by the sail, which is a radiation effect, or on its cooling by radiation, but it would make it a better cooling device by increasing *convection* over the skin. (Moving air cools bodies better than it heats them, as we all know from personal experience in breezes and winds.)

Bennett also found that *Edaphosaurus* had a smaller sail than a *Dimetrodon* of the same body size, implying that the *Edaphosaurus* sail was more efficient. The only way in which it differs is in the knobs, and the only way it looks more efficient is in its cooling effect. This implies even more strongly that the major function of the sail (in both animals) was cooling.

Bennett also thinks that early morning basking was probably inefficient in both animals (though it was better than nothing!). Early in the morning, their circulation systems would have been sluggish. The deep body was cool, the heart would have pumped slowly, and the blood would have been thick. They may not have been able to pump blood round the sail fast enough to transfer all the solar energy that was shining on the sail. However, once they were hot, the heart would have pumped strongly, and the blood would have flowed more freely, allowing them to cool efficiently by pumping hot blood from their overheated bodies round the sail to be cooled in the breeze. Altogether, Bennett sees the major problem for these big pelycosaurs as overheating in the heat of the day.

If pelycosaurs with sails thermoregulated, then other pelycosaurs (caseids, for example) probably thermoregulated too, in behavioral ways that left no traces on the skeleton. It is easy to imagine that the fine-tuning of the internal enzyme system that accompanied the first attempts at solar thermoregulation encouraged the evolution of internal control systems, such as those that varied blood flow to the surface. After the Permian, as advanced pelycosaurs evolved into therapsids and the rest disappeared, we see little sign of thermoregulatory devices preserved in the skeleton. There is indirect evidence, however, that therapsids had limited thermoregulation; that evidence is presented later in this chapter.

PERMIAN CHANGES

Shifting continental geography resulted in major biogeographic changes in the Permian (Chapter 6). The large southern continent Gondwana moved north to collide with Euramerica, and by the Middle Permian these blocks formed a continuous land mass. A little later, Asia crashed into Euramerica from the east, buckling up the Ural Mountains to complete the assembly of the continents into the global supercontinent Pangea (Figure 6.3).

These tectonic events put an end to the wet climates that had fostered the system of large lakes, swampy deltas, and shorelines along the south coast of Euramerica, where the Carboniferous coal forests had flourished (Figure 10.9).

Worldwide, Permian floras were dominated by gymnosperms, mostly gingkoes, conifers, and cycads. Conifers had evolved in Carboniferous times. Compared with other Late Paleozoic plants, they were better adapted for drought resistance, and they probably evolved in much drier uplands, because they are rare on lowland floodplains. Tree-sized lycopods disappeared from coal swamps in the Late Carboniferous as climates became drier. As the drying trend continued into the Permian, conifers extended into lowlands at the expense of the swamp plants that had dominated Carboniferous floras.

Total plant diversity dropped in this turnover. The drop was spread over a long time, however. It was not catastrophic or even abrupt. It was a response to changing climate, perhaps based on geographic changes, and it can be explained by normal evolutionary processes.

THE INVASION OF GONDWANA

Geological evidence from Gondwana shows that a huge ice sheet was centered on the South Pole (Figure 10.11) in Late Carboniferous and Early Permian times. Ice sheets moving northward scoured rock surfaces and deposited stretches of glacial debris on a continental scale.

The continental collisions that formed Pangea allowed land animals to walk into Gondwana. But pelycosaurs, which had always had a narrow tropical distribution, remained in the tropical areas, much reduced in diversity; Late Permian pelycosaurs are found only in North America and Russia. Instead, Gondwana was invaded by their descendants, the synapsid reptiles called therapsids. Therapsids had larger skull openings than pelycosaurs did, indicating that they had more powerful jaws. The whole skull was strengthened and thickened, and the jaw always had prominent canine teeth. Therapsids also had much better locomotion than pelycosaurs.

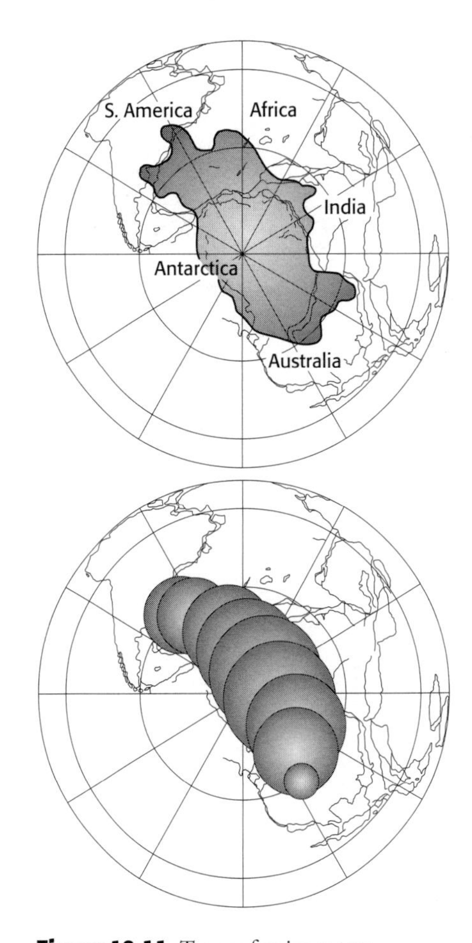

Figure 10.11 Traces of an ice age are widespread over the continents that formed Gondwana in the Late Paleozoic. The edges of the ice sheets can be marked with confidence, but not the precise time at which the ice reached those edges. The simplest explanation is that an ice cap formed over the South Pole of the time. Over millions of years, Gondwana slowly drifted over the pole, and the edges of the migrating ice cap left an irregular trace. The traces ended as the last little Permian ice cap melted.

Thermoregulation in Therapsids

Dimetrodon, with its sail and its great skin surface area, must have been able to maintain thermoregulation over a reasonable range of external environmental variation, and must have lived in an environment where such thermoregulation was both required and possible. It must also have lost very little water through its skin.

These adaptations of a sphenacodont may have been preadaptive for the therapsid invasion of more challenging habitats. *Dimetrodon* itself survived a climatic change in the Permian of Texas that eliminated most other pelycosaurs from the region, possibly because it could tolerate greater temperature ranges and drier conditions than they could. But in the end *Dimetrodon* and the other pelycosaurs became extinct, to be replaced in tropical latitudes by diapsid reptiles rather than synapsids (Chapter 11).

Therapsids evolved from sphenacodont pelycosaurs, and lived mainly at middle or high latitudes rather than in tropical regions: almost all therapsid clades evolved in Gondwana and spread outward from there. It's not clear whether they were outcompeted in the tropics by other reptile groups. The restriction, or adaptation, of therapsids to drier and more seasonal habitats may have encouraged their success in southern Gondwana, away from the tropics and toward higher latitudes. Thousands of specimens of therapsids have been collected from Late Permian rocks in South Africa, for example, and someone with time on his hands estimated that these beds contain about 800 billion fossil therapsids altogether! There are literally dozens of species, and we have a good deal of evidence about their environment. The glaciations were over and vegetation was abundant, with mosses, tree ferns, horsetails, true ferns, conifers, and a famous leaf fossil, *Glossopteris*. The climate may well have been mild considering that South Africa was at 60° S latitude. But it must have been seasonal, so the supply of plant food would have been seasonal too.

When we find large extinct synapsids at such high latitudes, we can be reasonably sure that they were unlike living reptiles in their metabolism. Their thermoregulation must have been more sophisticated than simple behavioral reactions. We know that mammals evolved from late therapsids in the Late Triassic. Did Permian therapsids already have a mammalian style of thermoregulation, with automatic internal control, a furry skin, and a high metabolic rate? We have too little evidence to say, but the scrappy evidence available suggests that the answer is no.

Therapsids had sprawling forelimbs and did not move very efficiently compared with later reptiles and mammals. Many of them lacked the secondary palate that

Figure 10.12 The body plan of therapsids looks as if it was good for conserving heat. This animal is *Keratocephalus*, a rhino-sized tapinocephalian from Gondwana. (The skeleton is displayed in the university museum in Tübingen, Germany.)

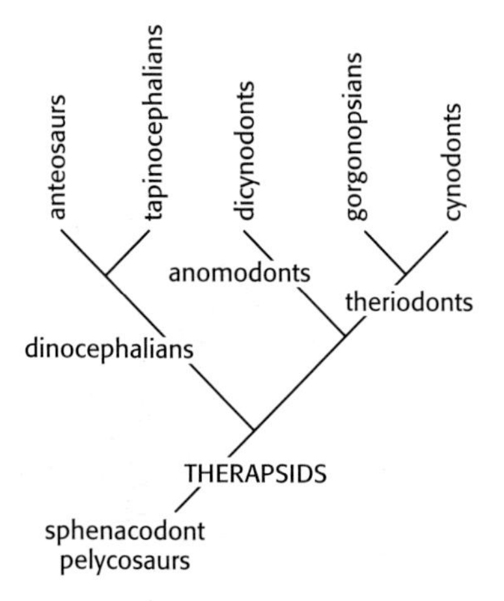

Figure 10.13 One possible hypothesis (phylogram) of the evolution of the major groups of therapsids. Much simplified from Rubidge and Sidor (2001). The only surviving clade of therapsids is the cynodonts: a group of advanced cynodonts evolved toward mammals in the Late Triassic.

would have allowed them to chew and breathe at the same time. Unlike other reptiles, many therapsids had short, compact, stocky bodies, with short tails: good adaptations for conserving body heat if not generating it (Figure 10.12). They may also have had hair or thick hides for conserving heat, but there is no way of detecting that from their fossils. All this suggests that therapsids did not have a large energy budget. They may have had some moderate form of internal temperature control, but nowhere near as good as that in living mammals.

THERAPSID EVOLUTION

Therapsids evolved quickly and are present in both Russia and South Africa in Late Permian rocks: in other words, they spread globally. Their evolutionary history has not yet been properly worked out, and the classification is still changing rapidly. Figure 10.13 shows one hypothesis about therapsid evolution.

The earliest therapsids are poorly known, though there is little doubt that they evolved from one lineage of sphenacodont pelycosaurs, as relatively small- to medium-sized carnivores. They are much outnumbered by the more advanced dinocephalians, the earliest abundant therapsids.

Dinocephalians

Dinocephalians moved much better than pelycosaurs. Their spine was quite stiff, and limb length and stride length were longer than in pelycosaurs. The forelimbs were still sprawling, but the hind limbs were set somewhat closer to the vertical, accentuating the wheelbarrow mode of walking we have already described for pelycosaurs.

Dinocephalians became very large. They ranged from Russia to South Africa. They had large skulls and, like all therapsids, they had strong canines. They also had well-developed incisors that seem to have been both efficient and important in feeding. Dinocephalians had unusual front teeth: their upper and lower incisors, and sometimes the canines too, interlocked along a line when the mouth closed, forming a formidable zigzagged array that would bite off pieces of food as well as piercing and tearing (Figure 10.14).

The earliest dinocephalians, the **anteosaurs**, were carnivores with skulls up to a meter long. Like sphenacodonts, they killed prey mainly by slamming the long, sharp front teeth into them, then tearing and piercing. Apparently the back teeth were not used very much; they were fewer and smaller than in sphenacodonts.

Most other dinocephalians, the **tapinocephalians**, look carnivorous at first sight (Figure 10.14), with large canines and incisors in the front of the jaw. But they had a broad, hippo-like muzzle, a large array of flattened back teeth, and massive bodies with a barrel-like rib cage that must have contained a capacious gut. These animals may have been omnivorous, but more likely the incisors were cropping, cutting teeth used on vegetation, and the canines were fighting tusks, not carnivorous weapons. (Look inside the mouth of a hippo sometime.) The jaw exerted most pressure when closed, for efficient chewing rather than slamming.

Some tapinocephalians were particularly bizarre, with horns; some of them, probably males, had great bony flanges on the cheeks (Figure 10.14). All tapinocephalians had thick skull bones, sometimes up to 11 mm (half an inch) thick. Herbert Barghusen suggested in 1975 that individuals butted heads, presumably to establish dominance within a group. Large vegetarians today tend to

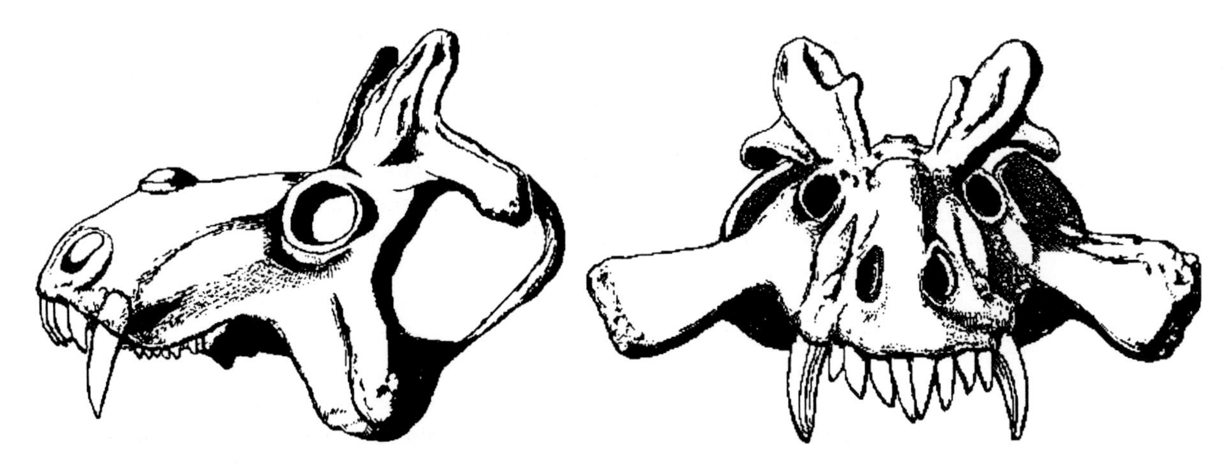

Figure 10.14 Tapinocephalians had large canines, probably for display and fighting rather than feeding. *Estemmenosuchus*, from the Late Permian of Russia, also had bizarre flanges on the skull that may have distinguished males. (After Chudinov.)

fight by head butting or pushing, while carnivores today are quick and agile and tend to use claws and teeth as they fight. Early therapsids, even the carnivores, were heavy and clumsy; they had sprawling limbs that were so committed to supporting their weight that they could not have used claws as weapons.

Advanced Therapsids

The rapid evolution of therapsids brought a new wave of advanced forms across the world in the Late Permian. These are the **theriodonts** and **anomodonts**. Theriodonts are all carnivores, with low flat snouts and very effective jaws. The **gorgonopsians** are theriodonts named for their ferocious appearance. They were the dominant large carnivores of the Late Permian. Gorgonopsians were specialist carnivores on large prey. Their sabertoothed killing action (Figure 10.15) clearly involved a very wide gape of the jaw and a slamming action that drove the canine teeth deep into the prey. The incisors were strong, but the back teeth were small and must have been practically useless. The snout was rather short, but deep enough to hold the roots of the canines. The limbs were long and fairly slender, and gorgonopsians may have been comparatively agile. The skull is only 50 cm (20 in) long in the largest known gorgonopsian, and even in adults the limbs were not required to be purely load-bearing as they had been in earlier therapsids. The joints could therefore be more lightly built, and the whole locomotion was improved. The hind limb could be swung into an erect position, stride length was greater, and the foot was lighter, altogether indicating greater speed. They didn't go in for much chewing, but simply tore large chunks from prey that was too large to eat in one bite (sharks and crocodiles do that too). Their front teeth had serrations on them to slice through muscle and tendon.

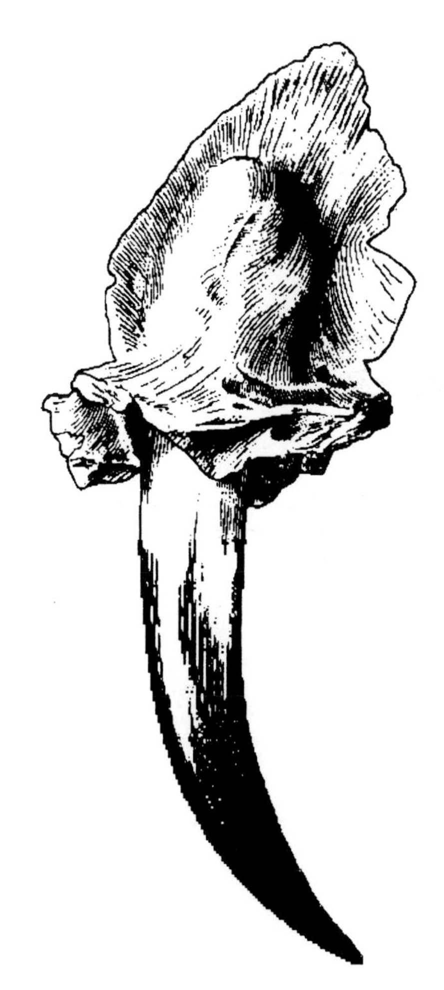

Figure 10.15 *Ivantosaurus* is an early theriodont from Russia with saberteeth about 10 cm (4 in) long. The character continued into the gorgonopsians. (After Chudinov.)

Cynodonts

Cynodonts are a clade of theriodonts. They appeared near the end of the Permian and two lineages of cynodonts survived and flourished in the Triassic, with one of them evolving into ancestors of mammals at the end of the Triassic. In their own right, cynodonts were important small- and medium-sized animals in the Triassic throughout Gondwana. Their innovations in food processing (especially an increase in "chewing") led to the evolution of many jaw, tooth, and skull features that we now see as "mammalian." Cynodonts used to be described as the most "mammal-like" of reptiles, for good reason. They are discussed in Chapter 15.

Anomodonts

Anomodonts are a group of therapsids that evolved in the Late Permian. The **dicynodonts** are by far the most important anomodonts and were the dominant herbivores of the Late Permian: they were the first truly abundant worldwide herbivores. They provide 90% of therapsid specimens and much of the therapsid diversity preserved in Late Permian rocks. The earliest dicynodonts were already so specialized as herbivores at their first appearance that they show no close resemblance to other therapsid groups and are difficult to classify. Early dicynodonts already had a secondary palate, so they could breathe and chew at the same time.

At their peak in the Late Permian, dozens of species of dicynodonts were living in Gondwana, and they survived long into the Triassic. They differed from other therapsids in having very short snouts, and they had lost practically all their teeth except for the tusk-like upper canines, which were probably used for display and fighting rather than for eating (Figure 10.16). Because there were no chewing teeth, the jaws must have had some sort of horny beak (like that of turtles) for shearing off pieces of vegetation at the front and grinding them on a horny secondary palate while the mouth was closed. The jaw joint was weak, and moved forward and back in a shearing action instead of sideways or up and down. As part of this system, the jaw musculature was unusual, set far forward on the jaw, and took up a good deal of space on the top and back of the skull. These unusual jaw characters had their effect on the whole shape of the skull, which was short yet high and broad, almost box-like. The extensive muscle attachments resulted in the eyes being set relatively far forward on a short face. Dicynodonts look as if they cropped relatively tough vegetation with their beaks, and then ground it up in a rolling motion in the mouth. As in other herbivores, the body was usually bulky, with short, strong limbs (Figure 10.12).

The success of dicynodonts is astonishing. Most dicynodonts were rather small, though they ranged from rat-sized to cow-sized. Presumably the fact that the horny feeding structures of dicynodonts were replaced continuously throughout life had a great deal to do with their success. Reptiles with teeth replace them throughout life, but intermittently, so it is difficult for them to achieve continuously effective tooth rows. Other therapsids evolved effective cutting and grinding teeth, but teeth do wear out with severe and prolonged use.

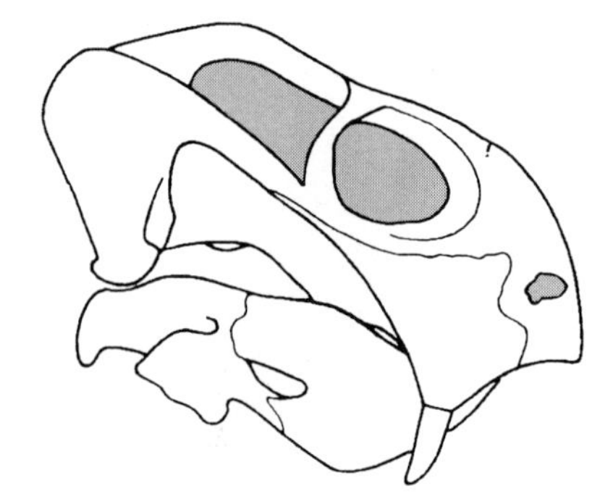

Figure 10.16 Dicynodonts had strange-looking faces because of their style of jaw construction. This is *Dicynodon*, with a skull about 15 cm (6 in) long. (After Cluver and Hotton.)

Dicynodont jaws varied a lot, presumably because of their diet. There were dicynodonts with cropping jaws and with crushing jaws (perhaps for large seeds), and many browsers and grazers. Some dicynodonts were specialized for grubbing up roots, and some for digging holes, although they remained vegetarian. In a spectacular discovery in South Africa, skeletons of a little dicynodont were found at the bottom of sophisticated spiral burrows (Figure 10.17).

The extent of specialization among dicynodonts suggests that the climate was reasonably mild and food supply reasonably reliable at the time, in spite of the high latitude and inevitable seasonal changes. Most Permian dicynodonts were small, with skulls about 20 cm (8 in) long. Possibly many of them were small so that they could burrow to avoid seasonal changes in temperature and food supply.

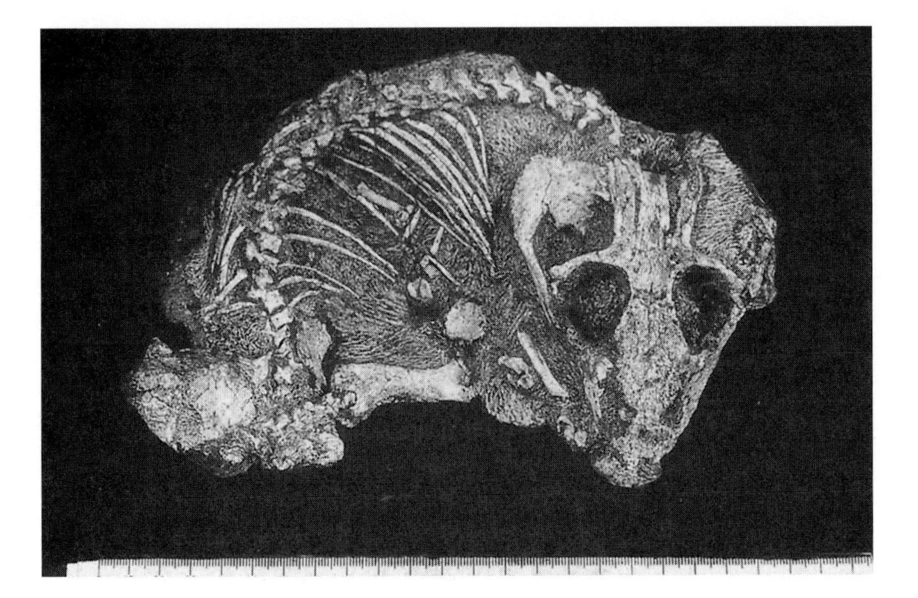

Figure 10.17 Several specimens of the little Permian dicynodont *Diictodon* have been found fossilized inside their burrows. The markings that look suspiciously like hair on this specimen are actually scratches from the needles used to clean the rock away from the bones. The scratches add an eerie realism to the fossil! (Courtesy of Dr. R. M. H. Smith of the South African Museum.)

Dicynodonts declined abruptly at the end of the Permian, but a few lineages persisted, often in great numbers. The best-known dicynodont of all is a very specialized Early Triassic form, *Lystrosaurus*. It has been found in India, Antarctica, South Africa, and south China; its distribution helped and is still helping to identify fragments of Gondwana.

Other Triassic dicynodonts became unusually massive, with short legs and barrel-like trunks. Late in the Triassic, some grew to the size of rhinos (Figure 10.12) and must have been very slow-moving. Perhaps they were ecological analogues of giant pandas, gorillas, or ground sloths. The last of the dicynodonts even lost their tusks, which may once have been used for visual display. Possibly, therefore, the last clumsy dicynodonts were nocturnal or hid in thick undergrowth in the forest, driven there by competition with more advanced reptiles. In a gloomy scene reconstructed in 1978 by C.E. Gow: "Thus died out the Dicynodontia, lurking in semi-obscurity in the depths of the bushes and only venturing out to feed at night. The archosaurs had come into their own."

Synapsids and Diapsid Replacements

With their generally large bodies, their radiation into herbivores and carnivores of varying sizes, and their experimentation with horns, fangs, and fighting, a Late Permian therapsid community viewed from a long way away would not seem totally strange to a modern ecologist, especially one familiar with the large mammals of the African savanna. However, the comparison would not stand close examination. The Permian carnivores were larger and much more numerous than modern African carnivores, and the locomotion of all Permian therapsids was very clumsy.

But a Late Permian community would not have looked as foreign as a Triassic one. Faunas did not evolve to look more mammal-like. Instead, therapsids were replaced by archosaurian diapsid reptiles, which had evolved from quite different Permian ancestors.

Further Reading

Barghusen, H. R. 1975. A review of fighting adaptations in dinocephalians (Reptilia, Therapsida). *Paleobiology* 1: 295–311.

Bennett, S. C. 1996. Aerodynamics and the thermoregulatory function of the dorsal sail of *Edaphosaurus*. *Paleobiology* 22: 496–506.

Gow, C. E. 1978. The advent of herbivory in certain reptilian lineages during the Triassic. *Palaeontologica Africana* 21: 121–41.

Cruickshank, A. R. I. 1978. Feeding adaptations in Triassic dicynodonts. *Palaeontologia Africana* 21: 121–32.

Haack, S. C. 1987. A thermal model of the sailback pelycosaur. *Paleobiology* 12: 450–8.

Reisz, R. R. 1981. A diapsid reptile from the Pennsylvanian of Kansas. *Special Publications of the Museum of Natural History of the University of Kansas* 7. [*Petrolacosaurus*.]

Ray, S., and A. Chinsamy. 2002. Functional aspects of the postcranial anatomy of the Permian dicynodont *Diictodon* and their ecological implications. *Palaeontology* 46: 151–83. [Adaptations for burrowing.]

Ruben, J. A., et al. 1987. Selective factors in the origin of the mammalian diaphragm. *Paleobiology* 13: 54–9.

Rubidge, B. S., and C. A. Sidor. 2001. Evolutionary patterns among Permo-Triassic therapsids. *Annual Review of Ecology and Systematics* 32: 449–80.

Smith, R. M. H. 1987. Helical burrow casts of therapsid origin from the Beaufort Group (Permian) of South Africa. *Palaeogeography, Palaeoclimatology, Palaeoecology* 60: 155–69.

Sues, H.-D., and R. R. Reisz. 1998. Origins and early evolution of herbivory in tetrapods. *Trends in Ecology & Evolution* 13: 141–5. [Review.]

CHAPTER ELEVEN

The Triassic Takeover

Chapter 10 may have given the impression that the only significant evolution among Permian and Triassic amniotes took place among synapsids. This, of course, is not true. Permian amniote faunas were dominated ecologically and numerically by synapsids, first by pelycosaurs and then by therapsids. But a good deal of evolution was going on among diapsid reptiles, and in the Triassic they came to replace synapsids as the dominant land vertebrates. The replacement was so dramatic that it has come to be a debating ground for the general question of replacement of one vertebrate group by another. The variety of Triassic diapsids leads to a mass of unfamiliar names, and I have tried to keep the list as simple as I can. Questions and some possible answers relating to the diapsid takeover form the major themes of this chapter.

DIAPSIDS

Basal Diapsids

Diapsids are probably descended from *Petrolacosaurus*, and most basal diapsids were basically lizard-like in size, structure, and behavior. But basal diapsids (Figure 11.1) also evolved into some interesting ways of life as early as the Late Permian. *Heleosaurus* from South Africa was a small terrestrial carnivore that may have been able to run bipedally. In Madagascar, *Coelurosauravus* (Chapter 13) was a glider and *Hovasaurus* was aquatic.

Three hundred specimens of *Hovasaurus* make it one of the best-known Permian diapsids. Overall, it was lizard-like, perhaps only 30 cm (1 ft) long from snout to vent. But the tail was exceptionally long, strong, and deep (Figure 11.2), so the whole animal was close to 1 m (3 ft) in length. The tail had at least 70 vertebrae and certainly looks like a swimming appendage. Inside the fossils, the abdominal cavity consistently contains a mass of small quartz pebbles. The pebbles often have a characteristic shape, tapering at both ends. Presumably they were swallowed by the animal during life. They are too small to be food-grinding pebbles and too far back in the abdomen to have occupied the stomach in life. Probably they were contained in a specially adapted abdominal sac. *Hovasaurus* almost certainly swallowed the stones as ballast for diving. Living Nile crocodiles do the same thing, and the extinct plesiosaurs may have done so too.

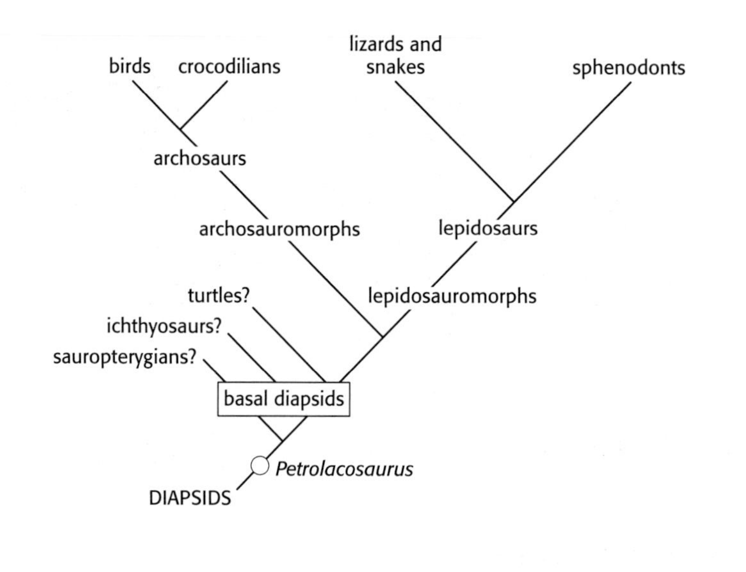

Figure 11.1 Cladogram of major groups of diapsids. The placement of some marine reptiles is unclear, though they probably arose early in diapsid evolution, perhaps, as shown here, from basal "stem diapsids."

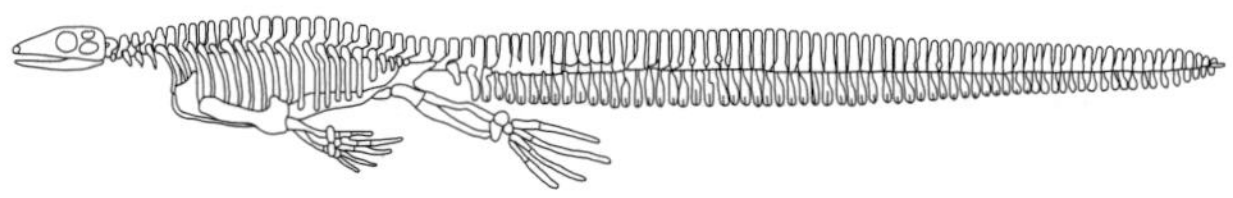

Figure 11.2 *Hovasaurus* was an aquatic diapsid from the Late Permian of Madagascar. The tail was very long and strong, and its abdominal cavity was fossilized with pebbles inside; they can be interpreted as ballast. Total length about 1 m (3 ft). (From Currie.)

The basal diapsids probably include the ancestors of the dominant marine reptiles of the Mesozoic, the ichthyosaurs and plesiosaurs that I will discuss in Chapter 14. Turtles may belong here too (Figure 11.1).

A radiation into major diapsid clades on land began in the latest Permian but was truly spectacular in the Triassic. The diapsid takeover from synapsids during that time was an astonishing series of events.

There are two major surviving clades of diapsids: the **lepidosaurs** (lizards, snakes, and the tuatara of New Zealand) and the **archosaurs** (crocodiles and birds) (Figure 11.1). We use a crown-group definition of the clades Lepidosauria and Archosauria: all those diapsids that are more closely related to the living survivors than to anything else. As usual, there are extinct clades that branched off below the base of these crown groups, and these are placed in larger clades called Lepidosauromorpha and Archosauromorpha. The scheme is logical though the names are clumsy.

Lepidosauromorphs

On land, the crown-group lepidosaurs have been the dominant group of small-bodied reptiles since the Mesozoic. They consist of two major clades (Figure 11.1). **Squamata** are the numerous and diverse smaller living reptiles, including lizards and snakes. **Sphenodontia** include only one living form, the tuatara, *Sphenodon* (Figure 11.3), an outwardly lizard-like animal that survives today only on a few islands off the coast of New Zealand (Chapters 17 and 21). Its skull characters show that it is not a true lizard. Sphenodonts are known as far back as the Triassic.

Figure 11.3 The tuatara, *Sphenodon* from New Zealand. (From Lydekker.)

Archosauromorphs

Archosauromorphs include the largest aerial and terrestrial animals that have ever lived, and they rose to dominate land ecosystems by the Late Triassic. They include a number of groups that will receive significant discussion (Figure 11.4). **Turtles** are certainly diapsids and may be basal archosauromorphs. **Rhynchosauria** are basal archosauromorphs that were the dominant large herbivores for a brief period during the Triassic.

The **Archosauria** evolved in the Middle to Late Triassic and include the clades **Crurotarsi** (crocodiles, alligators, and related extinct forms), **Pterosauria**, and **Dinosauria**. The pterosaurs evolved true flapping flight much earlier than birds did, and they dominated the skies throughout the Mesozoic (Chapter 13). The **Dinosauria** (Chapter 12) consist of prosauropods, sauropods, ornithischians, and theropods (which include birds). Birds (Chapter 13), of course, are derived diapsid, archosauromorph, archosaurian, dinosaurs.

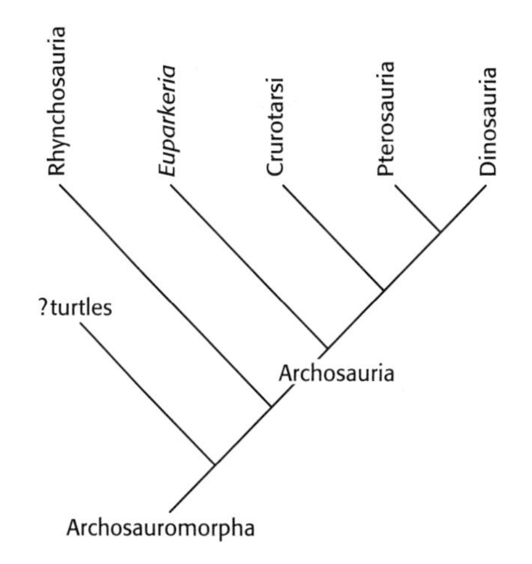

Figure 11.4 Cladogram of archosauromorphs. Turtles may be basal archosauromorphs rather than basal diapsids. Crurotarsi is the clade that includes living crocodilians and extinct forms that are closely related to them.

THE TRIASSIC DIAPSID TAKEOVER: THE PATTERN

Over 600 Permo-Triassic reptiles have been described. Their history is best preserved for those that lived in lowland faunas. Upland faunas are not often preserved, and aquatic reptiles are not as numerous as terrestrial ones.

Large pelycosaurs dominated the tropical regions of Euramerica in the Early Permian. The only amniote outside this area was aquatic, the little fish-eating *Mesosaurus* (Figure 11.5), which lived in and around the African and Brazilian parts of Gondwana.

By the Late Permian, land animals could walk into Gondwana. Therapsids had

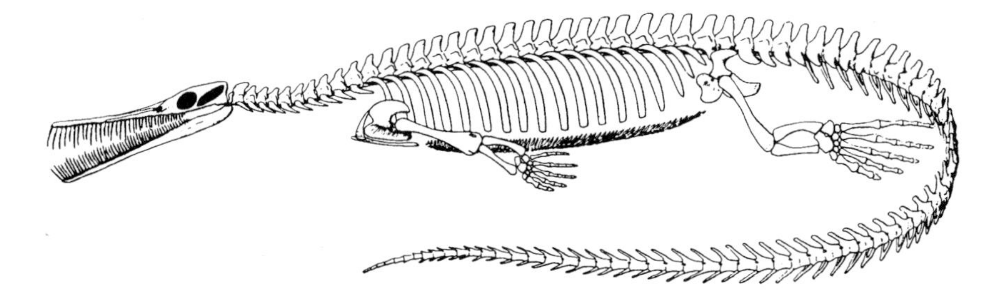

Figure 11.5 *Mesosaurus* was the first amniote to reach Gondwana. Its discovery in fresh-water Permian sediments in South Africa and Brazil helped to confirm that Africa and South America had once been joined in Gondwana. It looks like a fisheater. About 1 m (3 ft) long. (After MacGregor.)

replaced pelycosaurs as the dominant land reptiles. New advanced therapsids dominated the Late Permian, particularly dicynodonts. But in some areas, diapsids were also radiating into new ways of life. As we have seen, there were terrestrial, aquatic, and gliding diapsids in Madagascar; more important, the success of *Heleosaurus* in South Africa suggests that terrestrial diapsids were invading cooler regions for the first time.

Gondwana had rich Triassic faunas, and land animals were free to disperse throughout Pangea. Therapsid diversity dropped sharply in the Permo-Triassic extinction, although the species that did survive were widespread and numerous. Dicynodonts were extraordinarily abundant at larger sizes, and cynodonts were medium-sized herbivores. There were few therapsid predators: most of them were small- and medium-sized cynodonts such as *Cynognathus* (Chapter 15). Some of the early archosauromorphs were small and carnivorous, although therapsids outnumbered them 65 to 1 at first. But by the end of the Early Triassic, some archosauromorphs were 5 m (16 ft) long, with massive skulls a meter long. In South Africa, *Euparkeria* was a fast, lightly built carnivore that in retrospect is very close to the ancestry of the Archosauria.

Therapsids were the dominant herbivores well into the Late Triassic. But Middle Triassic diapsids showed marked improvements in running ability over earlier forms, and by the end of the Middle Triassic, rhynchosaurs became abundant vegetarians alongside the dicynodonts. There were even greater changes among the carnivores. Archosauromorphs of various sizes became abundant in the Middle Triassic, ranging from large quadrupedal rauisuchians to small bipedal ornithosuchids, with crocodile-like forms also present. Among therapsids, cynodont carnivores were at most medium-sized but were still abundant and diverse.

Therapsids and rhynchosaurs declined distinctly in the Late Triassic, though they were still important ecologically. By the latest Triassic most of the therapsids had disappeared, along with rhynchosaurs and many other archosauromorphs. The vegetarians of the latest Triassic were almost all prosauropod dinosaurs; diapsid carnivores were larger, more diverse, and more mobile than before, and they were joined by the first theropod and ornithischian dinosaurs. The first mammals were few and small.

Finally, at the end of the Triassic, dinosaurs quickly overwhelmed terrestrial ecosystems throughout the world, replacing thecodonts and therapsids alike in every medium- and large-bodied way of life, to form a land fauna dominated by dinosaurs that lasted throughout the Jurassic and Cretaceous Periods.

The replacement of therapsids by archosaurs was worldwide. Therapsid carnivores were replaced first, gradually during the Middle Triassic. Carnivorous archosaurs became gradually larger, more diverse, and more abundant with time. They also came to have a more erect limb structure, which indicates better locomotion, including bipedal running in many cases. The herbivore replacement was much more rapid, taking place in the early part of the Late Triassic.

There are arguments about the relative suddenness of some of the replacements. Detailed studies in Triassic rocks which extend from North Carolina to Nova Scotia have shown that here at least there was a sudden, major extinction of Late Triassic land vertebrates. Some people have wondered whether a meteorite impact was involved, pointing to the huge crater of Manicouagan in Québec, Canada (70 km [40 mi] across), which is dated with some uncertainty as Late Triassic. Others deny that the replacements were sudden at all, or, if they were, that there were at least two separate episodes of extinction and replacement. The arguments mostly reflect the difficulty of judging relative time in terrestrial deposits, and will eventually be resolved by careful field research.

There are more fundamental arguments about the relationship between archosaurs and therapsids. Some people argue that the therapsids largely died out for environmental reasons (climatic change, for example, or an asteroid impact) and were replaced by archosaurs that radiated after the therapsids had largely gone. Others suggest more direct competition, in which carnivorous archosaurs first outcompeted therapsid carnivores and then hunted out the remaining therapsid vegetarians. Vegetarian archosaurs then evolved to take advantage of the abundant plant life. In this model, the archosaur success resulted from competitive superiority.

When it comes to real evidence, we can show clearly that archosaurs were much superior to other contemporary land vertebrates in respiration and locomotion. The difference between Triassic archosaurs and synapsids is so clear and so profound that it provides all the reason we need to account for the decline of the synapsids, except for their shifty, vicious, nocturnal survivors, the mammals.

RESPIRATION, METABOLISM, AND LOCOMOTION

In 1987, David Carrier put together some simple but powerful ideas about the links between respiration, locomotion, and physiology.

Fishes have no problem maintaining high levels of exercise. Many sharks swim all their lives without rest, for example. Gill respiration gives all the oxygen exchange needed for such exercise levels, and with good hunting skills, the necessary food supply is readily available.

The evolution of lungs did not change that relationship between anatomy and physiology. Air has to be pumped in and out of an internal body cavity, however, and living lungfishes may have rather low exercise levels because their oxygen exchange is not geared for high performance.

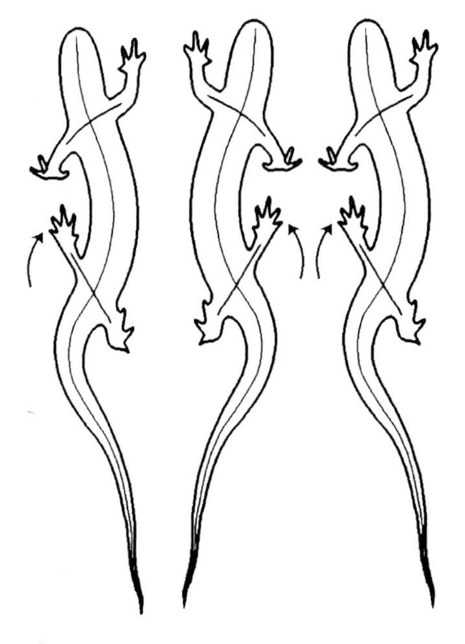

Figure 11.6 David Carrier pointed out that the sprawling locomotion of a lizard or salamander forces it to compress each lung alternately as it moves (see text).

Tetrapods moving about on land face a much more serious problem. The shoulder girdle and the forelimbs in particular, powered in part by the muscles of the trunk, are largely devoted to supporting and moving the body over the ground. In the sprawling gait of amphibians and living reptiles, the trunk is twisted first to one side and then the other in walking and running. As the animal steps forward with its left front foot, the right side of the chest and the lung inside it are compressed while the left side expands (Figure 11.6). Then the cycle reverses with the next step. This distortion of the chest interferes with and essentially prevents normal breathing, in which the chest cavity and both lungs expand uniformly and then contract. If the animal is walking, it may be able to breathe between steps, but sprawling vertebrates cannot run and breathe at the same time. I shall call this problem **Carrier's Constraint**.

Animals can run for a while without breathing: for example, Olympic sprinters usually don't breathe during a 100-meter race. Animals can generate temporary energy by anaerobic glycolysis, breaking down food molecules in the blood supply without using oxygen. But this process soon builds up an oxygen debt and a dangerously high level of lactic acid in the blood. Mammalian runners (cheetahs and humans, for example) often use anaerobic glycolysis even though they can breathe while they run; it's a useful but essentially short-term emergency boost, like an afterburner in a jet fighter.

Figure 11.7 This lizard has stopped to take a breath.

Living amphibians and reptiles, then, can hop or run fast for a short time, first using up the oxygen stored in their lungs and blood, then switching to anaerobic glycolysis. They cannot sprint for long, however. If lizards want to breathe, they have to stand still with feet symmetrical (Figure 11.7). Lizards run in short rushes, with frequent stops. By attaching recorders to the body, Carrier showed that the stops are for breathing, and that lizards don't breathe as they run. Therefore, all living amphibian and reptilian carnivores use ambush tactics to capture agile prey: chameleons and toads flip their tongues at passing insects, for example.

The giant varanid lizard, the ora or Komodo dragon, which eats deer, pigs, and tourists (most notably, Baron Rudolf von Reding on 18 July, 1974), goes a little way toward solving Carrier's Constraint by pumping air into its lungs from a throat pouch; but that only gives it a small improvement in performance. The Komodo dragon has a short sprinting range, but it prefers to ambush prey from 1 m away.

Amphibians and most living reptiles have a three-chambered heart, which has usually been regarded as inferior to the four-chambered heart of living mammals and birds. But the three-chambered heart is useful to a lizard. Lizards run to catch food or to get away from danger, so they must use their resources most efficiently at this time. In a run, it is useless and perhaps dangerous for the lizard to waste energy pumping blood to lungs that cannot work. The lizard thus uses all the heart and blood capacity it has to circulate its store of oxygen around the whole body. The price it pays is a longer recharging time when it has to resupply oxygen to the blood, but it is usually able to do this at a less critical moment.

Early tetrapods all had sprawling gaits and faced a great problem. Their respiration and locomotion used much the same sets of muscles, and both systems could not operate at the same time. Imagine the laborious journey of *Ichthyostega* from the water to its breeding pools, with a few steps and a few gasps repeated for the whole journey. One can understand why so many early tetrapods remained adapted to life spent largely in water, and why many early amniotes often looked amphibious. *Eryops*, for example, swam with its tail (Figure 9.3) and would have had no major difficulty in devoting its rib-cage muscles to taking deep breaths at the surface.

When we see land animals such as pelycosaurs, with stiffened backbones and teeth designed for carnivorous and vegetarian diets rather than fish eating, we have to conclude that the problem had at least partially been solved. It's no good, for example, to raise metabolic rate by solar thermoregulation if there is no reliable oxygen supply to the tissues.

I suggest that the secret of the pelycosaurs was the stiffening of the backbone. They simply did not twist the body much as they moved. They had long bodies and relatively short limbs compared with lizards, and in any case a short step would not have rotated the trunk very much or distorted the lungs. The stiffening of the body also meant that most of the forelimb rotation was taken up at the shoulder joint, rather than being transmitted to the trunk. Furthermore, pelycosaurs had wheelbarrow locomotion, and the front limbs were mainly reactive support props, so the muscles operating them did not exert forces on the chest wall except to support the shoulder joint. However, the driving muscles of the pelvic girdle attached far from the chest wall.

The pelycosaurs thus had a special synapsid mitigation of Carrier's Constraint: they evolved adaptations that went some way toward reducing its consequences. (If you understand that point, look back at the skeleton of *Ichthyostega* [Figure 8.17]. Perhaps there is an analogous reason for its peculiar rib structure?) But pelycosaurs could not solve Carrier's Constraint. There is no way that they were running freely, or breathing while they ran.

Fishes can swim in water with sustained energy because Carrier's Constraint does not apply to gill breathing. The same is probably true for the lung breathing of turtles, because their shell does not allow the lungs to be distorted as they swim and come to the surface to breathe.

Many living land vertebrates have evolved a beautiful answer to Carrier's Constraint. They have freed the mechanics of respiration from the mechanics of locomotion by evolving an erect stance. The body is suspended more freely from the shoulders, allowing the thorax to make its breathing movements with hardly any twisting.

The evolutionary solution to Carrier's Constraint that resulted from erect stance is shown best today in mammals. Mammals evolved the diaphragm, a set of muscles to pump air in and out of the chest cavity. Air is sucked in as the diaphragm contracts, and forced out by the reaction of the elastic tissues of the lung. At the same time, the locomotion in most mammals has evolved to encourage breathing on the run. The backbone flexes and straightens in an up-and-down direction with each stride, alternately expanding and compressing the rib cage evenly (Figure 11.8). This rhythmic pumping of the chest cavity in the running action can be synchronized with the action of the diaphragm to move air in and out of the lungs with little effort. Thus quadrupeds running at full speed—gerbils, jackrabbits, dogs, horses, and rhinoceroses—take one breath per stride, and wallabies take one breath per hop (Figure 11.8). Trotting is far more complex, but that doesn't harm the line of argument presented here. Human runners usually take a breath every other stride. It is such a natural action that we don't notice it: runners should try to breathe out of phase to get some idea of the mechanism.

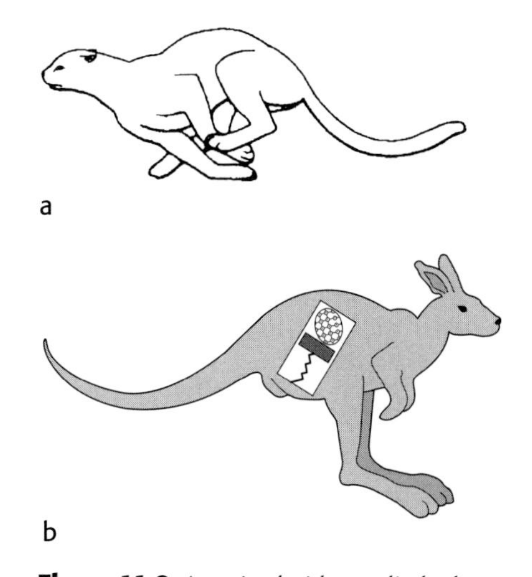

Figure 11.8 An animal with erect limbs does not rotate its chest as it runs as much as a sprawling animal does. The cat (a), other mammals, and certain other groups of tetrapods can breathe as they run, alternately squeezing air in and out of the lungs with each stride. The wallaby (b) forces air in and out of its lungs with each hop.

Animal locomotion often involves cyclic movements such as the strides and strokes of running or swimming limbs, or wingbeats in flight. Breathing may be made more efficient if it is synchronized with certain phases of limb movement. This is particularly important in human swimming, but it is a general principle. Flying insects synchronize their respiration with their wingbeats: the same muscular actions that raise and lower the wings also act to expand and compress the body, forcing air in and out of the spiracles. Birds do much the same thing (Chapter 13).

These principles are pieces of basic animal physiology, and they should be as true for extinct animals as they are for modern ones. Therefore, erect stance might be necessary for sustained running in any land animal, and its evolution should represent a great breakthrough in any tetrapod lineage, giving the basis for greatly improved running speed and stamina. Living reptiles are successful, but they are limited in the ecological roles they can perform because they have a sprawling gait and cannot sustain fast movement for very long.

Diapsids living today, such as lizards, don't have erect stance or sustained energy output, but we must not be fooled into thinking that all diapsids always lacked those capabilities. David Carrier suggested that Triassic diapsids, and the archosauromorphs in particular, were the first amniotes to make the breakthrough to erect gait and rapid, sustained locomotion. That breakthrough is preserved in the fossil record in the structure of the limbs and shoulder girdles of the early archosauromorphs. Erect gait and sustained locomotion was most likely the key innovation that made possible the diapsid, and in particular the archosaur, takeover of the Late Triassic.

I suggest here that an accident of history played an important role in forming the differences between Triassic diapsids and synapsids. Therapsids evolved largely in cool climates of the Late Permian, in northern Laurasia and southern Gondwana, while Permian diapsids evolved in warmer climates. Part of a heat-retaining syndrome in cool climates is to have stocky, compact bodies and short limbs and appendages, and therapsids are characteristically built that way (for example, Figure

10.12). Diapsids, however, had long, strong tails, and much of their body weight was on the hind limbs. It was relatively easier for diapsids to evolve to become partly or totally bipedal, and therefore to evolve erect limbs from a bipedal stance. Therapsids, with short tails, did not have that option: all of them were quadrupeds with a good deal of weight on the front feet. It may have been difficult to escape from the wheelbarrow locomotion that the therapsids inherited from the pelycosaurs, especially at larger body size. Truly erect gait, the solution to Carrier's Constraint, did not evolve among synapsids until the tiny mammals of the Early Jurassic.

RHYNCHOSAURS

Rhynchosaurs evolved in the Middle and Late Triassic with the decline of most large vegetarian therapsids and the disappearance of some. They were all herbivores, pig-sized animals with hooked snouts bearing a powerful cutting beak and hind limbs that look as if they might have been used for digging (Figure 11.9). Strong jaws bore batteries of slicing teeth, which are unusual among reptiles in that they were fused to bone at the base, not set into normal sockets. The teeth were ever-growing and were not replaced during life. As rhynchosaurs grew, they simply added more bone and more teeth at the back of the growing jaw as the teeth at the front became worn out. This style of tooth addition allowed rhynchosaurs great precision in tooth emplacement, so their bite was very effective for slicing vegetation with a scissor-like action.

Rhynchosaurs have been difficult to classify because of their peculiar features. They are probably a basal archosauromorph group. They were abundant and widespread in the Middle and Late Triassic and may have replaced therapsid groups because they too evolved an erect gait. However, rhynchosaurs rapidly became extinct at the end of the Triassic.

LOCOMOTION AND TRIASSIC ARCHOSAUROMORPHS

There was a repeated evolution of advanced locomotion among different early archosauromorphs.

Most early archosauromorphs were impressive carnivores, but they were large and dominantly quadrupedal, like *Archosaurus* from the Late Permian of Russia (Figure 11.10).

The best-known basal archosaur is *Euparkeria*, from South Africa. It was a larger, even more agile predator than *Heleosaurus*. It was about 1 m long, very lightly built,

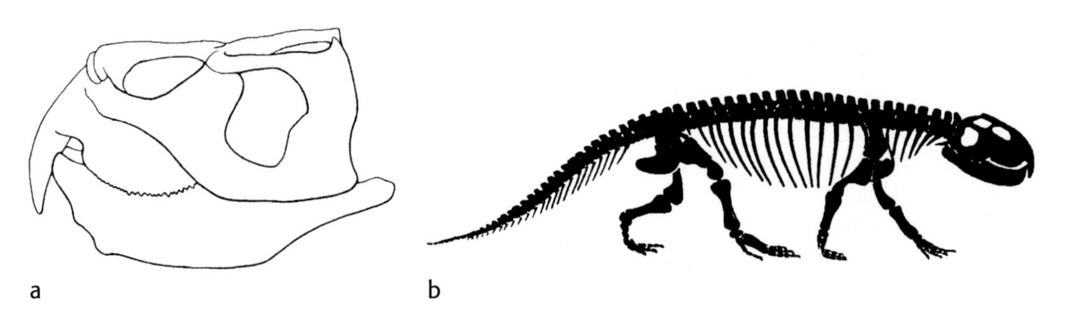

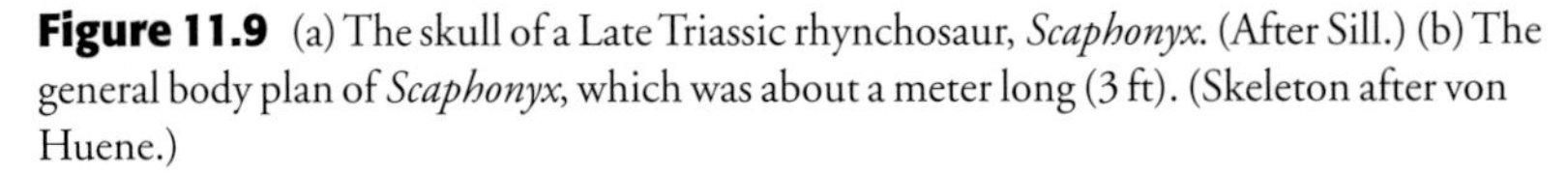

Figure 11.9 (a) The skull of a Late Triassic rhynchosaur, *Scaphonyx*. (After Sill.) (b) The general body plan of *Scaphonyx*, which was about a meter long (3 ft). (Skeleton after von Huene.)

Figure 11.10 *Archosaurus*, from the Late Permian of Russia. It's one of the wonderful legalisms of cladistics that *Archosaurus* is NOT an archosaur, only an archosauromorph. Whatever it is called, it was clearly an impressive sight to its Late Permian contemporaries.

and had a long, strong tail to give balance in running. Its skull was long and light, with many long, sharp stabbing teeth (Figure 11.11). *Euparkeria* was evidently fast-running, probably a bipedal runner but perhaps walking on all four feet. Its speed and agility may have promoted its success in comparison to contemporary therapsids. The limbs were set directly under the body, reducing the mechanical stresses of fast running and allowing sustained running. *Euparkeria* could be the common ancestor of crocodiles, pterosaurs, and dinosaurs (Figure 11.12).

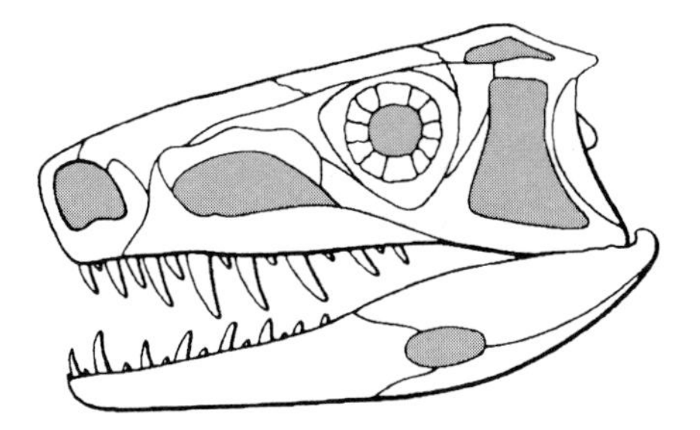

Figure 11.11 The advanced thecodont *Euparkeria* is a reasonable candidate for a generalized archosaur and dinosaur ancestor.

To walk (and run) bipedally, legs have to move dominantly forward, and that means that the ankle joint should not only be hinged in a forward-and-backward direction, but should be well braced so that it does not flop around sideways. There are many bones in the ankle region, and possibly because of this skeletal legacy, there was more than one joint that could be reconstructed into an efficient ankle for bipedal running. Two lineages of archosaurs can be distinguished on this basis: the Crurotarsi, which evolved the joint still seen today in crocodiles, and the Ornithodira, which evolved the joint seen today in birds (Figure 11.12). Each lineage exploited the ankle to achieve more erect gait, culminating not only in erect-limbed dinosaurs, but erect-limbed crurotarsians too.

For most of the Middle and Late Triassic, the largest carnivorous groups were rauisuchians, and ornithosuchians, a group of basal archosaurs typified by *Ornithosuchus* (Figure 11.13). Each included bipeds and quadrupeds, and large,

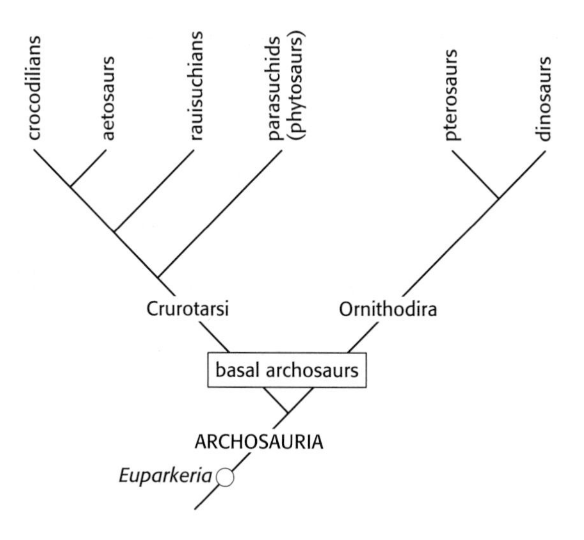

Figure 11.12 More detailed cladogram to show the relationships of Archosauria.

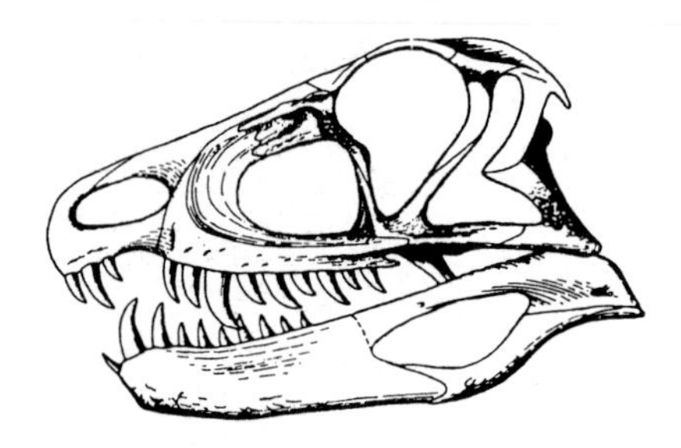

Figure 11.13 The basal archosaur *Ornithosuchus* evolved many adaptations in parallel not only with rauisuchians such as *Postosuchus* but also with later carnivorous dinosaurs. Skull about 25 cm (1 ft) long. (After Walker.)

geographically widespread animals. The rauisuchian *Postosuchus*, from the Late Triassic of Texas, was about 4 m long including the tail, and stood 2 m high. It was lightly built and walked (and ran) bipedally. But it was a hunter, with a heavy killing head, impressive wide-opening jaws, and serrated stabbing and cutting teeth. The eyes were large, set for forward stereoscopic vision, with bony eyebrows to shade them. *Postosuchus* is uncannily like a small version of the much larger and later carnivorous dinosaurs *Allosaurus* and *Tyrannosaurus* in overall body plan and presumably in ecology.

These carnivores were perhaps close in ecology to living monitor lizards, which have a semi-erect gait and are active predators with a preferred body temperature close to 37°C (98°F). The Komodo dragon of Indonesia is the top predator in its ecosystem, weighing over 100 kg (200 lb). Many of the basal Triassic archosaurs were mostly about the same size and, if anything, were more active, because they had erect gait and could probably run faster and further.

Some crurotarsians explored a way of life that we associate with living crocodiles: ambush hunting at the water's edge. Parasuchids (sometimes called phytosaurs) were large, long-snouted carnivores from the tropical belt of the Late Triassic. They evolved toward a crocodilian appearance and ecology (Figure 11.14). Two parasuchids from India more than 2 m (7 ft) long had stomach contents that included small bipedal archosaurs, and one had eaten a rhynchosaur.

Living crocodiles may well be some guide to the physiology, locomotion, and ecology of parasuchids. Crocodiles have a good circulatory system, with more advanced heart and lung modifications than other living reptiles. Although they normally walk slowly on land, in a sprawling stance, they are also capable of a faster run in which the limbs are nearly vertical. The little freshwater crocodile of Australia can gallop (briefly) at 16 kph (10 mph), but some parasuchids could probably have done even better. Crocodiles show that it's artificial to categorize animals as having only one possible gait and stance: even so, many animals specialize in one gait or another.

Parasuchids disappeared at the end of the Triassic, with many other archosaur groups. They were replaced in that ecological niche by true crocodilians. Early crocodilians had been small, long-legged, terrestrial predators. Some, from the Late Triassic of western Europe and eastern North America, look like lightly built running carnivores (Figure 11.15). Later some crocodilians adapted to water, replacing the parasuchids, and only then did they become much larger. They evolved a secondary palate so that they could bite and chew under water without flooding their nostrils, and they also lost some of the features of their terrestrial gait, becoming secondarily sprawling.

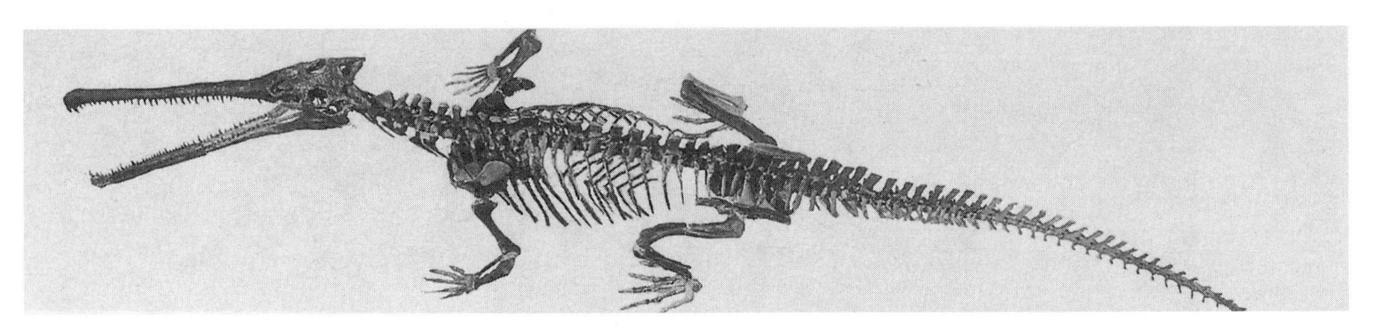

Figure 11.14 The Triassic parasuchid *Rutiodon*. (Negative 319167. Photograph by Edward Bailey. Courtesy of the Library Services Department, American Museum of Natural History.)

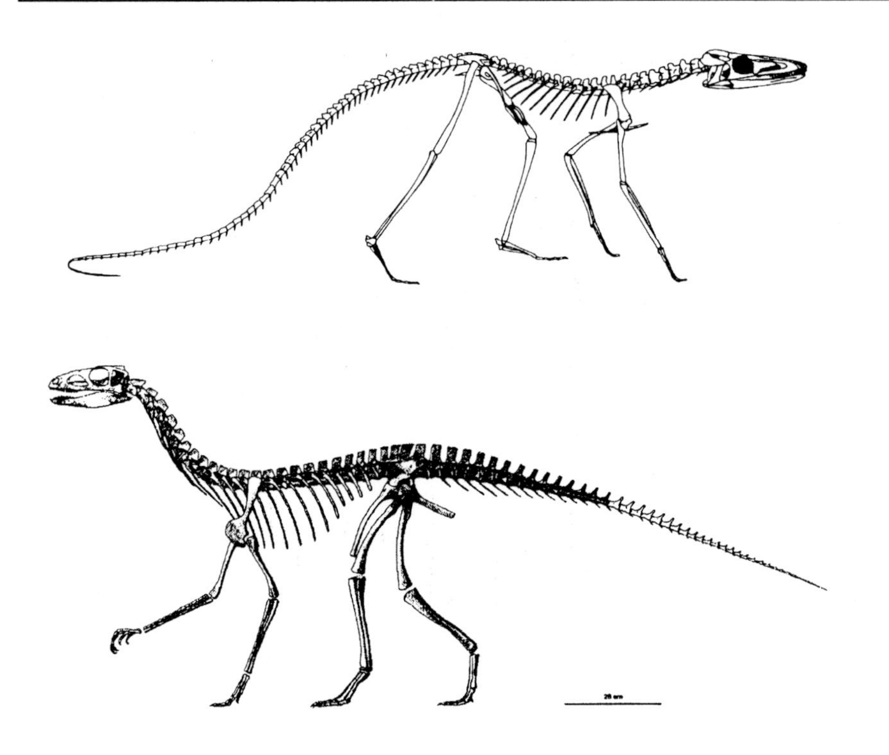

Figure 11.15 A reconstruction by P. J. Crush of a fast-running early terrestrial crocodile from the Late Triassic of Britain. It was very lightly built and only about 50 cm (18 in) long.

Figure 11.16 *Silesaurus*, a herbivore from the Late Triassic of Poland, is very closely related to true dinosaurs. Scale bar is 25 cm (close to a foot). (Published in Dzik, 2003. Courtesy Jerzy Dzik of the Polish Academy of Sciences, and the Society of Vertebrate Paleontology.)

DINOSAUR ANCESTORS

The first dinosaurs appeared in the Late Triassic, around 225 Ma. They were small, agile, and bipedal at first (their large to enormous sizes evolved later). Although the first dinosaurs were almost certainly small bipedal running carnivores, they quickly evolved into different feeding styles. *Silesaurus* (Figure 11.16), from the Late Triassic of Poland, was *almost* a dinosaur, lacking only a couple of skull features: it was a small herbivore with a beak for cropping vegetation, like many later dinosaurs.

Herrerasaurus and *Eoraptor*, the best-known of the earliest dinosaurs, were carnivores living in Argentina alongside a fauna dominated by rhynchosaurs, with synapsids present too. The community seems to have been stable for at least 10 m.y. or so, with the dinosaurs forming perhaps one-third of the carnivores. After that, the ecology seems to have changed rapidly, and dinosaurs became dominant, as we shall see in Chapter 12.

Further Reading

Benton, M. J. 1993. Late Triassic extinctions and the origin of the dinosaurs. *Science* 260: 769–70. [Benton tries to argue that dinosaurs replaced Triassic reptiles, but didn't outcompete them.]

Brochu, C. A. 2001. Progress and future directions in archosaur phylogenetics. *Journal of Paleontology* 75: 1185–201.

Carrier, D. R. 1987. The evolution of locomotor stamina in tetrapods: circumventing a mechanical constraint. *Paleobiology* 13: 326–41. [In my view, one of the real breakthrough papers of the 1980s.]

Carrier, D. R., and C. G. Farmer. 2000. The evolution of pelvic aspiration in archosaurs. *Paleobiology* 26: 271–93. [Another astounding paper from David Carrier, this time with Colleen Farmer. A novel and convincing reconstruction of respiration in archosaurs, with major application to early crocodilians.]

Dzik, J. 2003. A beaked herbivorous archosaur with dinosaur affinities from the early Late Triassic of Poland. *Journal of Vertebrate Paleontology* 23: 556–74. [*Silesaurus*.]

Farmer, C. G., and W. J. Hicks. 2000. Circulatory impairment induced by exercise in the lizard *Iguana iguana*. *Journal of Experimental Biology* 203: 2691–7. [More proof of, and more amazing implications of, Carrier's Constraint.]

Gatesy, S. M. 1991. Hind limb movements of the American alligator (*Alligator mississippiensis*) and postural grades. *Journal of Zoology, London* 224: 577–88.

Gower, D. J., and E. Weber. 1998. The braincase of *Euparkeria*, and the evolutionary relationships of birds and crocodilians. *Biological Reviews* 73: 367–411. [*Euparkeria* is basal to advanced archosaurs.]

Olsen, P. E., et al. 1987. New Early Jurassic tetrapod assemblages constrain Triassic Jurassic tetrapod extinction event. *Science* 237: 1025–9; see also discussion, v. 241, 1358–60.

Owerkowicz, T., et al. 1999. Contribution of gular pumping to lung ventilation in monitor lizards. *Science* 284: 1661–3. [Monitor lizards avoid Carrier's Constraint (a little) by pumping air from a throat pouch.]

Padian, K. (ed.). 1987. *The Beginning of the Age of Dinosaurs.* Cambridge, England: Cambridge University Press.

Parrish, J. M. 1987. The origin of crocodilian locomotion. *Paleobiology* 13: 395–414.

Zardoya, R., and A. Meyer. 2001. The evolutionary position of turtles revised. *Naturwissenschaften* 88: 193–200. [Summary of the problem, early 2001.]

CHAPTER TWELVE

Dinosaurs

We are familiar with dinosaurs in many ways: they have been with us since kindergarten or before, in comic strips, toys, stories, movies, nature books, TV cartoons, and advertising. Yet it's still not easy to understand them as animals. The largest dinosaurs were more than ten times the weight of elephants, the largest land animals alive today. Dinosaurs dominated land communities for 100 million years, and it was only after they disappeared that mammals became dominant. It's difficult to avoid the suspicion that dinosaurs were in some way competitively superior to mammals and confined them to small body size and ecological insignificance. We would dearly love to know the basis for that superiority.

The earliest dinosaurs were small bipedal carnivores and evolved from small, bipedal archosaurs (Chapter 11). They appeared in the Late Triassic of Gondwana at about the same time as the first mammals. All the spectacular variations on the dinosaur theme came later, but all four major dinosaur groups (Figure 12.1) had evolved by the end of the Triassic.

Most people know now that birds are highly evolved dinosaurs. That means that in cladistic terms, the clade Dinosauria includes birds as well as those dinosaurs that are not birds (the "nonavian dinosaurs"). That is a clumsy phrase. I shall use dinosaurs with a small d to refer to nonavian dinosaurs, and Dinosauria to mean dinosaurs plus birds.

THEROPODS

Theropods were all bipedal carnivores, retaining the body plan and the ecological character of ancestral dinosaurs. The earliest known dinosaurs were theropods: *Eoraptor* and *Herrerasaurus*, both from the Late Triassic of Argentina. *Eoraptor* was a very small animal with a skull only about 8 cm (3 in) long (Figure 12.2). Even so, it was well adapted as a fast-running carnivore, with sharp teeth and grasping claws on its forelimbs. *Herrerasaurus* was very like *Eoraptor* but much larger, between 3 and 6 m (10 and 20 ft) long.

A small, agile, bipedal body form was successful throughout theropod history, and small bipedal theropods formed the only dinosaur clade that still survives (as birds). However, at least four lineages of theropods evolved to giant size, in fact: the Jurassic allosaurs and in three Cretaceous groups: North American tyrannosaurs, in

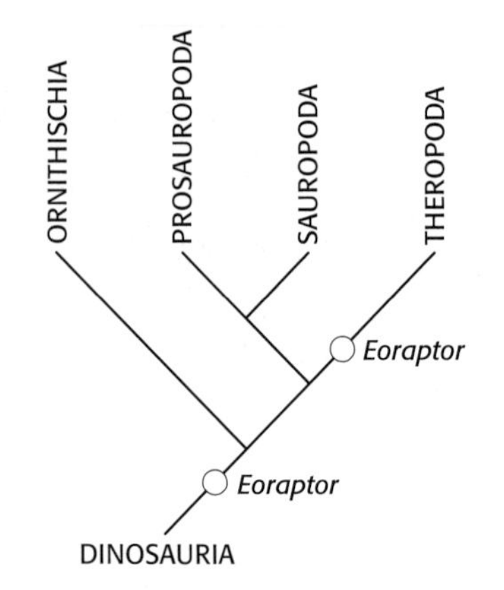

Figure 12.1 A phylogram showing the major groups of dinosaurs. The earliest well-known dinosaur, and the earliest well-known theropod, is *Eoraptor*. It is simple enough to be the ancestor of one, or both, so I've shown it in both places. Prosauropods are known from Late Triassic rocks, and sauropods from the Triassic/Jurassic boundary. Therefore, the cladogram predicts that all four groups of dinosaurs had diverged by that time, and that there are "ghost ornithischians" and "ghost sauropods" still to be discovered somewhere in Late Triassic rocks. New fossils will clean up these uncertainties (or make them worse!).

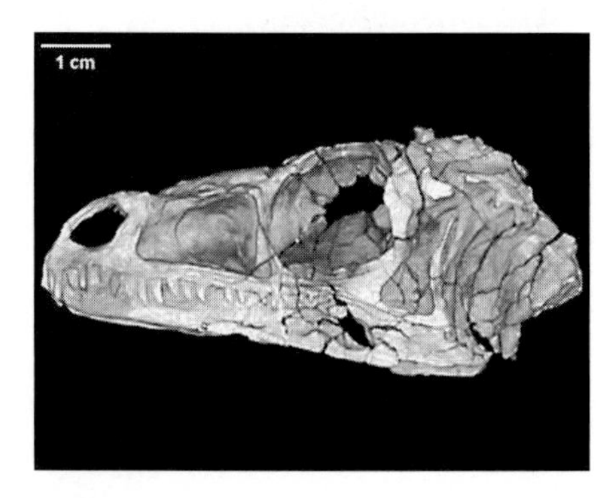

Figure 12.2 CT scan of the skull of *Eoraptor*. (Courtesy Tim Rowe and the Digimorph project at the University of Texas, Austin.)

Giganotosaurus from Argentina, and in *Carcharodontosaurus* from North Africa. These theropods were the largest land carnivores of all time, each weighing about 6 or 7 tons (more than an elephant), and standing about 6 m (20 ft) high, with a total length around 12 m (40 ft). At present, *Giganotosaurus* seems to be the largest, with an estimated length close to 14 m (40+ ft), and a weight of 6–8 tons. All these giant theropods must have relied on massive impact from the head for killing, aided by huge stabbing teeth that would have caused severe bleeding, usually lethal, in a prey animal.

Early theropods included *Coelophysis* from the Late Triassic of North America (Figure 12.3). At 2.5 m (8 ft) long, and lightly built (perhaps only 20 kg or 45 lb), it was clearly adapted for fast running. The bones of its skeleton were more extensively fused into stronger units than in the earliest theropods, so *Coelophysis* is placed into the first of the derived theropod groups, the coelophysoids (Figure 12.4). Jurassic ceratosaurs (Figure 12.5) and allosaurs included large powerful predators up to 6 m long.

The large clade of **coelurosaurs** (Figure 12.4) includes more advanced theropods of all sizes. *Compsognathus* is a basal coelurosaur that was small but an active predator with long arms and clawed fingers (Figure 12.6). **Ornithomimids** are the so-called ostrich dinosaurs. Their body plan much like that of a living ostrich, except that they had long arms and slim, dexterous fingers instead of wings. Ornithomimids had long legs and necks, large eyes, and rather large brains, but no teeth. They could have been formidable carnivores, of course, but perhaps they specialized on smaller prey animals. Many of them lacked large claws and had long fingers that could have been used to manipulate objects. **Tyrannosaurs** are well known, of course (Figure 12.7). They had enormous heads and tiny arms, a combination of characters that is still not understood. It is still a debate whether tyrannosaurs were giant predators or giant scavengers.

The third major lineage of coelurosaurs is made up of small- to medium-sized, agile carnivores, the **maniraptorans** (Figures 12.4; 12.8). **Oviraptors** are the nest-building dinosaurs from Mongolia, to be discussed later. **Therizinosaurs** are large, superficially bird-like theropods that were and are sometimes thought to be birds. *Mononykus*, from the Late Cretaceous of Mongolia, is a small theropod with a rather long tail, but it has a breastbone like a bird. Its arms were much modified, so the hand had only one strong, blunt, clawed finger. The scientists who described *Mononykus* wondered whether it dug with these strange hands, but they recognized that burrowing would not suit a rather tall, long legged theropod. I suggest that it used the hands for digging out its prey (small mammals?) or for molding its nest. Ecologically, I would compare it with the big, flightless, megapod birds of Australia and New Guinea, which build a huge nest from piles of leaves that they collect and shape by kicking backwards with their enlarged feet. This is a truly comical sight, especially as the bird has to keep looking over its shoulder to see what it is doing. *Mononykus*, I suspect, had a much easier time making its nest. The arms of *Mononykus* seem especially well designed for adduction (moving together under a load) so I envisage *Mononykus* digging a shallow nest, then sweeping together vegetation or sand to cover its eggs.

The other major clade of maniraptors includes **birds**, **troodonts**, and **dromaeosaurs**. Troodonts are the least specialized of the three (Figure 12.9). Dromaeosaurs include *Velociraptor*, famous star of the movie *Jurassic Park*, and *Deinonychus*, from the Early Cretaceous, one of the most impressive carnivores that ever evolved (Figures 12.10; 12.11). It was about 3.5 m long, it was clearly fast and agile, and had murderous slashing claws on both hands and feet, and a most impressive set of teeth.

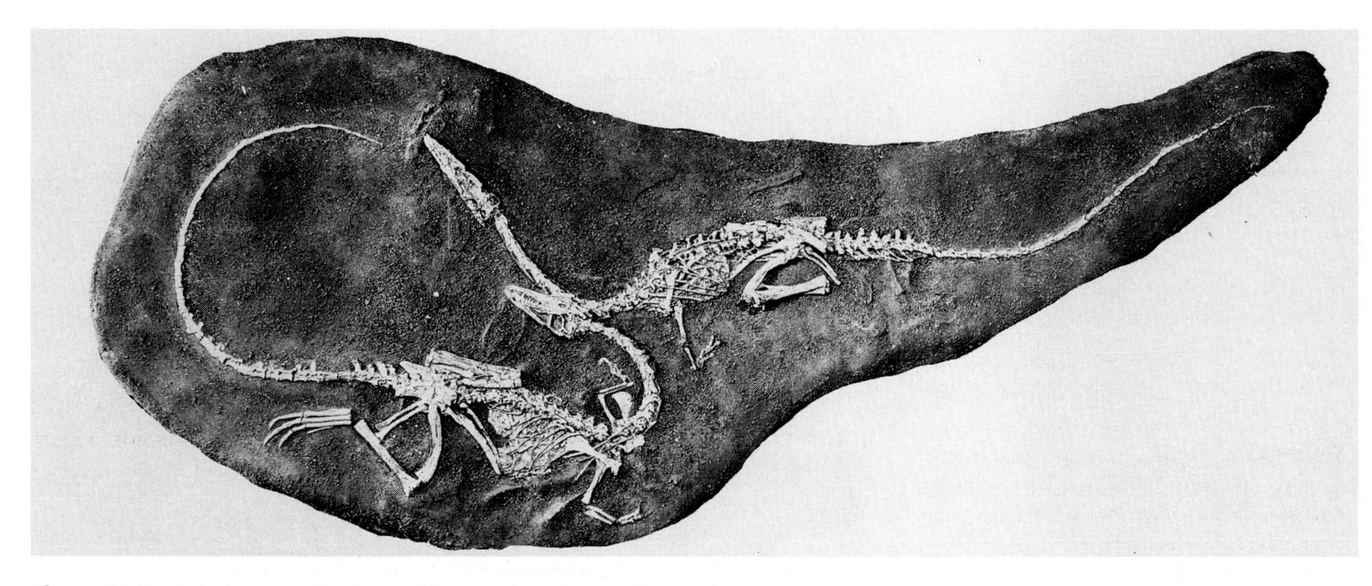

Figure 12.3 *Coelophysis*, an early theropod dinosaur from the Late Triassic of New Mexico. (Negative 329319. Photograph by Boltin. Courtesy of the Library Services Department, American Museum of Natural History.)

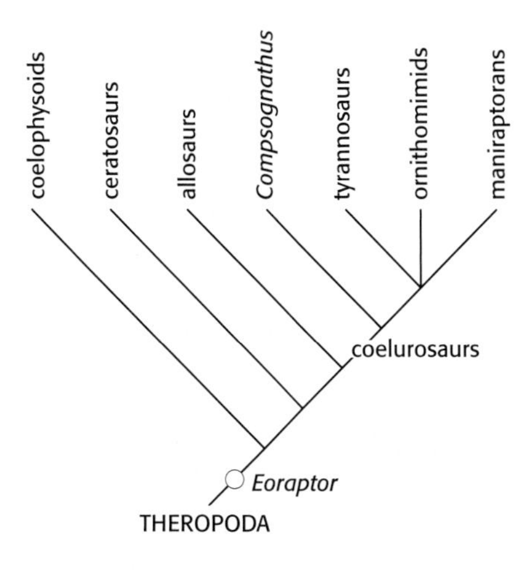

Figure 12.4 A phylogram of the theropods (simplified from Brochu, 2001).

Figure 12.5 A ceratosaur. Ceratosaurs were up to 6 m (20 ft) long. (From Marsh.)

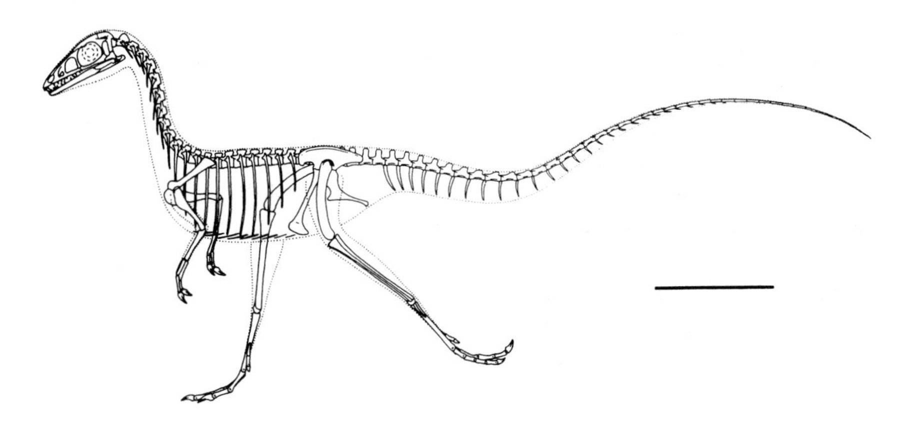

Figure 12.6 *Compsognathus*. (Courtesy of John Ostrom, Yale University.)

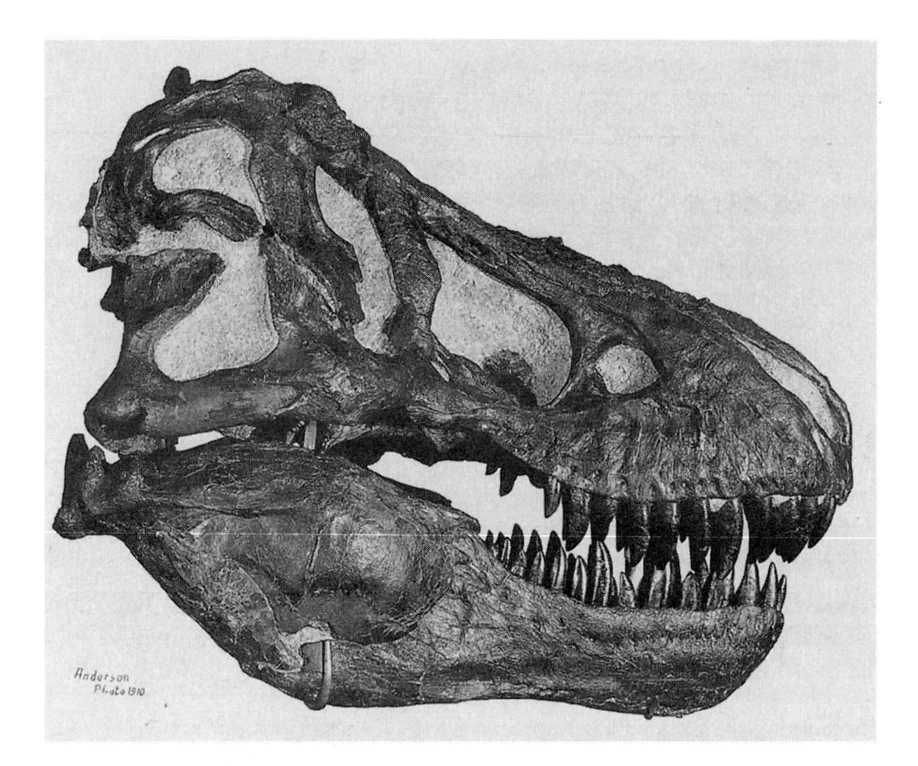

Figure 12.7 *Tyrannosaurus rex* had a massive skull. These jaws must have been the primary killing weapons. (Negative 35491. Photograph by A. E. Anderson. Courtesy of the Library Services Department, American Museum of Natural History.)

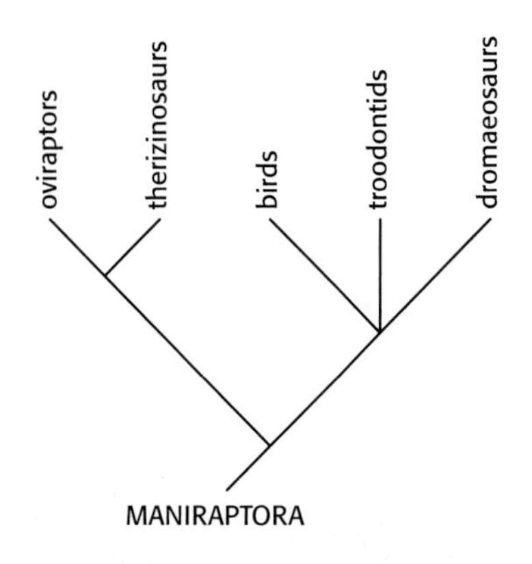

Figure 12.8 One of several alternative cladograms for maniraptoran theropods.

Figure 12.9 *Troodon*, a lightly built maniraptoran with a relatively large brain. (Reconstruction by Bob Giuliani. © Dover Publications Inc., New York. Reproduced by permission.)

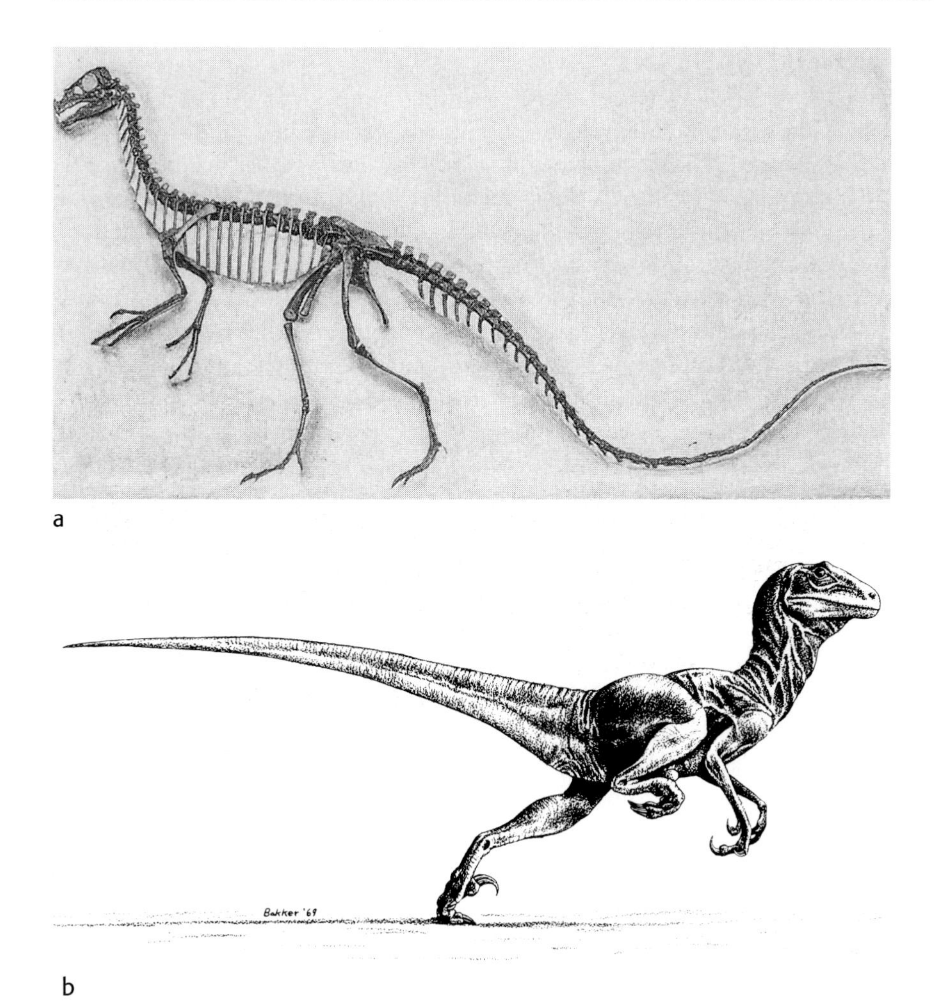

Figure 12.10 *Deinonychus.* (a) A skeletal reconstruction by John Ostrom, showing the animal as a powerful, fast-running predator. The tail was so tied by tendons that it acted as a wonderful counterweight for agile turning. (b) Flesh reconstruction of *Deinonychus* by Robert Bakker. Even apart from the physical weaponry of this dinosaur, Bakker's reconstruction of muscles and ligaments gives a vivid image of its power. (Both images courtesy of John Ostrom, Yale University.)

Figure 12.11 The skull of *Deinonychus* is long, light, and strong, and the teeth are designed for seizing, slashing, and cutting. The "velociraptors" of the movie *Jurassic Park* owed more to *Deinonychus* than they did to the real *Velociraptor.* (Courtesy of John Ostrom, Yale University.)

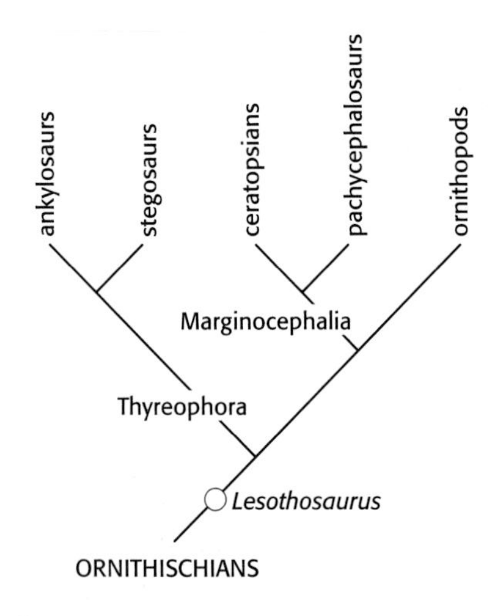

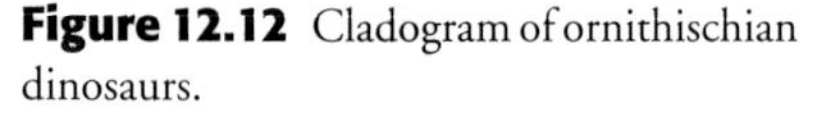

Figure 12.12 Cladogram of ornithischian dinosaurs.

ORNITHISCHIANS

The earliest ornithischians were the first herbivorous dinosaurs, and they gave rise to a spectacular radiation of dinosaurs, all of which were also herbivorous. Judging by their teeth, most ornithischians ate rather coarse, low-calorie vegetation, so many of them tended to be at least medium-sized. They were the most varied and successful herbivorous animals of the Mesozoic era, and they were abundant in terrestrial ecosystems right to the end of the Cretaceous. Small bipedal basal ornithischians, such as *Lesothosaurus*, gave rise to derived groups that were much heavier, with some animals weighing 5 tons or more (Figure 12.12). The armored dinosaurs (Thyreophora) form one derived clade that includes stegosaurs and ankylosaurs, and another major clade is the Marginocephalia, which includes the horned dinosaurs or ceratopsians and the heavy skulled pachycephalosaurs. However, most of the larger ornithischians were ornithopods, which include iguanodonts and the so-called "duck-bill" dinosaurs, the hadrosaurs.

The best-known early ornithischians are small bipedal forms dinosaurs from the Early Jurassic of Gondwana. *Lesothosaurus* (Figure 12.13a), was small, agile, and fast-running, but it clearly had vegetarian teeth. *Heterodontosaurus* had teeth that were even more specialized for a vegetarian diet (Figure 12.13b). Small teeth at the front of the upper jaw bit off vegetation against a horny pad on the lower jaw. The back teeth evolved into shearing blades for cutting vegetation. (The sharp incisors were for display or fighting.) The cheek teeth were set far inward, with large pouches outside them to hold half-chewed food for efficient processing.

The earliest ornithopods were less than a meter long, but they soon increased significantly in size. A general theme of ornithopod evolution was the successive appearance of groups that in different ways evolved toward the 5- to 6-ton size that seems to have been a weight limit for most terrestrial herbivores. Even at this size, many ornithopods remained bipedal. Others probably walked most of the time on all fours but raised themselves up on two limbs for running, or browsing on high vegetation (like goats and gerenuks today [Figure 12.14]).

Ornithopods evolved large batteries of teeth, and newly evolved modifications of the jaws and jaw supports allowed complex chewing motions. **Iguanodonts** were particularly abundant in the Early Cretaceous: they reached 9 m (30 ft) in length and stood perhaps 5 m (16 ft) high. They cropped off vegetation with powerful beaks before grinding it. Most iguanodonts were replaced ecologically by a variety of **hadrosaurs** (duck-billed dinosaurs) in the Middle and Late Cretaceous (Figures 12.14; 12.15). Hadrosaurs were about the same in size and body plan as iguanodonts, but had tremendous tooth batteries, with several hundred teeth in use at any time.

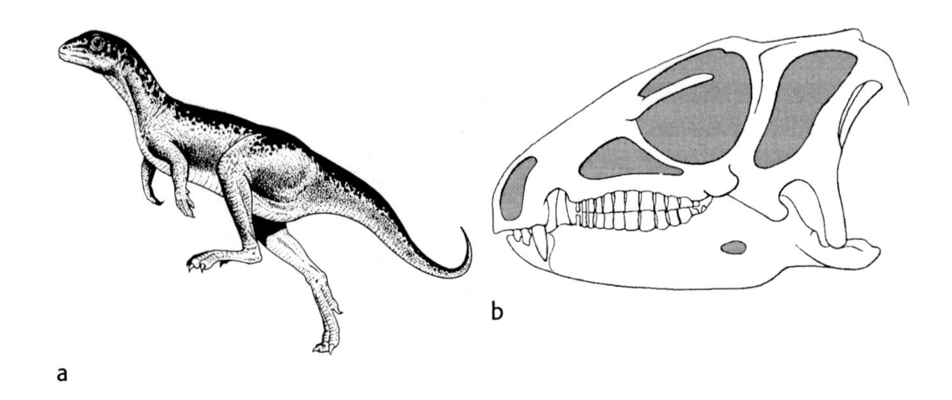

Figure 12.13 Two early ornithischians, both from southern Gondwana. Like all basal dinosaurs, they were erect, bipedal, and small. (a) *Lesothosaurus*. (Reconstruction by Bob Giuliani. © Dover Publications Inc., New York. Reproduced by permission.) (b) *Heterodontosaurus* had teeth that varied greatly along the jaw in size, shape, and presumed function. Skull about 10 cm (4 in) long. (After Charig and Crompton)

The other ornithischians were dominantly quadrupeds, but they betrayed their bipedal ornithopod ancestry with hindlimbs that were usually longer and stronger than the forelimbs. **Stegosaurs** (Figure 12.16), with their characteristic plates set along the spine, were the major quadrupedal ornithischians in the Jurassic, but they were replaced in the Early and Middle Cretaceous by the armored **ankylosaurs** (Figure 12.17). Later in the Cretaceous, the ornithischians were particularly abundant and varied in their body styles. Many quadrupedal forms lived alongside the hadrosaurs, including the **ceratopsians**, or horned dinosaurs (Figure 12.18).

Figure 12.14 The hadrosaur *Corythosaurus*, an advanced ornithopod, showing its ability to rear on its hind legs to reach vegetation. (Reconstruction by Bob Giuliani. © Dover Publications Inc., New York. Reproduced by permission.)

SAUROPODOMORPHS

Some early dinosaurs evolved to become very large, heavy quadrupedal vegetarians with broad feet and strong pillar-like limbs. The sauropodomorphs had an early radiation as prosauropods and a later radiation as the famous sauropods with which we are all familiar. **Prosauropods** were abundant, medium to large dinosaurs of the Late Triassic (Figure 12.19). They were typically about 6 m (20 ft) long, but *Riojasaurus* was unusually large at 10 m (over 30 ft). Prosauropods lived on all continents except Antarctica, with rich faunas known from Europe, Africa, South America, and Asia. They ranged into the Early Jurassic, when they were replaced by sauropods.

Prosauropods were all browsing herbivores. The teeth were generally good for cutting vegetation but not for pulping it. Opposing teeth did not contact one another, and all the grinding must have been done in a gizzard. (Masses of small stones have been found inside the skeletons of several prosauropods.) Prosauropods have particularly long, lightly built necks and heads, and light forequarters. They were clearly adapted to browse high in vegetation, perhaps reaching up from the tripod formed by the hind limbs and heavy tail. Only *Riojasaurus*, the largest, was always quadrupedal because of its weight, but it had a very long neck to compensate. Prosauropods were the first animals to browse on vegetation high above the ground,

Figure 12.15 The skeleton of a hadrosaur, the duck-billed dinosaur *Edmontosaurus*. Ossified tendons helped to hold the tail rigid over the pelvis. (From Marsh.)

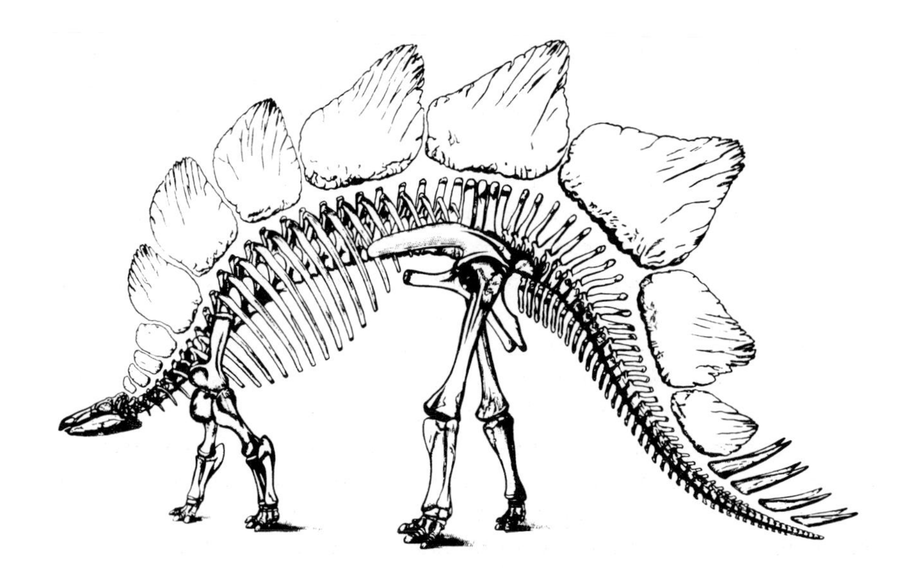

Figure 12.16 The skeleton of a stegosaur. People still argue about placement of the plates and the number of pairs of spikes on the tail. (From Marsh.)

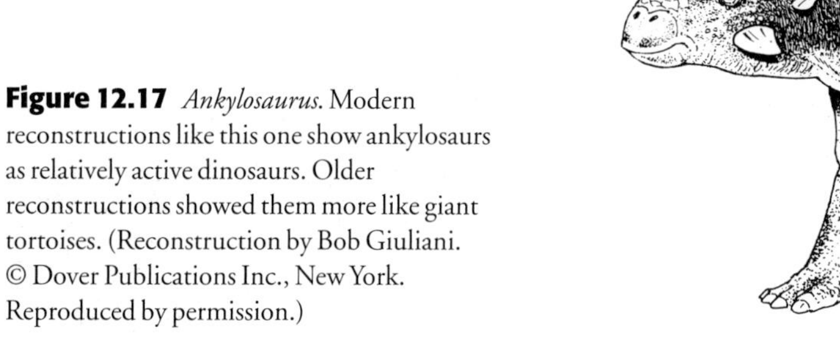

Figure 12.17 *Ankylosaurus.* Modern reconstructions like this one show ankylosaurs as relatively active dinosaurs. Older reconstructions showed them more like giant tortoises. (Reconstruction by Bob Giuliani. © Dover Publications Inc., New York. Reproduced by permission.)

Figure 12.18 *Triceratops* is the most famous of the horned dinosaurs or ceratopsians. (Reconstruction by Bob Giuliani. © Dover Publications Inc., New York. Reproduced by permission.)

Figure 12.19 *Anchisaurus*, a small prosauropod from the Early Jurassic of New England, about 2.5 m (8 ft) long. Clearly, it would have been equally comfortable in this bipedal pose, or on four feet. (From Marsh.)

and they represent a completely new ecological group of herbivores exploiting an important new resource in the zone up to perhaps 4 m (13 ft) above ground. The same adaptation was reevolved later in sauropods, and again in mammals such as the giraffe.

Prosauropods began small, like other dinosaur groups. *Anchisaurus* was (only) about 2.5 m (8 ft) long (Figure 12.19). But soon prosauropods were the largest and heaviest members of their communities, and they were abundant. *Plateosaurus* accounts for 75% of the total individuals in a well-collected site in Germany, and probably over 90% of the animal mass in its community.

Sauropods are the largest land animals that ever evolved. Remember that there is a natural human ambition to discover the largest or the oldest of anything; we must be cautious in assessing claims about the size and weight of dinosaurs without also surveying the evidence. But even a cautious person must admit that well-documented sauropod body weights are at least 50 tons; *Argentinosaurus*, from Argentina of course, may have been close to 100 tons. Famous names and enormous numbers are associated with sauropod anatomy: *Seismosaurus*, from New Mexico, was at least 28 m (90 ft) long, and maybe much more (it is probably a huge individual of *Diplodocus*). *Brachiosaurus* has long forelimbs carrying it over 12 m (40 ft) high, as tall as a four-story building, and with a weight estimated at 50 to 80 tons.

Sauropods were all herbivores, of course; no land animals that size could have been carnivorous. They had curiously small heads and very long necks that allowed them to browse on anything within 10 m (33 ft) of ground level. The tails were long also, but the body was massive, with powerful load-bearing limb bones and pelvis. All sauropods were quadrupedal, though perhaps the earliest forms could briefly have been bipedal. The major body mass was centered close to the pelvis, which was accordingly more massive than the shoulder girdle. Within sauropods, there were two major lineages. One, the Diplodocoidea, obviously includes *Diplodocus* itself, and other sauropods with fairly long skulls and peg-like teeth (Figure 12.20). The other is the Macronaria, named for the fact that the compact skull has huge gaps for nostrils (Figure 12.21). This group includes the camarasaurs and titanosaurs. The nostrils may have been associated with sound production, but it is difficult to see how to test that.

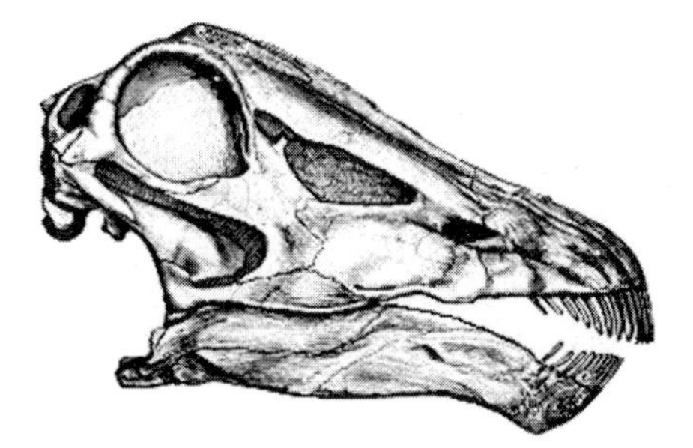

Figure 12.20 The head of a large sauropod, *Diplodocus*, showing the projecting teeth. (From Marsh.)

DINOSAUR PALEOBIOLOGY: LIFE AT LARGE SIZE

There are small dinosaurs: *Compsognathus* was only the size of a chicken, for example. But the dominant feature of dinosaurs, and the dominant aspect of their paleobiology, has to be the enormous size of the largest ones. Ornithischian dinosaurs are easier to understand than the others because they were vegetarians in the 5-ton range, comparable with living elephants or rhinos, perhaps. However, there are no 5-ton carnivores alive today on land that we can compare with carnosaurs such as *Tyrannosaurus*, and there are no 50-ton vegetarians that we can compare with the sauropods. Despite this, we can make some reasonable inferences about dinosaur biology.

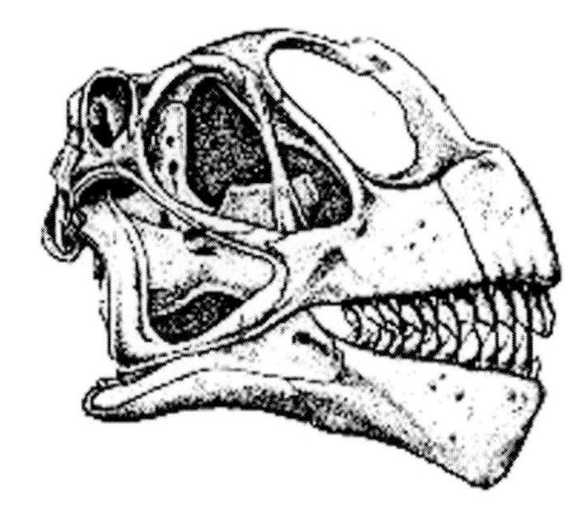

Figure 12.21 The head of a macronarian sauropod, *Camarasaurus*. It is difficult to test any hypothesis about the function of such huge nostrils. (From Marsh.)

Vegetarian Dinosaurs

Animals on purely vegetarian diets almost always have bacteria in their guts to help them break down cellulose. Large animals have slower metabolic rates than small ones, and for vegetarians this means a slower passage of food through the gut and more time for fermentation. Alternatively, a large vegetarian can digest a smaller percentage of its food and live on much poorer quality forage. Vegetarians usually grind their food well so that it can be digested faster, and this was accomplished in two different ways in the two major groups of vegetarian dinosaurs, the ornithischians and the sauropods.

Sauropods had very small heads for their size. This has sometimes been thought to indicate a soft diet that did not need much chewing. However, sauropods probably had a small head for the same mechanical reason that giraffes do: the head sits on the end of a very long neck. In both sets of animals the food is gathered by the mouth and teeth, then swallowed and macerated later.

Giraffes are ruminants, and boluses of food are regurgitated and chewed at leisure with powerful batteries of molar teeth. Sauropods probably used a different system for grinding food. They probably had enormous, powerful gizzards in which food was ground up between stones that the dinosaurs swallowed. Wild birds seek and swallow grit to help them grind food, and poultry farmers can increase egg production by feeding grit to their chickens. (A fossil moa of New Zealand was found with 2.5 kg [over 5 lb] of stones in its stomach area.) There are literally millions of dinosaur gizzard stones or gastroliths in Cretaceous rocks in the western interior of the United States. A high proportion of these are made of very hard rocks, often colored cherts. William Stokes made the irresistible suggestion that dinosaurs specifically sought brightly colored, rounded pebbles in stream beds for swallowing, providing themselves with perfectly shaped and very hard grinding stones. (Dinosaurs as the first rockhounds!)

Ornithischians generally had impressive teeth, especially in advanced hadrosaurs and ceratopsians, and they would have chewed up their food thoroughly, as living mammals do. Even so, there are gizzard stones associated with fossils of the little ceratopsian *Psittacosaurus.*

Food gathering and processing do not seem to have posed problems difficult enough to prevent vegetarian dinosaurs from reaching enormous size. Dinosaurs evolved to be much larger than other land vertebrates for reasons not connected with diet.

Posture and Habitat

We can use our knowledge of the mechanics of bone, muscle, and ligament to interpret dinosaur skeletons. Early workers had difficulty coming to grips with the size of dinosaurs, mostly because they did not understand biomechanics. Thus, early reconstructions show dinosaurs sitting in trees or jumping high in the air. In fact, they would have broken bones on impact after even a small fall. Older reconstructions showing large dinosaurs with bent knees and bent elbows, usually near water, are equally inappropriate, because they are based on comparison with lizards. A 5-ton stegosaur and especially a 50-ton sauropod must have had erect limbs to support their weight, like elephants or rhinos rather than lizards.

Early workers, who suspected that sauropods would have found it difficult to support their weight with bent limbs, suggested that instead they spent much of their life in swamps, buoyed up by water. The long necks were interpreted as snorkels (Figure 12.22). Even today, many drawings of sauropods show water nearby. The snorkel idea is very unlikely: in any depth of water, the pressure would have been great enough to prevent the rib cage from expanding, and the sauropod would have suffocated. All sauropods and most other dinosaurs should be envisioned as having lived in drier plains country. Duck-billed dinosaurs did not live in water either; their jaws had enormous batteries of teeth with large grinding surfaces, and they ate coarse vegetation in dry country.

DINOSAUR BEHAVIOR

Dinosaur behavior can be judged by footprints; for example, a dinosaur stampede has been discovered (Figure 12.23). Rocks laid down about 90 Ma as sediments near a Cretaceous lakeshore in Queensland, Australia, bear the track of a large carnosaur heading down toward the lake with a 2 m (6 ft) stride. Superimposed on this track are thousands of small footprints made by small, bipedal, lightly built dinosaurs, running back up the creek bed away from the water. More than 3000 footprints have now been uncovered, showing all the signs of a panicked stampede. At least 200 animals belonging to two species were stampeding. One of the species, probably a coelurosaur, ranged up to about 40 kg (90 lb), and the other, probably an ornithopod, ranged up to twice as large. Juveniles and adults of both species were digging in their toes as they tried to accelerate: 99% of the footprints lack heel-

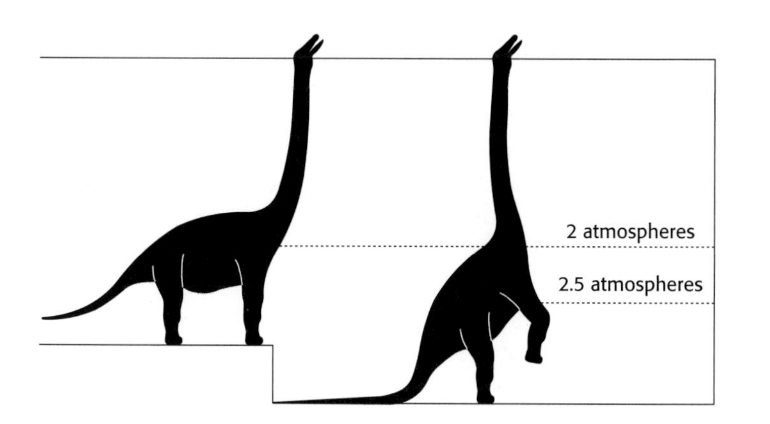

Figure 12.22 Sauropods could not have used their necks as snorkels. Water pressure on the rib cage would have prevented it from expanding, and the animal would have suffocated. (Sauropod postures after Bakker.)

Figure 12.23 A dinosaur stampede from Cretaceous rocks in Queensland. A large theropod walked or stalked from left to right along an ancient creek bed toward a shallow lake. Thousands of little footprints were made by small dinosaurs all stampeding the other way, across and over the tracks the carnosaur made on its way down to the lake. (Courtesy of the Queensland Museum, Australia.)

> **Limerick 12.1**
> Impelled by their dinosaur greed,
> They went down to the lake for a feed.
> But a theropod near
> Made them scamper with fear
> In the first recorded stampede.

marks. The footprints show slipping, scrabbling, and sliding, and the smaller species usually avoided the tracks of the larger one. They must have been hemmed against the lakeshore, breaking away in a terrified group.

The stampede sheds light on ecology as well as behavior, telling us that at least some dinosaurs gathered in herds and behaved just as African plains animals do today at waterholes on the savanna, responding immediately and instinctively to the approach of dangerous predators.

Trackways from the Middle Jurassic of Britain document a group of titanosaurs traveling in a group: a good idea, since the tracks of large theropods occur on the same bed!

Other indicators of behavior are preserved in dinosaur skeletons. The ornithischian dinosaur *Pachycephalosaurus* had a dome-shaped area of bone on the top of its skull (Figure 12.24). It is not solid bone but has air cavities in it, with an internal structure very much like that found in the skulls of sheep and goats that fight by ramming their heads together. Did *Pachycephalosaurus* do that? It weighed eleven times as much as a bighorn sheep, so the most recent interpretation of this dinosaur is that it probably butted opponents from the side rather than head-to-head. *Triceratops* and other ceratopsians, however, also had air spaces between the horns and the brain case, which may indicate that they competed by direct head-to-head impact.

Some hadrosaurs had huge crests on the head (Figure 12.25). The crests were not solid but contained tubes running upward from the nostrils and back down into the roof of the mouth (Figure 12.26). Only large males had large crests; females had smaller ones, and juveniles had none at all. The tubes are unlikely to have evolved for additional respiration or thermoregulation. (If so, adults would have needed large tubes whether they were female or male.) When the tubes are reconstructed

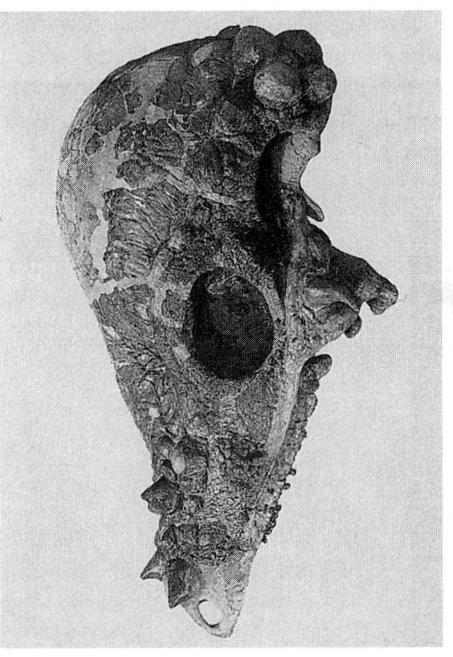

a

b

Figure 12.24 (a) Skull of the dome-headed dinosaur *Pachycephalosaurus*, arranged in head-butting position. (Negative 319456. Photo by C. H. Coles. Courtesy of the Library Services Department, American Museum of Natural History.) (b) *Pachycephalosaurus*. (Reconstruction by Bob Giuliani. © Dover Publications Inc., New York. Reproduced by permission.)

they look like medieval horns and can be blown to give a note (Figure 12.26). The varying sizes of crests allow us to infer differences between the sounds produced by young, by adult females, and by adult males, to go with the different visual signals provided by the crests. These hadrosaurs may have had a sophisticated social system, perhaps as complex as those we take for granted in mammals and birds.

Figure 12.25 Large (probably male) *Parasaurolophus* with enormous crest on the head. (Reconstruction by Bob Giuliani. © Dover Publications Inc., New York. Reproduced by permission.)

DINOSAUR EGGS AND NESTS

Major finds of fossilized dinosaur eggs and nests (Figure 12.27) have been made in Cretaceous rocks in Montana, Alberta, Mongolia, Spain, and Argentina. We now have eggs from all three major dinosaur groups.

The Montana nests are carefully constructed bowls of mud or sand lined with vegetation. In each nest the eggs were laid or arranged in a neat pattern so they would not roll around. Embryonic dinosaurs are preserved in some nests. Many of the nests are clustered together at regular close intervals, suggesting that they were in communal breeding grounds: nesting colonies, if you like. The nests had sometimes been remodeled and reused, perhaps in successive seasons. Many large (long-lived) birds do this today, including the red-tailed hawks in my pine tree.

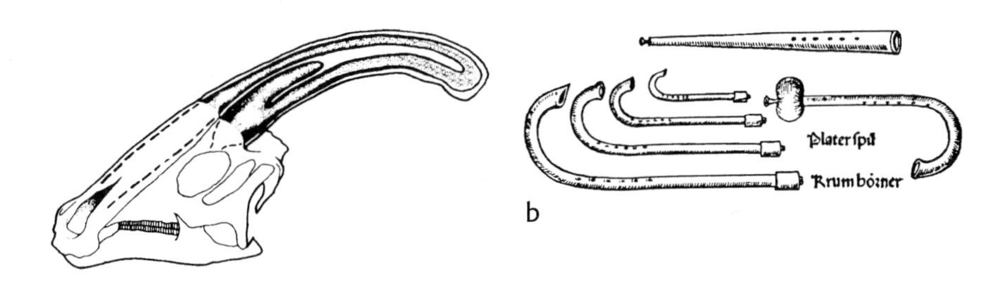

Figure 12.26 (a) A cross-section of a *Parasaurolophus* crest shows that it carried hollow tubes that were part of the nostril system. The left and right tubes from each nostril met in a central chamber above the roof of the mouth. Females and juveniles had smaller crests or none at all. (After Weishampel and Norman, 1981.) (b) Medieval German musical instruments that are very much like the air passages of an adult male *Parasaurolophus.* David Weishampel made this striking analogy in 1981.

Figure 12.27 Dinosaur eggs. (Negative 410765. Photo by Shackelford. Courtesy of the Department of Library Services, American Museum of Natural History.)

There is good evidence for long-term parental care by dinosaurs. One nest from Montana contained 15 baby duck-billed dinosaurs. We know that these baby dinosaurs were not new hatchlings because they are about twice too large for the eggshells found nearby, and because their teeth had been used long enough to have wear marks. But they were together in the nest when they died and were buried and fossilized, with an adult close by—*Maiasaura*, the "good mother."

> **Limerick 12.2**
> Makela and Horner have stressed
> That maiasaur mothers were best
> Their chicks' teeth are worn
> So they're fairly well grown
> But they're still in a group in the nest!

This interpretation has been challenged on the grounds that hatchling dinosaurs had well-developed limbs and pelvis, showing that they were capable of moving about as soon as they hatched. Therefore, they did not receive parental care. Of course, this argument does not work. At a very different body size, partridges hatch a large number of chicks, which are taken by a parent on trips from the nest to forage very soon after hatching. Ostriches run a crèche system for the care of foraging young.

Dinosaur eggs and nests were found in Mongolia in the 1920s. They were naturally associated with the most abundant dinosaur in the area, *Protoceratops*. To everyone's surprise, when an embryo was finally discovered inside one of the eggs, it was well enough preserved to be identified not as *Protoceratops*, but as the little theropod *Oviraptor* (Figure 12.28). The irony here is that an adult *Oviraptor* had originally been identified (and named) as a nest-robbing, egg-eater, preying on an innocent *Protoceratops*!

A block of rock collected in Mongolia in 1993 turned out to contain an adult *Oviraptor* that had been buried in a sandstorm while it was crouched over a nest of *Oviraptor* eggs (Figure 12.29). The only reasonable explanation of this find is that the dinosaur was brooding its eggs, just as most living birds do. By 1996, three of the seven known *Oviraptor* adults had been discovered on or near nests.

So, in terms of their posture and reconstructed behavior, including parental care, complex social structure, and intelligence (needed to run a complex society), dinosaurs should be compared not with living reptiles, but with living mammals and birds.

Mammals and birds are warmblooded, with high metabolic rates, and they are much more intelligent than other living vertebrates. But do these characters also apply to dinosaurs? The question is usually worded, "Were dinosaurs warmblooded?" The arguments are intense and interesting, and they are not yet resolved.

WERE DINOSAURS WARMBLOODED?

There are several aspects of this question. First, what does "warmblooded" mean? Did dinosaurs control their body temperature precisely? that is, were they

Figure 12.28 *Oviraptor*, misnamed as an egg thief when actually it was a builder and guardian of its own nests. (Reconstruction by Bob Giuliani. © Dover Publications Inc., New York. Reproduced by permission.)

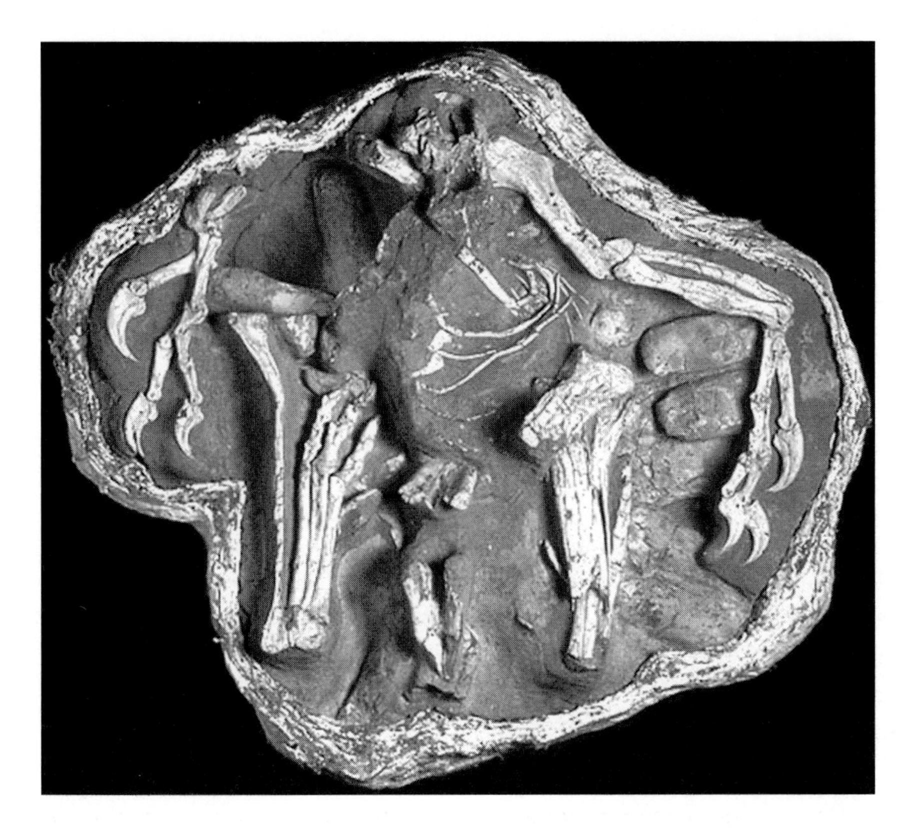

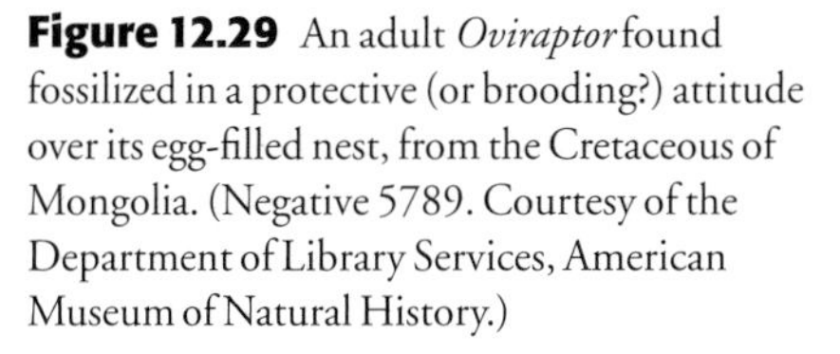

Figure 12.29 An adult *Oviraptor* found fossilized in a protective (or brooding?) attitude over its egg-filled nest, from the Cretaceous of Mongolia. (Negative 5789. Courtesy of the Department of Library Services, American Museum of Natural History.)

homeotherms? Or were they **heterotherms**, allowing their body temperature to vary quite widely, depending on their needs? Were they **endotherms**? did they produce a high body temperature by generating a lot of metabolic heat, raising their overall energy budget? Or were they **ectotherms**, controlling their body temperature dominantly by behavior, by clever use of sunlight, shade, shelter, activity level, and so on? If they were homeotherms, did they control their body temperature at a much higher level than their average environment, say 35°–40°C (95°–104°F), like most mammals and birds? Or at a lower level, say 30°C (86°F), more like a platypus or a hedgehog? Or was it lower still, hardly any warmer than the average temperature of an equatorial habitat, say 27°C (81°F)? Along with these answers come different estimates of the resting metabolic level, the energy output, and the activity level of dinosaurs. Were they comparable with elephants, ostriches, or lions, or more like living reptiles, or like none of the above? Were all dinosaurs alike in their physiology?

We are still in the process of learning the differences between living reptiles and mammals. Any vertebrate cells are capable of high-energy output if they are kept fueled with oxygen and food. Thus, the secret of evolving thermoregulation at high levels and at high resting metabolic rates lies in the engineering around the cells rather than in their biochemistry. Respiration and circulation systems, which transport oxygen and food, are the crucial factors. For example, the hearts of living mammals and reptiles are very different, and David Carrier has shown how and why their respiratory systems and locomotory systems are different too (Chapter 11).

To understand dinosaurs, we need to be able to reconstruct the circulation and respiration of these fossil vertebrates, as well as their locomotion. It is certainly not easy, but it's a better prospect than trying to infer the biochemistry of their cells!

Then there is an ecological factor. Whatever the advantages of high resting metabolic rate, it has a cost: more food must be eaten. The higher the metabolic rate, the greater the cost.

For nearly 30 years, professional and amateur paleontologists alike have been increasingly persuaded that dinosaurs were warmblooded, with connotations of high levels of activity. The novel and movie *Jurassic Park* represent the acme of this wave of thought. The author Michael Crichton was very much influenced by the work of John Ostrom and Robert Bakker and the immense success of the movie in turn set the popular image of dinosaurs. But we need to look at evidence rather than a movie screen.

DINOSAURS WITH FEATHERS

Feathers have always been regarded as structures unique to birds: in fact, they have been used for 200 years as one of the most important characters that define birds. Recently discoveries have shown that some dinosaurs had feathers too.

Six theropods from Early Cretaceous beds in China are of great interest because they are so well preserved. *Sinosauropteryx* and *Beipiaosaurus* have a halo of very fine structures on the body surface that look like down. *Protarchaeopteryx* has down feathers on its body, tail, and legs, and a fan-shaped bunch of long feathers, several inches long, at the very end of its tail. *Caudipteryx* also has down and strong tail feathers, but it has feathers on its arms as well. They are shorter toward the fingers, and longest toward the elbow, in contrast to the feathers on the wings of flying birds. Finally, and most important, *Microraptor* and *Sinornithosaurus* have true branching feathers on all four limbs and on the tail.

Caudipteryx is closely related to oviraptors, and *Beipiaosaurus* is a therizinosaur. *Microraptor* and *Sinornithosaurus* are both dromaeosaurs. We can use this information to assess how many theropods had feathers. Look at the cladogram of Figure 12.8. If oviraptors and dromaeosaurs and birds have feathers, and if feathers only evolved once (which seems almost inevitable, given their complexity), then all maniraptorans and perhaps even more theropods could have had feathers too.

Were all dinosaurs feathered? The answer is "No." We have direct evidence from skin impressions that some adult ornithischians and sauropods had scales. But baby elephants have hair, so the evidence about scaly adult dinosaurs may not tell the whole story. The small ornithischian *Psittacosaurus* had long bristles around its tail, for example, while the rest of its skin had scales.

Altogether, the Chinese theropods confirm that birds evolved from small ground-running predatory theropods, and that feathers evolved before birds and before flight. Of the Chinese theropods, *Protarchaeopteryx* is the closest relative yet discovered to birds, especially to the most primitive bird, *Archaeopteryx*.

So how do we now distinguish a bird from a small theropod? With difficulty, and certainly not by its feathers! (By now there is no significant feature of the skeleton that can be used with confidence.)

This is perhaps a good opportunity to illustrate how insignificant the transition can be from one group to another. The first bird hatched out of an egg laid by a theropod dinosaur, but unless there were many other hatchlings at almost exactly the same evolutionary stage to form a breeding population, the lineage would have gone extinct. Even if you had been there, watching those feathery chicks hatch and grow up, you would not have been aware that you were seeing the transition from theropod to bird in that generation of that evolving population. (And, of course, that is true of any other evolutionary transition that has ever occurred.) That means that if the fossil record is relatively complete, the change that defines the transition will necessarily be one so trivial that it will look artificial—and, of course, it is trivial and it is artificial!

The Origin of Feathers

The proteins that make feathers in living birds are completely unlike the proteins that make reptilian scales today. Feathers originate in a skin layer deep under the outer layer that forms scales. It is very unlikely that feathers evolved from reptilian scales, even though that thought is deeply embedded in the minds of too many paleontologists. Feathers probably arose as new structures under and between reptile scales, not as modified scales. Many birds have scales on their lower legs and feet where feathers are not developed, and penguins have such short feathers on parts of their wings that the skin there is scaly for all practical purposes. So there is no real anatomical problem in imagining the evolution of feathers on a scaly reptilian skin. But feathers evolved in theropods as completely new structures, and any reasonable explanation of their origin has to take this into account. Obviously, feathers did not evolve for flight. They evolved for some other function and were later modified for flight.

Feathers may have evolved to aid thermoregulation. The feathered Chinese theropods all have down, probably as insulation to keep their bodies at an even temperature. It would not matter whether they used their feathers to conserve heat in cold periods, or to keep heat out in hot periods, or both. In either case, insulation would have been useful.

The thermoregulatory theory for the origin of feathers is probably the most widely accepted one today, but it does have problems. Why feathers? Feathers are more complex to grow, more difficult to maintain in good condition, more liable to damage, and more difficult to replace than fur. Every other creature that has evolved a thermoregulatory coat—from bats to bees and from caterpillars to pterosaurs—has some kind of furry cover. There is no apparent reason for evolving feathers rather than fur even for heat shielding.

Limerick 12.3
Our store of good data is slight:
We don't know which answer is right.
A cutaneous feather
May be better than leather,
But for warmth? or for show? or for flight?

Furthermore, thermoregulation cannot account for the length or the distribution of the long feathers on the Chinese theropods. Short feathers (down) can provide good thermoregulation, but thermoregulation does not require long feathers, and it would not help thermoregulation very much to evolve long strong feathers on the arms and tail. So it is difficult to suggest that feathers evolved for thermoregulation alone. It would be better to think of another equally simple explanation.

I naturally prefer an idea that I developed years ago, with my colleague Jere Lipps. In living birds, feathers are for flying, for insulation, but also for camouflage and/or display. Lipps and I suggested that feathers evolved for display. The display may have been between females or between males for dominance in mating systems (sexual selection), or between individuals for territory or food (social selection), or directed toward members of other species in defense.

Living reptiles and birds often display for one or all of these reasons, using color, motion, and posture as visual signals to an opponent. Display is often used to increase apparent body size; the smaller the animal, the more effectively a slight addition to its outline would increase its apparent size. Lipps and I therefore proposed that erectile, colored feathers would give such a selective advantage to a small displaying theropod that it would encourage a rapid transition from a scaly skin to a feathery coat. Display would have been advantageous as soon as any short feathers appeared, and it would have been most effective on movable appendages, such

Figure 12.30 Forearm display would have drawn attention to the powerful claws of the first bird, and derived theropod, *Archaeopteryx*. (From Heilmann.)

as forearms and tail. (Display on the legs would not be so visible or effective.) Forearm display by a small theropod would also have drawn particular attention to the powerful weapons the theropod carried there, its front claws (Figure 12.30). *Caudipteryx* carried long feathers on its middle finger, between the two outside claws, and it could fold that middle finger away, with the feathers out of harm's way.

The display hypothesis explains more features of the feathered theropods and the first bird *Archaeopteryx* than other hypotheses, with fewer assumptions. It explains completely the feather pattern: the evolution of long strong feathers on arms and tail.

Once they evolved, feathers could quickly have been coopted for thermoregulation, and the down coat on the Chinese theropods may show that process. Down can only be for thermoregulation. Although down is not proof of warm blood, it is very strong evidence in favor of it. In living birds, down feathers are associated with the problem of heat loss for hatchlings.

Egg Brooding

There are still other implications. The little theropod *Oviraptor* brooded its eggs, just as living birds do. A few desert-dwelling birds protect their eggs from the sun by sitting on them, but most birds brood their eggs to keep them warm: even the desert-dwellers keep their eggs warm at night. If *Oviraptor* had feathers, then it was even more bird-like than we had imagined, and it is very unlikely that it was cold-blooded!

It now becomes a matter of judgment how far to extend *Oviraptor*'s body temperature to other dinosaurs. We do not know whether other dinosaurs brooded their eggs, but we do know that all dinosaur groups built nests. Among the whole array of dinosaurs, it is clear that we can make the best case for homeothermy, perhaps even endothermy, among lineages of theropods. A completely different line of evidence can be applied to other dinosaur groups.

High-Latitude Dinosaurs

It is fairly easy to imagine a large adult dinosaur having a fairly even temperature during small temperature swings from day to night, especially in mild tropical climates. But there are rich plant floras and rich dinosaur faunas containing hadrosaurs and tyrannosaurs in Late Cretaceous rocks of the north slope of Alaska, which at the time was at high northern latitude: at least 70°N and possibly as much as 80°N. Equally rich cool-climate floras and Early Cretaceous dinosaurs flourished in southern Australia, which was then at about 80°S, and dinosaurs have been found in Antarctica too. In these high latitudes, vegetarian dinosaurs had to survive strong seasonal changes in light, temperature, and plant food supply.

Cretaceous polar temperatures were rather mild: luxuriant floras grew at high latitudes in both hemispheres. But luxuriant growth in polar latitudes can only occur in summer. Winter darkness prevents plant growth, and winter forage would have been scarce even if average Cretaceous polar winter temperatures were above freezing. (Summer growth in Arctic plants today is often limited by low temperatures even though there is a lot of light.) Late Cretaceous floras of North America were dominated by deciduous plants everywhere north of Montana, presumably because light levels were too low to make it worthwhile for plants to maintain leaves all through the winter. There must have been at least occasional frosts in polar latitudes, and if it's dark, there's no way of reheating a region once a winter storm cools it down.

There are no large ectotherms in high latitudes today: crocodiles and turtles live mostly in the tropics and only occasionally range into temperate climates. There are no reptiles at all in Alaska. Yet large endotherms survive successfully even near today's ice caps. Carnivores such as polar bears, and large swimming mammals such as walruses, penguins, seals, and whales live year-round in polar regions, in spite of the large amount of food they need for body heating. Terrestrial omnivores such as brown bears and grizzly bears avoid the worst period of low food supply and cold by hibernating. But large herbivores dare not hibernate: they would be vulnerable to attack by insomniac carnivores. They migrate instead; caribou herds cover hundreds of kilometers between seasons.

If living animals are reliable ecological guides, the big vegetarian dinosaurs of Alaska were endotherms and migrated with the seasons. Today the great migrants are fliers (birds), swimmers (whales), or walkers (large mammals with large, precocious young that can walk and run very quickly after birth—caribou, or African plains animals). The Alaskan dinosaurs included many juvenile hadrosaurs, which implies that hatchlings were reasonably precocious and grew quickly. All herbivores 5 tons or larger must be able to forage over a lot of ground. Probably the foraging of Cretaceous hadrosaurs involved significant northward and southward movements with the seasons. Some hadrosaurs may have migrated annually to nesting grounds in the north, just as pregnant caribou migrate northward today. Young hadrosaurs would then have had the advantage of feeding on the growth of the Arctic summer. Meanwhile, tyrannosaurs and smaller predators followed the migration, just as wolves follow the caribou herds today.

The same argument for migration does not apply in southern Australia, where the land mass inhabited by the dinosaurs was separated from the rest of the continent. There is no escaping the conclusion that the dinosaurs lived there year round. And many of them were small: the little ornithischian *Leaellynasaura* was only chicken-sized, another was dog-sized, and even an Australian *Allosaurus* was much smaller than its North American counterparts. *Leaellynasaura* had very big eye-sockets for its size, suggesting large eyes, perhaps to see during the dark polar winter. It is difficult to avoid the inference that these dinosaurs were endothermic.

DOUBTS ABOUT ENDOTHERMY?

David Carrier showed us a link between erect posture and endothermy (Chapter 11). Erect posture allows birds and mammals to breathe enough oxygen to run endothermy, but it does not require them to do so. If dinosaurs were able to migrate, they could have avoided the polar winters that require endothermy to survive. *Oviraptor* could have successfully brooded its eggs as long as it was homeothermic: it did not have to be endothermic. Down feathers might have been used to regulate temperature in an ectothermic theropod, not necessarily in an endothermic one. The little Australian dinosaurs may have been able to run endothermy at a cheap rate (say heating to what we might call cool temperatures, around 70°F or so). All these lines of evidence suggest endothermy, sometimes strongly, but they do not require it.

Research on dinosaur bone structure has yielded ambiguous evidence on metabolic levels, and therefore on body temperature and behavior. For example, a hypsilophodont (small ornithischian herbivore) from southern Australia had continuous bone deposition, while an ornithomimosaur (small theropod "ostrich dinosaur") from the same beds had cyclic bone deposition. Normally, one would interpret continuous bone deposition as signifying year-round activity and/or homeothermy, and cyclic bone deposition as signifying strongly seasonal bone deposition and/or heterothermy. But these dinosaurs were too small to migrate, and the continuous deposition in the hypsilophodont suggests that it at least did not hibernate. We are left with the uncomfortable feeling that dinosaur bone structure is not easy to interpret, in terms of endothermy at least, in any simple way.

Behavioral and Passive Thermoregulation

Temperature can be controlled to some extent by behavior (basking, bathing, seeking shade) as in living lizards and crocodiles. Many of the advantages of thermoregulation are gained even if only the extremes of environmental temperature fluctuation are avoided. And behavioral thermoregulation is cheap: cooling does not involve sweating away valuable water, and solar heating does not burn food.

Most living reptiles are comparatively small, and they gain and lose heat quickly. This may be good or bad for them, but it follows the laws of physics. However, most dinosaurs were large enough that they would have had passive thermoregulation (thermoregulation that needs no action). Think how long it takes to defrost a frozen chicken, and even more, a frozen turkey. It would have taken most of the day to heat up even a one-ton dinosaur, and most of the night to cool it down. The largest dinosaurs would have had practically uniform temperature whether they chose to or not: they would have warmed and cooled so slowly relative to day and night that their inner temperature would not have changed much at all. Passive thermoregulation is also called inertial thermoregulation or **gigantothermy**. It would have maintained medium to large dinosaurs at a body temperature close to that of the daily average temperature of their surroundings: homeothermy, perhaps, but not endothermy.

For example, a large passively thermoregulating dinosaur in a mild tropical climate like that of modern-day Amazonia would have had a body temperature around 27°C (80°F), day and night, season after season. So in some habitats it is easy to believe that many large dinosaurs (say most ornithischians and sauropods) could function well with this sort of thermoregulation, at a body temperature somewhat lower than ours. As we have seen, one could possibly argue that polar dinosaurs are evidence for thermoregulation at comparatively cool temperatures.

Living Crocodiles

Insight came in 1999 with new research on Australian crocodiles. Crocodiles can bask in the sun to gain heat (Chapter 8). They can also walk into the water to cool down: overall, they use behavioral thermoregulation to maintain fairly even body temperatures. Frank Seebacher and his colleagues found, as expected, that 1-ton crocodiles could regulate their body temperatures more easily than smaller crocodiles (because of the effects of passive thermoregulation). But they also found, to their surprise, that large crocodiles were on average warmer than small ones (see Seebacher et al., 1999). There is no explanation for this (yet), but it means that if dinosaurs worked the same way, that large dinosaurs could have maintained high body temperatures (say 38°C, 100°F) simply by passive and behavioral thermoregulation.

Crocodiles are archosaurs, and are the nearest living relatives of dinosaurs (except for birds). I shall summarize the implications of Seebacher's team as the archosaurian model for dinosaur thermoregulation. In the archosaurian model, environmental heating and cooling costs only time, not energy. Dinosaurs, then, would have had a resting metabolic level about the same as crocodiles, much lower than living mammals. However, all vertebrates have a maximum exercise metabolism about seven times their resting value, so the archosaurian model would allow an impressive performance by dinosaurs, though not up to the level of living mammals and birds.

Remember that dinosaurs, with erect posture, could and did run fast. I have seen an Australian crocodile lunging for prey, and I have no illusions about the speed of movement of living reptiles. Add the ability to sustain that speed (dinosaurs had solved Carrier's Constraint), and dinosaur performance becomes formidable.

The archosaurian model also accounts for dinosaurs in high latitudes. Life in cool climates could have been possible, given that the animals would need food only at reptilian levels. The archosaurian model accounts for the fact that many dinosaurs are large: size alone would give relatively high body temperature and its advantages, at little cost.

The archosaurian model also accounts for some of the cooling adaptations of dinosaurs that seem puzzling at first sight. Larger dinosaurs probably had more problems cooling down than they did heating up.

Some paleontologists have been arguing for years that dinosaurs had a different, dinosaurian, thermal biology. And they may be right. James Farlow (in Weishampel et al., 1999) suggested that dinosaurs had the ability to turn their metabolic rate up and down, possibly with the seasons. This would be almost an automatic result of the archosaurian model, especially in seasonal climates outside the tropics. But is it true?

Size and Metabolic Rates

Metabolic rates among living vertebrates fall with increasing size in any group of similar animals; it's cheaper to feed an elephant than the same weight of horses. To put it simply, an animal ten times the size of another will require only six times the food and oxygen. Exceptions usually make sense. For example, hummingbirds have a higher metabolic rate than one would expect because of the (high-energy) way in which they hover to collect nectar, their (high-energy) food. So even if sauropods operated at the metabolic rates of mammals, they would have needed perhaps six times the food of an elephant, not ten times as much. But if sauropods operated on the archosaurian model, they would have needed even less food. The problems of sauropod life at very large size thus may not be as awesome as they seem.

Cooling Large Herbivorous Dinosaurs

The macerated plant material inside the gut of a herbivorous dinosaur must have produced an enormous amount of heat as it fermented. Anyone familiar with garden compost will realize that this could have raised body temperature considerably, especially in a 50-ton sauropod. This is more "free" heat to add to solar energy. However, the high body temperature may have been difficult to control and would have responded to how often, how much, and what the dinosaur ate. Without any other control, rapid fermentation would have raised body temperature, which would have increased fermentation rate, and so on, in a potentially dangerous runaway reaction. So herbivorous dinosaurs may have had more problems in controlling their temperatures than carnivorous dinosaurs.

However, sauropods have two features that may have made heat shedding easier than one would imagine. First, the larger the dinosaur, the easier the problem may have been, because a very large body size would have been slow to heat up, and night would have been a good time to shed unwanted heat. Second, sauropods progressively evolved air spaces in their skeleton, connected with the lung system. Not only did this lighten the skeleton, it allowed air to circulate in deep body tissues, including the vertebrae. The sauropod would then have shed heat by evaporative cooling as it breathed out. The same system is used by birds (look at a crow on a hot day). Other things being equal, this may help to explain why the sauropod dinosaurs evolved to such extraordinary size.

There is a corollary to the discovery of this air system. If sauropods had the same respiratory system as birds, they may have had a similar metabolic regime. The bird's breathing system is more efficient than that of mammals, and provides enough oxygen to power a warmblooded creature. Sauropods grow very fast in their early years: their bone structure suggests that they reach sexual maturity in about 10 years, and full size around 20. Such enormous growth rates can be found in mammals and birds, but not in any living reptile. Again, the implication is one of warm blood.

Stegosaurus may provide a completely different way of addressing the question of dinosaur metabolism. It is a large (5-ton) herbivorous dinosaur, famous for the plates on its back (Figure 12.16). They are not spines, and they are offset from the spinal column. The plates are staggered right and left along the back, and their surfaces are corrugated, as if in life they were covered in skin and had large blood vessels running up and down them. Engineering experiments with aluminum and steel models of stegosaurs, heated by electricity and put into wind tunnels, have shown that the plates are practically ideal for shedding excess heat to a breeze. They are not designed for basking. This implies that *Stegosaurus* had heat-stress problems, and needed to shed heat, at least at times. Of course, when it did not need cooling, the blood supply to the plates would have been shut off.

A complicated system like this implies that *Stegosaurus* needed efficient cooling but did not need warming. Is that because it was already warmblooded, yet at the same time was fermenting large amounts of heat-producing vegetation? *Stegosaurus* is also famous for having a large cavity in its spinal column that was once interpreted as the site of a large posterior "brain." This cavity is much better explained by comparing it with the air cavities found in bird skeletons, which are used to pass cooling air through the body before it reaches the lungs.

Triceratops (Figure 12.18) has a very large bony frill on its skull, and large horns. Recent research by Reese Barrick and colleagues suggests that the frill, largely covered by skin, could have been an effective heat-shedding surface for the whole body. The horns may also have been used to shed heat to keep the brain cool! (see Barrick et al., 1988).

Respiration

David Carrier and Colleen Farmer have documented astonishing similarities between the respiration systems of birds and crocodiles (the surviving archosaurs). Both groups, in different ways, use bone and muscle systems to aid breathing by moving the pelvis and guts to generate a pumping action that in turn affects the lungs. Carrier and Farmer argue convincingly that if one reconstructs the muscle systems in the pelvic region of theropod dinosaurs, there is a very high likelihood that theropods were able to run an analogous respiration-boosting system. Furthermore, the pumping system is almost inextricably linked with locomotion, allowing the respiration to be linked with the active motion of dinosaurs in an analogous way to that seen in mammals and birds (Chapter 10). They even extend the argument to pterosaurs, and to ornithischian dinosaurs, which are also active bipeds. If all this is true, then the oxygen supply system would certainly have been capable to generating high energy flows in the metabolism of dinosaurs (see Carrier and Farmer, 2000).

Small Dinosaurs

In the archosaurian model, thermoregulation is much easier for large animals than small ones. Yet *all* dinosaurs were small as babies, and many dinosaurs were small as adults, especially theropods. Even on Seebacher's archosaurian model, small dinosaurs were the ones that most needed warm blood. And small dinosaurs, whether they are babies in nests, or adults faced with tending nests, would certainly have done better if they were warmblooded, or had warmblooded parents sitting on the nest. The evolution of egg brooding by dinosaurs can perhaps be seen as an important innovation that crocodiles and other reptiles did not achieve. And perhaps it would seem reasonable that small theropods would also have evolved a thermoregulatory coat (of feathers) that was not evolved in other dinosaur lineages. Those two innovations (feathers and brooding) were critical theropod innovations (to make them better theropods) that led to the evolution of a completely new group of dinosaurs: the birds.

The Answer?

So (finally!), what's the answer? At least small theropod dinosaurs had thermoregulation (down feathers). All dinosaurs could breathe while they ran (erect gait), so had an ability for fast sustained running (confirmed by footprints). Sauropods had a bird-like respiration system, and the others probably did too (especially the theropods that were so closely related to birds). Sauropods also had very rapid growth. None of this *proves* that dinosaurs were warmblooded (endothermic and homeothermic), but all of it points that way. In short, there is a lot of evidence suggesting strongly that dinosaurs were active, warmblooded animals.

Further Reading

Ackerman, J. 1998. Dinosaurs take wing: new fossil finds from China provide clues to the origin of birds. *National Geographic* 194 (1): 74–99.

Alexander, R. McN. 1989. *Dynamics of Dinosaurs and Other Extinct Giants.* New York: Columbia University Press.

Alexander, R. McN. 1991. How dinosaurs ran. *Scientific American* 264 (4): 130–6.

Barreto, C., et al. 1993. Evidence of the growth plate and the growth of long bones in juvenile dinosaurs. *Science* 262: 2020–3.

Barrick, R. E., et al. 1998. The thermoregulatory function of the *Triceratops* frill and horns: heat flow measured with oxygen isotopes. *Journal of Vertebrate Paleontology* 18: 746–50. [Cooling devices.]

Brochu, C. A. 2001. Progress and future directions in archosaur phylogenetics. *Journal of Paleontology* 75: 1185–201.

de Buffrénil, V., et al. 1987. Growth and function of *Stegosaurus* plates: evidence from bone histology. *Paleobiology* 12: 459–73.

Carpenter, K. 1999. *Eggs, Nests, and Baby Dinosaurs: a Look at Dinosaur Reproduction.* Bloomington: Indiana University Press.

Carrier, D. R., and C. G. Farmer. 2000. The evolution of pelvic aspiration in archosaurs. *Paleobiology* 26: 271–93. [A novel and convincing reconstruction of respiration in archosaurs, with major application to dinosaurs.]

Carrier, D. R., and C. G. Farmer. 2000. The integration of ventilation and locomotion in archosaurs. *American Zoologist* 40: 87–100. [Some overlap with the previous paper.]

Chen, P.-J., et al. 1998. An exceptionally well-preserved theropod dinosaur from the Yixian Formation of China. *Nature* 391: 147–52. [*Sinosauropteryx*.]

Chiappe, L. 1998. Dinosaur embryos. *National Geographic* 194 (6): 34–41.

Chiappe, L. M., et al. 1998. Cranial morphology of the avian Alvarezsauridae: evidence from a new relative of *Mononykus*. *Nature* 392: 275–8.

Chiappe, L. M., et al. 2001. Embryonic skulls of titanosaur sauropod dinosaurs. *Science* 293: 2444–6.

Chin, K., et al. 1998. A king-sized theropod coprolite. *Nature* 393: 680–2.

Currie, P. J. 1989. Long-distance dinosaurs. *Natural History* June 1989: 60–5.

Day, J. J., et al. 2002. Sauropod trackways, evolution, and behavior. *Science* 296: 1659. [A group of sauropods: also see Day et al., 2002. *Nature* 415: 494.]

Desmond, A. J. 1975. *The Hot-Blooded Dinosaurs*. London: Blond and Briggs; issued in paperback by Hutchison, 1990. [A lively historical account.]

Erickson, G. M., et al. 2001. Dinosaurian growth patterns and rapid avian growth rates. *Nature* 412: 429–33. [See also Padian et al., 2001.]

Farlow, J. O., and M. K. Brett-Surman (eds). 1997. *The Complete Dinosaur*. Bloomington: Indiana University Press. [Many fine essays on all aspects. See especially Part 4: Biology of the Dinosaurs.]

Farlow, J. O., et al. 2000. Theropod locomotion. *American Zoologist* 40: 640–63.

Fastovsky, D. E., and D. B. Weishampel. 1996. *The Evolution and Extinction of the Dinosaurs*. New York: Cambridge University Press. [The best text book.]

Forster, C. A. 1996. New information on the skull of *Triceratops*. *Journal of Vertebrate Paleontology* 16: 246–58. [Head-butting?]

Forster, C., et al. 1998. The theropod ancestry of birds: new evidence from the Late Cretaceous of Madagascar. *Science* 279: 1915–19, and comment, 1851–2. [*Rahonavis*, a Cretaceous bird with a slashing dromaeosaur claw.]

Gillette, D. D., and M. G. Lockley (eds). 1989. *Dinosaur Tracks and Traces*. Cambridge: Cambridge University Press.

Gillette, D. G., and M. Hallett. 1994. *Seismosaurus, the Earth Shaker*. Princeton University Press. [A good read; explains very well how paleontologists work.]

Hargens, A. R., et al. 1987. Gravitational haemodynamics and oedema prevention in the giraffe. *Nature* 329: 59–60, and comment 13–14. [Now think about sauropod necks.]

Horner, J. R., and J. Gorman. 1988. *Digging Dinosaurs*. New York: Workman. [An excellent read and a persuasive insight into the scientific investigation of dinosaurs.]

Horner, J. R., and D. Lessem. 1994. *The Complete T. rex*. New York: Simon and Schuster. [Paperback. Light, bright and breezy.]

Horner, J. R. 2000. Dinosaur reproduction and parenting. *Annual Reviews of Earth & Planetary Sciences* 28: 19–45.

Hutchinson, J. R., and S. M. Gatesy. 2000. Adductors, abductors, and the evolution of archosaur locomotion. *Paleobiology* 26: 734–51.

Ji, Q., et al. 1998. Two feathered dinosaurs from northeastern China. *Nature* 393: 753–61, and comment 729–70.

Ji, Q., et al. 2001. The distribution of integumentary structures in a feathered dinosaur. *Nature* 410: 1084–8.

Lessem, D. 1992. *Kings of Creation*. New York: Simon and Schuster. [A journalist's book, not a scientist's, but a very good read.]

Mayr, G., et al. 2002. Bristle-like integumentary structures at the tail of the horned dinosaur *Psittacosaurus*. *Naturwissenschaften* 89: 361–5.

Milner, A. 2003. *Dinobirds*. London: Natural History Museum. [Short, concise, clear survey of feathered dinosaurs.]

Norell, M., et al. 1993. New limb on the avian family tree. *Natural History* 102 (9): 38–42. [*Mononykus*.]

Norell, M. A., et al. 1994. A theropod dinosaur embryo and the affinities of the Flaming Cliffs dinosaur eggs. *Science* 266: 779–82, and comment, 731. See comment also in *Nature*, v. 372, 130. [*Oviraptor* embryo.]

Norell, M. A., et al. 1995. A nesting dinosaur. *Nature* 378: 774–6, and comment 764–5.

Norell, M., et al. 2002. 'Modern' feathers on a non-avian dinosaur. *Nature* 416: 36–7. [On a dromaeosaur.]

Norman, D. B., and P. Wellnhofer. 2000. *The Illustrated Encyclopedia of Dinosaurs*. London: Salamander. [The best of the "coffee-table" books, with good science. This new edition includes pterosaurs.]

Novacek, M. J., et al. 1994. Fossils of the Flaming Cliffs. *Scientific American* 271 (6): 60–9. [Some of the new discoveries in Mongolia.]

Novacek, M. J. 1996. *Dinosaurs of the Flaming Cliffs*. New York: Anchor Books. [Entertaining account of the Mongolian expeditions of the 1990s.]

Padian, K., et al. 2001. Dinosaurian growth rates and bird origins. *Nature* 412: 405–8. [See also Erickson et al. 2001.]

Paladino, F. V., et al. 1990. Metabolism of leatherback turtles, gigantothermy, and thermoregulation of dinosaurs. *Nature* 344: 858–60.

Paul, G. S. 1988. Physiological, migratorial, climatological, geophysical, survival, and evolutionary implications of Cretaceous polar dinosaurs. *Journal of Paleontology* 62: 640–52.

Prum, R. O., and A. H. Brush. 2002. The evolutionary origin and development of feathers. *Quarterly Review of Biology* 77: 261–95.

Rayfield, E. J., et al. 2001. Cranial design and function in a large theropod dinosaur. *Nature* 409: 1033–7, and comments 987–8, and *Science* 291: 1475–6. [The bite of an *Allosaurus*.]

Rogers, R. R., et al. 2003. Cannibalism in the Madagascan dinosaur *Majungatholus atopus*. *Nature* 422: 515–18.

Rowe, T., et al. 2001. Dinosaur with a heart of stone. *Science* 291: 783. [A supposed fossilized heart of *Thescelosaurus* is not: it is a mineral concretion.]

Ruben, J. A., et al. 2003. Respiratory and reproductive paleophysiology of dinosaurs and early birds. *Physiological and Biochemical Zoology* 76: 141–64. [For years Ruben and his students have argued for reptile-like physiology for dinosaurs and early birds. This is a recent summary. Other papers are 1996. *Science* 273: 1204–7; 1997. *Science* 278: 1267–70, and comment 1229–30; 1999. *Science* 283: 514–16, and comment, 468.]

Sanz, J. L., et al. 1995. Dinosaur nests at the sea shore. *Nature* 376: 731–2.

Seebacher, F., et al. 1999. Crocodiles as dinosaurs: behavioural thermoregulation in very large ectotherms leads to high and stable body temperatures. *Journal of Experimental Biology* 202: 77–86.

Sereno, P. C., and F. E. Novas. 1992. The complete skull and skeleton of an early dinosaur. *Science* 258: 1137–40. [*Herrerasaurus*.]

Sereno, P. C., et al. 1993. Primitive dinosaur skeleton from Argentina and the early evolution of Dinosauria. Nature 361: 64–6. [*Eoraptor*.]

Sereno, P. C., et al. 1996. Predatory dinosaurs from the Sahara and Late

Cretaceous faunal differentiation. Science 272: 986–91, and comment 971–2. [*Carcharodontosaurus.*]

Sereno, P. C., et al. 1998. A long-snouted predatory dinosaur from Africa and the evolution of spinosaurids. *Science* 282: 1298–302, and comment 1276–7. [*Suchomimus.*]

Sereno, P. C. 1999. The evolution of dinosaurs. *Science* 284: 2137–47.

Shreeve, J. 1997. Uncovering Patagonia's lost world. *National Geographic* 192 (6): 120–37.

Stevens, K. A., and J. M. Parrish. 1999. Neck posture and feeding habits of two Jurassic sauropod dinosaurs. Science 284: 798–800. [This is only as good as the computer model it was done with: the results are probably wrong.]

Stokes, W. L. 1987. Dinosaur gastroliths revisited. *Journal of Paleontology* 61: 1242–6.

Thomas, D. A., and J. O. Farlow. 1997. Tracking a dinosaur attack. *Scientific American* 277 (6): 74–9.

Thulborn, R. A. 1990. *Dinosaur Tracks.* London: Chapman and Hall.

Thulborn, R. A., and M. Wade. 1979. Dinosaur stampede in the Cretaceous of Queensland. *Lethaia* 12: 275–9.

Vickers-Rich, P., and T. H. Rich. 1993. Australia's polar dinosaurs. *Scientific American* 269 (1): 50–5.

Wedel, M. J. 2003. Vertebral pneumaticity, air sacs, and the physiology of sauropod dinosaurs. *Paleobiology* 29: 243–55.

Weishampel, D. B. and C. Norman. 1981. Acoustic analysis of potential vocalization in lambeosaurine dinosaurs (Reptilia: Ornithischia). *Paleobiology* 7: 252–61.

Weishampel, D. B., et al. (eds). 1999. *The Dinosauria.* Berkeley: University of California Press.

Wilson, J. A. 2002. Sauropod dinosaur phylogeny: critique and cladistic analysis. *Zoological Journal of the Linnean Society* 136: 215–75.

Witmer, L. M. 2001. Nostril position in dinosaurs and other vertebrates and its significance. *Science* 293: 850–3.

Xu, X., et al. 1999. A therizinosauroid dinosaur with integumentary structures from China. *Nature* 399: 350–4. [*Beipiaosaurus.*]

Xu, X., et al. 1999. A dromaeosaurid dinosaur with a filamentous integument from the Yixian Formation of China. *Nature* 401: 262–6. [*Sinornithosaurus.*]

Xu, X., et al. 2000. The smallest known non-avian theropod dinosaur. *Nature* 408: 705–8. [*Microraptor.*]

Xu, X., et al. 2000. Branched integumental structures in *Sinornithosaurus* and the origin of feathers. *Nature* 410: 200–4. [Real feathers, not just "integumental structures."

Xu, X., et al. 2003. Four-winged dinosaurs from China. *Nature* 421: 335–40, and comment, 323–4; also comment in *Science* 299 491. [*Microraptor gui.*]

Zhou, Z., et al. 2003. An exceptionally preserved Lower Cretaceous ecosystem. *Nature* 421: 807–14. [Convenient review. No new science.]

CHAPTER THIRTEEN

The Evolution of Flight

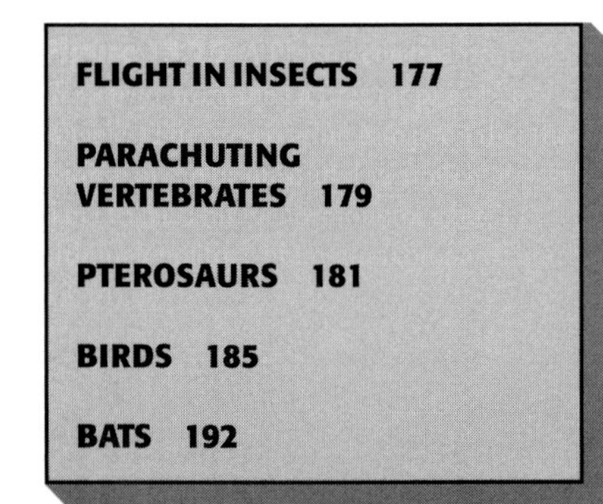

There are four kinds of flight: passive flight, parachuting, soaring, and powered flight. Passive flight can be used only by very tiny organisms light enough to be lifted and carried by natural winds and air currents, and light enough to suffer no damage on landing. Tiny insects, baby spiders, frogs' eggs, and many kinds of pollen, spores, and seeds can be transported this way. But their "flight" duration, direction, and destination is entirely at the mercy of chance events.

The first land organisms and the first aerial organisms were microscopic. As reproduction adjusted to the problems of life in air, spores were evolved by fungi, plants, and other small organisms—their soft reproductive cells were protected by a dry, watertight coating, rather than the damp slime that is sufficient in water. Dry spores could then be spread as passive floaters on the wind—Earth's first fliers. Plant spores occur in Ordovician rocks (Chapter 8), and they are numerous and widespread enough in Devonian rocks to be used as guide fossils in relative age dating. Apart from anything else, this suggests that some plant species had such a large area of tolerable habitats open to them that a long-range dispersal method had become worthwhile.

Gliding flight includes parachuting, in which the flight structures slow a fall, and soaring, in which the flight structures allow an organism to gain height by exploiting natural air currents. Parachuting and soaring may seem to grade into one another, but their biology is very different, and the two flight modes probably have distinctly different origins. Parachuting organisms have simpler flight structures and much less control over the direction, speed, and height of flight than soarers. They seek short-range travel from one point to another, and their landing point is reasonably predictable because they do not seek external air currents for lift. Parachuting is used in habitats where external air currents are minimal, especially in forests. Wind gusts and air currents are potentially disastrous to animal parachutists, just as they are to human paratroops.

Powered flight is usually accomplished by some sort of flapping motion with special structures (wings). It needs a lot of energy, but gives much greater independence from variations in air currents, and it is usually accompanied by a high level of control over flight movements. Because powered flight is achieved by controllable appendages, almost all powered fliers can glide to some extent, some very poorly (no better than parachutists) and some very well indeed. Raptors and soaring seabirds are examples of powered fliers that glide well.

Soaring is used by flying organisms that range widely over a broad habitat. It is a low-energy flight style because the lift comes from external air currents rather than muscular expenditure by the flier. Energy costs are mainly related to the maintenance and adjustment of gliding surfaces in the air flow. Soarers may need occasional bursts of flapping flight if there are no up-currents, or in transferring from one up-current cell to another. Flapping is sometimes needed for takeoff, until airspeed exceeds stalling speed, or for final adjustments of attitude and speed in landing. Because flapping flight is needed occasionally by all soarers today (especially in emergencies), soaring probably cannot evolve from parachuting but only from powered flight.

Flight of all kinds demands a light, strong body. Soaring especially emphasizes lightness in muscle mass as well as overall structure. Powered flight has more requirements, including a significant output of energy and strength. Even the best soarers among living birds, albatrosses and condors, cannot flap for long before they are exhausted, because their flight muscles are small relative to their size and total weight. It might be difficult for a specialized soarer to reevolve the ability to sustain flapping flight.

One would imagine that powered flight could evolve easily from gliding. Any evolving wing should be a fail-safe device, allowing a gliding fall during flight training. But an animal that has already evolved efficient gliding would not easily improve its flight by flapping in mid-glide, because that would disturb the smooth airflow over the gliding surfaces. Aerodynamic analysis shows that an evolutionary transition is possible from gliding to flapping, but only in very special circumstances. The glider must add fairly large, rapid wing beats, not little flutters, and because wing beats require considerable expenditure of energy, there must be a corresponding payoff in energy saved (for example, the animal must save some walking or climbing, or must reach a larger food supply).

The evolution of flight demands lightening and strengthening of the whole body structure, and the evolution of a flight organ from a preexisting structure (a limb, for example) that could otherwise perform some other function. Flight may have strong advantages in locomotion—for food-gathering, escape, rapid travel between base and food supply, or migration—but it also has costs, not just in energy but also in constraints on body form and function that may have accompanying drawbacks. Flight has evolved many times in spite of all these problems.

Flight usually involves relatively large lifting structures; in almost every case a small lifting structure is no better than none at all. Lifting structures must already be present before flight can evolve, and they must therefore have evolved for some different function. This theme dominates the discussion of flight origins in this chapter.

FLIGHT IN INSECTS

Primitive insects are known from Devonian rocks, but flying insects are not found until the Late Carboniferous, as insects radiated in the Carboniferous forest canopy. Many insects of all sizes are known from the coal beds of the Carboniferous, and half of all known Paleozoic insects had piercing and sucking mouthparts for eating plant juices. In turn, these smaller insects were a food source for giant predatory dragonflies (Figure 9.14) and for early amniotes (Chapter 9).

In living insects (except mayflies), only the last molt stage, the adult, has wings, and there is a drastic metamorphosis between the last juvenile stage (the nymph) and the flying adult. Wings have to be as light and strong as possible; in living insects this is achieved by withdrawing as much live tissue as possible. Most of the wing is left as a light mass of dead tissue that cannot be repaired. This gives great flying efficiency, though it usually means a short adult lifespan. The automatically short life expectancy of flying insects has played a strong part in the evolution of social behavior among some insects, in which the genes of a comparatively few breeding but nonflying adults are passed on with the aid of a great number of cheap, throwaway, sterile flying individuals (worker bees, for example). Some insects shed their wings. In many (ants, for example,) the wings are functional only for a brief but vital period during the mating flight. Insects do not have a long enough life expectancy to have the luxury of learning, so they carry with them a "read-only memory" that seems to govern their behavior entirely by "instinct."

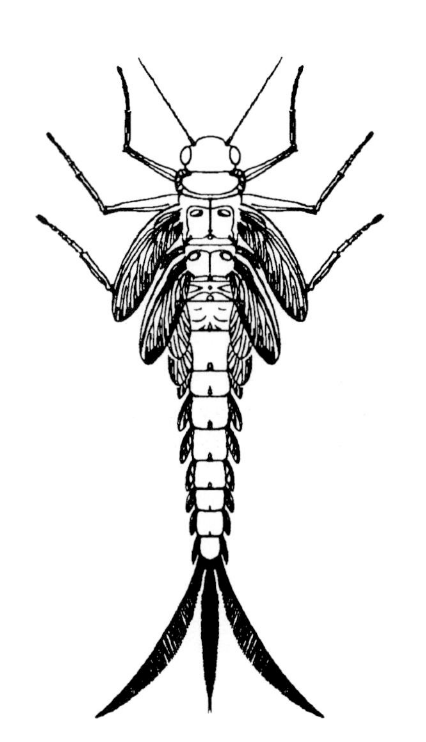

Figure 13.1 A mayfly nymph from the Lower Permian of Oklahoma. The wings on the thorax are big and look functional, but they are for underwater rowing rather than flight. There are also smaller winglets on all the segments of the abdomen. (Redrawn after a reconstruction by Jarmila Kukalová-Peck.)

But these are characters of living insects, and they would not necessarily apply to early insects that had not yet evolved flight. However, there is now good evidence from Carboniferous insects on the evolution of flight in this group, the first animals to take to the air under control. Jarmila Kukalová-Peck (1987) has pointed out that insects (and angels) are the only flying creatures that evolved flapping flight without sacrificing limbs to form the wings. Insects have thus lost little of their ability to move on the ground.

Many living insects are good gliders. Dragonflies, which were among the earliest insects to evolve flight, have wings arranged so that they are very stable in a gliding attitude, but dragonflies use flapping flight to chase rapidly and expertly after their prey. We still do not understand dragonfly flight. Somehow, complex eddies are produced between the two sets of wings, which beat out of phase. At some phases in the wing cycle, dragonfly wings produce lift forces that are 15 to 20 times the body weight.

Other insects have complex locking devices to hold their wings in a gliding position without energy expenditure. They need these locking devices for gliding because in powered flight their wings flap freely during complex movements. This line of reasoning suggests that insects evolved flight as flappers and later adjusted in a complex way to gliding.

The critical fact about the evolution of insect wings is that arthropod limbs consisted originally of two branches: a walking leg and another jointed unit—the exite—that was used as a filtering device or a gill. These structures are still found in most marine arthropods, but at first sight they seem to have been lost in insects, which have only walking legs. They were not lost: exites disappeared because they evolved into wings. We can see some of the stages in this evolution. In the young water-dwelling stages—nymphs—of living mayflies, the exites along the abdomen are shaped into plate-like gills. The same structure is found on the thoracic exites of larval dragonflies, some beetles, and several other groups of insects.

There are fossilized nymphs of Late Carboniferous insects, and many of them had exites modified into plate-like gills (Figure 13.1). The plates were probably also used for swimming (living mayfly nymphs use their plated gills for swimming in the same way). However, the plates were also preadapted to flight (short at first, of course), and this pathway to flapping flight is the leading hypothesis for the origin of insect flight.

An intermediate stage in the evolution of insect flight may still exist in some primitive living insects, mayflies and stoneflies. James Marden and Melissa Kramer showed in 1994 that these primitive insects "skim" across water surfaces, using a wing action that is exactly like flying, but they also receive some lift from the legs, which remain in contact with the water surface.

Figure 13.2 A parachuting vertebrate, the flying frog. (From Lydekker.)

PARACHUTING VERTEBRATES

Several living forest-dwelling vertebrates have evolved parachuting flight, using skin flaps as flight surfaces. They include flying squirrels, three different lineages of Australian gliding marsupials (greater gliders, squirrel gliders, and feathertail gliders), *Draco* the flying lizard, flying geckos, flying frogs (Figure 13.2), and even flying snakes. This suggests that parachuting adaptations evolve in animals of the forest canopy that habitually jump from branch to branch, from tree to tree, or from trees to the ground. Any method of breaking the landing impact or of leaping longer distances would be advantageous and might evolve rapidly.

None of these parachuting animals has powered flight, however. The energy for gliding flight is gravitational, generated as the animal climbs in the tree and released as it parachutes off the branch. Parachuting can evolve in animals with rather low metabolic rates. It does not require the high metabolic rate of birds and bats, which have powered flight. Characteristically, parachuting animals have short limbs, long trunks, flexible spines, and quadrupedal stance.

Early Gliding Vertebrates

The earliest known gliding vertebrate is the Late Permian reptile *Coelurosauravus*. Its fossils have been found in Germany, Britain, and Madagascar, so it was widespread across Pangea. All these areas were near tropical shores at the time. *Coelurosauravus* is an ordinary, small diapsid reptile in the structure of its skull and body, about the size of a small squirrel. But the trunk is dominated by 20 or more long, curving, lightly constructed rod-shaped bones that extended outward and sideways from the body. They supported a skin membrane that was close to an ideal airfoil in shape, 30 cm (1 ft) across, and could only have been used for gliding

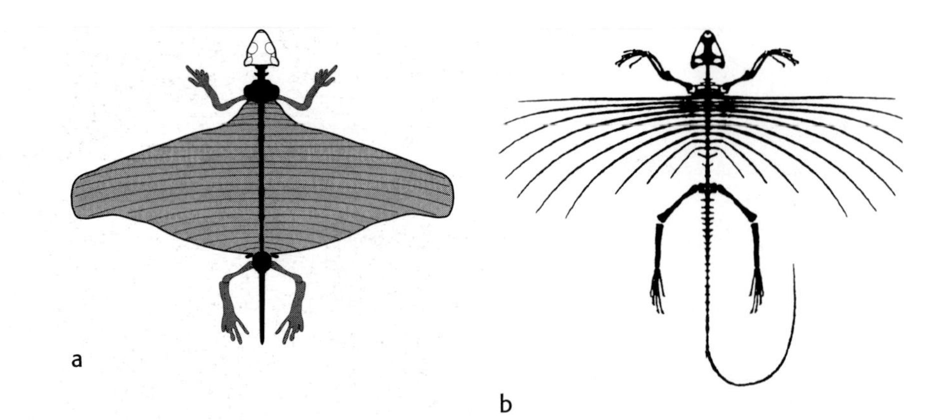

Figure 13.3 (a) *Coelurosauravus*, a gliding reptile from the Permian. (Airfoil reconstruction based on a skeletal diagram by Robert Carroll.) (b) *Icarosaurus*, a gliding reptile from the Triassic of New Jersey, represents another independent evolution of gliding flight among vertebrates. (After Colbert, 1970.)

Figure 13.4 Reconstruction of a Triassic gliding reptile from Kazakhstan, *Sharovipteryx*, from the single specimen discovered. (After Sharov.)

(Figure 13.3a). More impressive still, the bones are jointed so that the airfoil could have been folded back along the body when it was not in use. Extra-long vertebrae allowed space for this folding along the spinal column. These bones are not ribs, but must have evolved specifically under the skin as a gliding structure. Because of this unique character, *Coelurosauravus* is placed in a major basal diapsid group of its own, the Weigeltisauria.

We can judge how well *Coelurosauravus* was adapted for gliding flight by comparing it with Triassic and living reptilian gliders. Kuehneosaurids are a family consisting of two gliding reptiles from the Late Triassic, *Kuehneosaurus* from Britain and *Icarosaurus* from New Jersey (Figure 13.3b). They too had effective airfoils, but since they were stretched out on elongated ribs, gliding must have evolved independently in kuehneosaurids and *Coelurosauravus*.

No gliding lepidosaur is known from the fossil record after the Triassic, so the living lizard *Draco*, which also uses elongated ribs to support an airfoil, must have evolved gliding independently. Ligaments and muscles between the ribs of *Draco* give precise control of the gliding surface, and this was probably true also in the fossil gliders. In Permian, Triassic, and living gliders, all four limbs remain free for walking, grasping, and climbing. All three groups can fold up the airfoil when it is not in use. But there are interesting differences in the way the airfoil is organized.

In *Draco* and in the Triassic gliders, the ribs are single, unjointed bones. When *Draco* folds its airfoil, the spinal ends of the ribs have to be moved between the back muscles, which means that the ribs cannot be very big or very strong. The Triassic gliders had long levers mounted on the spinal ends of their ribs to get around this problem, and of course *Coelurosauravus* avoided the problem altogether by having a separate jointed airfoil that was not made from ribs. In some ways, then, *Coelurosauravus* was better designed than its later counterparts.

Sharovipteryx was discovered by accident in a search for fossil insects in Late Triassic rocks in Central Asia. It was a small reptile, and preserved skin clearly indicates that it had a gliding membrane. *Sharovipteryx*, however, was unique in that the membrane was stretched between very long, strong hind limbs and a long tail, so a large, broad wing surface was set well behind the trunk and head (Figure 13.4), rather like some modern aircraft—the supersonic Concorde, for example.

Carl Gans and his Russian colleagues suggested in 1987 that *Sharovipteryx* glided very well, perhaps with the aid of a canard, or accessory membrane, associated with the short, normal-looking forelimbs (Figure 13.5). Sensitive control over flight could have been maintained by slight backward-and-forward motion of the hind limbs, as in many gliding birds today. A similar but cruder system is used in swing-wing aircraft.

Longisquama is a strange reptile from the same Triassic rocks as *Sharovipteryx*. Its remains include a series of long, flattened bones with flared, curved tips. Susan Evans suggested that these were ribs from a gliding airfoil. The lightness and flattening of the bones and the curvature of their tips would all make sense if that were true. *Longisquama* was probably a glider, very much like the kuehneosaurids.

An airfoil does not appear by magic, especially a folding one. Robert Carroll points out that there may have been other good reasons for evolving a folding, extended rib structure. The great area of exposed skin could have been used in thermoregulation, for example. If the extinct reptiles behaved like *Draco*, they may have used their airfoils for display as well as for flight, and may have evolved them first for display.

We recognize these fossil reptiles as gliders because they had specialized skeletons. By comparison with small, insect-eating vertebrates in forests today, there were probably many other jumping and gliding reptiles in Permian and Triassic forest canopies, with skin flaps unsupported by bones. The forest canopy was probably rich in many species of small insectivorous amphibians and reptiles.

Figure 13.5 Reconstruction of *Sharovipteryx* suggested by Carl Gans and his Russian colleagues. They were able to fly a successful model by adding a canard to the leading edge of the flight surface. (Courtesy of Carl Gans of the University of Michigan.)

PTEROSAURS

Pterosaurs are the most famous flying reptiles. They too evolved in the Triassic, but they are archosaurs, most likely closely related to dinosaurs, and quite unrelated to gliding lizards. The earliest pterosaurs so far discovered are already fully adapted for flight, so their distant or direct ancestors are not yet known.

Pterosaurs have very lightly built skeletons, with air spaces in many of the bones. Their forelimbs were extended into long struts that supported a wing, as in birds and bats. Pterosaurs were unique, however, in that most of the wing membrane was supported on one extraordinarily long finger, while three other fingers were normal and bore claws. The fourth finger was about 3 m (10 ft) long in the largest pterosaurs. In contrast, birds support the wing with the whole arm, and bats use all their fingers as bony supports through their wing membranes (Figure 13.6). Pterosaurs thus have a unique wing anatomy, but as the largest flying creatures ever to evolve and as a group that flourished for more than 140 m.y., they can't be dismissed as primitive or poorly adapted.

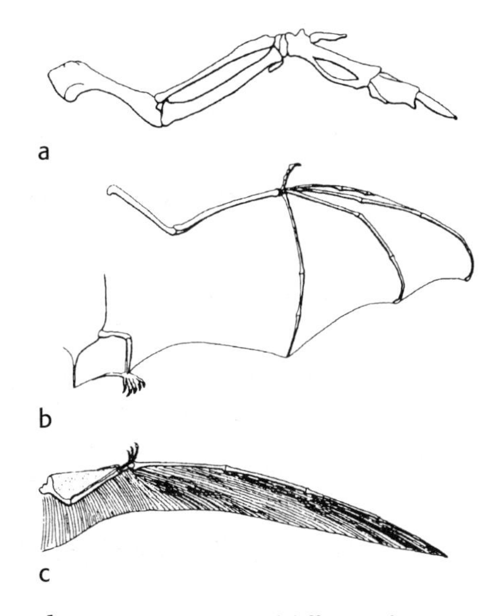

Figure 13.6 The skeletal differences between the wings of a bird (a), a bat (b), and a pterosaur (c). I have reconstructed the pterosaur wing as narrow, in line with Kevin Padian's interpretation (see text).

Most pterosaurs had large eyes sighting right along the length of long, narrow, lightly-built jaws. The teeth were usually thin and pointed, often projecting slightly outward and forward (Figure 13.7). This is most likely an adaptation for catching fish. Almost all pterosaur fossils are preserved in sediments laid down on shallow seafloors, and where stomach contents have been preserved with pterosaur skeletons, they always contain fish remains such as spines and scales. Some pterosaurs may have fished on the wing, like living birds such as gadfly petrels or skimmers, which fly along just above the water surface and dip in their beaks to scoop up fish or crustaceans. One can imagine *Anhanguera* doing this (Figure 13.8). Other pterosaurs may have fed like terns, which dive slowly so that only the head, neck, and front of the thorax reach under the water, while the wings remain above the surface. Some pterosaurs with long sharp beaks may have fished standing in the water like herons, or sitting on the water. It seems unlikely that pterosaurs crash-dived into water like pelicans or gannets, or swam underwater like penguins: pterosaur wings were too long and too fragile. At least one pterosaur, *Pterodaustro* from Argentina, had teeth that were so fine, long, and numerous that it must have been a filter feeder, perhaps like a flamingo; and *Ctenochasma* looks like a filter-feeder too

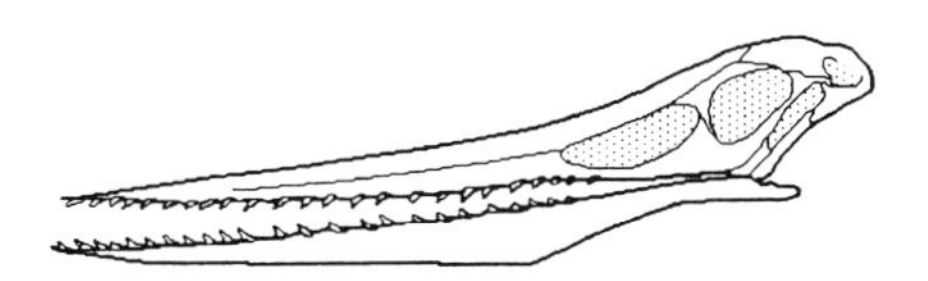

Figure 13.7 The skull of a typical fishing pterosaur, *Rhamphorhynchus*, in side view, with sharp pointed teeth that protruded slightly to impale a fish in a strike. (Figure by C. McQuilkin, provided courtesy of Larry Witmer of Ohio University.)

Figure 13.8 *Anhanguera*, from the Cretaceous of Brazil, reconstructed by Larry Witmer. (Figure by C. McQuilkin, provided courtesy of Larry Witmer of Ohio University.)

(Figure 13.9). Some short-jawed pterosaurs may have eaten shore crustaceans or insects.

If I had to choose one modern ecological analog for most pterosaurs, I would choose frigate birds. These tropical fishing birds have the lightest wing loading of any bird, and they spend days at a time out at sea, soaring above the ocean, presumably half-sleeping on the wing. Occasionally they descend to the surface to pick off a fish, but they do not dive in: they cannot even land on the water because their feathers are not waterproof. They take longer to raise a chick than any other bird, and they can live for more than 30 years.

There are two main groups of pterosaurs. **Rhamphorhynchoids** (Late Triassic to Late Jurassic) are the stem group of early pterosaurs, rather than a clade. Most of them had wingspans under 2 m (6 ft), and some were as small as sparrows. At least some of them had crests on the head, and long, thin, stiff tails that carried a vertical vane on the tip. Perhaps these were dynamic stabilizers in flight.

Pterodactyloids are a clade of advanced pterosaurs that replaced rhamphorhynchoids in the Late Jurassic and flourished until the end of the Cretaceous. Pterodactyloids had no tails, and many were much larger than rhamphorhynchoids. The large forms were adapted for soaring rather than continuous flapping flight, although they all flapped for takeoff. *Pterodactylus* itself was sparrow-sized, but *Pteranodon*, from the Cretaceous of North America, had a wingspan of about 7 m (22 ft); and the gigantic pterosaur from Texas, *Quetzalcoatlus*, was 10–11 m (30–5 ft) in wingspan, the largest flying creature ever to evolve. (An incomplete set of pterosaur fossils from Romania may be pieces of an even larger form: guesses vary around 12 m.)

Although pterosaur bones were light and fragile, several examples of outstanding preservation have shown us many details of their structure. Black shales in Lower Jurassic rocks of Germany have shown details of rhamphorhynchoids; Late Jurassic members of both pterosaur groups have been found exquisitely preserved in the Solnhofen Limestone of Germany and in lake deposits in Kazakhstan in Central Asia; from the Lower Cretaceous of Brazil we have partial skeletons preserved without crushing; and the Upper Cretaceous chalk beds of Kansas have yielded huge specimens of *Pteranodon*. Discoveries of skin, wing membranes, and stomach contents allow biological interpretations of these exciting animals.

However, those interpretations vary widely. Pterosaurs have no living descendants that we can study, and we have not found their ancestors. It is difficult to choose a living analog: some people look at bats for anatomical and functional guidance, others look at birds. A few things are very clear: all pterosaurs, including the giant forms, were capable of powered, flapping flight. All had a large bony plate on

Figure 13.9 The skull of *Ctenochasma*, which probably scooped water and filtered small prey from it. (Courtesy of Christopher Bennett of the University of Bridgeport; © S. C. Bennett.)

the front of the rib cage to which powerful flight muscles were attached. At the other end, the flight muscles attached to a wide flange at the top of the arm (Figure 13.10). But controversy has raged over five issues: what were the wings of pterosaurs like? how well did they fly?; how did pterosaurs stand, and walk, on the ground? what were pterosaur ancestors? and how did pterosaurs evolve flight? I present here the interpretations that I think are more plausible: the alternatives are discussed on the Web page.

The pterosaur wing was attached to the hindlimb. Opinion varies (and pterosaurs may vary) in where that attachment was: high on the hip, or low, at the knee or ankle. The disagreement continues because the wing is often preserved lying across the skeleton after death, and it is difficult to tell whether the contact between wing and leg is anatomical or simply accidental.

David Unwin and Peter Wellnhofer argue that pterosaurs were unique, with a complex of skin components working to form a compound airfoil that involved all four limbs: the "broad-wing" reconstruction (Figure 13.11a) (see, for example, Norman and Wellnhofer, 2000, and Unwin and Henderson, 2002). Kevin Padian (see, for example, Ryaner and Padian, 1993) has argued that there is no evidence that any pterosaur had a wing attached any further down the leg than mid-thigh. If they are correct, pterosaurs had long, narrow wings, much like a falcon or an albatross (Figure 13.11b).

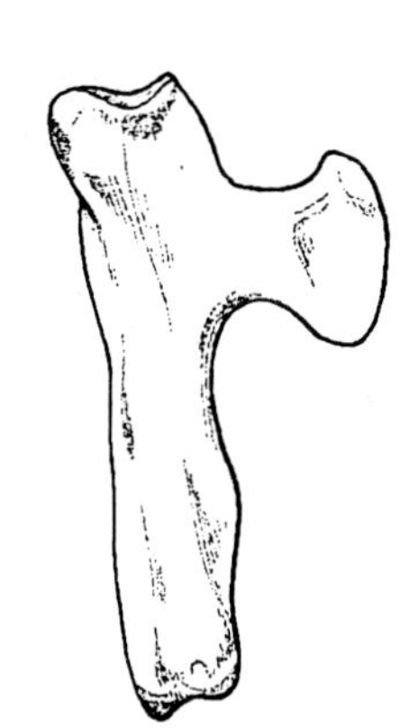

Figure 13.10 The humerus of all pterosaurs has a special flange where powerful flight muscles attached. This bone is from *Nyctosaurus*. (After Williston.)

The wing itself was not simply a giant skin membrane: that would have been too weak to power flapping flight. Furthermore, with bones, joints, and ligaments only on the leading edge of the wing, a pterosaur needed a way to control the aerodynamic surface of the wing. Beautifully preserved specimens show that the wing had special adaptations. It was stiffened by many small, cylindrical fibers, which were probably tied together by small muscles. The combination of structural stiffeners and muscles allowed fine control over the surface, and at the same time made the wing reasonably strong, not easily damaged or warped, and not likely to billow in flight like the fabric of a hang glider.

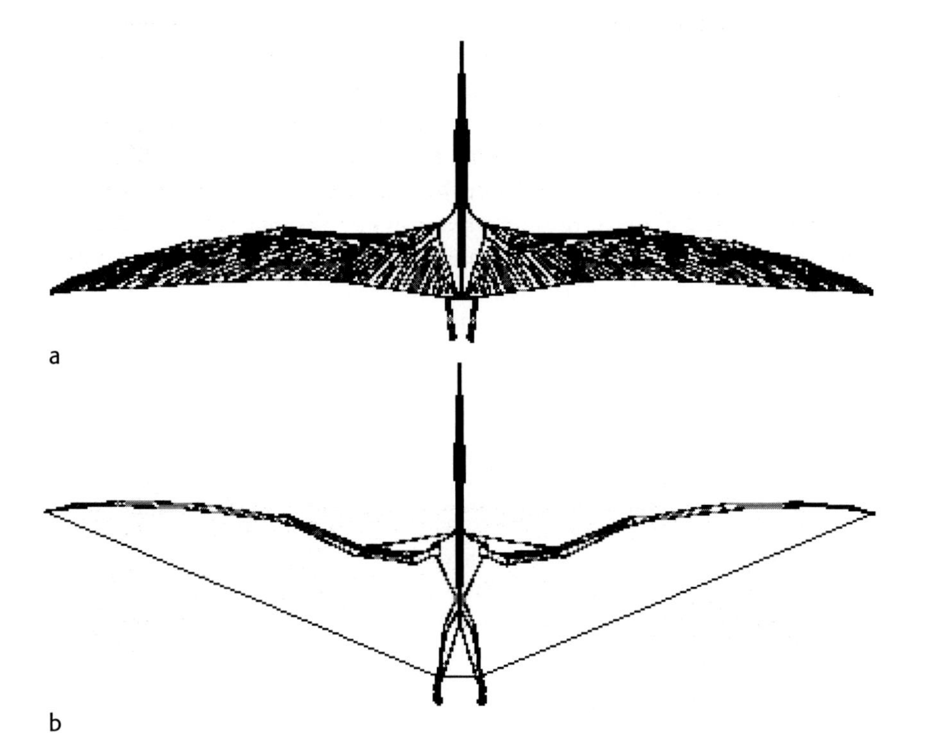

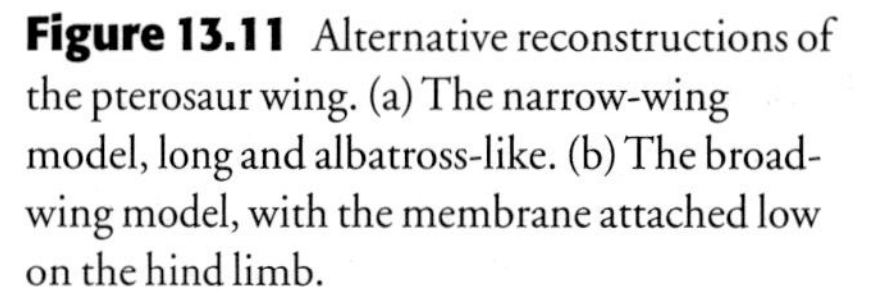

Figure 13.11 Alternative reconstructions of the pterosaur wing. (a) The narrow-wing model, long and albatross-like. (b) The broad-wing model, with the membrane attached low on the hind limb.

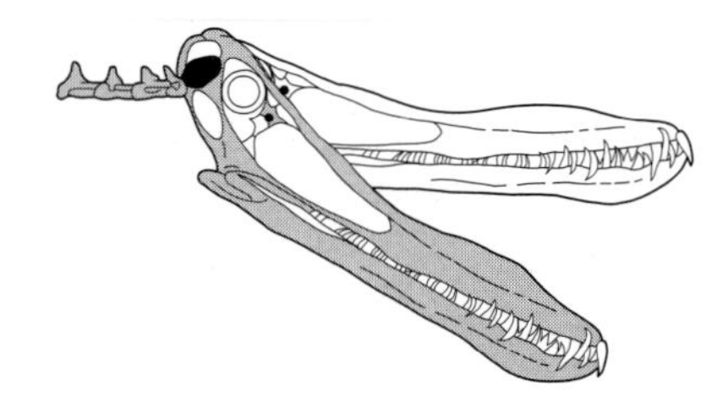

Figure 13.12 A CT scan of the skull of *Anhanguera* suggests that it held its head angled downward, rather than in the expected horizontal position. Probably the angled position is the one used as the pterosaur scanned the water surface for fish. (Figure by C. McQuilkin, provided courtesy of Larry Witmer of Ohio University.)

A team of researchers led by Larry Witmer made CT scans of two uncrushed pterosaur skulls (see Witmer et al., 2003). The scans revealed the size and shape of the pterosaur brain. In both brains, the lobes associated with balance were very large, and this allowed the researchers to reconstruct the head to be arranged in the usual, or preferred, attitude it had in life. While the little Early Jurassic pterosaur *Rhamphorhynchus* apparently held its head horizontally (as birds do in normal flight), the later and larger Cretaceous pterosaur *Anhanguera* seems to have held its head angled downward, perhaps in fishing position (Figures 13.8; 13.12). This is not unreasonable. Herons spend hours in this kind of attitude as they stand waiting for fish, even though they fly with their heads horizontal. In completely different ways of life, pelicans and kites (at least the white-tailed kite and the white pelican of California) hold their heads "normally" as they fly from place to place, but kites hover over potential prey sites, and pelicans go into slow searching flight mode, both with their heads tilted dramatically downward.

All the small rhamphorhynchoids and many of the pterodactyloids had active, flapping flight. Naturally, the gigantic pterosaurs could not have flapped for long, and they probably spent most of their time soaring, as does the living albatross. Aerodynamic analysis shows that pterosaurs were the best slow-speed soaring fliers ever to evolve.

Flapping flight involves very high energy expenditure. Birds are warmblooded, as are bats and many large insects when they are in flight: dragonflies, moths, and bees are examples. Thus one might guess that pterosaurs too were warmblooded. Several Jurassic pterosaurs have fur preserved on the skin: if pterosaurs had fur, they were probably warmblooded. Flapping flight has evolved only three times among vertebrates (in pterosaurs, in birds, and in bats) and in each case the animal was apparently warmblooded before or just as it achieved flight. Pterosaur bones had air spaces running through them in the same way that living bird bones do. In birds, this system helps to provide air cooling, and it is reasonable to interpret pterosaur bone structure in the same way.

The second unresolved question is about the leg structure of pterosaurs and their walking ability. Padian thinks that small pterosaurs could have moved well on the ground. If pterosaurs had broad wings, they would have been very clumsy on the ground, like living bats. This question is under debate (to put it mildly), but should be resolved soon.

If most pterosaurs ranged widely over the ocean searching for fish, it would have been impossible for pterosaur nestlings to feed themselves until they had reached a fairly advanced stage of growth and flight capability. Nesting behavior and care of the young would therefore have been mandatory. Kevin Padian has described a "pterosaur nursery" preserved in Cretaceous rocks in Chile.

The social behavior of pterosaurs may have been complex. Many pterosaurs were dimorphic. Males were larger, with long crests on the back of the head and with relatively narrow pelvic openings. Females were smaller, with smaller crests but larger pelvic openings. New discoveries show that the soft tissues associated with some crests were extravagantly large, and were much more likely to have been display structures than aids to flight (Figure 13.13).

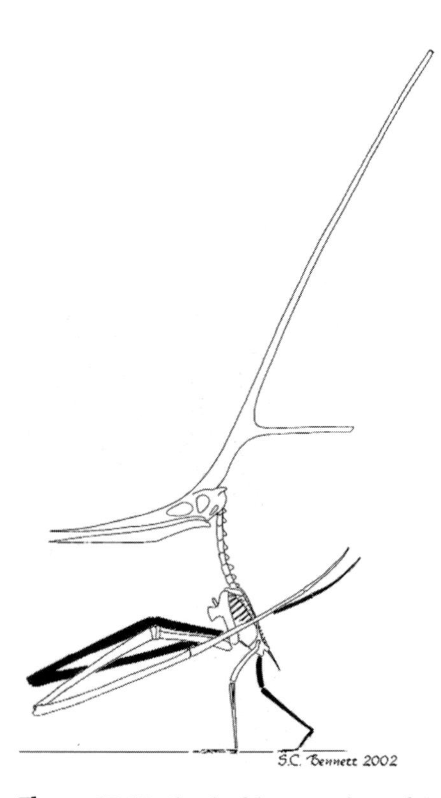

Figure 13.13 Sketch of the astounding soft-tissue crest discovered in the Cretaceous pterosaur *Nyctosaurus*. (Courtesy of Christopher Bennett of the University of Bridgeport; © 2002 S. C. Bennett.)

Pterosaurs were the first vertebrates to achieve flapping flight. Padian has argued that pterosaurs were terrestrial bipeds as well as good fliers. If so, then pterosaur flight evolved among ground-running animals, not among tree dwellers.

The earliest pterosaurs known are Late Triassic. They were already well-evolved flying creatures, and we have not yet found plausible ancestors. The earliest well-preserved pterosaur, *Austriadactylus*, already had a bony crest on its skull and a very long tail.

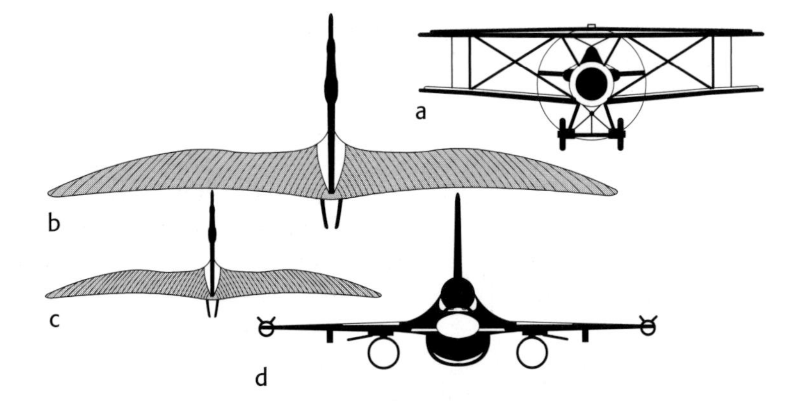

Figure 13.14 The wingspans of various fliers. (a) A Sopwith Camel F1, wingspan 8.5 m (28 ft); (b) The pterosaur *Quetzalcoatlus*, the largest flying animal ever to evolve, conservatively estimated to have a wingspan of 10 m (33 ft); (c) *Pteranodon*, a large pterosaur, conservatively estimated to have a wingspan of 7.5 m (25 ft); (d) A General Dynamics F16A Fighting Falcon, wingspan 10 m (33 ft). Judging the sizes of the pilots in the airplanes will serve as an extra reminder of the size of the pterosaurs.

The largest pterosaur, *Quetzalcoatlus* (Figure 13.14), was also the latest. It and related forms lived right at the end of the Cretaceous. Its bones are found in nonmarine beds, deposited perhaps 400 km (250 mi) inland. Perhaps it was the ecological equivalent of a vulture, soaring above the Cretaceous plains and scavenging on carcasses of dinosaurs. *Quetzalcoatlus* did have a strangely long, strong neck, but its beak seems too lightly built for this method of feeding. Perhaps it was more like a gigantic heron, standing and fishing in inland lakes and swamps, or picking up frogs, turtles, or arthropods such as crayfish from shallow water.

We do not know why pterosaurs became extinct. As we have seen, they were most likely active, warmblooded animals with flapping flight much like that of birds. Yet pterosaurs became extinct at the end of the Cretaceous, at the same time as the dinosaurs disappeared, while birds did not. We shall return to that question in Chapter 18.

BIRDS

Living birds are warmblooded, with efficient thermoregulation that maintains body temperatures higher than our own. Birds breathe more efficiently than mammals, pumping air through their lungs rather than in and out. They have better vision than any other animals. Birds build extraordinarily sophisticated nests: bowerbirds are second only to humans in their ability to create art objects. New Caledonian crows learn to make tools faster than chimpanzees do. And above all, birds can fly better, farther, and faster than any other animals, an ability that demands complex energy supply systems, sensing devices, and control systems.

Birds include ostriches and penguins, which cannot fly, and hummingbirds, which can hardly walk. But birds share enough characters for us to be sure that they form a single clade, descended from (i.e., part of) archosaurs. The skull, pelvis, feet, and eggs of birds and reptiles are so clearly archosaurian that Darwin's friend T. H. Huxley called birds "glorified reptiles."

The earliest known bird is *Archaeopteryx* from Upper Jurassic rocks in Germany, perhaps the most famous fossil in the world (Figure 13.15). Only seven specimens, plus a single feather, have been found, and one of those is currently missing, probably stolen. The first complete *Archaeopteryx* was immediately recognized as a fossil bird, because it had feathers on its wings and tail. But without feathers, it looks very much like a small theropod dinosaur. Two of the six known specimens lay unrecognized for a long time, labeled as small theropods.

Figure 13.15 There are only a few known skeletons of *Archaeopteryx*, the first bird. This is the "Berlin" specimen, which belongs to the Humboldt Museum in Berlin. (From an early engraving.)

Archaeopteryx has a theropod pelvis, not the tight, box-like structure of living birds. It has a long, bony tail, clawed fingers, and a jaw full of savage little teeth.

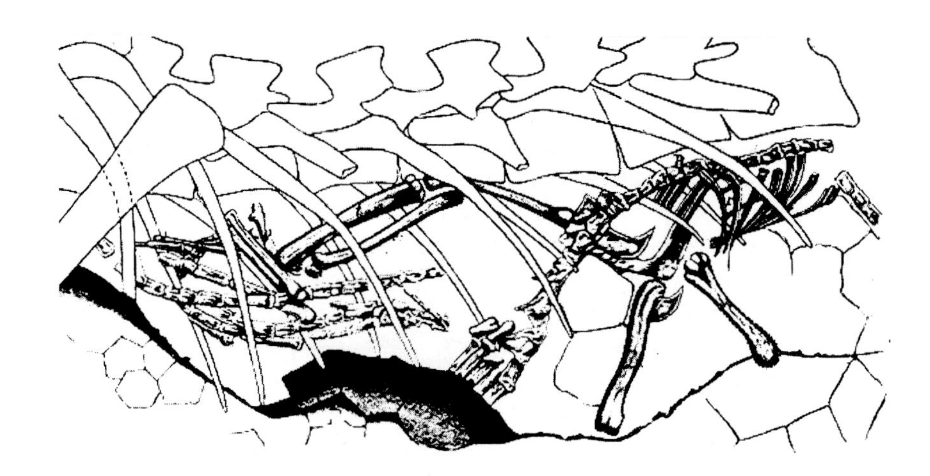

Figure 13.16 The best specimen of *Compsognathus* turned out to have a lizard folded up inside it: its last meal, and possibly an ill-judged one. *Compsognathus* bones in outline; lizard bones emphasized. (From a sketch by Nopsca.)

These are all theropod features. *Archaeopteryx* lacks many features of living birds; in fact, the only birdlike features on the entire bony skeleton of *Archaeopteryx* are a few characters of the skull. Even the wishbone or furcula is also found in theropod dinosaurs (perhaps *all* of them!).

Compsognathus is a small theropod dinosaur that was also preserved in the Solnhofen Limestone (Figure 12.6). It was a fast-running predator a little larger than *Archaeopteryx.* One specimen has the skeleton of a long, slim, fast-running lizard neatly folded up inside it (Figure 13.16). Both *Compsognathus* and *Archaeopteryx* are always preserved in an unusual body attitude, with the neck severely ricked back over the body. We know why this happens. If an animal dies today on or near the beach or on a desert salt pan, it may be mummified by wind and salt spray before it rots or is eaten by predators. The muscles slacken and the tendons dry out. The long tendons that support the head contract severely, dragging the skull backwards over the spine. At the same time, any body feathers on a bird usually drop off, but the stronger wing and tail feathers stay fixed in position.

Occasionally, mummified birds may be washed out to sea on a high tide, or blown into the water by a gale. They may float for several weeks before becoming waterlogged, and even when they finally sink, they retain their peculiar body attitude. *Compsognathus* did not have feathers, and it was without question a terrestrial biped. *Archaeopteryx* and *Compsognathus* were fossilized in the same way, as mummified bodies that floated out from a shoreline some distance away. There is no need to suggest that *Archaeopteryx* could fly because it sank and was buried at sea.

The Origin of Flight in Birds

Almost all paleontologists are now convinced that birds evolved from theropods. There are now many transitional fossils between theropods and the first birds. In particular, feathers have been found on several theropods that were certainly ground-running animals (Chapter 12).

Since ground-running theropods had feathers, the origin of flight in birds has nothing to do with the appearance of feathers. I argued in Chapter 12 that display and thermoregulation may have been involved in the origin of feathers, but not flight. (Flight evolved in bats and pterosaurs without feathers.) There have been three important hypotheses for the origin of bird flight, and I shall add a fourth.

The arboreal hypothesis

The arboreal hypothesis suggests that ancestral birds evolved flight by jumping out of trees. The arboreal theory was the most favored until recently, and it still has supporters. But it must be abandoned in the face of the new theropods from China. With long, erect limbs, a comparatively short trunk, and bipedal locomotion, *Archaeopteryx* and the feathered theropods are exactly the opposite in body plan of all living mammals and reptiles that jump and glide from tree to tree. There is nothing in the ancestry of birds as we now know it to suggest any arboreal adaptations at all.

Flapping arms or proto-wings (in fact, any feathers at all on wings or tail) increases drag. Aerodynamically, the transition from gliding to flapping is difficult: there is only a narrow theoretical window through which the transition could have been made. Such a transition would have been especially difficult for *Archaeopteryx* or any other known theropod candidate for the ancestry of birds, because they had long, bony tails with long feathers on them. This kind of tail adds much more drag than it adds lift.

The cursorial hypothesis

Perhaps some adaptations in a ground-dwelling theropod could provide some of the anatomy and behavior necessary for flight, such as lengthening the forearms, especially the hands, placing long, strong feathers in those areas, and evolving powerful arm movements. Early in the twentieth century, Baron Nopsca, perhaps impressed by the apparent speed and dexterity of *Compsognathus* (Figure 13.17), suggested that a fast-running reptile might evolve long scales on the arms. In this theory, the scales generated lift as the arms were actively flapped on the run (Figure 13.18). The animal could now take long leaps, perhaps encouraging the scales to evolve into feathers and the leaps to evolve into powered flapping flight.

Feathers did not evolve from scales, but in any case this original version of the cursorial hypothesis does not work. Any lift generated by a flapping arm decreases the ground traction given by the feet, and acceleration is lost. A racing car is held down on the track by its airfoils for good traction, and an aircraft cannot be driven through its wheels on the takeoff run. A running theropod that flapped its arms would increase drag: the faster the run, the greater the drag. Only a very small amount of thrust would have been generated by the arms in the early stages of the process.

The running raptor

More recent versions of the cursorial hypothesis are much better: they are mechanically sounder and include behavior that involved strong, synchronized arm strokes and the evolution of strong pectoral muscles.

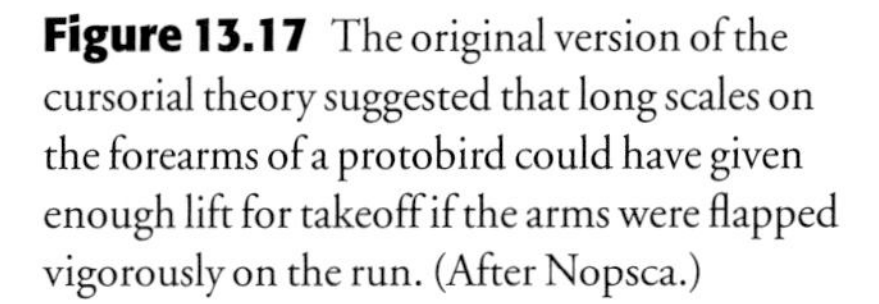

Figure 13.17 The original version of the cursorial theory suggested that long scales on the forearms of a protobird could have given enough lift for takeoff if the arms were flapped vigorously on the run. (After Nopsca.)

Figure 13.18 Gerhard Heilmann reconstructed the wing of *Archaeopteryx* in 1926. Instead of primary feathers on the ends of the wing, the fingers carried, displayed, and no doubt used, claws.

The most plausible of the recent versions is that of Phillip Burgers and Kevin Padian in 2001. They envisage flight evolving from running, with the early advantage of flight being greater speed over the ground. The achievement of flight would replace thrust on the ground by foot traction by aerodynamic forward thrust from the wing. In the take-off run, energy expended by the forelimbs would replace energy expended by the hindlimbs, after a transition period in which all limbs would be contributing to forward thrust.

Unlike the early versions, lift is not important at first. The first stages of this fast, low-level flight would be aided by the phenomenon of ground effect. Essentially, the eddies generated by the wings interact with the ground immediately under the wings, providing enough lift at very low altitude to achieve take-off. Thus the wing stroke would not have to produce much lift as long as there was no advantage in acquiring height.

In the scenario, the bird is now capable of fast-flapping low-level flight, but its advantage ends if it ascends out of the shallow zone of ground effect. All the wing action is energetically expensive, especially in the early stages of lift-off. Rapid flapping is essential throughout the scenario. And finally, none of this scenario begins to work until (unless) wing thrust is powerful enough to replace the (powerful) leg thrust of a running theropod. (The earliest feathered wings would not have been very effective as thrust devices.)

The display and fighting hypothesis

Jere Lipps and I suggested that display was involved in the evolution of flight as well as feathers. Theropods had long, strong display feathers on arms and tail (Chapter 12). Successful display was increased by lengthening the arms, especially the hand, and by actively waving them, perhaps flapping them rapidly and vigorously. Flapping in display would have encouraged the evolution of powerful pectoral muscles, and the supracoracoideus system.

Display can be very effective, and not just for sexual ends. Frigate birds and bald eagles often try to rob other birds of food instead of catching prey themselves. Because the penalty for wing injury is high, many birds can be intimidated by display into giving up their catch rather than fighting to defend it.

But a threat display cannot always be an empty bluff. Fighting is the last resort. Living birds often fight on the ground, even those that fly well. The wings no longer have claws but are still used as weapons in forward and downward smashes (steamer ducks are particularly deadly at this). Beaks and feet can be used as weapons too, and are most effective when used in a downward or forward strike.

A strong wing flap, directed forward and downward, is also the power stroke that gives lift to a bird in takeoff. Lipps and I suggest that strong wing flapping is a simple extension of display flapping, honed in fighting behavior. Powerful flapping used to deliver forearm smashes could have lifted the bird off the ground, allowing it also to rake its opponent from above with its hind claws. The more rapidly the wings could be lifted for another blow, the more effective the fighting. This would rapidly encourage an effective wing-lifting motion that minimized air resistance, so the wing action would then be almost identical to a takeoff stroke.

Kevin Padian (e.g., see Padian, 2001) also sees the wing stroke evolving from the arm strike used by a theropod in predation (rather than competition). It is not clear (to me) how this could have led easily to whole-body takeoff, however: predators try to avoid situations in which they would need a prolonged struggle to subdue a prey. (Sexual competition is another matter: you may be fighting for your posterity, not just for another meal.)

Archaeopteryx fits our display-and-fighting hypothesis well. It was well adapted for display. Like any small theropod, it was well equipped for fighting with its teeth and the strong claws on hands and feet. *Archaeopteryx* did not have long primary feathers on its fingers (Figure 13.18), probably because they would have hidden the claws in display and would most likely have broken in a fight.

Archaeopteryx could not fly well; I suspect that it hardly flew at all. It may have been able to glide a short distance, but it could not have sustained flapping flight. There is no breastbone, and no hole through the shoulder joint through which to pass the large tendon that gives the rapid, powerful, twisting wing upstroke in living birds. This tendon passes through the shoulder joint, and as well as raising the wing, it twists it. On the upstroke, the twist arranges the wing and feathers so that they slip easily through the air, with little drag. At the top of the upstroke, the wing is in exactly the right position to give a powerful downbeat. Without the supracoracoideus system (Figure 13.19), which is easily identified because it leaves a strong trace on the shoulder joint, a bird cannot fly by wing flapping. In fact, it cannot even take off and land, because the greatest power from rapidly beating wings is required during slow flight.

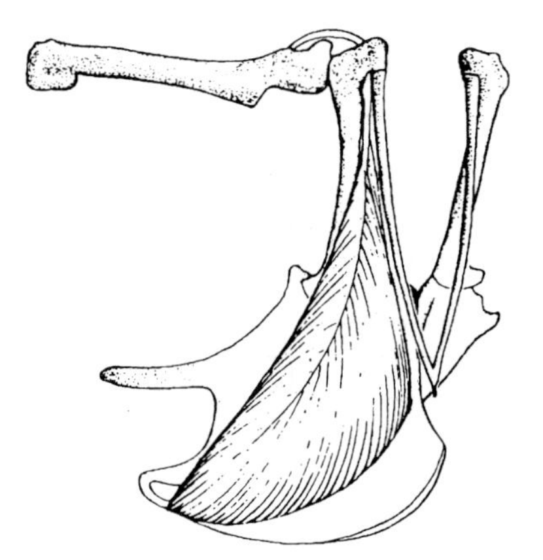

Figure 13.19 *Archaeopteryx* lacked many features of living birds. All living birds have a powerful muscle, the supracoracoideus, shown here passing from the breastbone up through a special hole in the shoulder joint to the top side of the humerus. It acts to fold up (and streamline) the wing as it is raised. Meanwhile the flexible wishbone acts as a spring, to prevent the shoulder girdle from distorting too much as the wing muscles pull on the shoulder joints. *Archaeopteryx* had not evolved this system. There is no hole in the shoulder joint, no muscle attachment for the supracoracoideus on the humerus, and its wishbone was big, solid, and rigid rather than flexible and spring-like.

In small flying birds today, the wishbone acts as a spring that repositions the shoulder joints after the stresses of each wing stroke. It is needed to give the rapid flaps necessary for flight (a starling flies with 14 complete wing beats per second). The wishbone also helps to pump air in and out of the lungs, and to recover some of the muscular energy put into the downstroke. But the wishbone in *Archaeopteryx* and the wishbones of theropods are U-shaped and strong and solid; they could not have acted as effective springs. Furthermore, *Archaeopteryx* did not have the long primary feathers on the wing tips, or the breastbone anchoring the muscles that are needed for routine takeoff and landing. It could not have raised its arms high above its body for an effective downstroke. In fact, *Archaeopteryx* evolved structures that were active deterrents to flight. Its tail was long and bony, with long feathers. Among living birds with display feathers, this sort of tail is aerodynamically the worst of all possible tail styles, adding a lot of drag and little lift.

Archaeopteryx, then, was a fierce little fast-running, displaying bird, which probably spent its life scurrying around the Solnhofen shore, hunting for small prey such as crustaceans, reptiles, and mammals. In hunting style, *Archaeopteryx* was probably much like the roadrunner of the dry country of the American Southwest, but its ecological setting was like that of a steamer duck: on a shoreline with year-round food supply. If so, then *Archaeopteryx* did not compete in the air with the pterosaurs that are also found in the Solnhofen Limestone, and the first bird did not fly.

From fight to flight

Display and fighting in birds takes a lot of energy, whether it is for territory, dominance, or food, but it provides an enormous payoff in survival and selection. Sexual display in most living birds must be done correctly, or no mating takes place. New behaviors can be evolved rapidly, and they are often evolutionarily cheap, because they usually don't require important morphological changes in their early stages.

Our scenario stresses lift as well as thrust. It suggests that the earliest birds evolved flight behavior, anatomy, and experience at low ground speed and low height: ideal preflight training. The selective payoff for successful mastery of the flight motions gave significant advantages, even before flight itself was possible. Short-lived but intense activity could provide major adaptive advantage, an advantage that would begin as soon as the feathered surfaces of the wings could generate any lift at all.

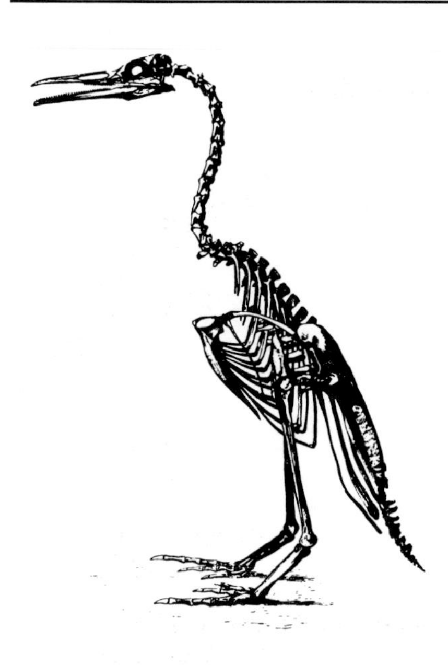

Figure 13.20 A reconstruction of a flightless, diving Cretaceous bird, *Hesperornis*. It had a flying ancestor, because its shoulder and breastbone have the structures associated with rapid and powerful upward wingbeats. It had teeth, however, a primitive character for birds. (From Marsh.)

Figure 13.21 A reconstruction of a Cretaceous bird, *Ichthyornis*. It appears to have been an agile flier that ate fish, with a way of life rather like a living tern. Note that *Ichthyornis* had teeth. (From Marsh.)

Rapid wingbeats would not be essential, as they are in the cursorial hypotheses: the wingbeats would simply have to be faster than those of the competition.

From the stage exemplified by *Archaeopteryx*, the many advantages of flight were added to those of social or sexual competition. I do not think it is a coincidence that the males of the Early Cretaceous Chinese birds *Confuciusornis* and *Changchengornis* had extravagantly long (display) feathers on the tail! In more advanced birds than *Archaeopteryx*, the supracoracoideus tendon system evolved in the shoulder, while the wishbone evolved into a spring. The breastbone evolved as the anchor for the flight muscles. The forearms became longer, lighter, and more fragile in bone structure, becoming specialized as wings, and losing the finger claws. Meanwhile, the feet and beak became the dominant fighting weapons, as in most living birds.

Cretaceous Birds

The radiation of birds was very rapid. Early Cretaceous rocks have yielded bird remains in all the northern continents and in Australia. *Sinornis*, a sparrow-sized bird from the Early Cretaceous of China, had many features directly related to much better flight and perching than was possible in *Archaeopteryx*. The body and tail were shorter, and the tail had fused vertebrae at its end that provided a firm but light base for strong tail feathers. The center of mass of the body was much farther forward, closer to the wings. *Sinornis* had a breastbone, a shoulder joint that allowed it to raise its wings well above the horizontal, and fingers that were adapted to support feathers rather than grasping and tearing claws. The wrist could fold much more tightly forward against the arm than the 90° seen in *Archaeopteryx*, so the wing could be folded away cleanly in the upstroke or on the ground, reducing drag. The foot was much better adapted for perching. Even so, *Sinornis* still had some very primitive features: the skull and pelvis were much like those of *Archaeopteryx*.

New Early Cretaceous fossils from Spain and from China tell the same story. Rapid evolution among Early Cretaceous birds dramatically improved their flying and perching ability; perhaps this is why most of them were small and light. *Confuciusornis* and *Changchengornis*, from China, had lighter bones than *Archaeopteryx*, and had genuine beaks rather than jaws with teeth. *Eoalulavis* from Spain had evolved the alula, the arrangement of feathers near the wing tip that allows slow flight without stalling.

Most Cretaceous bird fossils are from shoreline habitats, but that may reflect preservation bias rather than ecological reality. We have good fossils of Late Cretaceous diving birds such as *Hesperornis* (Figure 13.20), and *Ichthyornis* was tern-like in its adaptations (Figure 13.21).

Cenozoic Birds

When the dinosaurs died out at the end of the Cretaceous, there must have been a very interesting opportunity for surviving creatures to invade the ecological niches associated with larger body size on the ground. The two leading contenders were birds and mammals, and although mammals quickly became large herbivores, it was birds that became the dominant land predators in some regions in the Paleocene. These birds evolved to become flightless terrestrial bipeds once more. (This is the only way a bird can become heavy and powerful.)

Large, flightless birds called diatrymas (after *Diatryma*, one of the largest) lived across the Northern Hemisphere in the Paleocene and Eocene. They stood about 2 m (6 ft) tall, and they had massive legs with vicious claws and huge, powerful beaks that look like killing instruments as efficient as the heads of tyrannosaurs. A full-grown *Diatryma* could have killed many of the mammals in its community. Carnivorous birds with very similar adaptations, the phorusrhacids, dominated the plains ecosystem of South America somewhat later (Figure 13.22). Diatrymas became extinct at the end of the Eocene, but the phorusrhacids survived until the Late Pliocene. Some phorusrhacids were 2.5 m (7 ft) tall, and a spectacular late phorusrhacid, *Titanis*, crossed to Florida from South America less than 3 m.y. ago. It was larger than an ostrich and no doubt caused at least temporary consternation among the Floridian mammals of the time.

The southern continents have a number of large flightless birds. Living forms such as the ostrich, cassowary, rhea, and emu are familiar enough, but even more interesting forms are now extinct. All these birds are loosely grouped together as ratites (Figure 13.23). The moas of New Zealand reached well over 3 m (10 ft) in height. *Aepyornis*, the "elephant bird" of Madagascar (Figure 13.24a), was living so recently that its eggshells are still found lying loose on the ground. The eggs are unmistakable because they had a volume of 11 liters (2 gallons). Early Muslim traders along the African coast certainly saw these eggs, and they may even have seen living elephant birds in Madagascar, giving rise to folktales about the fearsome roc that preyed on elephants and carried Sinbad the Sailor on its back (Figure 13.24b). *Aepyornis* and *Dromornis*, a giant extinct Australian bird (Chapters 18 and 21), are close competitors for Heaviest Bird Ever to Evolve. The *Guinness Book of World Records* currently favors *Dromornis*, which was powerfully built and weighed perhaps 500 kg (1100 lb).

Figure 13.22 One of the large flightless carnivorous birds of the Cenozoic of South America, a phorusrhacid from Argentina. The original reconstruction by Sinclair was shown with a greatly curved neck so that it would fit onto the size of the page he was allowed; this is my attempt to straighten out the neck into a normal posture. (After Sinclair.)

The largest flying birds

The largest flying birds so far discovered are teratorns, immense birds from South America, now extinct, who reached North America during the Pleistocene. Hundreds of specimens have been found in the tar pits of La Brea in Los Angeles, California (Figure 13.25), and from Florida and Mexico. But the largest teratorn was *Argentavis* from the Late Miocene of Argentina: it had a wingspan of 7.5 m (24 ft). By contrast, the largest living bird is the royal albatross, just over 3 m (10 ft) in wingspan.

The beak of *Argentavis* suggests that it was a predator, not a scavenger. It probably stalked prey on the ground. With a skull 55 cm (2 ft) long and 15 cm (6 inches) wide, it could have swallowed prey animals 15 cm across. Its bones are associated with other vertebrate fossils, but 64% of those are from *Paedotherium*, a

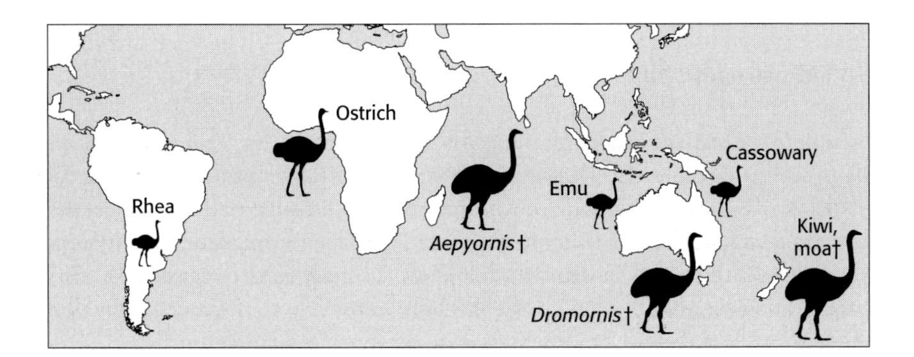

Figure 13.23 The distribution of some living and fossil ratites.

Figure 13.24 (a) *Aepyornis*, the elephant bird of Madagascar that became extinct only a few hundred years ago (an ostrich is next to it for scale). On the right (b) is the roc, Smizurgh of Arab and Persian legend, carrying off three elephants at once. The legend of the roc may have been based on ostriches, but ostriches must have been familiar to Arab traders visiting East Africa. Most likely the much larger and rarer elephant bird was the real inspiration for the stories. (From a nineteenth-century story book.)

Figure 13.25 A Pleistocene teratorn, *Teratornis*, from the La Brea tar pits of Los Angeles. The preservation is so good that the wingspan can be measured accurately at nearly 4 m (13 ft). This is enormous for a bird, although not for a pterosaur (compare Figure 13.14).

little mammal about the size of a jackrabbit (in other words, an easy swallow for *Argentavis*).

In the same size range as teratorns were pelagornithids, gigantic marine birds that must have spent most of their time soaring over water. They ranged worldwide from the Eocene to the Late Miocene. They were lightly built, but the wingspan was close to 6 m (nearly 20 ft) in the largest specimens. Their beaks were very long, with tooth-like projections built into their edges, presumably to help them hold squirming prey. More than any other living birds, pelagornithids were the ecological equivalents of pterosaurs, and it will be fascinating when further research allows us to reconstruct their mode of life accurately.

BATS

The latest evolution of flapping flight among vertebrates took place among bats. In all bats, the wing is stretched between arm, body, and leg, with the fingers of the hand splayed out in a fan toward the wingtip (Figure 13.6). The wing membrane has little strength of its own, but it is elastic, and tension has to be maintained in it by muscles and tendons. The hind leg is used as an anchor for the trailing edge of the membrane, which means that the limb is not free for effective walking and running. Bats therefore are forced into unusual habits, which include roosting in inaccessible places where they hang upside down. Because bats are placental mammals, they have evolved special adaptations to maintain flight during pregnancy and nursing. For example, the pelvis has features that allow the body to be streamlined yet still have a rather large birth canal. Baby bats have needle-sharp milk teeth that allow them to hold tightly to the mother's fur in flight (ouch at feeding time!).

Figure 13.26 The earliest known fossil bat, *Icaronycteris* from the Early Eocene of Wyoming. The fossil is shown here, with the shadowy outline of the wings reconstructed. (Courtesy of the late Glenn Jepsen.)

The earliest known bat, *Icaronycteris*, is an extremely well-preserved fossil from Early Eocene lake beds in Wyoming (Figure 13.26). The Messel Oil Shale, in Middle Eocene rocks of Germany, has yielded dozens of bat skeletons. Some of them still contain the bats' last meals (primitive moths). Even the smallest ear bones are preserved, and they tell us that these bats were already equipped with the echolocating sonar that all insect-hunting, fishing, and frog-eating bats have today. The baby bats at Messel already had sharp milk teeth.

Current opinion is that bats evolved flight in trees, through a parachuting stage, though there is no evidence at all from bat ancestors, because we have not identified them. Bat sonar presumably evolved from the acute hearing of little, nocturnal, insect-hunting mammals in the forest canopy of the Cretaceous (fruit bats have lost their sonar). Obviously, bats must already have had an eventful evolutionary history before the Eocene, but we still have to find these fossils.

Further Reading

Flight (but not birds)

Averof, M., and S. M. Cohen 1997. Evolutionary origin of insect wings from ancestral gills. *Nature* 385: 627. [Genetic evidence from living arthropods.]

Carrier, D. R., and C. G. Farmer. 2000. The evolution of pelvic aspiration in archosaurs. *Paleobiology* 26: 271–93. [Relevant to the evolution of pterosaur and bird respiration.]

Clark, J. M., et al. 1998. Foot posture in a primitive pterosaur. *Nature* 391: 886–9. [*Dimorphodon* could not have run, in this interpretation.]

Colbert, E. C. 1970. The Triassic gliding lizard *Icarosaurus*. *Bulletin of the American Museum of Natural History* 143: 85–142.

Dalla Vecchia, F. M., et al. 2002. A crested rhamphorhynchoid pterosaur from the Late Triassic of Austria. *Journal of Vertebrate Paleontology* 22: 196–9. [A very well-preserved and very early pterosaur.]

Dudley, R. 2000. The evolutionary physiology of animal flight: paleobiological and present perspectives. *Annual Reviews of Physiology* 62: 135–55. [Good discussion of the evolution of insect flight (by jumping, says Dudley). He tries to make all other instances of flight evolution fit the same model, but I don't believe that (yet). As with everything Dudley does, this is packed with good ideas.]

Frey, E., et al. 1997. Gliding mechanism in the Late Permian reptile *Coelurosauravus*. *Science* 275: 1450–2, and comment 1419.

Gans, C., et al. 1987. *Sharovipteryx*, a reptilian glider? *Paleobiology* 13: 415–26.

Haubold, H., and E. Buffetaut. 1987. A new interpretation of *Longisquama insignis*, an enigmatic reptile from the Upper Triassic of Central Asia. *Comptes rendus de l'Académie des Sciences de Paris, Série II* 305: 65–70.

Jones, T. D., et al. 2000. Nonavian feathers in a Late Triassic archosaur. *Science* 288: 2202–5, and comment 2124–5. [*Longisquama*. I think this is nonsense (and I've seen the [single] specimen). See discussions in *Science* 291: 1899–901; *Nature* 408: 428.]

Kellner, A. W. A., and D. de A. Campos. 2002. The function of the cranial crest and jaws of a unique pterosaur from the Early Cretaceous of Brazil. *Science* 297: 389–92. [*Thalassodromeus* seems to have been functionally and ecologically like the living skimmer.]

Kingsolver, J. G., and M. A. R. Koehl. 1994. Selective factors in the evolution of insect wings. *Annual Review of Entomology* 39: 425–51.

Kramer, M. G., and J. H. Marden. 1996. Almost airborne. *Nature* 385: 403–4.

Kukalová-Peck, J. 1987. New Carboniferous Diplura, Monura, and Thysanura, the hexapod ground plan, and the role of thoracic side lobes in the origin of wings (Insecta). *Canadian Journal of Zoology* 65: 2327–45.

Marden, J. H., and M. G. Kramer. 1994. Surface-skimming stoneflies: a possible intermediate stage in insect flight evolution. *Science* 266: 427–30, and comment, v. 270, 1685. See also Marden's nontechnical 1995 article in *Natural History* 104 (2): 4–8.

Naish, D., and D. M. Martill. 2003. Pterosaurs—a successful invasion of prehistoric skies. *Biologist* 50: 213–16. [A review; same message as Unwin and Henderson, 2002.]

Norman, D. B., and P. Wellnhofer. 2000. *The Illustrated Encyclopedia of Dinosaurs*. London: Salamander. [The best of the "coffee-table" books, with good science and beautiful illustrations. Despite the title, includes pterosaurs.]

Rayner, J. M. V., and K. A. Padian. 1993. The wings of pterosaurs. *American Journal of Science* 293-A: 91–166. [The narrow-wing interpretation.]

Sigé, B., et al. 1998. The deciduous dentition and dental replacement in the Eocene bat *Palaeochiropteryx tupaiodon* from Messel: the primitive condition and beginning of specialization of milk teeth among Chiroptera. *Lethaia* 31: 349–58.

Socha, J. J. 2002. Gliding flight in the paradise tree snake. *Nature* 418: 603–4.

Springer, M. S., et al. 2001. Integrated fossil and molecular data reconstruct bat echolocation. *Proceedings of the National Academy of Sciences* 98: 6241–6. [Echolocation. evolved with the earliest bats, and was lost later on in the "megabats," the fruit bats.]

Tintori, A., and D. Sassi. 1992. *Thoracopterus* Bronn (Osteichthyes: Actinopterygii): a gliding fish from the Upper Triassic of Europe. *Journal of Vertebrate Paleontology* 12: 265–83.

Unwin, D. M., and D. M. Henderson. 2002. On the trail of the totally integrated pterosaur. *Trends in Ecology & Evolution* 17: 58–9. [Pterosaurs were clumsy quadrupeds on land, but excellent fliers in the air.]

Weimerskirch, H., et al. 2003. Frigatebirds ride high on thermals. *Nature* 421: 333–4. [A wonderful analog for pterosaurs, in my opinion.]

Witmer, L. M., et al. 2003. Neuroanatomy of flying reptiles and implications for flight, posture and behaviour. *Nature* 425: 950–3, and comment by Unwin, 910–11.

Zimmer, C. 1998. Into the night. *Discover* 19 (11): 110–15. [Bat evolution and echolocation.]

Birds

Ackerman, J. 1998. Dinosaurs take wing: new fossil finds from China provide clues to the origin of birds. *National Geographic* 194 (1): 74–99.

Balmford, A., et al. 1993. Aerodynamics and the evolution of long tails in birds. *Nature* 361: 628–31. [*Archaeopteryx*-type tails make flight worse, not better.]

Burgers, P., and K. Padian. 2001. Why thrust and ground effect are more important than lift in the evolution of sustained flight. In Gauthier, J., and L. F. Gall (eds), *New Perspectives on the Origin and Early Evolution of Birds*. New Haven, CT: Peabody Museum of Natural History, Yale University, 351–61.

Campbell, K. E., and E. P. Tonni. 1981. Preliminary observations on the paleobiology and evolution of teratorns (Aves, Teratornithidae). *Journal of Vertebrate Paleontology* 1: 265–72.

Carrier, D. R., and C. G. Farmer. 2000. The evolution of pelvic aspiration in archosaurs. *Paleobiology* 26: 271–93. [Relevant to evolution of pterosaur and bird respiration.]

Chatterjee, S. 1997. *The Rise of Birds*. Baltimore: Johns Hopkins University Press.

Chiappe, L. M., et al. 1999. Anatomy and systematics of the Confuciusornithidiae (Theropoda: Aves) from the Late Mesozoic of northeastern China. *Bulletin of the American Museum of Natural History* 242.

Chiappe, L. M., and L. M. Witmer (eds). 2002. *Mesozoic Birds: Above the Heads of Dinosaurs*. Berkeley, California: University of California Press. [Massive overview in 20 chapters, with full references. See especially chapters by Witmer, Clark et al., Chiappe, and Gatesy.]

Chiappe, L. M., and G. J. Dyke. 2002. The Mesozoic radiation of birds. *Annual Review of Ecology and Systematics* 33: 91–124.

Cowen, R., and J. H. Lipps. 1982. An adaptive scenario for the origin of birds and of flight in birds. *Proceedings of the 3rd North American Paleontological Convention, Montréal*, 109–12. [Also posted on the Web site.]

Dantzker, M. S., et al. 1999. Directional acoustic radiation in the strut display of male sage grouse *Centrocercus urophasianus*. *Journal of Experimental Biology* 202: 2893–909.

Dial, K. P. 2003. Wing-assisted incline running and the evolution of flight. *Science* 299: 402–3, and comment 329. [I don't believe it.]

Dunn, P. O., et al. 2001. Mating systems, sperm competition and the evolution of sexual dimorphism in birds. *Evolution* 55: 161–75. [Sexual selection (display) can drive the evolution of sexual dimorphism in birds.]

Earls, K. D. 2000. Kinematics and mechanics of ground take-off in the starling *Sturnus vulgaris* and the quail *Coturnix coturnix*. *Journal of Experimental Biology* 203: 725–39. [Both these small birds take off with a powerful jump: this is fully compatible with our display hypothesis.]

Feduccia, A. 1996. *The Origin and Evolution of Birds*. New Haven: Yale University Press. [Contains a lot of data and a lot of opinion, and it's often difficult to tell them apart.]

Garner, J. P., et al. 1999. On the origins of birds: the sequence of character acquisition in the evolution of avian flight. *Proceedings of the Royal Society of London B* 266: 1259–66, and enthusiastic comment by. Hedenström, *Trends in Ecology & Evolution* 14: 375–6. [Garner et al. suggest a clumsy complex series of changes, e.g., they require a perching theropod that would climb trees or bushes, then pounce on its prey. Perch-and-pounce predators today (e.g., some owls and hawks) are superbly aerobatic and are powerful fliers, quite unlike a very early bird just evolving flight.]

Houde, P. 1986. Ostrich ancestors found in the Northern Hemisphere suggest new hypothesis of ratite origins. *Nature* 324: 563–5, and comment 516. [Since supported by molecular evidence.]

Jenkins, F. A. 1993. The evolution of the avian shoulder joint. *American Journal of Science* 293-A: 253–67.

Ji, Q., et al. 1999. A new Late Mesozoic confuciusornithid bird from China. *Journal of Vertebrate Paleontology* 19: 1–7. [*Changchengornis*, with display tail feathers just like *Confuciusornis*.]

Marshall, L. G. 1994. The terror birds of South America. *Scientific American* 270 (2): 90–5.

Norell, M. A., and J. A. Clarke. 2001. Fossil that fills a critical gap in avian evolution. *Nature* 409: 181–4. [*Apsaravis*.]

Nudds, R. L. & Bryant, D. M. 2000. The energetic cost of short flights in birds. *Journal of Experimental Biology* 203: 1561–72. [Short flights are even more expensive in terms of energy than we had thought. I think that supports a display-and-fighting idea, because it places the origin of flight as a short-flight, intense-effort situation.]

Padian, K., and L. M. Chiappe. 1998a. The origin and early evolution of birds. *Biological Reviews* 73: 1–42. [Major survey, with extensive reference list.]

Padian, K., and L. M. Chiappe. 1998b. The origin of birds and their flight. *Scientific American* 278 (2): 28–37, and correspondence, v. 278 (6): 8–8A.

Padian, K. 2001. Stages in the origin of bird flight: beyond the arboreal–cursorial dichotomy. In Gauthier, J., and L. F. Gall (eds), *New Perspectives on the Origin and Early Evolution of Birds*. New Haven, CT: Peabody Museum of Natural History, Yale University, 255–72.

Padian, K. 2003. Four-winged dinosaurs, bird precursors, or neither? *BioScience* 53: 450–3. [*Microraptor*: comment on Xu et al., 2003.]

Poore, S. O., et al. 1997. Wing upstroke and the evolution of flapping flight. *Nature* 387: 799–802.

Rich, P. V. 1980. The Australian Dromornithidae: a group of extinct large ratites. *Contributions in Science of the Los Angeles County Natural History Museum* 330: 93–104.

Sanz, J. L., et al. 1996. An Early Cretaceous bird from Spain and its implications for the evolution of avian flight. *Nature* 382: 442–5.

Sanz, J. L., et al. 1997. A nestling bird from the Lower Cretaceous of Spain: implications for avian skull and neck evolution. *Science* 276: 1543–6, and comment 1501.

Shipman, P. 1998. *Taking Wing: Archaeopteryx and the Evolution of Bird Flight*. New York: Simon and Schuster.

Tykoski, R. S., et al. 2002. A furcula in the coelophysid theropod *Syntarsus*. *Journal of Vertebrate Paleontology* 22: 728–33. [If *Syntarsus*, an early basal theropod, has a furcula, all theropods except maybe *Eoraptor* and *Herrerasaurus* had one, and eventually it may be found in them too.]

Witmer, L. M., and K. D. Rose. 1991. Biomechanics of the jaw apparatus of the gigantic Eocene bird *Diatryma*: implications for diet and mode of life. *Paleobiology* 17: 95–120.

Xu, X., et al. 2003. Four-winged dinosaurs from China. *Nature* 421: 335–40, and comment, pp. 323–4; also comment in *Science* 299, 491. See also Padian, 2003. [*Microraptor gui*.]

Zhou, Z., and F. Zhang. 2002. A long-tailed, seed-eating bird from the Early Cretaceous of China. *Nature* 418: 405–9. [*Jeholornis*: early, fairly primitive, seed-eating bird.]

Zhou, Z., et al. 2003. An exceptionally preserved Lower Cretaceous ecosystem. *Nature* 421: 807–14. [Convenient review of the Jehol Biota from China that includes feathered dinosaurs and early birds. No new science.]

Zimmer, C. 1992. Ruffled feathers. *Discover* 13 (5): 44–54. [Sankar Chatterjee and the *Protoavis* affair.]

Zimmer, C. 1997. Terror, take two. *Discover* 18 (6): 68–74. [*Titanis* and friends.]

CHAPTER FOURTEEN

The Modernization of Land and Sea

At the end of the Permian, the world's biology was decimated. The Mesozoic is the time (era, if you like) when that biology was not only reconstituted as a diverse global fauna and flora, but it took on many of the characteristics of the modern world. A SCUBA diver in Permian seas would not have seen many familiar creatures, even if they were playing familiar ecological roles. However, a SCUBA diver in the Late Cretaceous would have found a much more familiar world.

In this chapter we will look at some of the marine and terrestrial organisms that display this major change. I will concentrate on the top predators of the Mesozoic oceans, chiefly marine reptiles, and on land I will concentrate on the engine that drove the change: the transition from a conifer-dominated land flora to an angiosperm-dominated one that gave not only a different structure to land ecosystems, but filled them with beautiful blossoms and fragrant scents.

MESOZOIC OCEAN ECOSYSTEMS

Paleozoic oceans had fishes that must have been exploiting food sources that are not well preserved in the fossil record: the plankton of the ocean surface. But those Paleozoic fishes were neither numerous nor diverse. Lobefins were in decline after some of them had invaded the land, but the rayfins are not numerous or obvious fossils in most collections of marine animals from the Paleozoic. Instead, the larger carnivores in open water were various lineages of cephalopods, ammonoids that were essentially squids with shells. They were relatively slow-moving and clumsy.

But after the P–T extinction, fishes become the mid-sized predators of the ocean. Ammonites were still abundant, but the major additions to the global oceans in terms of large-bodied predators were not fishes, but fisheaters, and they are dominated by marine reptiles. This says (to me) that Mesozoic oceans were now productive enough that the ecosystem could sustain a level of large predators that could not have succeeded in Paleozoic oceans. One could argue about the root cause: additional nutrients reaching the oceans as plants colonized more land surface is one probable process that has been called the "seafood hypothesis" (Chapter 6). Whatever the reason, Mesozoic ecosystems differ dramatically from Paleozoic ones because large-bodied animals on land (dinosaurs) and in the air (pterosaurs) had their oceanic counterparts in large marine reptiles. Most of these reptile groups evolved

Figure 14.1 The giant Cretaceous marine turtle, *Archelon*, evolved a carapace that was lightened so that it could maintain buoyancy in the water. (Courtesy of the Peabody Museum of Natural History, Yale University.)

in Triassic times, but reached their greatest abundance in the Jurassic and Cretaceous. Several different clades of reptiles evolved spectacular adaptations to life at sea.

Turtles

We have already seen that turtles are diapsids, though it is debatable whether they are basal diapsids or basal archosauromorphs (Chapter 11). The first well-known turtle is from the Late Triassic of Europe, and it already had bony plates on its surface, though it had not yet accomplished the turtle trick of having the shoulder blades inside the ribs.

Turtles were widespread and successful in Jurassic and Cretaceous seas and estuaries. Perhaps the most famous is the giant Cretaceous turtle *Archelon*, which was 3 m (10 ft) long and nearly 4 m (13 ft) in flipper span. It was so large that it couldn't have swum with a complete solid carapace, so it had only a bony framework (Figure 14.1). Large marine turtles are anything but primitive in their biology. Their limbs are modified into hydrofoils, and they "fly" underwater. Marine turtles can navigate precisely over thousands of kilometers and they are warmblooded, maintaining their body temperatures at levels significantly higher than the water around them.

Crocodiles

Crocodiles are archosaurs, and their ancestry is clearly terrestrial (Chapter 11). All crocodiles were terrestrial predators in the Late Triassic. There were large, powerful

Figure 14.2 An ichthyosaur, beautifully preserved with skin, from the Jurassic of Germany. (Courtesy of the Library Services Department, American Museum of Natural History.)

crocodile-like aquatic parasuchids in the Triassic, and true crocodiles did not become aquatic until these others became extinct. *Terrestrisuchus*, a small crocodile from the Late Triassic of Britain, had long, slim, erect limbs, and with a length of less than 1 m, probably ran quite fast on land (Figure 11.15). Terrestrial crocodiles lived on well into the Jurassic, and in the end may have been outcompeted on land by bipedal theropod dinosaurs.

Ever since the Early Jurassic, most crocodiles have been amphibious. Many of them are predators at or near the water's edge. Some became almost entirely aquatic, and others returned yet again to land in the Cenozoic to become powerful terrestrial predators. Crocodiles that became amphibious or aquatic evolved to large size and were reasonably common in Mesozoic seas and rivers. *Deinosuchus* from the Late Cretaceous of Texas is the largest crocodile that ever lived. It had a skull 2 m (6 ft) long. It was about 10 m (33 ft) long, and weighed about 5 tons. It may have taken duck-billed dinosaurs as prey (they are found in the same rock formations) in the same way that living Nile crocodiles take hippos.

Crocodiles today are not equipped to kill large prey quickly. They usually kill large prey by holding them under the water until they drown. There's no reason to suppose that *Deinosuchus* did anything more sophisticated as it hunted large dinosaurs. *Stomatosuchus* was a gigantic duck-billed crocodile from the Middle Cretaceous of Africa. It is difficult to reconstruct its ecology because there are no living duck-billed crocodiles.

Figure 14.3 The hand of an ichthyosaur from the Jurassic of Britain. This specimen is in the Oxford University Museum.

Ichthyosaurs

Ichthyosaurs are not easily related to any other reptile groups, but the best guess is that they are highly derived basal diapsids. They were shaped much like dolphins, except that the tail flukes are horizontal in dolphins and vertical in ichthyosaurs. The most advanced ichthyosaurs had a continuation of the spine running into the lower tail fin (Figure 14.2). The main propulsion would then have been a side-to-side body motion, like a fish rather than a dolphin. The limbs were modified into small, stiff fins for steering and attitude control, again like dolphins, so that ichthyosaurs would have been very maneuverable up and down in the water as well as sideways (Figure 14.3). The tail fin was usually very deep, which is characteristic of swimmers that use fast acceleration in hunting prey. Ichthyosaurs were beautifully streamlined, but would have been unable to move on land.

Beautiful ichthyosaur fossils have been known for 200 years, and they figured in many early discussions of evolutionary theory because everyone could recognize

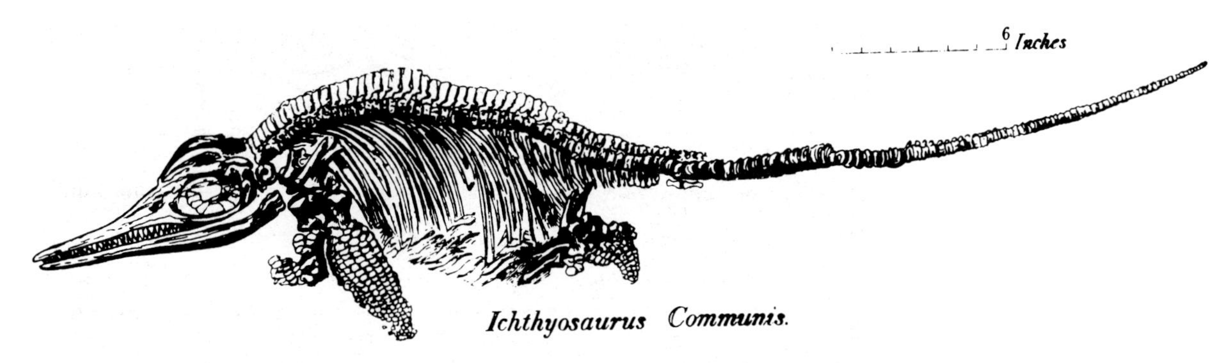

Figure 14.4 Beautiful fossils of ichthyosaurs were discovered early in the nineteenth century and caused much discussion about evolution. (From Buckland.)

their exquisite adaptations for life in water (Figure 14.4). Ichthyosaurs all had good vision, with large eyes looking right along the line of the jaw. In advanced ichthyosaurs the jaw was long and thin, with many piercing conical teeth that were well designed for catching fish. Preserved stomach contents include fish scales and hooklets from the arms of cephalopods, possibly soft-bodied squids. One spectacular Jurassic ichthyosaur, *Eurhinosaurus*, had a sword-like upper jaw projecting far beyond the lower, with teeth all along its length. *Eurhinosaurus* was probably an ecological equivalent of the swordfish, using its upper jaw to slash its way through a school of fish, then spinning around to catch its crippled victims.

Most early ichthyosaurs had blunt, shell-crushing teeth and may have hunted and crushed ammonites and other shelled cephalopods in a way of life that did not demand high levels of hydrodynamic performance. The earliest ichthyosaurs found to date, from the Early Triassic of the Northern Hemisphere, were small, about 1 m (3 ft) long, but they were already specialized for marine life. *Mixosaurus* is a typical small, early ichthyosaur, from Middle Triassic rocks ranging from the Arctic to Nevada to Indonesia; but the best-preserved specimens come from the Alps. The spine had not yet turned down to form the lower tail fin, but almost all the other features show excellent adaptation to swimming, with the limbs totally modified into effective fins.

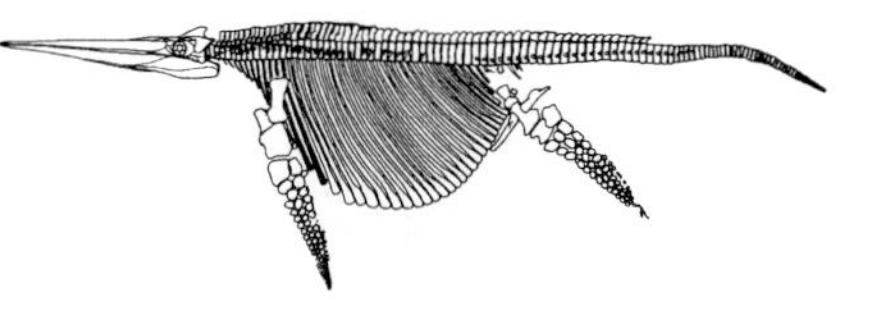

Figure 14.5 A huge, powerful, deep-bodied early ichthyosaur, *Shonisaurus*, up to 15 m (50 ft) long, from the Triassic of Nevada. (After Merriam.)

Shonisaurus from the Late Triassic of Nevada, at 15 m (50 ft) long, is one of the largest ichthyosaurs known. It was robust too, with a huge, strong deep body and long, powerful fins (Figure 14.5). Thirty-seven specimens of *Shonisaurus* were fossilized together, most of them facing in the same direction—probably the result of a mass stranding, like the calamity that happens occasionally to whales and dolphins today.

Jurassic ichthyosaurs were abundant and varied, but there was only one Cretaceous ichthyosaur, *Platypterygius*. It had lost the large tail for fast acceleration, and instead its limbs were modified into large fins. This suggests that it had more of a cruising style of hunting than most ichthyosaurs did. It may have used the limb fins as underwater wings for propulsion rather than steering, in the style used by sea turtles and penguins and reconstructed for some plesiosaurs.

Sauropterygians

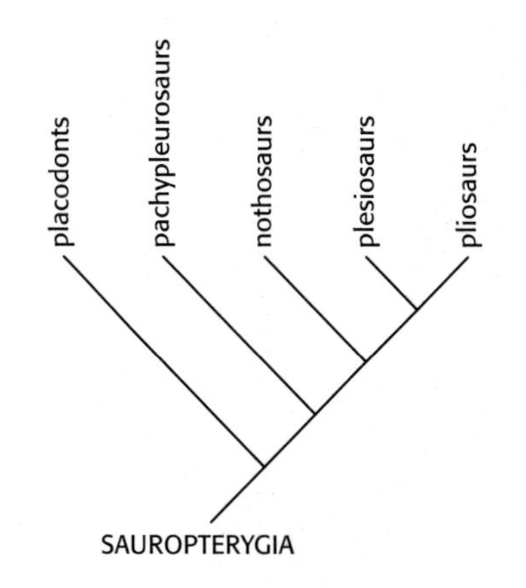

Figure 14.6 A cladogram of the sauropterygians, which included some of the more spectacular marine reptiles of the Mesozoic. (Simplified from Rieppel, 2002.)

Sauropterygians (Figure 14.6) are a large clade of reptiles whose ancestry can be traced back to basal diapsids of the Permian (Chapter 11). Sauropterygians had unusually large limbs for land animals that had evolved toward life in water (compare

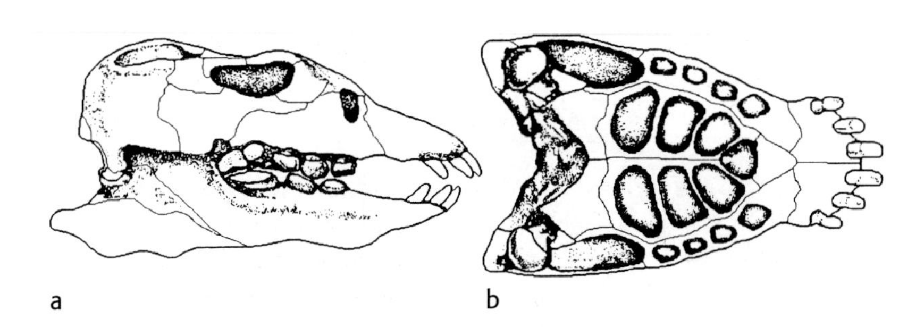

Figure 14.7 (a) The skull of *Placodus*, a marine reptile from the Triassic of Germany that may have had an ecology much like the living walrus. (b) The lower jaw of *Placodus*, showing the clam-crushing teeth of the palate. (What did it do with its tongue during this process?) (After Broili and Romer.)

crocodiles, seals, and whales). Most had small heads and comparatively long necks for their body size, so their prey (presumably fishes) must have been relatively small.

Placodonts are early but very specialized sauropterygians known only from the Triassic of Europe: in other words, they were an early offshoot from the basal part of the sauropterygian clade (Figure 14.6). They had their own set of adaptations that may reflect a specialized ecology like the living walrus, which dives down to shallow seafloors to dig and crush clams. *Placodus* itself had unusual teeth that suited it for this way of life. The large comb-like teeth at the front of the jaw (Figure 14.7a) were probably used to dig into the seafloor to scoop up clams, and sediment could be washed off them by shaking the head with the mouth open. The clean clams were then crushed between flat molar teeth in the lower jaw and flat plates on the roof of the mouth (Figure 14.7b). Placodonts did not need great maneuverability or speed, and many had heavy plated carapaces that covered them dorsally and ventrally, rather like a turtle.

Pachypleurosaurs were the simplest early sauropterygians in structure, though we have good specimens only from the Middle Triassic. The intermediate forms between them and their Permian ancestors remain to be discovered. Pachypleurosaurs are small marine reptiles that are best known from Middle Triassic rocks of the Alps, but they also occur in China. Hundreds of specimens are known, but only a few are well preserved. Their limbs were not very strong, and they were modified for life in water in the sense that they could be folded back against the body for extremely low water resistance; the tail was powerful. Pachypleurosaur swimming style accentuated power and flexibility of the body and tail, though this did not include the rib cage: pachypleurosaurs had thick ribs that presumably made up a very stiff thorax. This adaptation is a solution to Carrier's Constraint (Chapter 11) in active, air-breathing swimmers. I suspect that pachypleurosaurs swam like living monitor lizards, with the front limbs tucked away against the rib cage and the hind limbs used as rudders. The forelimbs were relatively short but quite powerful, perhaps for dragging the animal out onto land for breeding and egg-laying.

Nothosaurs were more advanced Late Triassic sauropterygians. All nothosaurs were large compared with their pachypleurosaur ancestors. They extended the rigid thorax of pachypleurosaurs by evolving ribs far back along the body. With their bodies stiffened in this way, nothosaurs probably relied less on the tail for propulsion than pachypleurosaurs did, and they had strong forelimbs that may have contributed compensating swimming power. The nothosaur forelimb was not really wing-like and may have used a rowing action to give propulsion. The hindlimbs were quite strong but not particularly well adapted for a swimming stroke (Figure 14.8).

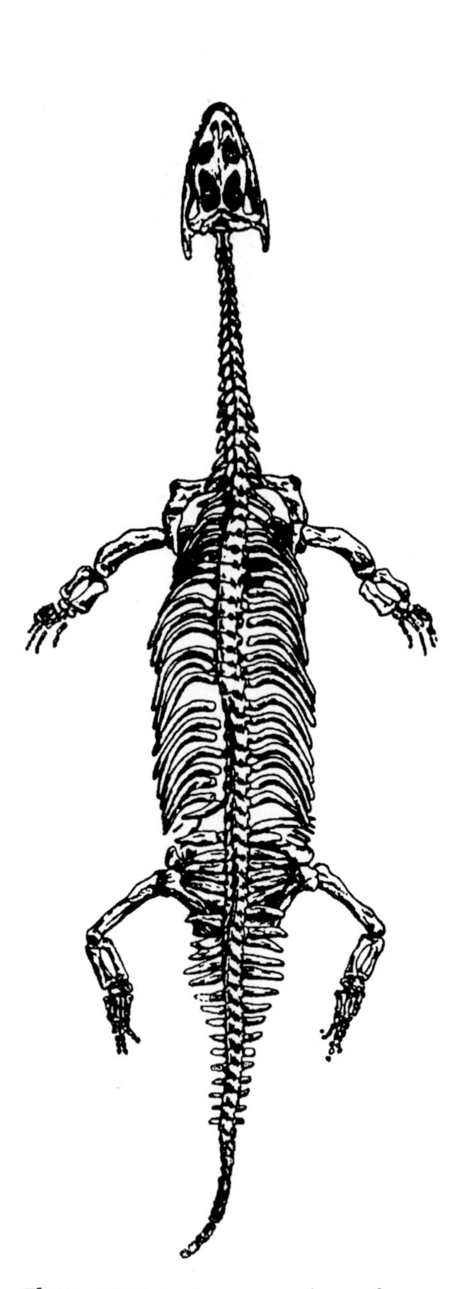

Figure 14.8 *Lariosaurus*, a nothosaur from the Triassic of Switzerland, about 60 cm (2 ft) long. (Idealized after Peyer.)

One nothosaur group evolved in Early Jurassic times into the largest and best-known sauropterygian clade, the **Plesiosauria**. The Plesiosauria had large bodies, and limbs that were very strong, equally well-developed front and back, and highly modified for swimming. They swam with all four limbs that used the stiffened body as a solid mechanical base, in a further extension of the swimming style of nothosaurs. The limbs were strengthened and further modified for efficient swimming strokes, eventually becoming much more important in swimming than the tail. The jaws have modifications that look very well evolved for fish eating.

The Plesiosauria flourished worldwide in marine ecosystems from the Early Jurassic until the end of the Cretaceous. They came in two versions, pliosaurs and plesiosaurs. **Pliosaurs** had short necks and long, large heads, and they looked rather like powerful, long-headed ichthyosaurs. They swam mainly with the strong limbs, however, all four of which were large, paddle-shaped structures, shaped into effective hydrofoils. Some pliosaurs were huge: *Leiopleurodon* reached 20 m (65 ft) long!

Plesiosaurs had the same limb structure but had very long necks and small heads. An average adult was about 3 m (10 ft) long, with a neck that had 40 vertebrae. Some plesiosaurs were very large too. *Elasmosaurus* from the Cretaceous of Kansas was 12 m (40 ft) long, with 76 neck vertebrae!

Plesiosaurian limbs were jointed to massive pectoral and pelvic girdles (Figure 14.9), presumably by very strong muscles and ligaments. In 1976 Jane Robinson suggested that these structures could be explained if all four limbs were used in an up-and-down power stroke, in underwater "flying" like that of penguins—except, of course, that four limbs were involved instead of two. She realized that the plesiosaurian body had to be tightly strung with powerful ligaments to transmit the propulsion generated by the limbs to the body that they pulled through the water; and she found grooves in the skeleton where the ligaments had run.

But plesiosaurian limbs were not jointed well enough to the shoulder and pelvic girdles to allow strictly "flight" power strokes, and they could not have been lifted above the horizontal. They also show no sign of powerful muscle attachments. Steven Godfrey suggested instead in 1984 that the propulsion stroke was downward and backward in a combination of "flying" and rowing; living sea lions swim this way. However it worked, plesiosaurian swimming required precise coordination between the limb strokes.

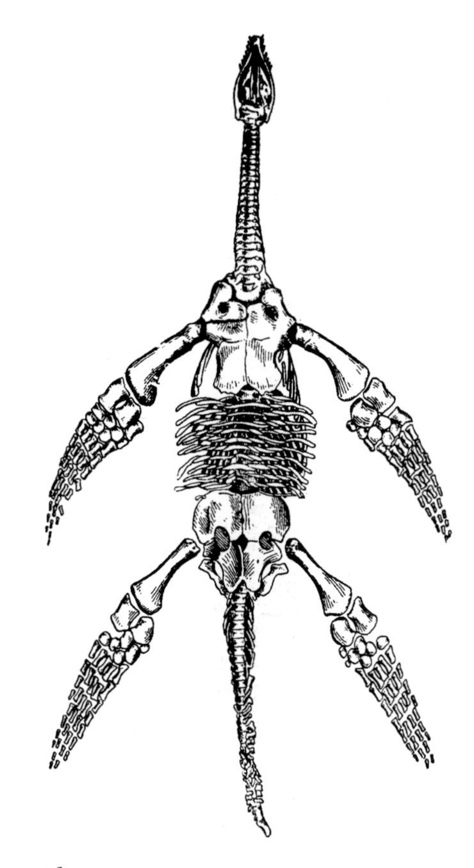

Figure 14.9 The plesiosaur *Rhomaleosaurus*, viewed from underneath the skeleton. The limbs must have dominated the swimming—but how? (From Andrews.)

But here is an unresolved question: how did the limb strokes coordinate? Did all four limbs work in synchrony? Did the power stroke of both front limbs alternate with the power stroke of both back limbs? Or did right front and left back limb strokes coincide with left front and right back? Most people favor the first technique of synchronous strokes, which is also used by sea lions. The second option would involve too much stress on the trunk, which would be alternately extended and compressed if power strokes alternated between front and back limbs. The third option, however, would require only resistance to trunk twisting, which could easily be accomplished by the ligaments along the spine and those connecting the large bony masses along the underside (Figures 14.9): this could potentially make a plesiosaurian much more maneuverable than the other techniques would.

It is difficult to envisage how plesiosaurians hunted. Perhaps, with their large heads, pliosaurs hunted large fishes at fairly high speed. But plesiosaurs are different. They have large bodies but small heads and long necks. Perhaps they stalked smaller prey and used sustained underwater "flight" mainly for migration or for cruising to feeding grounds.

Mosasaurs

Mosasaurs were essentially very large Late Cretaceous monitor lizards, up to 10 m (30 ft) in length, the largest lizards that have ever evolved. Their evolution of aquatic adaptations in parallel with ichthyosaurs and plesiosaurs is astonishing.

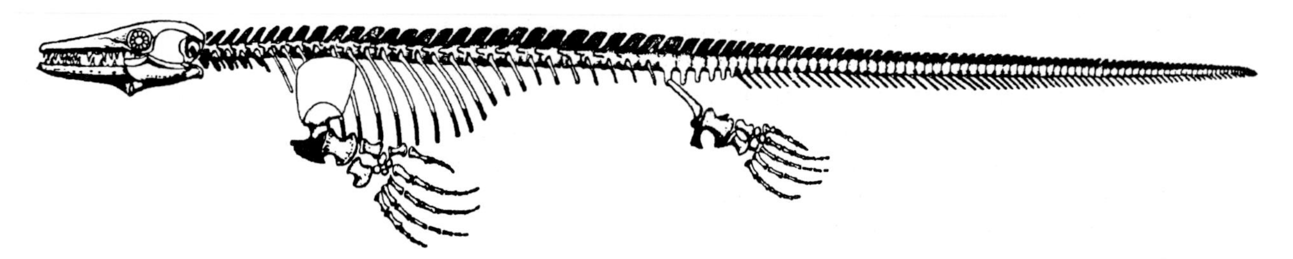

Figure 14.10 The Cretaceous mosasaur *Platecarpus*, essentially a gigantic lizard adapted for carnivorous life in the sea. About 4 m (13 ft) long. (After Merriam.)

Mosasaur bodies were long and powerful, with tails and limbs adapted for swimming. The main propulsion came from flexing the body and sculling with the tail, which was flat and deep, as it is in living crocodiles. But, in addition, the limbs were modified into beautiful hydrofoils. The elbow joint was rigid, and the shoulder joint was designed for up-and-down movement. Although the forelimbs could have given some lift, most mosasaurs probably used them as steering surfaces, as dolphins do.

Some forms like *Platecarpus*, however, had well-developed forelimbs (Figure 14.10), and may have used them in a kind of underwater flying, like penguins. The hind limbs were like the forelimbs, though smaller, with the major muscle attachments also giving up-and-down movement. Because the pelvis was not fixed to the backbone, the hind limb strokes cannot have delivered much power. The hind limbs could rotate, and probably worked like aircraft elevators to adjust pitch and roll.

Mosasaurs had long heads set on a flexible but powerful neck. The large jaws often had a hinge halfway along the lower jaw, which may have served as a shock absorber as the mosasaur hit a large fish at speed. This hinge and the powerful stabbing teeth suggest that most mosasaurs ate large fishes (Figures 14.11; 14.12). Other mosasaurs had large, rounded, blunt teeth like those of *Placodus* (Figure 14.7), and they probably crushed mollusc shells to reach the flesh inside.

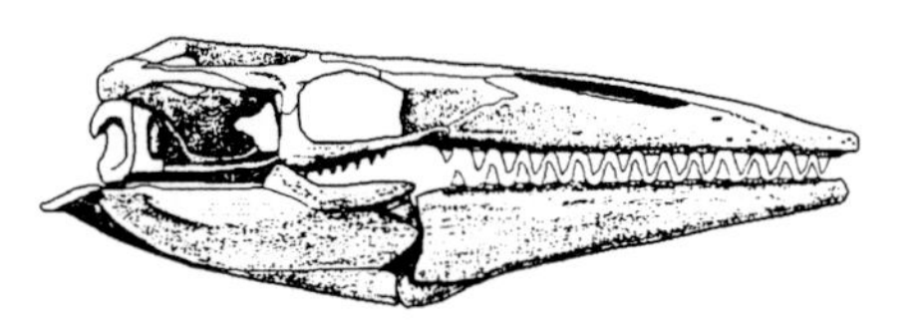

Figure 14.11 The jaws and teeth of the mosasaur *Clidastes*. (After Russell.)

Air Breathers at Sea

All these Mesozoic reptiles were air breathers and therefore faced special problems for life in the sea. Precisely the same problems are faced today by marine mammals. The major one, of course, is the fact that air breathers must visit the surface for air, but there are also problems in introducing young to a complex and dangerous world where they must be prepared to use sophisticated skills immediately after birth. Many marine reptiles, mammals, and birds return to the shore for reproduction. Turtles simply lay large clutches of eggs and leave them buried in the sand, a method that results in horrific mortality but has obviously worked successfully for 200 million years. Seals, sea lions, and penguins have their young on shore in safe nurseries, so that they can breathe air, be fed, and grow for a while before they take to a swimming and foraging life at sea.

But living dolphins and whales never come ashore. They have special adaptations for air breathing, breeding, giving birth and caring for the young at sea. The young are born tail first, and mothers and other related adults will push them to the surface until they learn to breathe properly. The young must be able to dive immediately to suckle, and whales feed their babies milk under high pressure.

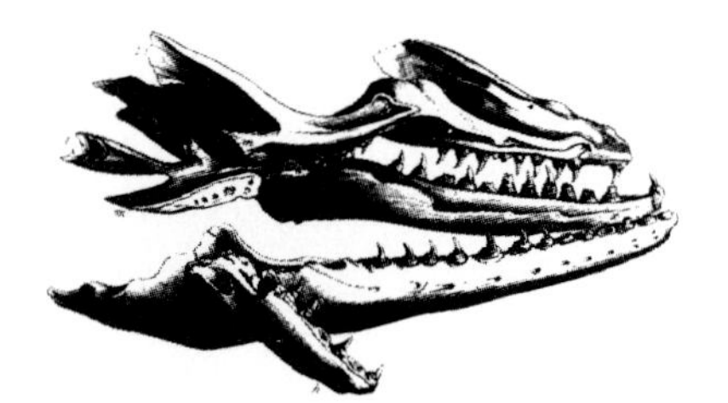

Figure 14.12 *Mosasaurus*, from Maastricht, the original mosasaur. The tremendous jaws of this creature were the talk of Europe in the late 1700s (along with the American Revolution, the French Revolution, and the Industrial Revolution). (From Buckland.)

There is evidence that Mesozoic marine reptiles solved the same kinds of problems in spectacular fashion. Several fossils of ichthyosaurs have been found with young preserved inside the rib cage of adults, evidence that ichthyosaurs at least had evolved live birth. The preserved fetuses have long, pointed jaws, showing that they would have been able to feed for themselves immediately after birth, and they were born tail first as whales are. All this implies that ichthyosaurs had special mechanisms for training the young to swim and feed, and it also suggests parental care on a scale comparable with that of dinosaurs (Chapter 12).

Mosasaurs do not look as if they would readily have come ashore to lay eggs, and no fetuses or even juveniles have been found associated with adults. There is some indirect evidence that they had live birth at sea: the pelvis is very unusual. This may have resulted simply from adaptation to swimming, but perhaps the normal pelvis was expanded to give birth to live offspring much bigger than any normal egg.

The early sauropterygians had limbs that would have allowed them to haul themselves out onto a beach to lay eggs or to give birth, rather like sea lions. The fairly strong limbs of placodonts and the small body size of pachypleurosaurs make them particularly easy to imagine on the shore. Nothosaurs and plesiosaurs are usually much larger, and would have had to work much harder to drag themselves up a beach. Plesiosaurs, with limbs modified into long hydrofoils, may have been totally sea-going, with live birth at sea as in ichthyosaurs, whales, and dolphins.

Air breathers at sea are subject to Carrier's Constraint (Chapter 11): they cannot swim fast if they flex the body side-to-side. As in their terrestrial counterparts, marine mammals and birds do not have a problem: their bodies flex up and down as they swim. But mosasaurs, as lizards, certainly could not have swum at speed for long. As nothosaurs evolved into plesiosaurs, they also evolved stiffened trunks that avoided Carrier's Constraint (compare Figure 14.8 with Figure 14.9), and their underwater flight is a reflection of that evolutionary breakthrough.

What about ichthyosaurs? They certainly look fast, yet their tail fin flexes sideways, and the body does not look stiff (Figure 14.3). Ryosuke Motani tells me that the size of the centers of the vertebrae imply considerable stiffness of the backbone, and that in turn implies that ichthyosaurs had solved Carrier's Constraint.

Many large and powerful swimming creatures today are warmblooded to some extent: many sharks, tuna, and several turtles, as well as dolphins. The metabolic effort of swimming contributes to a warm body. Thus, one could guess that ichthyosaurs and plesiosaurs were warmblooded, and that ichthyosaurs at least had live birth as well. This does not make them mammals, but it does suggest that they were most impressive creatures.

Almost all these magnificent marine reptiles became extinct at the end of the Cretaceous, along with dinosaurs, pterosaurs, and a significant number of marine invertebrates. Only crocodiles and turtles have survived to give us some clues about the mode of life of large reptiles. Unfortunately, these survivors are far from being typical Mesozoic reptiles!

THE MODERNIZATION OF LAND PLANTS

As plants invaded drier habitats from Devonian times onward, they evolved ways to retain water and protect their reproductive stages from drying out. The major advance was the perfection of seeds, which are fertilized embryos packed in a reasonably watertight container filled with food. The embryo can survive in suspended animation within the seed until the parent plant arranges for its dispersal. Germination can be delayed until after successful transport to a favorable location. The seedling then bursts its seed coat and grows, using the nutrition in the seed until its roots and leaves have grown large and strong enough to support and maintain the growing plant.

Seeds had evolved in Late Devonian times, and seed ferns were a successful component of Late Paleozoic floras, including the coal forests; they flourished into the Triassic. But Mesozoic gymnosperms perfected the seed system, making up 60% of Triassic and 80% of Jurassic species. Gymnosperms include conifers, cycads, and gingkoes. Mesozoic forests had trees up to 60 m (200 ft) high, forming famous fossil beds such as the Petrified Forest of Arizona. Conifers were the dominant land plants during the Jurassic and Early Cretaceous, and they are still by far the most successful of the gymnosperms. Finally, around the Jurassic–Cretaceous boundary, the flowering plants or angiosperms evolved and eventually came to dominate land floras.

Seed plant reproduction has two phases, fertilization and seed dispersal. The plant must be pollinated, and after the seed has formed it must be transported to a favorable site for germination. A major factor in the evolution of angiosperms is their manipulation of animals to do these two jobs for them.

MESOZOIC PLANTS AND POLLINATION

Conifers and many other plants are pollinated by wind. They produce enormous numbers of pollen grains, which are released to blow in the wind in the hope that a grain will reach the pollen receptor of a female plant of the same species. Wind pollination works, just as scattering sperm and eggs into the ocean works for many marine invertebrates. But the process looks very expensive. The pollen receptor in conifers is only about one square millimeter in area, so to achieve a reasonable probability of fertilization, the female cone must be saturated with pollen grains at a density close to one million grains per square meter.

Parent plants do some things to cut the costs of wind pollination. Male cones release pollen in dry weather in just the right wind conditions, for example, and female cones are aerodynamically shaped to act as efficient pollen collectors. But for practical purposes, wind pollination is consistently successful only if many individuals of the same species live in closely packed groups: conifers in temperate forests or grasses in prairies and savannas. An ecological setting like a tropical rain forest, where many species have well-scattered individuals, is not the place for wind pollination.

We can imagine Jurassic floras dependent on wind pollination, with plants ready to release large supplies of pollen. Insects then, as now, probably foraged for the food offered by plentiful pollen and soft, unripe female organs waiting for fertilization. We know that there were large, clumsy beetles in the Jurassic, and they probably visited plants for food. As they moved from plant to plant, they may have visited the same species frequently, collecting and transferring pollen by accident. Insects could help even by visiting one plant or one sex: in some living cycad gymnosperms, wind carries pollen only to the surface of the female cone, but insects clustering around the cone carry it into the pollen receptors.

Over time, the plant structure may have evolved toward cooperation with insects in certain ways. Perhaps delicate structures were protected, but pollen was made easier to gather, and female pollen collectors were moved closer to the male pollen emitters. Such changes would have made pollen transfer by insects more likely, and less costly to the plant. Devices to attract insects—strong scents at first, then brightly colored flowers—perhaps evolved side by side with rewards such as nectar.

Those plants that successfully attracted insects would have benefited by increasing their chances of fertilizing and being fertilized. Insects deliver pollen much more efficiently than wind.

An ideal pollinator should be able to exist largely on pollen and nectar, so that it can gather all its food requirements by visiting plants. It should visit as many (similar) plants as possible, so it should be small, fast-moving, and agile. A nocturnal pollinator should have a good sense of smell, and a daytime pollinator should have good vision or a good sense of smell, or both.

The only Jurassic candidates to fit this job description were insects. Birds and bats had not yet evolved, and small mammals were probably too sluggish. Insect pollinators had an increasing incentive to learn and remember certain smells and sights, and those that evolved rapid, error-free recognition of pollen sources, and clever search patterns to find them, would have become superior food gatherers and probably superior reproducers. Today, insects discriminate strongly between plant species, even between color varieties of particular species. Some insects congregate for mating around certain plant species.

Figure 14.13 Beautiful flowers had already evolved in the Early Cretaceous. This figure is a reconstruction of *Archaeanthus*, a flower discovered in the Early Cretaceous of China. (Courtesy of David Dilcher of the University of Florida.)

Magnolias and Moths, Cycads and Beetles

Early Cretaceous flowers, though small, had relatively large petals (Figure 14.13), and the flowers could have produced many small seeds. The living family Winteraceae are primitive, medium-sized trees related to magnolias, and their fossils date back to Early Cretaceous times. About 50 or so species of Winteraceae are found today, in moist, tropical forests.

The Winteraceae have an intriguing pollination system. The trees bloom all summer, but there are never more than a few flowers open at once (usually only one per tree). Each flower lasts for two days. On the first day it displays female organs, and on the second morning it extends male organs, thus avoiding inbreeding. On the second day the male phase produces pollen in a sticky, nutritious oil. A primitive moth is attracted to the flower and feeds on the oily pollen, getting much of it entangled and dried on its body. The moths arrive in dozens and are strong fliers. Presumably they go off to search for another tree when they have stripped off all the male pollen.

The female phase extrudes a strong scent, but it provides no food for the moths. Instead, the scent seems to act as a mating stimulant. Both female and male moths are attracted in large numbers to the flowers, and in the display and rapid movement involved, pollen previously collected from another tree is delivered to the female stigma.

The moths involved in the pollination are among the most primitive known. Instead of sucking mouth parts for collecting nectar, they have grinding jaws that they use to chew pollen and spores. Their fossil record also dates back to the Early Cretaceous, so this style of pollination may be very ancient indeed, perhaps a good clue to the success of the early angiosperms. Many other magnolia-like angiosperms have large, fragrant flowers, where insects congregate to feed and mate (and pollinate).

Primitive angiosperms are not the only plants pollinated by insects: living gymnosperms such as cycads are often insect-pollinated. However, many people suspect that the success of angiosperms began as insects congregated to mate in and around their flowers, encouraged originally by scents.

Animal pollination can deliver a large mass of pollen on the stigma, rather than a few wind-blown grains. Competition between individual pollen grains to fertilize the ovule allows the female angiosperm more mate choice than in other plants

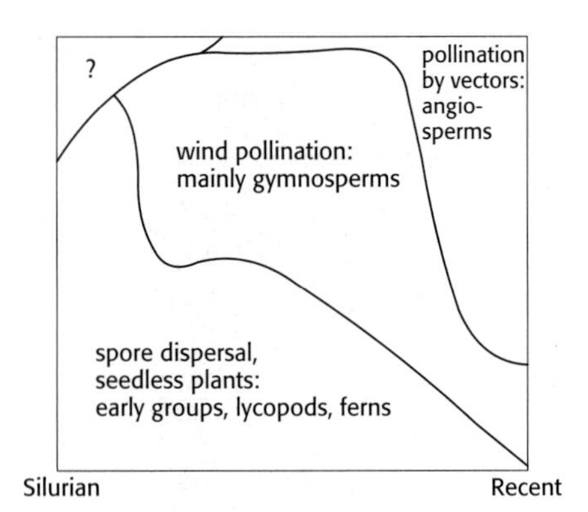

Figure 14.14 The percentage of plants pollinated by vectors increased sharply in the Cretaceous with the rise of angiosperms. But because many of those vectors (insects, for example) had been present since the Late Carboniferous, there must have been more to the rise of angiosperms than pollination by vectors. (Data simplified from Niklas, Tiffney, and Knoll.)

(remember Chapter 3). Pollen grains are haploid, so they cannot carry hidden recessive genes (as we do). A female plant could in theory select certain pollen grains over others by placing chemical or physical barriers between the stigma and the ovule; the first pollen grain to breach the barrier is selected over others for fertilization. Experimentally, plants that are allowed to exercise pollen choice in this way have stronger offspring than others. This aspect of angiosperm reproduction may have been one of the most important factors in their success.

Of course, pollination encouraged tremendous diversity among the pollinators as they came increasingly to specialize on particular plants. The astounding rise in diversity of beetles and bees began in Cretaceous times, and there are now tens of thousands of species of each. The bees and beetles associated with angiosperms are many times more diverse than those associated with gymnosperms.

Pollination cannot be the whole story, however. Insects help to pollinate cycads too, yet angiosperms are enormously successful while cycads have always been a relatively small group of plants. Several other Mesozoic plants experimented with ways of persuading organisms to transport pollen, and flower-like structures evolved more than once.

Furthermore, if pollination were the key to angiosperm success, flowers could have evolved as soon as flying insects became abundant in the Late Carboniferous. There are some signs that insect pollination began then as something of a rarity. But angiosperms appeared much later and rather suddenly in the Early Cretaceous (Figure 14.14). Therefore, angiosperm success is not related simply to their evolution of flowers. In fact, all Mesozoic plants that had insect pollination must have paid a significant price for fertilization, in insect damage.

MESOZOIC PLANTS AND SEED DISPERSAL

If seeds fall close under the parent plant, they may be shaded so that they cannot grow, or they may be eaten by animals or birds that have learned that tasty seeds are often found under trees. Many plants rely on wind to disperse their seeds. Sometimes seeds are provided with little parachutes or airfoils to help them travel far away from the parent; winged seeds evolved almost as soon as seeds themselves, in the Late Devonian.

But seeds dispersed by wind will often fall into places that are disastrous for them. Although wind dispersal works, it seems very wasteful: it can work only in plants that produce great numbers of seeds. Wind-dispersed seeds must be light, so cannot carry much energy for seedling growth. They have to germinate in relatively well-lit areas, where the seedling can photosynthesize soon after emerging above ground.

Alternatively, a plant could have its seeds carried away by an animal and dropped into a good place for growth. Many animals can carry larger seeds than wind can, and larger seeds can successfully germinate in darker places. As with pollination, animals must be persuaded, tricked, or bribed to help in seed dispersal.

Some animals visit plants to feed on pollen or nectar, and others browse on parts of the plants. Others simply walk by the plant, brushing it as they pass. Small seeds may be picked up accidentally during such visits, especially if the seed has special hooks, burrs, or glues to help to attach it to a hairy or feathery visitor. Such seeds may be carried some distance before they fall off. Small seeds may be eaten by a visiting herbivore, but some may pass unharmed through the battery of gnawing or grinding teeth, through the gut and its digestive juices, to be automatically deposited in a pile of fertilizer.

Plants face two different problems in persuading animals to disperse seeds and in persuading them to pollinate. In pollination there is often a payment on delivery: the pollinator collects nectar or another reward as it picks up and again as it delivers the pollen. There is no such payment on delivery of a seed. Any payment is made by the plant in advance, so that seed dispersers have no built-in payment for actual delivery of the seed. It would be better for them to cheat and to eat every seed. Thus plants often rely on tricks (burrs, for example) to fix seeds to dispersers. Velcro was evolved by plants long before the idea was copied by an astute human. Alternatively, plants may pack many small seeds into a fruit so that the disperser will concentrate on the fruit and swallow the seeds without crushing them (in strawberries, for example).

Some plants actually invite seed swallowing. They have evolved a tasty covering around the seed (a berry or fruit), and if the animal or bird eats the seed along with the fruit, every surviving seed is automatically planted in fertilizer. Tiny seeds are likely to be swallowed without being chewed, but can carry little food for the developing embryo. Large seeds loaded with nutrition are often protected by a strong seed coat or packed inside a nut.

Seed dispersal by animals is not a cost-free service. Many dispersers eat the seeds, passing only a few unscathed through their gut. So there is a significant wastage of seeds, depending on a delicate balance between the seed coat and the teeth and stomach of the disperser. Too strong a seed coat, and the disperser will turn to easier food or germination will be too difficult; too weak a seed coat, and too many seeds will be destroyed. Some plants are so delicately adjusted to a particular disperser that the seeds germinate well only if they are eaten by that disperser.

Figure 14.15 The very early angiosperm *Archaefructus*, from the Early Cretaceous of China. (Courtesy of David Dilcher of the University of Florida.)

Angiosperms evolved carpels as a new and unique protection for their ovules, and eventually for the developing seeds (Figure 14.15). Carpels probably evolved to protect against large, hungry insects. Soon, however, the angiosperm seed coat began to protect seeds as they passed through vertebrate guts. A seed with a strong coat was proof against many possible predators, but perhaps at the same time came to be desired food for one or a few animals that could break the seed coat. A plant could evolve to a stable relationship with a few such seed predators: the predators would receive enough food from the seeds to keep them visiting the plant regularly, but would pass enough seeds unscathed through the gut that the plant benefited too.

Seed dispersal by animals surely evolved after insect pollination. Jurassic insects may have become good pollinators, but they were too small to have been large-scale seed transporters. Jurassic reptiles were large enough, but often had low metabolic rates, so any seeds they swallowed were exposed to digestive juices for a long time. Reptiles do not even have fur in which seeds can be entangled (though feathered theropod dinosaurs might have done!).

Seeds were undoubtedly dispersed by dinosaurs to some extent, since the huge vegetarian ornithischians and sauropods ate great quantities of vegetation. But in spite of the size of the deposit of fertilizer that must have surrounded seeds passing through a dinosaur, browsing dinosaurs probably damaged and trampled plants more than they helped them. It's unlikely that any Mesozoic plant would have encouraged dinosaur browsing.

Effective transport over a long distance can take a seed beyond the range of its normal predators and diseases, and can allow a plant to become very widespread provided that there are pollinators in its new habitat. As angiosperms adapted to seed dispersal by animals, they probably dispersed into new habitats much faster than other plants. Other things being equal, we might expect a dramatic increase in the angiosperm fossil record as they adapted toward seed dispersal by animals rather than wind. (Some living angiosperms are pollinated by wind but have their seeds dispersed by animals. These include grasses, which did not evolve until well into the Cenozoic.)

We have seen that there were few effective animal seed-transporters in the Jurassic, and that dinosaurs were unlikely candidates in the Cretaceous. Philip Regal has suggested that birds and mammals triggered the radiation of angiosperms by aiding them in seed dispersal. Birds and mammals have feathers and fur in which seeds easily become entangled; seeds pass quickly through their small bodies with their high metabolic rates and are likely to be unharmed unless they have been deliberately chewed. Angiosperm seeds would have been especially suited to vertebrate transport because of their extra protective coating. Conifer seeds are usually small and light, designed to blow in the wind, and conifers depend on close clusters for pollination. Isolated conifers are likely to be unsuccessful reproducers, and additional transport would make little difference to their long-term success.

However, the early angiosperm radiation took place in the Early and Middle Cretaceous, when mammals and birds were still minor members of the ecosystem. This early success of angiosperms may be explained better by the "fast seedling" hypothesis. This idea is based on the fact that angiosperm seeds germinate sooner, and the seedlings grow faster and photosynthesize better than those of gymnosperms. Angiosperms may simply have outcompeted gymnosperms in the race for open spaces.

Regal's idea applies better to the later radiation of mammals and land birds in the Cenozoic, when angiosperms increased greatly in diversity, size, and abundance, and came to dominate most land floras. Today ferns are characteristic only of damp environments, conifers dominate mainly in temperate forests, and other ancient plants such as cycads and gingkoes are rare.

Bruce Tiffney showed that Cenozoic angiosperm seeds were much larger than Cretaceous ones. The ability of angiosperms to become dominant forest trees in ecosystems, and their successful evolution of large-seed dispersal aided by animals and birds, were Early Cenozoic events.

The rise to dominance of the angiosperms provided a food bonanza for seed dispersers. Birds and small mammals, especially early primates and bats, all joined the seed- and fruit-eating guilds in Early Cenozoic times. Some tropical flowers today still rely for pollination on bats, small marsupials, or lemurs.

We can imagine a whole set of pollinators and seed dispersers evolving together with the plants on which they specialized. For most plants, it would be best not only to be conspicuous, but also to be different from other plant species, to encourage pollinators and seed seekers to be faithful visitors.

Suppose that particular techniques are needed to extract seeds or pollen from a plant. A visitor that learns the secret has an advantage over others and will tend to visit that species rather than foraging at random, which might require learning several collecting techniques. Fewer strangers are likely to visit the plant and rob its regular visitors of their rewards. The plant is much more likely to be fertilized or dispersed by faithful visitors than random browsers. An insect, which has only a short adult life, a limited memory, and a limited learning capacity, is more likely to be a faithful pollinator to the first plant it learns to forage from, or to the species for which it is genetically programmed. It's easy to imagine the evolution of a great variety of bright and highly scented flowers and fruits, together with a great variety of their specialized pollinators and seed seekers. Again, the evolution of the faithful visitor may have been much later than the evolution of angiosperms themselves. Only in the Early Cenozoic do angiosperm flowers show evidence of pollination by faithful visitors such as bees, wasps, bats, and other small animals, and seed dispersal by birds, mammals, and large insects.

Living angiosperms have developed extraordinary devices for pollination as well as seed dispersal. One arctic flower provides its insect pollinators with a bowl of petals that forms a perfect parabolic sun-bathing enclosure. Orchids have petals shaped and colored like female insects, and they are pollinated by undiscriminating and optimistic males. It's quite by accident that we happen to sense and appreciate the scents and colors of the flowers around us, because most of them were selected for the eyes and senses of insects. (We probably have color vision to help us choose between ripe and unripe fruit.) But we can gain a scientific as well as an aesthetic kick from looking at flowers if we admire their efficiency as well as their beauty.

ANGIOSPERMS AND MESOZOIC ECOLOGY

The first angiosperms appeared around the Jurassic–Cretaceous boundary. The earliest angiosperm of all is from the famous Liaoning lake sediments in northern China that have also yielded feathery dinosaurs and early birds. *Archaefructus* (Figure 14.15) is preserved almost completely, and seems to be a water-dwelling weed. There are no petals, but the plant has closed carpels with seeds inside, a classic feature of angiosperms. Cladistic analyses of *Archaefructus* place it as the most primitive as well as the earliest angiosperm.

Angiosperms were diverse by the Middle Cretaceous, especially in disturbed environments such as riverbanks. But how does the rise of angiosperms fit into the larger picture of Mesozoic ecology?

At the end of the Jurassic, we see a reduction of the sauropod dinosaurs that probably had been high browsers, and the rise of low-browsing ornithischians. More seedlings would now have been cropped off before reaching maturity, and any plant that could reproduce and grow quickly would have been favored.

Conifers reproduce slowly. It takes two years from fertilization until the seed is released from the cone, and wind dispersal typically does not take the seed very far. The whole reproductive system of conifers depends on wind and works best in a group situation such as a forest.

However, most angiosperms are designed for pollination by animals, especially insects; for rapid germination and growth; and for rapid release of seeds (within the year). An angiosperm is much more likely to succeed as a weed, rapidly colonizing any open space, and is more likely to be widely distributed because of its dispersal method. The earliest angiosperms were small, weedy shrubs, exactly the kind of plant that could survive heavy dinosaur browsing. A conifer forest, once broken up by dinosaur browsing or natural accident, would most likely have been recolonized by shrubs and weeds that could invade and grow rapidly (look at the results of clear-cutting in a conifer forest today). The weeds themselves would have reproduced quickly, so would have been more resistant to browsing than were young conifer seedlings.

Even without dinosaur browsing, angiosperms would have found habitats where they would have been very successful. In Middle Cretaceous rocks, for example, angiosperm leaves dominate sediments laid down in river levees and channels. Shifting and changing riverbank areas favor weeds because large trees are felled by storms and frequent floods. Most Middle Cretaceous pollen, however, comes from sediments laid down in lakes and near-shore marine environments. This is the windblown pollen from stable forests on the shores and on lowland plains away from violent floods, and it is dominantly conifer pollen.

Angiosperms did not take over the entire Cretaceous world, however. They were very slow to colonize high latitudes. (I suspect this reflects their greater dependence on insect pollinators, which drop off in both number and diversity in higher latitudes.)

Furthermore, Late Cretaceous fossil floras preserved in place under a volcanic ash fall in Wyoming show that even if angiosperms dominate a local flora in diversity of species, they may make up only a small percentage of the biomass. In the Big Cedar Ridge flora, angiosperms made up 61% of the species, but covered only 12% of the ground. We have to be careful in distinguishing between the diversity, the abundance, and the ecological importance of angiosperms. They cannot really be said to have dominated the ecology of any Cretaceous area.

Nevertheless, one can argue, as Bruce Tiffney (e.g., in 1998) and others have done, that the angiosperm radiation provided the basis for the radiations of the 5 ton ornithischians of the Later Cretaceous. They seem to have lived in much larger herds than Jurassic dinosaurs, up to several thousand in the case of *Maiasaura*, and Later Cretaceous dinosaurs were much more diverse as well as more abundant than their predecessors.

Ants and Termites

The success of angiosperms benefited pollinators and seed dispersers, and vice versa, and the later evolution of angiosperms was related to the ecology of large animal browsers. But today, some of the most effective tropical herbivores are leaf-cutting ants, and most terrestrial vegetation litter is broken down by termites. One-third of the animal biomass in Amazonia is made up of ants and termites. In the savannas of West Africa there are more like 2000 ants per square meter! There may be 20 million individuals in a single colony of driver ants, but the world record is held by a supercolony of ants in northern Japan, which has 300 million individuals, including a million queens, in 45,000 interconnected nests spread over 2.7 sq. km (1 sq. mi).

The higher social insects (bees, ants, termites, and wasps) began a major evolutionary radiation in the Late Cretaceous, as angiosperms became dominant in terrestrial ecosystems. The earliest known bee, found in Cretaceous amber from New Jersey, is a female worker bee adapted for pollen gathering. Bee society already had a sophisticated structure.

Angiosperm Chemistry

As we have seen, many angiosperms attract animals to themselves for pollination and seed dispersal. The plant usually pays a price in the production cost of substances such as nectar and in the cost of seeds eaten. Browsing animals and plant- and sap-eating insects often eat more plant material than they return in the form of services to the plant, and attracting such creatures results in a net loss of energy.

Angiosperms have therefore evolved an amazing variety of structures and chemicals that act to repel herbivores. These can be as simple and as effective as spines and stings, they can be contact irritants as in poison ivy and poison oak, or they can be severe or subtle internal poisons. Cyanide is produced by a grass on the African savanna when it is grazed too savagely. Many of our official and unofficial pharmacological agents were originally designed not for human therapy but as plant defenses. More than 2000 species of plants are insecticidal to one degree or another. Caffeine, strychnine, nicotine, cocaine, morphine, mescaline, atropine, quinine, ephedrine, digitalis, codeine, and curare are all powerful plant-derived chemicals, and it is not a coincidence that many of them are important insecticides or act strongly on the nervous, reproductive, or circulatory systems of mammals (some are even contraceptive and would act directly to decrease browsing pressure). One hundred and fifty million pyrethrum flowers are harvested every day, to fill a demand for 25,000 tons of "natural" insecticide per year. A million tons of nicotine per year were once used for insect control, until it was found that the substance was extremely toxic to mammals (self-destructive humans still smoke it!). Other plant chemicals are powerful but can be used to flavor foods in low doses. All our kitchen flavorings and spices are in this category. Garlic keeps away insects as well as vampires and friends.

For paleobiologists, the problem of angiosperm chemistry is its failure to be preserved in the fossil record. Clearly, the increasing success of angiosperms in the Late Cretaceous and Early Cenozoic occurred in the face of intense herbivory by the radiating mammals and insects of that time. The chemical defenses of angiosperms probably evolved very early in their history.

LIMERICK 14.1
We're proud of humanity's powers,
But these potions and medicines of ours,
Coffee, garlic, and spices,
Evolved as devices
That insects would stop bugging flowers.

Further Reading

Marine reptiles

Buchholz, E. A. 2001. Swimming styles in Jurassic ichthyosaurs. *Journal of Vertebrate Paleontology* 21: 61–73.

Carroll, R. L. 1981. Plesiosaur ancestors from the Upper Permian of Madagascar. *Philosophical Transactions of the Royal Society of London B* 293: 315–83.

Cowen, R. 1996. Locomotion and respiration in aquatic air-breathing vertebrates. In D. Jablonski et al. (eds), *Evolutionary Paleobiology*, 337–52. Chicago: University of Chicago Press. [Great idea, probably wrong. Rats!]

Dobie, J. L., et al. 1986. A unique sacroiliac contact in mosasaurs (Sauria, Varanoidea, Mosasauridae). *Journal of Vertebrate Paleontology* 6: 197–9.

Erickson, G. M., and C. A. Brochu. 1999. How the "terror crocodile" grew so big. *Nature* 398: 205–6.

Godfrey, S. 1984. Plesiosaur subaqueous flight: a reappraisal. *Neues Jahrbuch für Geologie und Paläontologie Monatshefte* 11: 661–72.

Lingham-Soliar, T. 1992. A new mode of locomotion in mosasaurs: subaqueous flight in *Plioplatecarpus*. *Journal of Vertebrate Paleontology* 12: 405–21.

McGowan, C. 1988. Differential development of the rostrum and mandible of the swordfish (*Xiphias gladius*) during ontogeny and its possible functional significance. *Canadian Journal of Zoology* 66: 496–503. [Compares with ichthyosaurs.]

McGowan, C. 1991. *Dinosaurs, Spitfires, and Sea Dragons*. Cambridge, MA: Harvard University Press. [Chapters 8–10 deal with marine reptiles.]

Motani, R., et al. 1999. Large eyeballs in diving ichthyosaurs. *Nature* 402: 747.

Motani, R. 2000. Rulers of the Jurassic seas. *Scientific American* 283 (6): 52–9. [Ichthyosaurs.]

Rieppel, O. 2002. Feeding mechanics in Triassic stem-group sauropterygians: the anatomy of a successful invasion of Mesozoic seas. *Zoological Journal of the Linnean Society* 135: 33–63. [More convincing detail than I have space to discuss.]

Robinson, J. A. 1975. The locomotion of plesiosaurs. *Neues Jahrbuch für Geologie und Paläontologie Abhandlungen* 149: 286–32.

Robinson, J. A. 1976. Intracorporal force transmission in plesiosaurs. *Neues Jahrbuch für Geologie und Paläontologie Abhandlungen* 153: 86–128.

Taylor, M. A. 1987. A reinterpretation of ichthyosaur swimming and buoyancy. *Palaeontology* 30: 531–5.

Zardoya, R., and A. Meyer. 2001. The evolutionary position of turtles revised. *Naturwissenschaften* 88: 1943–2000.

Flowering plants

Balandrin, M. F., et al. 1985. Natural plant chemicals: sources of industrial and medicinal materials. *Science* 228: 1154–60.

Bernhardt, P. 1999. *The Rose's Kiss: A Natural History of Flowers*. Washington, D.C.: Island Press.

Bond, W. J. 1989. The tortoise and the hare: ecology of angiosperm dominance and gymnosperm persistence. *Biological Journal of the Linnean Society* 36: 227–49.

Buchmann, S. L., and G. P. Nabhan. 1996. *The Forgotten Pollinators*. Washington, D.C.: Island Press. [Read this book! You will look at the world around you in a different way.]

Crepet, W. L. 2000. Progress in understanding angiosperm history, success, and relationships: Darwin's abominably "perplexing phenomenon." *Proceedings of the National Academy of Sciences* 97: 12939–41. [Short snappy summary of recent research on angiosperm origins.]

Currie, C. R., et al. 1999. The agricultural pathology of ant fungus gardens. *Proceedings of the National Academy of Sciences* 96: 7998–8002.

Diamond, J. M. 1999. Dirty eating for healthy living. *Nature* 400: 120–1. [Parrots eat clay to detoxify the plant poisons in their diet.]

Dilcher, D. 2000. Major evolutionary trends in the angiosperm fossil record. *Proceedings of the National Academy of Sciences* 97: 7030–6. [Stresses reproductive mechanisms (pollination and seed dispersal) in angiosperm evolution.]

Doyle, J. A. 1998. Phylogeny of vascular plants. *Annual Reviews of Ecology and Systematics* 29: 567–99. [A long review of the classification and evolution of all land plants, including Mesozoic gymnosperms and angiosperms.]

Friis, E. M., et al. 2001. Fossil evidence of water lilies (Nymphaeales) in the Early Cretaceous. *Nature* 410: 357–60. [A tiny water-lily flower from the Early Cretaceous of Portugal.]

Futuyma, D. J., and M. Slatkin (eds). 1983. *Coevolution*. Sunderland, MA: Sinauer. [Excellent chapters by Futuyma, Feinsinger, and Janzen.]

Grimaldi, D., and D. Agosti. 2000. A formicine in New Jersey Cretaceous amber (Hymenoptera: Formicidae) and early evolution of the ants. *Proceedings of the National Academy of Sciences*, published online November 14, 2000. [Ants and other social insects did not reach their modern dominance in terrestrial insect ecosystems until well into the Cenozoic: compare Wing and Boucher, below.]

Labandeira, C., and J. J. Sepkoski. 1993. Insect diversity in the fossil record. *Science* 261: 310–15.

Lewis, A. C. 1986. Memory constraints and flower choice in *Pieris rapae*. *Science* 232: 863–5.

Mathews, S., and M. J. Donoghue. 1999. The root of angiosperm phylogeny inferred from duplicate phytochrome genes. *Science* 286: 947–50.

Michener, C. D., and D. A. Grimaldi. 1988. The oldest fossil bee: apoid history, evolutionary stasis, and antiquity of social behavior. *Proceedings of the National Academy of Sciences* 85: 6424–6.

Mulcahy, D. L., and G. B. Mulcahy. 1987. The effects of pollen competition. *American Scientist* 75: 44–50.

Paxton, R. J., and Tengš, J. 2001. Doubly duped males: the sweet and sour of the orchid's bouquet. *Trends in Evolution and Ecology* 16: 167–9. [Outrageous swindling of its pollinators by an orchid.]

Regal, P. J. 1977. Ecology and evolution of flowering plant dominance. *Science* 196: 622–9.

Ren, D. 1998. Flower-associated Brachycera flies as fossil evidence for Jurassic angiosperm origins. *Science* 280: 85–8, and comment by Labandeira 57–9. [Speculation, but not proof, that angiosperms, and insect pollination of angiosperms, had Jurassic origins.]

Soltis, P. S., et al. 1999. Angiosperm phylogeny inferred from multiple genes as a tool for comparative biology. *Nature* 402: 402–4, and comment 358–9.

Sun, G., et al. 2002. Archaefructaceae, a new basal angiosperm family. *Science* 296: 899–904, and comment 821.

Tang, W. 1982. Heat and odor production in cycad cones. *Fairchild Tropical Garden Bulletin* 42 (3): 12–14.

Tang, W. 1987. Insect pollination in the cycad *Zamia pumila* (Zamiaceae). *American Journal of Botany* 74: 90–9.

Tiffney, B. H. 1998. Land plants as food and habitat in the age of dinosaurs. In Farlow, J. O., and M. K. Brett-Surman (eds), *The Complete Dinosaur*. Bloomington: Indiana University Press, 352–70.

Wing, S. L., and L. D. Boucher. 1998. Ecological aspects of the Cretaceous flowering plant radiation. *Annual Reviews of Earth & Planetary Sciences* 26: 379–421.

CHAPTER FIFTEEN

The Origin of Mammals

The origin of mammals had practically no significance in Mesozoic ecology. Mammals were small, rare, members of Mesozoic land communities. Yet they evolved into us and the great array of mammals that dominate the large- and small-bodied vertebrate faunas of the world today.

Living reptiles and living mammals are very different, with no surviving intermediates, and this requires us to make some mental adjustments as we try to understand how their ancestral counterparts, the diapsids and synapsids, evolved in such divergent ways in the Triassic.

Living mammals suckle their young, and they are warmblooded: endothermic and homeothermic. They have hair, not scales. They have only one bone along their lower jaw, instead of the reptilian four bones, and the jaw hinges between this lower jaw, the dentary, and the squamosal, replacing the joint of earlier synapsids (and diapsids), which had been between the articular and quadrate (Figure 15.1).

Mammalian teeth are not replaced continuously during life. Typically, milk teeth are replaced only once, and other teeth, such as the big molars or wisdom teeth, are formed only once. Mammalian teeth meet very accurately and work very efficiently, at the cost of severe problems if teeth are damaged, lost, or worn out.

The three bones that are "missing" from the lower jaw evolved into the middle ear of mammals, giving mammals particularly acute hearing at high frequency (squeaks and insect buzzing). In addition, the mammal brain is enlarged and specialized. The forebrain has huge lobes that wrap around older parts of the brain and contain a completely new structure, the neocortex, found only in mammals. The parts of the brain that are greatly increased in volume provide improved sensitivity to hearing, smell, and touch, and they are divided into the left and right lobes that psychologists talk about so much.

It is impossible to imagine all these differences arising overnight, but we can see some of them evolving gradually within the therapsids that were the ancestors of mammals. The fossil record of the transition is richest in jaws and teeth. The dentary bone in the therapsid jaw, originally the small section at the front, came to dominate the jawbone until the rearmost three bones on each side were only little nubbins near the hinge. The teeth became even more differentiated, and, in particular, the teeth behind the canines became larger and more complex in their shape and structure. This may suggest that tooth replacement during life became slower, but that is difficult to judge from the fossil record. Later therapsids evolved the

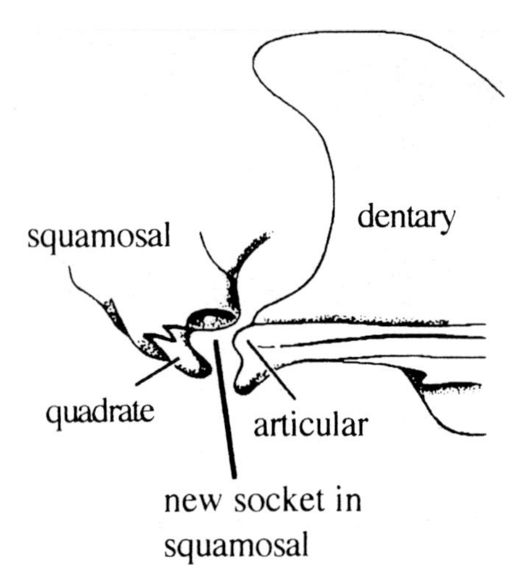

Figure 15.1 The structure of the back of the jaw in the advanced cynodont *Probainognathus*. Only a small transition would be needed to change the hinge from the articular and quadrate, as normal in early synapsids, to the dentary and squamosal, as normal in mammals. (After Romer.)

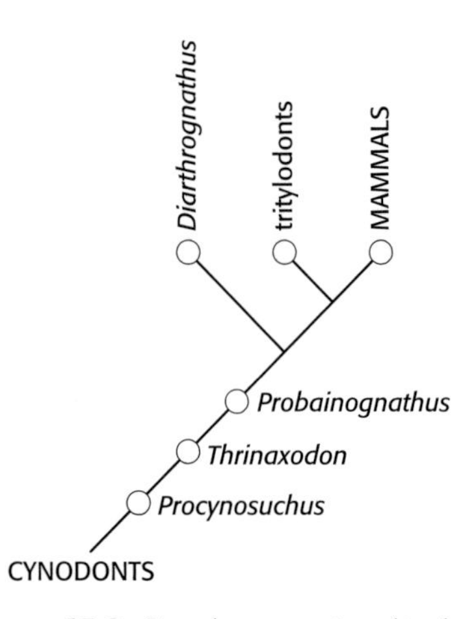

Figure 15.2 Cynodonts mentioned in the text, close to the ancestry of mammals.

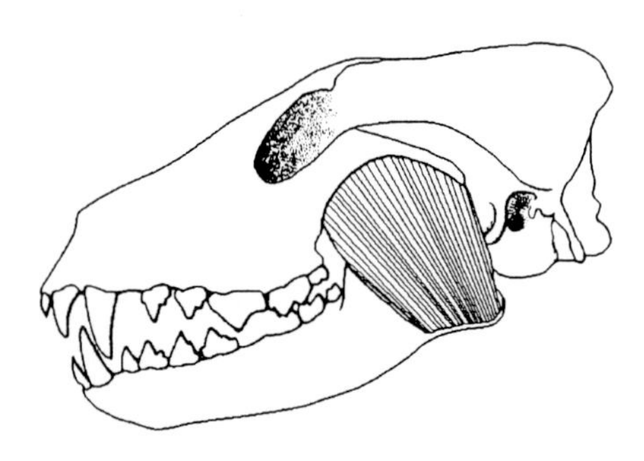

Figure 15.3 The masseter muscle is set into the angle of the jaw in mammals.

secondary palate, the division between the mouth and the nasal passages that allows mammals, including humans, to breathe and chew at the same time.

In terms of soft parts and thermoregulation, we have no direct evidence and must make indirect deductions. We've already seen evidence for seasonal climates in southern Gondwana, suggesting that temperature control of some sort probably evolved among therapsids long before their bony characters became mammalian.

Therapsids were abundant and diverse at the beginning of the Triassic (Chapter 10). But the larger therapsids gradually disappeared, and by the end of the Triassic the survivors were rather small. Cynodont therapsids (Chapter 10) need special attention because they evolved into mammals. They were the last major therapsid group to appear in the geological record, in the Late Permian, and are best known from Gondwana. At least six groups of carnivorous and herbivorous cynodonts evolved some mammalian characters during the Triassic, and a small-bodied carnivorous cynodont group evolved into the first mammals. I give only a general account of the evolution of mammalian characters in cynodonts, to show that the changes were gradual ones that produced more efficient cynodonts.

EVOLVING MAMMALIAN CHARACTERS

There's a paradox about the evolutionary transition from therapsid to mammal: it is too well known and complete. Everyone agrees that the therapsids are a clade, that cynodonts are a clade within therapsids, and that mammals are a clade within cynodonts (Figure 15.2). But there is controversy over which therapsid was actually the first mammal (and which animals the clade Mammalia includes).

The majority position among workers on this transition is to use a crown-group definition of Mammalia: thus the first member of Mammalia would be the common ancestor of *living* mammals. This definition would exclude from "mammals" a lot of Triassic and Jurassic creatures that had many "mammalian" characters such as single lower jaw bone, jaw joint between dentary and squamosal, separate middle ear bones, expanded brain, and so on. I suspect that these "nonmammals" also had fur, and would have looked and behaved like mammals. One would have to grit one's teeth and call these creatures Mammaliaforma to conform with a crown-group definition of Mammalia. Instead, in this Chapter I use a node-based definition of Mammalia. They are the first synapsids develop the jaw characters described above and in Figure 15.1, plus all of their descendants.

Jaws

The secondary palate, which allows chewing and breathing at the same time, evolved in other therapsids as well as cynodonts. However, cynodonts evolved a key innovation involving the rearrangement of the jaw: the **masseter**, a large muscle that runs from the skull under the cheekbone to the outer side of the lower jaw (Figure 15.3). In living mammals it is the most powerful muscle that closes the jaw. (Put your fingers on the angle of your own jaw, clench your teeth and relax again, and you will feel the masseter at work.) The evolution of the masseter had several important effects.

First, jaw movements were easier to control and could become more precise and complex. There was much more accurate lateral and back-and-forward movement of the lower jaw in chewing. Second, biting became more powerful. Third, the force of the bite was transmitted more directly through the teeth rather than indirectly by leverage around the jaw hinge. The lower jaw was slung in a cradle of muscles, and stresses acting on the jaw joint during chewing were much reduced.

In reptiles, the lower jaw is made of several bones, but as chewing efficiency improved, the dentary bone, the most forward bone in the lower jaw, became the largest. The other bones became smaller and were crowded back towards the jaw joint. Stresses on the jaw joint itself were reduced as the masseter evolved, and the three bones behind the dentary on each side became specialized for transmitting vibrations to the stapes rather than strengthening the back of the jawbone. Eventually the dentary became the only bone in the lower jaw and the others became part of the ear.

As this happened, the jaw joint was gradually remodeled. In reptiles the jaw is hinged between the articular and quadrate bones, but in living mammals the jaw hinges between the dentary on the lower jaw and the squamosal bone of the upper jaw. Many people have worried about the apparent jump of the jaw joint from one pair of bones to another, since evolution is a gradual process. However, all the relevant bones in cynodont skulls were small and close together, allowing a major structural shift without major displacement of the jaw hinge (Figures 15.1; 15.4).

Probainognathus, from the Middle Triassic of South America, is very close to the cynodont ancestor of mammals (Figure 15.1). Later changes still needed to complete the transition included smaller size; completion of the change in jaw structure to hinge only on the squamosal and dentary; completion of the middle ear from the three "excess" bones on each side of the lower jaw; enlargement of the brain; formation of definite premolars and molars in the jaw and reduction of tooth replacement to only two sets; better sculpture of the molars, with the mammalian jaw movements that go with it; and changes in the backbone that made it more flexible in curling up during mammalian springing and hopping. None of these changes would have been difficult or unlikely.

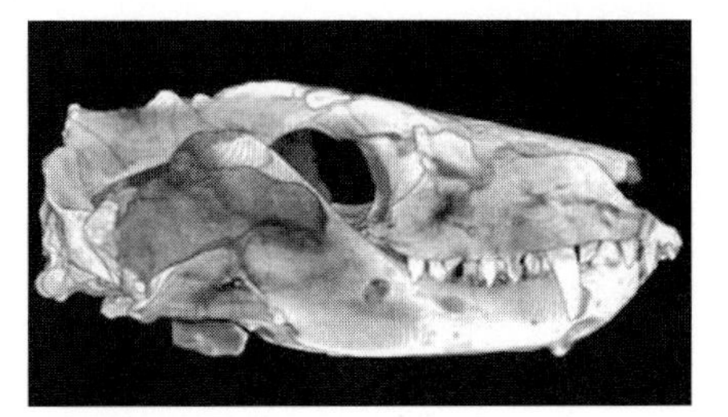

Figure 15.4 CT scan of the skull of *Thrinaxodon*, a cynodont from the Lower Triassic of Gondwana. Skull about 7 cm long. The small size and close packing of the structures around the jaw joint would allow substantial changes without major shifts in bony components. (Courtesy Timothy Rowe and DigiMorph Project at the University of Texas at Austin.)

Teeth and Tooth Replacement

Cynodonts had teeth as well differentiated as those of many later mammals. They had complex, multicusped teeth behind the canines, which implies more complex food processing than in other therapsids. The jaw changes gave greater biting forces near the hinge and smaller errors in occlusion. The teeth themselves, meeting their counterparts accurately, came to be exquisitely sculptured to perform their functions precisely. Different cynodonts, presumably with different diets, evolved shearing, crushing, or shredding actions. *Procynosuchus*, an early cynodont (Figure 15.2), may have been the first therapsid to chew insects rather than crush them and swallow them whole. Shearing is well seen in later carnivorous cynodonts, and there may have been limited self-sharpening of the teeth. Among herbivores, the teeth were organized for crushing; even here, slightly worn (self-wearing) opposing surfaces made a better crushing surface than new tooth surfaces. Look at a newly exposed permanent tooth of a child to see how irregular an edge it has when it first erupts.

Reptiles replace their teeth often during life, and although the process has some systematic pattern to it, any adult reptile has a mixture of larger, older teeth and smaller, newer teeth along its jaw. This means that top and bottom teeth cannot be relied upon to meet precisely against one another, so that tooth functions are comparatively crude. In advanced cynodonts, however, the jaw was slung in a rearranged set of muscles so that jaw control was more precise; the teeth also show precise occlusion between top and bottom jaws. Tooth replacement must have been more controlled and less frequent among cynodonts than in normal reptiles, and the fossil record confirms that. Cynodont teeth were replaced precisely, to maintain good occlusion of different, specialized teeth along a growing jaw. Thus the molar-like teeth of young animals were replaced by canines, while new molar-like teeth were added to the back of the jaw.

Hearing

Like its ancestors, the early cynodont *Procynosuchus* had a hearing system that transmitted ground-borne vibrations through the forelimbs and shoulder girdle to the brain, by way of the bones of the lower jaw and a massive stapes. As therapsid feeding came to emphasize chewing and slicing, it was important for teeth to be arranged all the way along the jaw (actually along the dentary bone), far back toward the hinge. The three bones on each side of the jaw behind the dentary became smaller, and so did the stapes, especially as therapsid body size became smaller. The hearing system evolved to detect and transmit airborne sound, and the posterior jaw bones.

Clearly, airborne sound was increasingly important to late cynodonts and early mammals. Perhaps they hunted insects at least partly by sound. The middle ear bones were linked to the jaw in very early mammals, but later they came to be suspended from the skull. As the hearing pathway was separated from the jaw, the mammal no longer had to listen to its own chewing so much, so would have had much better hearing. It took some time, probably into the Jurassic, to reorganize the other three bones into the "mammalian" middle ear. Only advanced mammals evolved the complex spiral inner ear.

Brains

The huge increase in brain size and complexity between advanced cynodonts and mammals occurred at the same time as the changes in the jaw and ear structure. Tim Rowe suggested in 1996 that these changes were connected. Essentially, he says, a growth clock was reset, allowing the brain to keep growing longer than the structures around it. As the skull and jaw adjusted to accommodate a bigger brain, other changes could occur. In the living opossum, the ear bones reach adult size after three weeks, while the brain grows for 12 weeks.

Rowe's suggestion does not explain the changes in jaw and ears, but it sets up an evolutionary situation in which the changes could happen. It provides an ecological and/or behavioral context in which a relatively large, more complex brain evolved, and it encourages us to ask why such a brain would have been important to a mammal.

Locomotion

Cynodonts still had wheelbarrow locomotion (Chapter 10): the hind limbs provided propulsion while the forelimbs gave only passive support. Cynodont hind limbs evolved to become semierect, whereas the forelimbs remained sprawling (Figure 15.5). The change in the hind limbs brought the feet closer together, and the ankle changed enough to give more direct propulsion along the line of travel. Some improvement in the shoulder joints allowed better locomotion, but it was only a

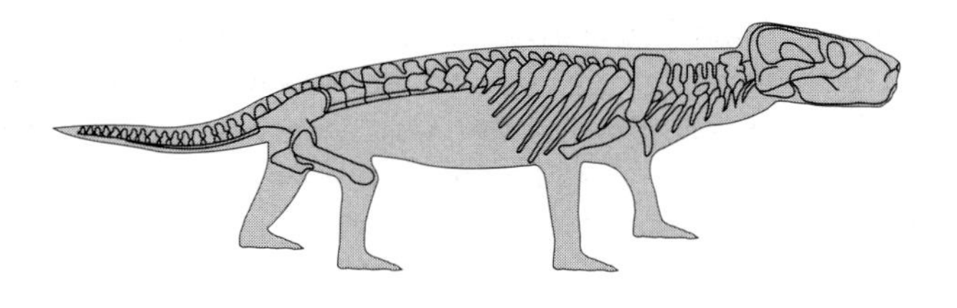

Figure 15.5 The cynodont *Thrinaxodon*, showing the limb structure. (Based on skeletal reconstruction by Farish Jenkins.)

better wheelbarrow style. The spine shows adaptations toward greater stiffness, so that power was transmitted more efficiently from the hind limbs. The cynodonts closest to mammals, the tritylodonts, evolved epipubic bones, which projected from the pelvis and interacted with muscles of the body wall to stiffen the trunk. Late cynodonts also evolved more flexible neck vertebrae, so that the head could swivel freely on the stiffened body. Even with these changes among cynodonts, truly erect limbs were not evolved by the first mammals but came much later.

LIMERICK 15.1
Early mammals suckled their brood,
They breathed in and out as they chewed.
Their molar tooth facets
Were masticatory assets,
But their locomotion was crude.

Thermoregulation and Metabolic Level

Because their jaws and teeth show such an emphasis on efficient food processing, cynodonts probably had higher metabolic rates than pelycosaurs. This does not mean that cynodonts reached the metabolic levels of modern mammals, especially as their limbs (especially the forelimbs) were semierect at best. The spine of therapsids still flexed laterally rather than up and down (but note the strangely widened ribs on *Thrinaxodon* in Figure 15.5, which perhaps were retrofitted devices that cut down on lateral flexing).

Several lines of evidence suggest that therapsids, and cynodonts in particular, were evolving toward endothermy. Mammals have a diaphragm as an important part of the breathing system. This sheet of muscle forces the lungs to expand, helping respiration. A diaphragm can work only when there are no ribs around the abdomen, so its evolution can be detected in fossil vertebrates. It seems to have evolved within the cynodonts, which lost their abdominal ribs (*Thrinaxodon*, Figure 15.5).

Primitive mammals today have comparatively low metabolic levels, and they thermoregulate at temperatures far below those of most mammals. Therapsids were mostly medium-sized, with stocky bodies. Perhaps they operated at a body temperature of 28°–30°C (82°–6°F), a little less than primitive mammals today. In other words, they could have been moderately warmblooded, with at least primitive thermoregulation.

Whatever therapsid body temperature was, they did not evolve great performance. They improved their breathing enough to maintain a fairly high basal metabolic rate (diaphragm, perhaps the ribs of *Thrinaxodon*), but they were not erect athletes the way that dinosaurs were, and they could not support sustained high speed because of Carrier's Constraint. Therapsids could have evolved limited endothermy without solving Carrier's Constraint. Therapsid physiology probably differed dramatically from that of living mammals, from that of living reptiles, and from that of dinosaurs.

As therapsids evolved into mammals, they became smaller. A therapsid with endothermy would have found this difficult, because small bodies lose heat faster than large ones. A possible solution is suggested by the thermal ecology of the little Australian marsupial *Pseudantechinus*, which forages for insects at night in the Australian desert. This is not a problem in the summer, but desert temperatures at night in the winter are usually below freezing. *Pseudantechinus* is so small that it cannot maintain its body temperature in freezing air. So in winter, it forages until it is cold, then goes into shelter and allows its body temperature to drop into torpor, 10°C or more below "normal." It wakes up, basks to regain body heat and digest its food, and ventures out at dusk to forage while it is still warm. This strategy may also have been used by the first mammals, until they achieved full homeothermy later in the Mesozoic.

Other Mammalian Characters

The bones of the cynodont snout have holes and grooves that suggest important blood vessels and nerve canals. Evidence from the early cynodont *Procynosuchus* suggests tight-fitting skin on the snout except immediately around the mouth. Perhaps there were well-developed lips in *Procynosuchus* and later cynodonts, to go with the extra chewing inferred from the teeth and jaws (not to mention suckling in baby *Procynosuchus*!). A well-developed snout blood supply might suggest important sensory organs such as whiskers and noses.

Procynosuchus had lower incisors arranged in a horizontal comb. A similar arrangement occurs in living lemurs, who use the incisors to groom the fur of other members of the troop. If that was true of *Procynosuchus*, it suggests strongly that all cynodonts had hair or fur.

MAMMALIAN REPRODUCTION

The major biological differences between living reptiles and living mammals are not in the skeleton, but in other characters. Reptiles have large eggs with a large energy store, and their young hatch as independent juveniles capable of living without parental care. Mammals have small eggs, and their young depend on parental care. Other major differences are physiological: most living mammals have high body temperatures and hair to insulate them, while reptiles lack hair and are cold-blooded.

Small, warmblooded animals have a high ratio of body surface to volume, and this is especially true for young (tiny) individuals. If therapsids were warmblooded, how did they deal with this problem? And how might the problem bear upon the origin of mammals, especially in view of the fact that the earliest mammals were tiny (smaller than mice)? Tiny, warmblooded animals must find and eat very large quantities of food compared with their body size.

We can find more clues from small living warmblooded vertebrates, the birds. Many nestling birds are helpless and cold-blooded. They depend on their parents for food and for warmth, and they have very low metabolic rates. But because they do not have to find their own food to keep warm, nestlings can devote all their food intake to growth. Helpless nestlings have very large digestive tracts for their size. Warm blood, temperature control, and the ability to make coordinated movement come later and gradually. This strategy avoids the energy problem of warm blood at small size, and nestlings are essentially cold-blooded until they have grown to considerable size. Furthermore, most birds cut down environmental temperature fluctuations in their nestlings by caring for them in nests designed to maintain a uniform temperature. But the system demands intensive care by one or both parents.

Most likely, some similar strategy was followed by late therapsids and mammals, but in burrows rather than nests. The little therapsid *Diictodon* was digging burrows by the end of the Permian (Figure 10.17). As therapsids evolved to very small size in the Late Triassic, the need for parental care would have become more and more acute. As the pelvis became smaller, eggs would necessarily have become smaller and smaller, with less and less yolk, and the young would have hatched earlier and been more helpless. Although freed from the anatomical problem of laying large eggs, the parent(s) were now committed to providing a steady supply of food to the young after hatching, like birds and unlike most reptiles. However, smaller eggs and the rapid growth of helpless hatchlings gave an opportunity for very rapid reproductive rates in closely spaced litters (or clutches).

Suckling

Living monotremes still have the kind of reproduction that we infer for advanced cynodonts and early mammals. The platypus lays and hatches tiny eggs in a nest inside a burrow. Monotremes also nourish their hatchlings by suckling, rather than collecting food for them. This behavior has advantages: the parent does not have to leave the hatchling to search for suitable food for it, because any normal adult food can be converted into milk. The hatchling digests milk easily, and its parent is never far away, providing protection and warmth.

Charles Darwin suggested how suckling might have evolved in mammals, even before Western science discovered monotremes. His theory survives with only minor modifications. Let's assume that mammalian ancestors were already caring for eggs by incubating them. A special gland may have secreted moisture to keep the eggs humid during incubation. Hatchlings that licked the incubation gland benefited by gaining water to help deal with the food brought back by the parents, and perhaps the secretions had the added advantage of being antibacterial. The adaptation was selective as long as the fluid helped hatchlings to survive and grow. Gradually, as the secretions came to contain mineral salts and trace elements and then nutritious organic compounds (milk) as well as water, the mother's excursions for food could be reduced and the hatchlings benefited even more by her increased attendance. Rapid evolution of full lactation from specialized nipples followed, with efficient suckling by the hatchlings.

The mammalian system is interesting because only the female parent is specialized to have milk glands, so that the male may take little or no role in caring for the young. Male mammals have nipples, of course, and there is no obvious biochemical reason why baby mammals should suckle only from the female, so the reason is probably genetic. The development of milk glands in mammals is controlled by a set of the Hox genes that are universal among metazoans, typically laying out nerves, vertebrae, segments, limbs and other body systems. Almost certainly, the lactation system is switched on, under genetic control, as the female goes through pregnancy and delivery. The switching system is complex, and has components from three of the four separate Hox gene clusters that mammals carry. Even in females, there are occasional mutations that upset this complex system. Since males do not go through pregnancy, they would not receive the signals to switch on lactation genes.

The development of suckling can be dated indirectly. Cynodonts had a secondary palate and could chew while still breathing, but even the tiniest baby cynodont had teeth and so probably did not suckle. Perhaps the parent brought food to the nest or into the burrow. But the first mammals had very limited tooth replacement, possibly related to their small size and short lifespans, and they probably suckled in some fashion. The evolutionary transition from licking to suckling was not as simple in baby mammals as it might seem: suckling demands full and flexible cheeks. Cheeks must have evolved, along with many other "mammalian" characters, among Triassic cynodonts. Certainly some sort of cheek would have been needed to cover the newly evolving masseter muscle.

Live Birth

Suppose mammals had reached the point of being reproductively like monotremes: they laid eggs but suckled their hatchlings. What would cause or encourage the evolution of live birth (viviparity)?

There is nothing unusual about live birth. It has evolved independently many times in fishes, amphibians, reptiles, and mammals: in fact, in every living vertebrate group but birds. It has evolved independently in at least 90 different groups of lizards and snakes, and some insects have evolved it too. But how and when did it evolve in the mammal lineage?

Laying eggs is a difficult proposition below a critical body size. The egg must be laid through a pelvic opening, and a shelled egg with a reasonable amount of yolk must have a certain minimum size to be viable. Constraints on the pelvis that would forbid laying a large, shelled egg may not apply to a fetus, which is structurally and physiologically more flexible than an egg. A fetus does not need a yolk or shell during its development; it can be squeezed through a birth canal more safely than can an eggshell. Inside a thermoregulating mother, a fetus develops at a more uniform temperature than in a nest. The growing fetus has an unlimited supply of water and oxygen, and an easy way of getting rid of CO_2 and other wastes, all of which are problems for an embryo inside an eggshell. There is far less chance of predation or infection. Finally, if suckling has already been evolved, the young never need be separated from the care and protection of the mother, even if they are helpless at birth.

Egg-laying monotremes survive today, proving that viviparity is not essential for mammals in spite of the list we have just compiled. But all other living mammals have live birth. The necessary evolutionary steps would include the gradual improvement of ways to transport material between mother and fetus, the beginnings of the placenta. The first viviparity would have been on the marsupial pattern, with or without a pouch to contain the young, but it need not have been as specialized a process as it is in living marsupials. We do not know when mammals evolved live birth, but indirect evidence suggests that it took place in the Cretaceous.

EARLY MAMMALS

Early mammals were tiny, and their fossils are rare and difficult to collect except by washing and sieving enormous volumes of soft sediment. But after years of effort we now have fragments of mammals (mostly teeth) from many localities in many continents, beginning close to the Triassic–Jurassic boundary. The problem with teeth is that there are specific advantages to having particular types of teeth, and it is becoming clear that tooth patterns have evolved more than once over time, much to the confusion of mammalian classification. We simply do the best we can . . .

Among Late Triassic mammals the family Morganucodontidae, named after

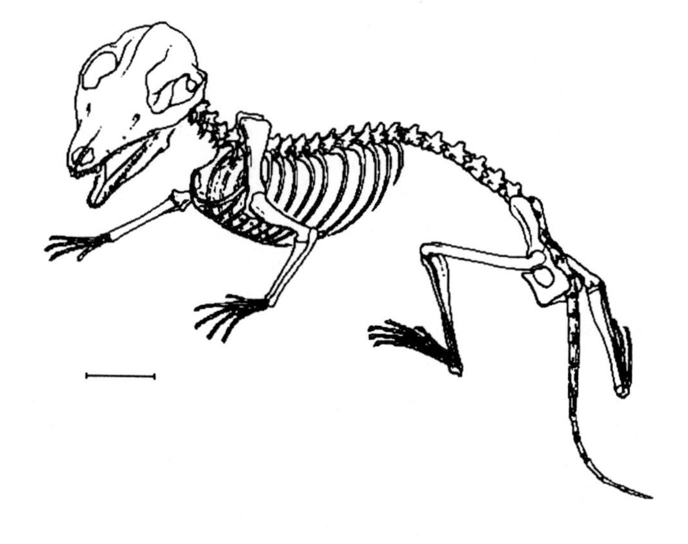

Figure 15.6 A morganucodont. The scale bar is 2 cm. (After Jenkins and Parrington, and Jenkins.)

Morganucodon, could be ancestral to most later groups. Morganucodonts are fairly well known from two nearly complete skeletons found in South Africa (Figure 15.6). They were small animals, perhaps only 10 cm (4 in) to the base of the tail, and weighing only about 25 g, about an ounce, much like modern shrews. They had small but nasty jaws and were obviously little carnivores, probably eating insects, worms, and grubs. They had relatively longer snouts and much larger brains than cynodonts. The skeletons show that they were agile climbers and jumpers. The neck was very flexible, as in living mammals, and the spine could have flexed up and down in addition to the lateral bending of therapsids.

The jaw joint was still like that of late cynodonts. But the teeth were fully differentiated, and the molars had double roots. As in most mammals today, the front teeth were replaced once, and there was only one set of molars. The molar teeth had three cusps in a line, so the name triconodont is used for this structure. Triconodont molars worked by shearing vertical faces up and down past one another (Figure 15.7), giving a zigzag cut exactly like that of pinking shears in dressmaking. This is efficient, especially for thin or soft material, but requires precise up-and-down movement. Triconodont teeth evolved more than once, confusing the picture of early mammal evolution.

The Late Triassic genus *Kuehneotherium* looks closer than *Morganucodon* to living mammals, but that is difficult to establish because we only have its jaws and teeth. Other fragmentary fossils remain puzzling. A tiny fossil from the Early Jurassic of China, named *Hadrocodium*, has some advanced features of the skull, despite its early age. It is the nearest mammal yet to the direct ancestry of the living mammals. All Jurassic mammals were small and probably nocturnal. They were carnivorous, insectivorous, or perhaps omnivorous: only a few had teeth that could chew up fibrous vegetation.

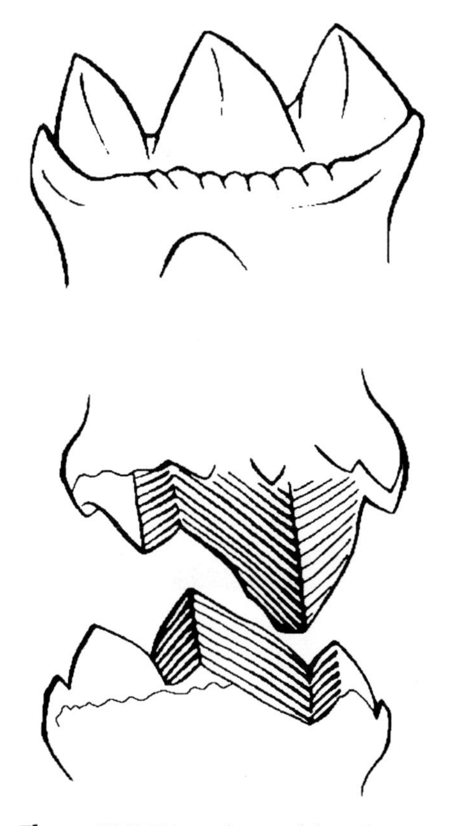

Figure 15.7 Triconodont teeth have three cusps in a row (above, after Simpson). They had an action rather like that of pinking shears (below, after Jenkins.).

Therians and Non-Therians

The three living clades of mammals are the monotremes, the marsupials, and the placentals. The easy differentiation between them today is reproductive: monotremes lay eggs and suckle their young, while marsupials and placentals have live birth. The marsupials and placentals are groups as therian mammals, the

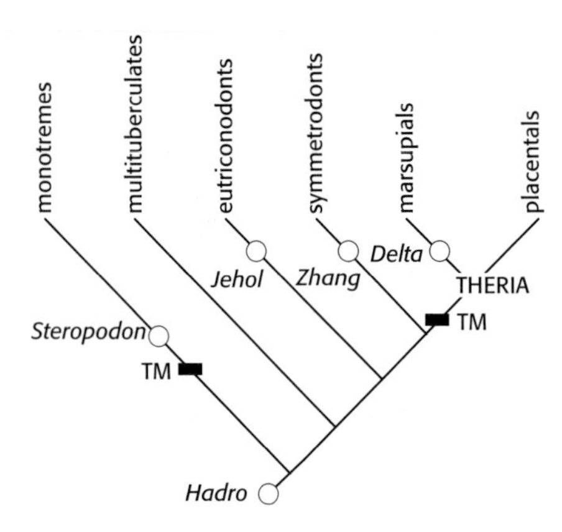

Figure 15.8 One hypothesis for the evolution of advanced mammal groups. TM marks the two independent evolutions of the tribosphenic molar tooth. *Hadro* is *Hadrocodium* from the Early Jurassic of China; *Zhang* is *Zhangheotherium*, and *Jehol* is *Jeholodens*, both from the Lower Cretaceous in China; and *Delta* is *Deltatheridium*, a basal marsupial from the Cretaceous of Mongolia. The monotreme branch is Gondwanan, but all the others are Laurasian, at our present state of knowledge.

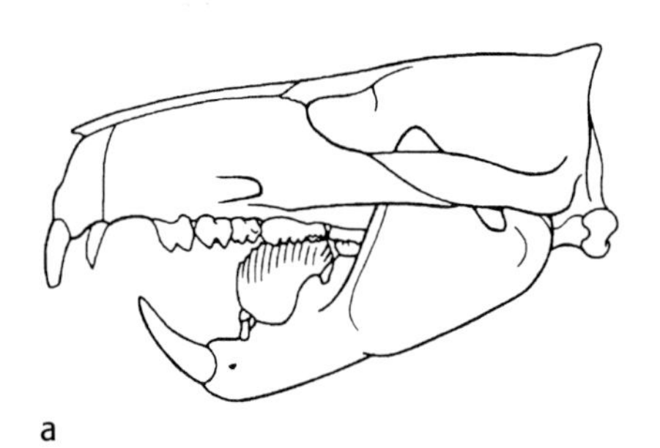

a

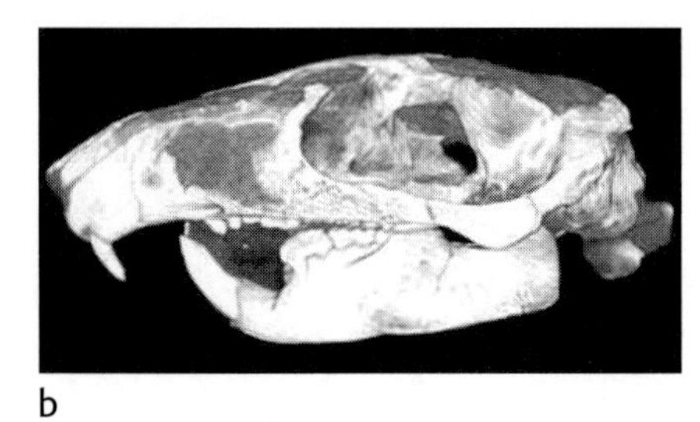

b

Figure 15.9 Skulls of two multituberculates. (a) *Ptilodus*, from the Paleocene of North America. (After Simpson, Krause, and Kielan-Jaworowska.). (b) CT scan of *Kryptobataar*, a smaller and older form from the Cretaceous of Mongolia. Skull only about 3 cm long. (Courtesy Timothy Rowe and DigiMorph Project at the University of Texas at Austin.)

Theria. But how do we deal with extinct clades that give us little clue about their reproduction?

Until recently, there was another basic character that distinguished therians from monotremes: therians have **tribosphenic** molar teeth. These are complex in shape and can perform a large variety of functions as upper and lower teeth interact. They evolved from simpler teeth by adding new surfaces that shear past one another as the jaw moves sideways in a chewing motion. Tribosphenic molars are particularly well suited for puncturing and shearing, and especially for grinding, superbly fitting mammals for a diet of insects and high-protein seeds and nuts. Living monotremes do not have tribosphenic molars.

It now turns out that the tribosphenic type of molar is so efficient that teeth like it evolved independently in two different groups. The ancestor of the monotremes, a little mammal called *Asfaltomylos*, evolved tribosphenic-type molars in South America (then part of Gondwana), as early as the Jurassic. In contrast, the mammals of the Northern Hemisphere (Laurasia) apparently did not evolve a tribosphenic molar until well into the Cretaceous, when it was evolved by the ancestor of the living therians.

Other extinct clades of northern creatures were part of the Laurasian radiation, and they did not have the tribosphenic molar. They were mammals but not therians. These clades included advanced triconodonts (or eutriconodonts), symmetrodonts, and multituberculates. We can draw a provisional evolutionary diagram that shows these new pieces of evidence (Figure 15.8).

The eutriconodonts were successful into Early Cretaceous times. Two well-preserved eutriconodonts come from the same remarkable rocks in China that also yielded feathered theropod dinosaurs (Chapters 12), many early birds (Chapter 13), and the earliest flowering plant (Chapter 14). *Gobiconodon* evolved to be the size of a possum. *Jeholodens* was smaller: strangely, it had sprawling hind limbs but erect forelimbs.

Symmetrodonts are Jurassic and Cretaceous mammals. The first complete symmetrodont, *Zhangheotherium*, also from the Lower Cretaceous of China, was only a few inches long. Its ear structure was primitive, and it had sprawling front limbs, though its shoulder joint was beginning to show advanced characters. Symmetrodonts are closer to therians than multituberculates.

Multituberculates are small mammals that evolved superficially rodent-like teeth, and were successful in the Late Jurassic, Cretaceous, and Early Cenozoic. They often make up more than half the mammals in Late Cretaceous faunas. They survived the great extinction at the end of the Cretaceous and reached their greatest diversity in the Paleocene, before being replaced by more modern mammals, especially the true rodents. Multituberculates are sometimes called the "rodents of the Mesozoic," but their ecology may not be so simple to reconstruct.

The incisor teeth of multituberculates were usually specialized for grasping and puncturing, rather than gnawing, but there were six in the upper jaw and only two in the lower. The very large, sharp-edged premolars were designed for holding and cutting, while the molars were grinding teeth. The system looks well suited for cropping and chewing vegetation with a back-and-forward jaw action (Figure 15.9). Although their radiation corresponds with the general rise of the flowering plants, many multituberculates were probably omnivores, like rats rather than guinea pigs. Specific forms can be interpreted more precisely. Some incisors were ever-growing and self-sharpening, well designed for gnawing. (Gnawing teeth may not have evolved for chewing nuts and seeds, but to open up wood to get at insects.) Other multituberculates had long, thin, saber-like incisors, like some modern insectivores that use them to impale insects. Still others probably used the shearing premolars and the crushing molars to eat fruits or seeds.

The range in body size (mouse- to rabbit-sized) indicates a fairly wide ecological range among multituberculates. Some later forms from the Early Cenozoic were clearly tree dwellers. *Ptilodus* had a prehensile tail and squirrel-like hind feet that could rotate backwards for climbing downward (Figure 15.10).

Kryptobataar, from the Late Cretaceous of Mongolia (Figure 15.9), is an important multituberculate because we can say confidently that it had live birth. It had a narrow, rigid pelvis that was incapable of widening during birth. Thus the birth canal would have been at most only 3–4 mm wide. The animal could not have laid any reasonable-sized egg, but it could have borne a very small fetus (newborn marsupials weigh about 1 g).

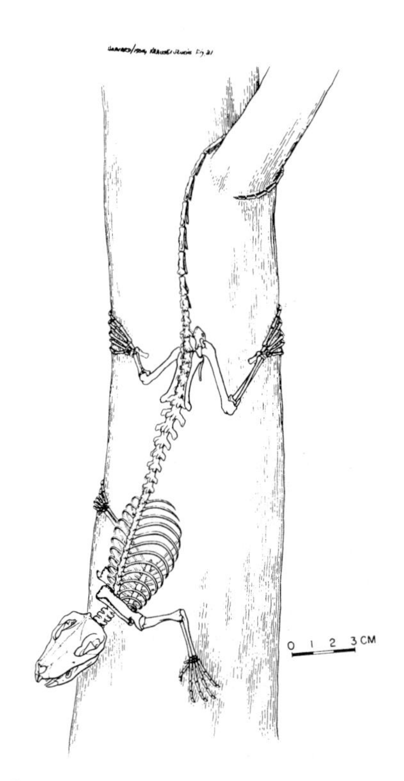

Figure 15.10 Reconstruction of *Ptilodus*, a tree-climbing multituberculate from the Early Cenozoic. (Courtesy of David Krause, SUNY Stony Brook.)

THERIAN MAMMALS

Formally, **therians** include **metatherians** and **eutherians**. Metatherians consist of the common ancestor of living marsupials and all its descendants; eutherians consist of the common ancestor of living placentals and all its descendants.

Early mammals probably reproduced by delivering small, helpless young once they had evolved beyond the monotreme stage of egg-laying. The divergence of mammals into separate marsupial and placental clades probably had taken place by the Early Cretaceous. Each style of reproduction, in its own way, solves some of the problems of the mammalian way of life. Marsupial and placental styles of reproduction are now quite distinct, but they probably both evolved from a state that we would now identify as simple but largely marsupial. Even so, it's usually impossible to infer the reproductive style of any given Cretaceous mammal, especially when the pelvic regions are not well preserved.

Even today more than 90% of all mammals weigh less than 5 kg (11 lb) as adults. All small mammals give birth to tiny helpless young, probably because they do not have enough body volume to pack into their babies all the requirements for fully independent mammal life. Tiny mammal babies are cold-blooded at first and absolutely dependent on parental care. Mammals with large bodies can accommodate and give birth to larger, more competent offspring. This factor may have been the key to the success of the large placental mammals as opposed to large marsupials, but it doesn't apply to small placentals and marsupials.

Living marsupials bear fetuses surrounded by a membrane like the eggshell membrane of a bird or a monotreme. Its most important component is the trophoblast, a cell layer that allows very close contact between fetal and maternal tissue yet prevents the passage of substances that would cause the mother to reject the foreign body growing within her. Only a limited amount of nutrition can be passed to the growing embryo from the mother, and after a certain gestation time it is better for the fetus to be born so that it can take nutrition more efficiently by suckling. Living marsupials, therefore, have short gestation periods followed by long suckling periods, often with the young in a pouch.

Sometime early in the Cretaceous, a line of small mammals evolved a new derived character, the true **placenta**. This is a specialized structure built in the uterus jointly by the fetus and the mother. The placenta has an enormous surface area (50 times the skin area of a newborn human), and it is used to supply the fetus with nutrition, oxygen, and hormones, and to pass waste products from the fetus to the mother for disposal. Essentially, the placenta is a large, discriminatory, two-way pump. The trophoblast of placental mammals is much more effective than it is in marsupials, allowing the placenta to support a growing fetus much longer. As a result, placental mammals can evolve a long gestation period, so they can have shorter lactation periods before the young reach a stage where they are independent of the mother.

Marsupials never evolved a placenta or a trophoblast as efficient as that of placental mammals, so they cannot supply the fetus with all its needs past a certain stage of development. Their trophoblast separates mother from fetus but allows only a limited range of materials to pass between them. As a result, marsupial newborns are fetuses that must be agile enough to reach the nipple.

None of this means that marsupials are inferior to placentals. A marsupial mother who experiences a natural crisis can easily abandon her young while they are fetuses, because she already carries them as an external litter. She may be ready to breed again quickly. A few placentals can absorb their fetuses, but most placental mothers must carry their internal young to term for a comparatively long gestation period, even during a flood, drought, or harsh winter, often at the risk of their own lives. Marsupial females can delay fetal development after implantation, whereas placental females rarely can.

The marsupial reproductive system stresses flexibility in the face of an unpredictable environment, so it may sometimes be superior to the placental system. Native marsupials and introduced placentals of the same body weight in the same environments in Australia (wallabies versus rabbits, for example) take on average about the same time to rear their young successfully. We still don't know the relative energy cost of the two methods. Placental and marsupial styles of reproduction, each in their own way, reduce the hazards of rearing young at small body size, but one is not always more efficient than the other.

Note that the flexibility of marsupials in abandoning their young is comparable with that of birds, who may abandon a nest in a crisis, even if there are eggs or young in it. Herons and storks will abandon a single chick if there is enough time left in the year for them to start another clutch of eggs that gives them a greater chance of rearing several chicks. A principle called the Concorde Fallacy seems to operate in human affairs. If a great deal has been invested in a project, then a great deal more will be invested in order to see it through to the bitter end, even after it is clear that the project will never repay its cost. The supersonic Concorde airliner was one case, but there are many others, such as the Vietnam War, nuclear power plants, and the Space Shuttle.

Animals operating under natural selection cannot afford to waste anything and must be ruthless in cutting their losses as soon as they detect eventual failure. Lions and cheetahs should (and do) abandon the chase as soon as they see they cannot catch their prey, and prospective parents should abandon their young if they cannot be reared successfully. In these terms, the allegedly superior placental reproductive system is more likely to result in wasteful expense than either the egg-laying of birds or the marsupial system. It is simply a bigger gamble than the others. In the long run, the three methods must be about equal in their results, because different animals practice them all successfully.

Other major differences between marsupials and placentals today are in thermoregulation and metabolic rate. Size for size, placentals thermoregulate at slightly higher temperatures and have slightly higher metabolic rates. They are "faster livers," as one writer has put it. This need not affect reproduction, because female marsupials increase their metabolic rate during pregnancy and lactation, up to placental levels. It is true that the brain grows faster in fetal placental mammals than in marsupials, and there is a small but significant difference in adult brain size, weight for weight, between the two groups. In turn, the metabolically active brain uses more oxygen in placentals, partly accounting for their higher energy budget. In spite of the metabolic differences, however, there is no systematic difference in at least one vitally important aspect: locomotion. Marsupials can run at about the same maximum speeds as equivalent placentals, and they have about the same stamina.

The lineages leading to marsupials and placentals diverged in the Cretaceous. The tiny mammal *Eomaia* from northern China is a eutherian, that is, it lies along the placental line, distinct from the marsupial lineage. It has kneecaps, for example! It probably had not yet evolved placental reproduction. *Eomaia* has a skeleton with climbing and tree-dwelling adaptations, unlike other early mammals. It was tiny, probably weighing less than 25 g, less than an ounce. But it is well enough preserved that we can see hair on it, the first known evidence for hair in the mammal lineage. (Endothermy had already evolved, because *all* living mammals have hair and endothermy.)

At first marsupials and placentals would not have been greatly different ecologically. Early placental mammals would still have had tiny, helpless young. The evolution of precocious young such as colts, calves, and fawns, which are large and can run soon after birth, had to wait until placental mammals reached large size; not until then did placental mammals become more successful in their distribution and diversity. Placental mammals may well have little or no advantage over marsupials when both are small, but large precocious young are not an option for marsupials, while they are for placentals.

THE INFERIORITY OF MAMMALS

If cynodonts were moderately warmblooded, their evolution to smaller size would almost automatically have produced adaptations such as insulation, parental care, and so on. But why did they evolve to smaller size? Given our ideas about dinosaur biology and physiology (Chapter 12), it was probably because of competition from archosaurs, which certainly had solved Carrier's Constraint. Ecologically squeezed between the first dinosaurs (fast-moving predators with sustained running) and the small, lizard-like reptiles of the Triassic (running on cheap solar energy with a low resting metabolic rate), late cynodonts may have escaped extinction only by evolving into a habitat suitable for small, warmblooded animals and no-one else: the night. In doing so, they underwent the radical changes in body structure, physiology, and reproduction that resulted in the evolution of mammals.

By the end of the Triassic, archosaurs had replaced and probably outcompeted the therapsids, driving them underground, deep into forests, or into nocturnal habits all over the world. And as the last few therapsids became extinct or were confined to tiny body size, the dinosaurs evolved into one of the most spectacular vertebrate groups of all time.

Burrowing in the dark, the mammals lived in a habitat that required much greater sensitivity to hearing, smell, and touch. This requirement may have selected for a relatively large, complex brain and sophisticated intelligence. So why didn't they take over the Cretaceous world? It may have depended on the competition.

With the spread of flowering plants in the Early Cretaceous, herbivorous dinosaurs, insects, and mammals all increased in diversity. The increase in food in the form of insects, seeds, nuts, and fruits provided a great ecological opportunity for small mammals. Mammals did increase in diversity through the Cretaceous, but not in a spectacular way, and probably in environments that do not yield many fossils: the forest canopy. Forest canopy ecosystems had flourished since

Carboniferous times (Chapter 9). Mesozoic mammals, small-bodied and insectivorous, were clear candidates to invade such ecosystems. By the end of the Cretaceous, it is easy to envisage a diverse set of mammals occupying many small-bodied ways of life in the forest, particularly at night. The ancestors of primates and bats most likely evolved their special characters in the forest canopy. Small mammals are very important in the canopy even today: the equatorial forest has many species of birds active by day and mammals at night, each with a small-bodied way of life, eating insects, seeds, nuts, and fruits. But mammals simply could not compete mechanically with dinosaurs: their locomotion and probably their metabolism simply did not allow them to survive on open ground by day.

Immediately after the end of the Cretaceous and the disappearance of the dinosaurs, mammals began a tremendous radiation into all body sizes and many different ways of life. The inverse relationship between the success of Mesozoic archosaurs, especially dinosaurs, and Mesozoic therapsids and mammals is probably not a coincidence. It reflects some real inability of the mammalian lineage to compete successfully in open terrestrial environments at the time. The extinction at the Cretaceous–Tertiary boundary that finally seems to have "released" the evolutionary potential of mammals must be seen in the context of the rest of the world's life in the Mesozoic, and we shall look at that in the next chapter.

Further Reading

Blackburn, D. G. et al. 1989. The origins of lactation and the evolution of milk: a review with new hypotheses. *Mammal Review* 19: 1–26.

Cifelli, R. L. 2001. Early mammalian radiations. *Journal of Paleontology* 75: 1214–26.

Duboule, D. 1999. No milk today (my Hox have gone away). *Proceedings of the National Academy of Sciences* 96: 322–3. [Short comment on an accompanying detailed paper.]

Geiser, F., et al. 2002. Was basking important in the evolution of mammalian endothermy? *Naturwissenschaften* 89: 412–14. [Yes. At least it's important for the tiny living Australian marsupial *Pseudantechinus*.]

Hillenius, W. J. 1994. Turbinates in therapsids: evidence for Late Permian origins of mammalian endothermy. *Evolution* 48: 207–29.

Hu, Y., et al. 1997. A new symmetrodont mammal from China and its implications for mammalian evolution. *Nature* 390: 137–42. [*Zhangheotherium*.]

Ji, Q., et al. 1999. A Chinese triconodont mammal and mosaic evolution of the mammalian skeleton. *Nature* 398: 326–30, and comment 283–4; also comment in *Science* 283: 1989–90. [*Jeholodens*.]

Ji, Q., et al. 2002. The earliest known eutherian mammal. *Nature* 416: 816–22, and comment 788–9. [*Eomaia*.]

Kemp, T. S. 1982. *Mammal-like Reptiles and the Origin of Mammals*. London: Academic Press.

Kielan-Jaworowska, Z., et al. 1987. The origin of egg-laying mammals. *Nature* 326: 871–3.

Luo, Z-X., et al. 2001. Dual origin of tribosphenic mammals. *Nature* 409: 53–7, and comment 28–31. Comment in *Science* 291, 26.

Luo, Z-X., et al. 2001. A new mammaliaform from the Early Jurassic and evolution of mammalian characteristics. *Science* 292: 1535–40, and comment 1496–7. [*Hadrocodium*.]

Luo, Z-X., et al. 2002. In quest for a phylogeny of Mesozoic mammals. *Acta Palaeontologica Polonica* 47: 1–78. [Long detailed paper which explains the problems in interpreting early mammal evolution.]

Madsen, O., et al. 2001. Parallel adaptive radiations in two major clades of placental mammals. *Nature* 409: 610–14.

Meng, J., and A. R. Wyss. 1995. Monotreme affinities and low-frequency hearing suggested by multituberculate ear. *Nature* 377: 141–4, and comment 104–5. [These may be ancestral similarities.]

Murphy, W. J., et al. 2001. Molecular phylogenetics and the origins of placental mammals. *Nature* 409: 614–18.

Rauhut, O. W. M., et al. 2003. A Jurassic mammal from South America. *Nature* 416: 165–8. [A new australosphenid, probably a monotreme ancestor.]

Reilly, S. M., and T. D. White 2003. Hypaxial motor patterns and the function of epipubic bones in primitive mammals. *Science* 299: 400–2. [The evolution of epipubic bones in advanced cynodonts.]

Rosowski, J. J., and A. Graybeal. 1991. What did *Morganucodon* hear? *Zoological Journal of the Linnean Society* 101: 131–68. [We don't know yet.]

Rougier, G. W., et al. 1998. Implication of *Deltatheridium* specimens for early marsupial history. *Nature* 396: 459–63.

Rowe, T. 1996. Coevolution of the mammalian middle ear and neocortex. *Science* 273: 651–4. [Good idea: may not be right. See Wang et al. 2001.]

Rowe, T., and J. Gauthier. 1992. Ancestry, paleontology, and definition of the name Mammalia. *Systematic Biology* 41: 372–8. [One side of the argument!]

Rubidge, B. S., and C. A. Sidor. 2001. Evolutionary patterns among Permo-Triassic therapsids. *Annual Review of Ecology and Systematics* 32: 449–80.

Rybczynski, N., and R. R. Reisz. 2001. Earliest evidence for efficient oral processing in a terrestrial herbivore. *Nature* 411: 684–7.

Savage, R. J. G., and M. R. Long. 1987. *Mammal Evolution: An Illustrated Guide*. London: British Museum (Natural History).

Sereno, P. C., and M. C. McKenna. 1995. Cretaceous multituberculate skeleton and the early evolution of the mammalian shoulder girdle. *Nature* 377: 144–7; and comment 104–5.

Sidor, C. A., and J. A. Hopson. 1998. Ghost lineages and "mammalness": assessing the temporal pattern of character acquisition in the Synapsida. *Paleobiology* 24: 254–73.

Szalay, F. S., et al. (eds). 1993. *Mammal Phylogeny: Mesozoic Differentiation, Multituberculates, Monotremes, Early Therians, and Marsupials.* New York: Springer-Verlag.

Wang, Y., et al. 2001. An ossified Meckel's cartilage in two Cretaceous mammals and origin of the mammalian middle ear. *Science* 294: 257–361.

CHAPTER SIXTEEN

The End of the Dinosaurs

Almost all the large vertebrates on Earth, on land, at sea, and in the air—all dinosaurs, plesiosaurians, mosasaurs, and pterosaurs—suddenly became extinct about 65 Ma, at the end of the Cretaceous Period. At the same time, most plankton and many tropical invertebrates, especially reef-dwellers, became extinct, and many land plants were severely affected. This extinction event marks a major boundary in Earth's history, the K–T or Cretaceous–Tertiary boundary, and the end of the Mesozoic Era. The K–T extinctions were worldwide, affecting all the major continents and oceans. There are still arguments about just how short the event was. It was certainly sudden in geological terms and may have been catastrophic by anyone's standards.

Despite the scale of the extinctions, however, we must not be trapped into thinking that the K–T boundary marked a disaster for all living things. Most groups of organisms survived. Insects, mammals, birds, and flowering plants on land, and fishes, corals, and molluscs in the ocean went on to diversify tremendously soon after the end of the Cretaceous. The K–T casualties included most of the large creatures of the time, but also some of the smallest, in particular the plankton that generate most of the primary production in the oceans.

There have been many bad theories to explain dinosaur extinctions. More bad science is described in this chapter than in all the rest of the book. For example, even in the 1980s a new book on dinosaur extinctions suggested that they spent too much time in the sun, got cataracts, and because they couldn't see very well, fell over cliffs to their doom. But no matter how convincing or how silly they are, any of the theories that try to explain only the extinction of the dinosaurs ignore the fact that extinctions took place in land, sea, and aerial faunas, and were truly worldwide. The K–T extinctions were a global event, so we look for globally effective agents to explain them: geographic change, oceanographic change, climatic change, or an extraterrestrial event (Chapter 6). The most recent work on the K–T extinction has centered on two hypotheses that suggest a violent end to the Cretaceous: a large asteroid impact and a giant volcanic eruption. I discussed these in outline in Chapter 6, but it is worth focusing more closely on this, the better-known of the two largest mass extinctions.

AN ASTEROID OR COMETARY IMPACT?

An asteroid hit Earth precisely at the time of the K–T extinction. The evidence for the impact was first discovered by Walter Alvarez and colleagues (see Alvarez et al, 1980), who found that rocks laid down precisely at the K–T boundary contain extraordinary amounts of the metal iridium (Figure 16.1). It doesn't matter whether the boundary rocks were laid down on land or under the sea. In the Pacific Ocean and the Caribbean the iridium-bearing clay forms a layer in ocean sediments; it is found in continental shelf deposits in Europe; and in much of North America it occurs in coal-bearing rocks laid down on floodplains and deltas. The dating is precise, and the iridium layer has now been identified in more than 100 places worldwide. Where the boundary is in marine sediments, the iridium occurs in a layer just above the last Cretaceous microfossils, and the sediments above it contain Paleocene microfossils from the earliest Cenozoic.

iridium, parts per trillion
10 100 1000
coal
5 mm intervals

Figure 16.1 A typical iridium spike at the K–T boundary. This data set is from New Mexico, where the K–T boundary lies in a coal bed. The scale for iridium is logarithmic. (Data from C. J. Orth et al., 1981.)

The iridium is present only in the boundary rocks and therefore was deposited in a single large spike: a very short event. Iridium occurs in normal seafloor sediments in microscopic quantities, but the iridium spike at the K–T boundary is very large. Iridium is rare on Earth. Although chemical processes in a sediment can concentrate iridium to some extent, the K–T iridium spike is so large that it must have arisen in some unusual way. Iridium is much rarer than gold on Earth, yet in the K–T boundary clay iridium is usually twice as abundant as gold, sometimes more than that. The same high ratio is found in meteorites. The Alvarez group therefore suggested that iridium was scattered worldwide from a cloud of debris that formed as an asteroid struck somewhere on Earth.

An asteroid big enough to scatter the estimated amount of iridium in the worldwide spike at the K–T boundary may have been about 10 km (6 mi) across. Computer models suggest that if such an asteroid collided with Earth, it would pass through the atmosphere and ocean almost as if they were not there and blast a crater in the crust about 100 km across. The iridium and the smallest pieces of debris would be spread worldwide by the impact blast, as the asteroid and a massive amount of crust vaporized into a fireball.

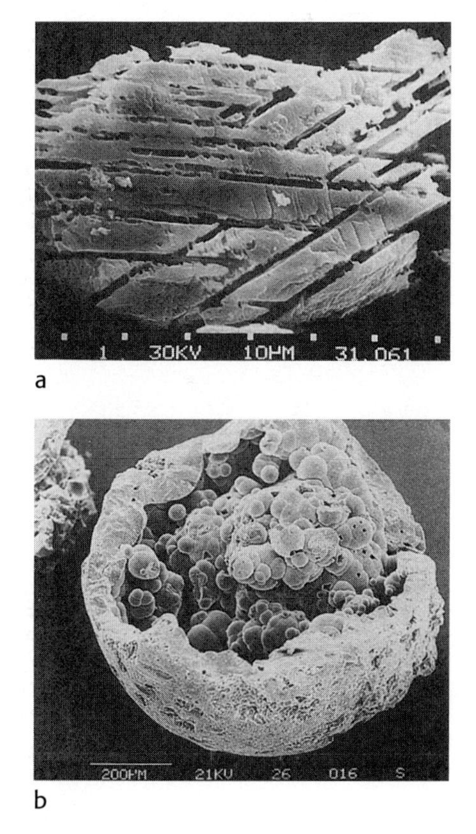

If indeed the spike was formed by a large impact, what other evidence should we hope to find in the rock record? Well-known meteorite impact structures often have fragments of shocked quartz and spherules (tiny glass spheres) associated with them (Figure 16.2). The glass is formed as the target rock is melted in the impact, blasted into the air as a spray of droplets, and almost immediately frozen. Over geological time, the glass spherules may decay to clay. Shocked quartz is formed when quartz crystals undergo a sudden pulse of great pressure, yet do not melt. The shock causes peculiar and unmistakable microstructures (Figure 16.2a).

All over North America, the K–T boundary clay contains glass spherules (Figure 16.2b), and just above the clay is a thinner layer that contains iridium along with fragments of shocked quartz. It is only a few millimeters thick, but in total it contains more than a cubic kilometer of shocked quartz in North America alone. The zone of shocked quartz extends west onto the Pacific Ocean floor, but shocked quartz is rare in K–T boundary rocks elsewhere: only some very tiny fragments occur in Europe. All this evidence implies that the K–T impact occurred on or near North America, with the iridium coming from the vaporized asteroid and the shocked quartz coming from the continental rocks it hit.

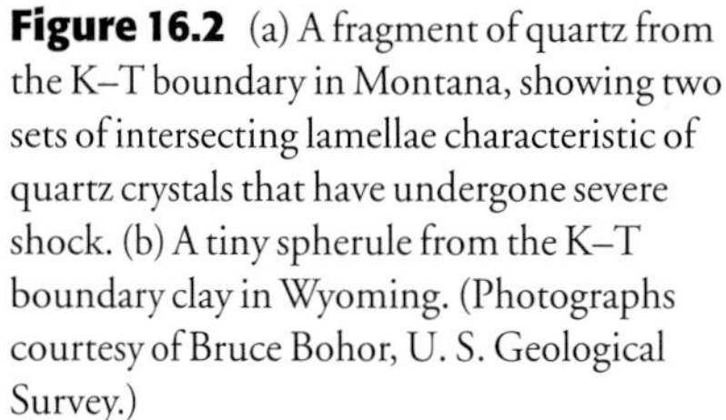

Figure 16.2 (a) A fragment of quartz from the K–T boundary in Montana, showing two sets of intersecting lamellae characteristic of quartz crystals that have undergone severe shock. (b) A tiny spherule from the K–T boundary clay in Wyoming. (Photographs courtesy of Bruce Bohor, U. S. Geological Survey.)

The K–T impact crater has been found. It is a roughly egg-shaped geological structure called Chicxulub, deeply buried under the sediments of the Yucatán peninsula of Mexico (Figure 16.3). The structure is about 180 km across, one of the largest impact structures so far identified with confidence on Earth. A borehole

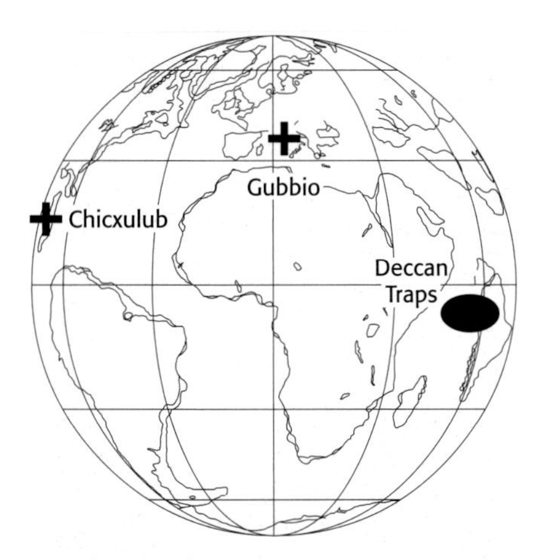

Figure 16.3 Important localities for discussing the K–T boundary, shown in the positions they occupied at the time, 65 Ma. Gubbio is the town in Italy where Walter Alvarez collected the sample that yielded the first evidence of the iridium spike. Note that Chicxulub is 120° away from the Deccan Traps plume, not 180°.

drilled into the Chicxulub structure hit 380 m (more than 1000 ft) of igneous rock with a strange chemistry that could have been generated by melting together a mixture of the sedimentary rocks in the region. The igneous rock contains high levels of iridium, and its age is 65 Ma, exactly coinciding with the K–T boundary.

On top of the igneous rock lies a mass of broken rock, probably the largest surviving debris particles that fell back on to the crater without melting, and on top of that are normal sediments that formed slowly to fill the crater in the shallow tropical seas that covered the impact area.

Well-known impact craters often have tektites associated with them as well as shocked quartz and tiny glass spherules. Tektites are larger glass beads with unusual shapes and surface textures. They are formed when rocks are instantaneously melted and splashed out of impact sites in the form of big gobbets of molten glass, then cooled while spinning through the air.

In Haiti, which was about 800 km from Chicxulub at the end of the Cretaceous, the K–T boundary is marked by a normal but thick (30 cm) clay boundary layer that consists mainly of glass spherules (Figure 16.2). The clay is overlain by a layer of turbidite, submarine landslide material that contains large rock fragments. Some of the fragments look like shattered ocean crust, but there are also spherical pieces of yellow and black glass up to 8 mm across that are unmistakably tektites. The tektites were formed at about 1300°C from two different kinds of rock; and they are dated precisely at 65 Ma. The black tektites formed from continental volcanic rocks and the yellow ones from evaporite sediments with a high content of sulfate and carbonate. The rocks around Chicxulub are formed dominantly of exactly this mixture of rocks, and the igneous rocks under Chicxulub have a chemistry of a once-molten mixture of the two. Above the turbidite comes a thin red clay layer only about 5–10 mm thick that contains iridium and shocked quartz.

One can explain much of this evidence as follows: an asteroid struck at Chicxulub, hitting a pile of thick sediments in a shallow sea. The impact melted much of the local crust and blasted molten material outward from as deep as 14 km under the surface. Small spherules of molten glass were blasted into the air at a shallow angle, and fell out over a giant area that extended northeast as far as Haiti, several hundred kilometers away, and to the northwest as far as Colorado. Next followed the finer material that had been blasted higher into the atmosphere or out into space and fell more slowly on top of the coarser fragments.

The egg-shape of the Chicxulub crater shows that the asteroid hit at a shallow angle, about 20°–30°, splattering more debris to the northwest than in other directions. This accounts in particular for the tremendous damage to the North American continent, and the skewed distribution of shocked quartz far out into the Pacific.

Other sites in the western Caribbean suggest that normally quiet, deep-water sediments were drastically disturbed right at the end of the Cretaceous, and the disturbed sediments have the iridium-bearing layer right on top of them. At many sites from northern Mexico and Texas, and at two sites on the floor of the Gulf of Mexico, there are signs of a great disturbance in the ocean at the K–T boundary. In some places, the disturbed seafloor sediments contain fossils of fresh leaves and wood from land plants, along with tektites dated at 65 Ma (Figure 16.4). Around the Caribbean and at sites up the eastern Atlantic coast of the United States, existing Cretaceous sediments were torn up and settled out again in a messy pile that also contains glass spherules of different chemistries, shocked quartz fragments, and an iridium spike. All this implies that a great tsunami or tidal wave affected the ocean margin of the time, washing fresh land plants well out to sea and tearing up seafloor sediments that had lain undisturbed for millions of years. The resulting bizarre mixture of rocks has been called "the Cretaceous–Tertiary cocktail."

Once Chicxulub was identified, it became possible to calculate some details of the fireball. The larger fragments, solid and molten, were blasted outward at lower angles, but not very far, and were deposited first and locally as rock fragments and tektites (about 15 min travel time to Colorado!). At the same time, smaller fragments, including shocked quartz, were lofted upward, but they fell out quickly and regionally (about 30 min to reach Colorado). The bulk of the mass in the fireball was vaporized and molten debris high above the atmosphere that was deposited last, and globally, slowly drifting downward as frozen droplets. The impact energy, for comparison with hydrogen bomb blasts, was around 100 million megatons.

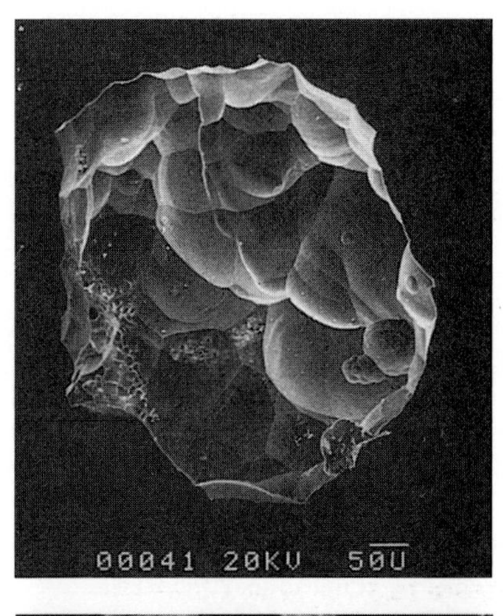

Figure 16.4 Two tektites from the K–T boundary at Mimbral, Mexico, including one referred to affectionately as the "Mimbral yoyo." (Courtesy of Phillippe Claeys of the Vrije Universiteit, Brussels, Belgium.)

A GIANT VOLCANIC ERUPTION?

Exactly at the K–T boundary, a new plume (Chapter 6) was burning its way through the crust close to the plate boundary between India and Africa. Enormous quantities of basalt flooded out over what is now the Deccan Plateau of western India to form huge lava beds called the Deccan Traps. A huge extension of that lava flow on the other side of the plate boundary now lies underwater in the Indian Ocean (Figures 16.3; 16.5). The Deccan Traps cover 500,000 sq. km now (about 200,000 sq. mi), but they may have covered four times as much before erosion removed them from some areas. They have a surviving volume of 1 million cu. km (240,000 cu mi) and are over 2 km thick in places. The entire volcanic volume that erupted, including the underwater lavas, was much larger than this (Figure 16.5).

The date of the Deccan eruptions cannot be separated from the K–T boundary. The peak eruptions may have lasted only about one million years, but that short time straddled the boundary. The rate of eruption was at least 30 times the rate of Hawaiian eruptions today, even assuming it was continuous over as much as a million years; if the eruption was shorter or spasmodic, eruption rates would have been much higher. The Deccan Traps probably erupted as lava flows and fountains like those of Kilauea, rather than in giant explosive eruptions like that of Krakatau. The Deccan plume is still active; its hot spot now lies under the volcanic island of Réunion in the Indian Ocean.

Thus the K–T boundary coincided with two very dramatic events. The Deccan Traps lie across the K–T boundary and were formed in what was obviously a major event in Earth history. The asteroid impact was exactly at the K–T boundary. Certainly something dramatic happened to life on Earth, because geologists have defined the K–T boundary and the end of the Mesozoic Era on the basis of a large extinction of creatures on land and in the sea. An asteroid impact, or a series of gigantic eruptions, or both, would have had major global effects on atmosphere and weather.

DID A CATASTROPHE CAUSE THE EXTINCTIONS?

Almost all the scientists directly involved in trying to explain the K–T extinctions are emotionally committed to one catastrophic hypothesis or the other, or are emotionally against both. This has resulted in claims that seem to overinterpret the evidence. One must be prepared to make one's own decision, and certainly all claims must be subject to close scrutiny.

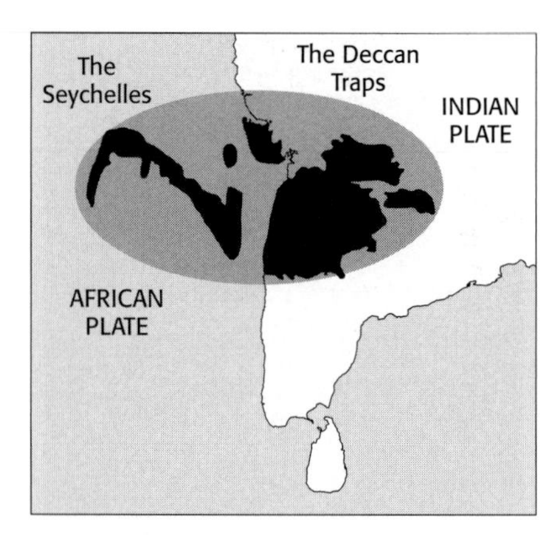

Figure 16.5 The Deccan Traps of India erupted as the Indian and African plates tore apart about 65 Ma. An enormous volume of lava was emplaced above a mantle hot spot (stippled), not only on land on the Indian plate, but offshore to form an enormous underwater mass on the African plate around what is now the Seychelles Islands and part of the Indian Ocean floor. The black areas mark surviving volumes of lava, shown as they were formed in the geography of the K–T boundary. Much more lava must have been erupted than has survived.

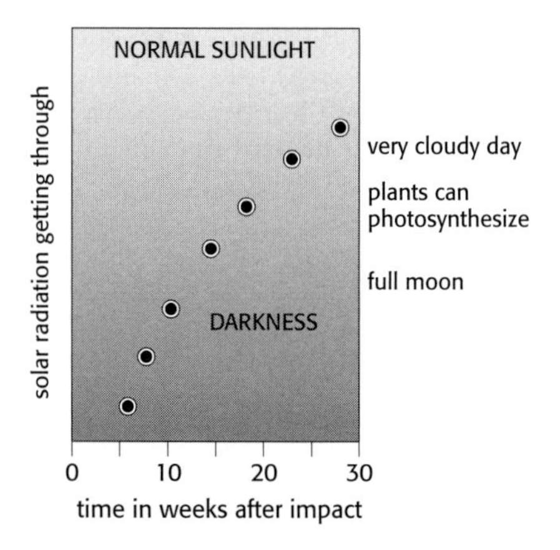

Figure 16.6 An early version of the "impact winter" scenario. The dots show the predictions of a computer program that calculates light levels at Earth's surface as dust in the atmosphere blocks the sun's rays after an impact. This model is more drastic than some. In this one, a lot of dust falls out slowly. Phytoplankton in the ocean or plans on land cannot photosynthesize under any circumstances for 20 weeks (five months) after the impact. (From data in *Geological Society of America Special Paper* 190).

In addition to the effects of a plume eruption and an asteroid impact described in Chapter 6, the impact at Chicxulub may have been particularly deadly because the target rocks contain high quantities of sulfur. The impact would have produced enormous amounts of sulfate aerosols in the atmosphere that acted as nucleation sites for acid rains much more intense and devastating than anything we have generated from industrial pollution. One model suggests rain with the strength of battery acid! The direct effect (in some versions of the impact) is enough to suffocate some air breathers, to destroy plant foliage, and to dissolve the shells of marine creatures living along shores and in the surface waters of the ocean. The balance of CO_2 between air and ocean is upset, and a chain of climatic events makes ocean surface waters barren for perhaps 20 years.

What do we do with these catastrophic scenarios? Naturally, we compare them with the evidence from the geological record. Birds, tortoises, and mammals live on land and breathe air: the evidence from the K–T boundary shows that they survived the K–T boundary event. Therefore they and the air they breathed weren't set on broil for several hours (Chapter 6); they were not inhaling battery acid either. To put it simply, these extreme scenarios did not happen.

The most persuasive scenarios of catastrophic extinction are quickly summarized. Regionally, there is little doubt that the North American continent would have been absolutely devastated. Globally, even a short-lived catastrophe among land plants and surface plankton at sea would drastically affect normal food chains. Pterosaurs, dinosaurs, and large marine reptiles would have been vulnerable to food shortage, and their extinction after a catastrophe seems plausible. Lizards and primitive mammals, which survived, are small and often burrow and hibernate; they would have found plenty of nuts, seeds, insect larvae, and invertebrates buried or lying around in the dark. In the oceans, invertebrates living in shallow water would have suffered greatly from cold or frost, or perhaps from CO_2-induced heating. But deeper-water forms are insulated from heat or cold shock and have low metabolic rates; they therefore would be able to survive even months of starvation. High-latitude faunas in particular were already adapted to winter darkness, though perhaps not to extreme cold. Thus, tropical reef communities could have been decimated, but deep-water and high-latitude communities could have survived much better. All these patterns are observed at the K–T boundary.

The problem with catastrophic hypotheses for the K–T extinctions is that the catastrophes must have been severe but not too severe, because so many creatures survived. Dust and soot must have fallen quickly (within a year) to satisfy some scenarios, but had to remain suspended longer in the atmosphere to produce other effects (Figure 16.6). Close examination of soot from the K-T boundary shows fungal decay in the original plant tissue. That means that the fires that produced the soot may have been ordinary in size, or they may have been gigantic, but they did certainly not form in wildfires immediately after the impact as the computer models would suggest: they formed at least weeks and probably months after the plant or plants died.

We don't yet know whether the K–T impact and eruption would have had catastrophic, severe, or only mild biological and ecological effects, or whether those effects would be local, regional, or global. In each scenario, however, the killing agent is transient: it would have operated for only a short time geologically.

PALEONTOLOGICAL EVIDENCE FROM THE K–T BOUNDARY

The paleontological evidence from the K–T boundary is ambiguous. While many phenomena are well explained by an impact or a volcanic hypothesis, others are not. The fossils do provide us with real evidence about the K–T extinction events, instead of inferences from analogy or from computer models.

The best-studied terrestrial sections across the K–T boundary are in North America. Immediately this is a problem, because we know that the effects of the asteroid impact were greater here than in most parts of the world. Perhaps this has given us a more catastrophic view of the boundary event that we would gather from, say, comparable careful research in New Zealand. Even so, it is obvious that life, even in North America, was not wiped out: many plants and animals survived the K–T event.

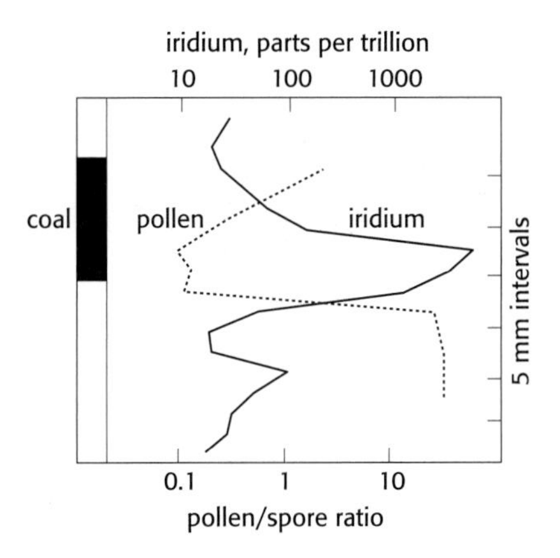

Figure 16.7 In many K–T boundary sites the iridium spike coincides with a spore spike. For some unknown period of time, hardly any angiosperm pollen was deposited, though fern spores were deposited in abundance. These data are from the same locality as those of Figure 16.1.

Land Plants

North American land plants were devastated from Alberta to New Mexico at the K–T boundary. The sediments below the boundary are dominated by angiosperm pollen, but the boundary itself has little or no angiosperm pollen and instead is dominated by fern spores in a spore spike analogous to the iridium spike (Figure 16.7). Normal pollen counts occur immediately after the boundary layer. The spore spike therefore coincides precisely with the iridium spike in time and is equally intense and short-lived. A fern spore spike also occurs in New Zealand, suggesting that the crisis was global in extent.

The spore spike could be explained by a short but severe crisis for land plants, generated by an impact or an eruption, in which all adult leaves died off (some mix of lack of light, or prolonged frost, or acid rain). Perhaps ferns were the first plants to recolonize the debris, and higher plants returned later. This happened after the eruption of Krakatau in 1883. Ferns quickly grew on the devastated island surfaces, presumably from windblown spores, but they in turn were replaced within a few decades by flowering plants as a full flora was reestablished.

Evidence from leaves confirms the data from spores and pollen. Land plants recovered from the crisis, but many Late Cretaceous plant species were killed off. The survivors probably remained safe during the crisis as seeds and spores in the soil, or even as roots and rhizomes.

Angiosperms were in the middle of a great expansion in the Late Cretaceous (Chapter 14), and the expansion continued into the Paleocene and Eocene. Climate (and plant diversity) were fluctuating before the K–T boundary, but not to the extent that the plant crisis at the boundary can be blamed on climate. Yet there were important and abrupt changes in North American floras at the K–T boundary. In the Late Cretaceous, for example, an evergreen woodland grew from Montana to New Mexico in a seasonally dry, subtropical climate. At the boundary the dominantly evergreen Late Cretaceous woodland changed to a largely deciduous Early Cenozoic swamp woodland growing in a wetter climate. The fern spike represents a period of swampy mire at the boundary itself. Deciduous trees survived the K–T boundary events much better than evergreens did; in particular, species that had been more northerly spread southward.

These changes could be explained in two different catastrophic scenarios: a catastrophe that wiped out all vegetation locally, with recolonization from survivors; or a catastrophe that selectively destroyed evergreen plants.

Freshwater Communities

Some ecological anomalies at the K–T boundary are not easily explained by a catastrophic scenario. Freshwater communities were less affected than terrestrial ones. For example, turtles and other aquatic reptiles survived in North Dakota while dinosaurs were totally wiped out. Freshwater communities are fueled largely by stream detritus, which includes the nutrients running off from land vegetation. It has been suggested that animals in food chains that begin with detritus rather than with primary productivity would survive a catastrophe better than others. That may be true generally and seems to be true for freshwater communities at the K–T boundary, but such communities would survive any ecological crises better, catastrophic or not.

Environmental Sex Determination

Most catastrophic scenarios are so severe that it's difficult to see how some groups of animals survived. Many living reptiles have environmental sex determination (ESD). The sex of an individual with ESD is not determined genetically, but by the environmental temperatures experienced by the embryo during a critical stage in development. Often, but not universally, in warmer temperatures, the sex that is larger as an adult develops. This pattern probably evolved because, other things being equal, warmer temperatures promote faster growth and therefore larger final size (at least for ectotherms). Female turtles are larger than males because they carry huge numbers of large eggs, so baby turtles tend to hatch out as females if the eggs develop in warm places and as males in cooler places. (This makes turtle farming difficult.) Crocodiles and lizards are just the reverse. Males are larger than females because there is strong competition between males, so eggs laid in warmer places tend to hatch out as males. ESD is not found in warmblooded, egg-laying vertebrates (birds and monotreme mammals), and it didn't occur in dinosaurs if they too were warmblooded.

ESD is found in such a wide variety of ectothermic reptiles today that it probably occurred also in their ancestors. If so, a very sudden change in global temperature should have caused a catastrophe among ectothermic reptiles at the K–T boundary. But it did not. Crocodilians and turtles were hardly affected at all by the K–T boundary events, and lizards were affected only mildly.

High-Latitude Dinosaurs

Late Cretaceous dinosaurs lived in very high latitudes north and south, in Alaska and in south Australia and Antarctica. These dinosaurs would have been well adapted to strong seasonal variation, including periods of darkness and very cool temperatures. An impact scenario would not easily account for the extinction of such animals at *both* poles, no matter what time of year the asteroid hit.

Birds

The survival of birds is the strangest of all the K–T boundary events, if we are to accept the catastrophic scenarios. Smaller dinosaurs overlapped with larger birds in size and in ecological roles as terrestrial bipeds. How did birds survive while dinosaurs did not? Birds seek food in the open, by sight; they are small and warmblooded, with high metabolic rates and small energy stores. Even a sudden storm or a slightly severe winter can cause high mortality among bird populations. Yet an impact scenario, according to its enthusiasts, includes "a nightmare of environmental disasters, including storms, tsunamis, cold and darkness, greenhouse warming, acid rains and global fires." There must be some explanation for the survival of birds, turtles, and crocodiles through any catastrophe of this scale, or else the catastrophe models are wrong.

Where Are We?

It is clear that at least the extreme models are wrong. It's not clear that impact hypotheses or volcanic hypotheses can explain satisfactorily the extinction patterns we see in the fossil record. There are nagging fears that we are overstating the effects of the impact because the results are so clear in North America.

An impact or a gigantic eruption that might otherwise have caused only a regional extinction might have caused the global K–T extinction by inducing longer-term climatic changes. These changes would be best recorded in ocean sediments and marine fossils. Tropical reef communities were drastically affected in the K–T extinctions, as were microplankton in the surface waters of the ocean. The pattern of marine K–T extinctions is consistent with a massive breakdown in normal marine ecology.

Oxygen isotope measurements across the K–T boundary suggest that oceanic temperatures fluctuated markedly in Late Cretaceous times and through the boundary events. Furthermore, carbon isotope measurements across the K–T boundary suggest that there were severe, rapid, and repeated fluctuations in oceanic productivity in the 3 m.y. before the final extinction, and that productivity and ocean circulation were suppressed for at least several tens of thousands of years just after the boundary, and perhaps for 1 or 2 m.y. afterward. These changes could have devastated terrestrial ecosystems as well as marine ones. Steve D'Hondt has suggested that climatic change is the connection between the impact and the extinction: the impact upset normal climate, with long-term effects that lasted much longer than the immediate and direct consequences of the impact (see D'Hondt, 1996, 1998).

There were survivors: hardly any major groups of organisms became entirely extinct. Even the dinosaurs survived in one sense (as birds). In particular, planktonic diatoms survived well, possibly because they have resting stages as part of their life cycle. They recovered as quickly as the land plants emerged from spores, seeds, roots, and rhizomes. The sudden interruption of the food chains on land and in the sea may well have been quite short, even if full recovery of the climate and full marine ecosystems took much longer. On the one hand, climatic modelers and paleobotanists have concluded that land plants recovered to full production in perhaps ten years; yet D'Hondt and his colleagues suspect that normal surface productivity took a few thousand years to reestablish in the oceans after a few thousand years. On the other hand, it took about three million years for the full marine ecosystem to recover, probably because so many marine predators (crustaceans, molluscs, fishes, and marine reptiles) disappeared, and had to be replaced by evolution among surviving relatives.

We still do not have an explanation for the demise of the victims of the K–T extinction, while so many other groups survived. We do not know whether it was the impact alone, or the combination of the impact and the plume volcanism, that caused the extinction, and we do not know the linkages between the physical events and the biological and ecological effects. It would be astonishing if the impact played no role, and it would be astonishing if the volcanism played no role.

The unusual severity of the K–T extinction, its global scope, and the sudden and dramatic biological features such as the fern-spore spike may have happened because an asteroid impact and a gigantic eruption occurred when global ecosystems were particularly vulnerable to a disturbance of oceanic stability. We will probably gain a better perspective on the K–T boundary as we gather more information about the P–T extinctions. Perhaps mass extinctions also require a tectonic or geographic setting that makes the global ecosystem vulnerable.

Further Reading

Alvarez, L. W., et al. 1980. Extraterrestrial cause for the Cretaceous–Tertiary extinction. *Science* 208: 1095–108. [The paper that started it all.]

Alvarez, W., et al. 1995. Emplacement of Cretaceous–Tertiary boundary shocked quartz from Chicxulub crater. *Science* 269: 930–5. [Why the shocked quartz overlies the rest of the impact layer.]

Alvarez, W. 1997. *T. rex and the Crater of Doom.* Princeton: Princeton University Press. [The best of many books on the extinction.]

Bourgeois, J., et al. 1988. A tsunami deposit at the Cretaceous–Tertiary boundary in Texas. *Science* 241: 567–70.

Bralower, T. J., et al. 1998. The Cretaceous–Tertiary boundary cocktail: Chicxulub impact triggers margin collapse and extensive sediment gravity flows. *Geology* 26: 331–4.

D'Hondt, S., et al. 1996. Oscillatory marine response to the Cretaceous–Tertiary impact. *Geology* 24: 611–14.

D'Hondt, S., et al. 1998. Organic carbon fluxes and ecological recovery from the Cretaceous–Tertiary mass extinction. *Science* 282: 276–9.

Emanuel, K. A., et al. 1995. Hypercanes: a possible link in global extinction scenarios. *Journal of Geophysical Research* 100: 13755–65.

Head, G., et al. 1987. Environmental determination of sex in the reptiles. *Nature* 329: 198–9.

Heinberg, C. 1999. Lower Danian bivalves, Stevns Klint, Denmark: continuity across the K–T boundary. *Palaeogeography, Palaeoclimatology, Palaeoecology* 154: 87–106. [No huge mass extinction among seafloor organisms in Denmark.]

Hildebrand, A. R., et al. 1995. Size and structure of the Chicxulub crater revealed by horizontal gravity gradients and cenotes. *Nature* 376: 415–17, and comment 386–7.

Hofmann, C., et al. 2000. $^{40}Ar/^{39}Ar$ dating of mineral separates and whole rocks from the Western Ghats lava pile: further constraints on duration and age of the Deccan traps. *Earth and Planetary Science Letters* 180: 13–27. [Confirms that most of the Deccan Traps were erupted within 1 m.y., at a time that cannot be distinguished from the K–T boundary.]

Johnson, K. R., and B. Ellis. 2002. A tropical rainforest in Colorado 1.4 million years after the Cretaceous–Tertiary boundary. *Science* 296: 2379–83.

Jones, T. P., and B. Lim. 2002. Extraterrestrial impacts and wildfires. *Palaeogeography, Palaeoclimatology, Palaeoecology* 164, 57–66.

Koeberl, C., et al. 1996. Impact origin of the Chesapeake Bay structure and the source of the North American tektites. *Science* 271: 1263–6. [An 90-km crater from 35 Ma under Chesapeake Bay.]

Kring, D. A. 2000. Impact events and their effect on the origin, evolution, and distribution of life. *GSA Today*: August 2000. [Snappy review of impact effects, concentrating on the K–T event. Good list of references.]

Li, L., and G. Keller. 1998. Abrupt deep-sea warming at the end of the Cretaceous. *Geology* 26: 995–8.

MacDougall, J. D. 1988. Seawater strontium isotopes, acid rain, and the Cretaceous–Tertiary boundary. *Science* 239: 485–7.

Max, M. D., et al. 1999. Sea-floor methane blow-out and global firestorm at the K–T boundary. *Geo-Marine Letters* 18: 285–91. [Competes for bad paper of the decade.]

Melosh, H. J., et al. 1990. Ignition of global wildfires at the Cretaceous/Tertiary boundary. *Nature* 343: 251–4. [Microwave summer.]

Morgan, J., et al. 1997. Size and morphology of the Chicxulub impact crater. *Nature* 390: 472–6.

O'Keefe, J. A., and T. J. Ahrens. 1989. Impact production of CO_2 by the Cretaceous/Tertiary impact bolide and the resultant heating of the Earth. *Nature* 338: 247–8.

Orth, C. J., et al. 1981. An iridium anomaly at the palynological Cretaceous–Tertiary boundary in northern New Mexico. *Science* 214: 1341–3.

Pardo, A., et al. 1999. Paleoenvironmental changes across the Cretaceous–Tertiary boundary at Koshak, Kazakhstan, based on planktic foraminifera and clay mineralogy. *Palaeogeography, Palaeoclimatology, Palaeoecology* 154: 247–73. [No huge mass extinction among marine organisms in Kazakhstan.]

Pope, K. O., et al. 1997. Energy, volatile production, and climatic effects of the Chicxulub Cretaceous–Tertiary impact. *Journal of Geophysical Research* 102: 21645–64.

Pope, K. O. 2002. Impact dust not the cause of the Cretaceous–Tertiary mass extinction. *Geology* 30: 99–102.

Rampino, M. R., and T. Volk. 1988. Mass extinctions, atmospheric sulphur and climatic warming at the K/T boundary. *Nature* 332: 63–5.

Rampino, M. R., et al. 1988. Volcanic winters. *Annual Reviews of Earth and Planetary Science* 16: 73–99.

Ryder, G., et al. (eds) 1996. *The Cretaceous–Tertiary Event and Other Catastrophes in Earth History. Geological Society of America Special Paper* 307. [A major international conference on the K–T event. Two previous conferences were published in volumes 190 and 247.]

Schuraytz, B. C., et al. 1996. Iridium metal in Chicxulub impact melt: forensic chemistry on the K–T smoking gun. *Science* 271: 1573–6.

Sharpton, V. L., et al. 1992. New links between the Chicxulub impact structure and the Cretaceous–Tertiary boundary. *Nature* 359: 819–21.

Schultz, P. H., and S. D'Hondt. 1996. Cretaceous–Tertiary (Chicxulub) impact angle and its consequences. *Geology* 24: 963–7.

Sheehan, P. M., and T. A. Hansen. 1986. Detritus feeding as a buffer to extinction at the end of the Cretaceous. *Geology* 14: 868–70.

Sigurdsson, H., et al. 1992. The impact of the Cretaceous/Tertiary bolide on evaporite terrane and generation of major sulfuric acid aerosol. *Earth and Planetary Science Letters* 109: 543–59.

Smit, J. 1999. The global stratigraphy of the Cretaceous–Tertiary boundary impact ejecta. *Annual Reviews of Earth & Planetary Sciences* 27: 75–113.

Stothers, R. B. 1984. The great Tambora eruption of 1815 and its aftermath. *Science* 234: 1191–8.

Turco, R. P. et al. 1990. Climate and smoke: an appraisal of nuclear winter. *Science* 247: 166–76. [Revised version of their original 1983 suggestion.]

Vajda, V., et al. 2001. Indication of global deforestation at the Cretaceous–Tertiary boundary by New Zealand fern spike. *Science* 294: 1700–2, and comment 1668–9. [Confirms the global extent of the catastrophe to plant life.]

White, R. S., and D. P. McKenzie. 1989. Volcanism at rifts. *Scientific American* 261(1): 62–71. [Why the Deccan Traps are so enormous.]

Wilf, P., et al. 2003. Correlated terrestrial and marine evidence for global climate changes before mass extinction at the Cretaceous–Paleogene boundary. *Proceedings of the National Academy of Sciences* 100: 599–604. [But not enough to *explain* the extinctions!]

Wolbach, W. S., et al. 1990. Fires at the K/T boundary: carbon at the Sumbar, Turkmenia, site. *Geochimica et Cosmochimica Acta* 54: 1133–46. [The *assumptions* in their previous papers have magically become *facts*!]

Wolfe, J. A., and G. R. Upchurch. 1986. Vegetation, climatic, and floral changes at the Cretaceous–Tertiary boundary. *Nature* 324: 148–52.

Wolfe, J. A. 1987. Late Cretaceous–Cenozoic history of deciduousness and the terminal Cretaceous event. *Paleobiology* 13: 215–26.

CHAPTER SEVENTEEN

Cenozoic Mammals: Origins, Guilds, and Trends

The end of the Cretaceous Period was marked by so many changes in life on the land, in the sea, and in the air that it also marks the end of the Mesozoic Era and the beginning of the Cenozoic Era (in which we live). Survivors of the Cretaceous extinctions built up into a very impressive and varied set of organisms, beginning in the Paleocene Epoch, the first 10 m.y. of the Cenozoic. In the marine fossil record, the Cenozoic Era is dominated by molluscs, especially by bivalves and gastropods, the clams and snails of beach shell collections.

On land, the Cenozoic is marked by the dominance of flowering plants, insects, and birds, and in particular by the radiation of the mammals from insignificant little insectivores into dominant large animals in almost all terrestrial ecosystems. Cenozoic mammals have a very good fossil record. There are thousands of well-preserved skeletons, and we understand their evolutionary history very well. I shall not try to give anything close to an overall survey of mammalian evolution. Instead, I shall use the mammal record to illustrate the ways in which evolution has acted on animals, because the same effects can be seen (more dimly) throughout the rest of the fossil record.

Evolution is the overall result of environmental factors acting on organisms through natural selection. But it is easier to understand evolutionary processes if we can isolate some of the different aspects involved. In this chapter and the next, I shall describe how successive groups of mammals evolved to replace dinosaurs, and discuss some of the major evolutionary events of the Cenozoic era. I shall look at four major aspects of evolution as I survey Cenozoic mammals, and in each case try to identify the various opportunities that allowed or encouraged evolutionary change to occur:

- the ecological setting of evolution;
- improving or changing well-defined adaptations;
- geographical influences on evolution;
- climatic influences on evolution.

Much of the turnover in the fossil record consists of the ecological replacement of one group of animals by another. An older group may disappear, for various reasons, offering an ecological opportunity for a new set of species that evolves and replaces the older set. Sometimes the new group outcompetes the older group, so that we see not just ecological replacement but ecological displacement. There are many examples of parallel evolution, in which certain body patterns that are apparently well suited for a particular way of life evolve again and again in different continents at different times. Understanding these processes helps us to sort through the complexity of catalogs of fossils.

Then we look at a smaller-scale phenomenon, evolution by improvement. Given that a particular body plan is well suited for executing a particular way of life, we often see changing morphology through time within a single group of organisms. These evolutionary changes can often be interpreted as a series of increasingly good adaptations to the characteristic way of life, or as a set of alternative adaptations within the general way of life. Coevolution, as in the arms race between predator and prey, in the relationship between plant and herbivore, or between plant and pollinator, can lead to increasingly efficient adaptation. In a successful, long-lived group that survives, one can trace the various adaptations that eventually led to the derived characters of the survivors.

Obviously, one must first have a good idea of the evolutionary relationships within the group (a reliable phylogram, in other words). In almost all cases, the evolutionary and adaptive pattern of a group is not a straight line but a winding pathway through time. But the attempt to trace a lineage through the complexity of evolution can be instructive, showing how the adaptations correspond to environmental opportunities.

THE EVOLUTION OF CENOZOIC MAMMALS

The surviving major groups of terrestrial creatures after the Cretaceous extinction were mammals and birds. Crocodilians were amphibious rather than terrestrial. Most Mesozoic mammals had been small insectivores, probably nocturnal, many of them tree-dwellers or burrowers, and usually with limbs adapted for agile scurrying rather than fast running. Flying birds must be small, but there is not the same constraint on terrestrial birds. There was probably intense competition between ground-dwelling birds and mammals in a kind of ecological race for large-bodied ways of life during the Paleocene, with crocodilians playing an important secondary role in some areas. The mammals evolved explosively, their diversity rising from 8 to 70 families.

The Radiation of Mammals: Molecular Studies

The fossil record suggests that there was "explosive radiation" among mammals in the early Cenozoic. This receives a ready explanation that I used in Chapter 15: the dinosaurs had been dominant in terrestrial ecosystems, worldwide, for over 100 million years, and had effectively suppressed any ecological radiation of Mesozoic mammals. With the disappearance of the (nonbird) dinosaurs, new ecological roles suddenly became available for mammals (and birds), and dramatic adaptive radiation was a predictable response to that ecological opportunity. We see very few mammals in Cretaceous rocks, and they are all small.

But is that explosive radiation an evolutionary (genetic) explosion, or is it an ecological explosion? Perhaps the different groups of mammals had already diverged genetically, at small body size, long before the end of the Cretaceous, but were ecologically released after the K–T extinction. (Note that this question is exactly the same as the one we asked in Chapter 4 about the radiation of the metazoans in the Late Pre-Cambrian relative to the Cambrian explosion.)

How would we detect and describe a Cretaceous radiation of major mammal lineages? We could look more carefully at Cretaceous mammals, to try to find advanced characters among them. But the record is so poor that this approach has been very difficult. In any case, if the ancestors of, say, horses were mouse-sized, they would not look like, eat like, run like, or behave like horses, so they would also lack most of the skeletal characters that we use to recognize horses. Genetics is not ecology. An alternative approach is to look at molecular evidence.

Molecular geneticists proposed that under certain circumstances, evolutionary changes in DNA and proteins might be selectively neutral, unaffected by natural selection. Such molecular changes should happen at a random rate that is fairly constant through time. In theory, such **molecular clocks** of evolutionary change, based on proteins, or specific genes, or DNA sequences from the nucleus or from mitochondria, could allow us to determine the times of divergence of living animals without ever having to look at or for their fossil ancestors.

The concept that there should be molecular clocks often disagreed with the facts of the fossil record, though it allowed geneticists to publish many papers quickly that essentially added nothing to our understanding but a lot to our confusion. Most geneticists now accept that molecular clocks do not run reliably, and have found ways to analyze their data that throws away that assumption. Their results have become much more compatible with the real fossil record, and finally genetics has come into its own as a wonderful complement to fossil morphology as we try to work out the history of groups of organisms.

So let us return to the question of the timing of the mammalian radiation. The Paleocene radiation of mammals that we see in the fossil record apparently occurred so fast that we cannot distinguish the very great number of branches that resulted in the great diversity of the major groups of mammals alive today. Meanwhile, molecular geneticists are painting a vivid picture of early branching events among Cretaceous placental and marsupial mammals. Are these two viewpoints compatible? The answer is "yes."

The Radiation of Mammals: Molecular Results

Stripped of "clock" assumptions, many molecular results are reasonable (that is, they agree with fossil evidence!), which gives confidence in the methods. Thus, marsupials and monotremes always fall outside the groups that form the placental mammals. However, some results within placentals came as a surprise. Some of these are very exciting, because they give insights into the mammalian radiation that had not been discovered by standard morphological comparison.

We saw in Chapter 15 that classical methods in paleontology defined a southern (Gondwana) origin of monotremes from stem mammals (mammaliaforms), and a northern (Laurasian) origin of the therians (marsupials and placentals). But molecular methods established another set of landmarks in mammal history. Early in the Cretaceous, marsupials and placentals arrived in Gondwana as it was breaking up, and founded lineages, one in Africa and two in South America, that evolved in those regions separately from mammalian evolution elsewhere.

Marsupials have been established in South America for a long time, and they almost certainly reached Australia across Antarctica before the South Pole froze up.

More important, a large group of African placentals, the Afrotheria, forms a clade separate from other placentals. The clade today includes elephants, sea-cows, hyraxes, aardvarks, elephant shrews, and golden moles: a tremendous array of different body plans, sizes, and ecologies. The reality of the clade is unanimously supported by all molecular evidence, yet we had not (and probably never would have) discovered this clade by analyzing fossil skeletons. The reality of the Afrotheria implies that Africa became isolated in Cretaceous times with a set of early placentals that evolved to fill all these ecological roles, separate from evolution on other continents, in an astounding case of an adaptive radiation. I shall return to this story in Chapter 20.

Second, a South American clade of placental mammals that has long been recognized turns out to have very deep roots. The Xenarthra, which includes living sloths and armadillos (and many extinct mammals) also shows prominently in molecular analyses.

The rest of the placentals are left as a northern clade, which, on molecular evidence, split into two groups, one the ancestors of ungulates, carnivores, and bats, the other the ancestors of primates and rodents.

When did these branches take place? Using assumptions that do not include a "clock," the molecular results imply that the Afrotheria became separate at perhaps 105 Ma (Middle Cretaceous) and the Xenarthra perhaps 95 Ma. These estimates coincide roughly with the major break-up of Gondwana to form the southern continents, and since the molecular evidence and the geological evidence are completely independent, this again adds credibility to the analyses.

The northern placentals split into their major clades during the Later Cretaceous, so that perhaps 20 or so separate lineages survived the K–T extinction to radiate in the Paleocene. While many paleontologists are happy with the *pattern* of these molecular results, they are dubious that the inferences on the *timing* of the branches are reliable.

The discussion may have a simple resolution. Molecular results measure gene changes. They do not, and cannot, measure ecological changes. So most likely there *were* deep genetic branches in the Cretaceous, but they only produced sets of Cretaceous mammals that were ecologically restricted, so would have been anatomically restricted too. There is no rule that Cretaceous mammals should look like, or have ecological roles like, their eventual descendants of today, or even of Cenozoic times. Simply stated, the major groups of placental mammals evolved (diverged from one another) in the latter half of the Cretaceous. Their radiation into many families and genera may have begun in the Late Cretaceous but was dominantly a Paleocene and Eocene process, taking place as an *ecological* radiation. Nevertheless, there are Cretaceous mammals still to be discovered, interpreted, and placed into the framework; and since world geography was changing in the Cretaceous as the continents moved about, there will be opportunities to test ideas about evolutionary divergence against geological evidence.

By the end of the Cretaceous there were mammals with varied sets of genes but muted variation in morphology. The principle is clear: now the scientists involved should tone down the rhetoric and try to work out what actually happened!

Now let us return to the fossil record, which, remember, contains many *extinct* clades that played major roles in evolution and ecology, but cannot be assessed by molecular analyses.

The Paleocene

By Paleocene times, placental mammals included recognizable ancestors of a great many living groups, including marsupials, shrews, rabbits, modern carnivores, elephants, primates, whales, and hedgehogs. The ancestors of the peculiar

Figure 17.1 *Chriacus*, a Paleocene arctocyonid condylarth that looks very much like the living coati. (Restoration by E. Kasmer, under the supervision of K. D. Rose. Courtesy of Kenneth D. Rose of The Johns Hopkins University.)

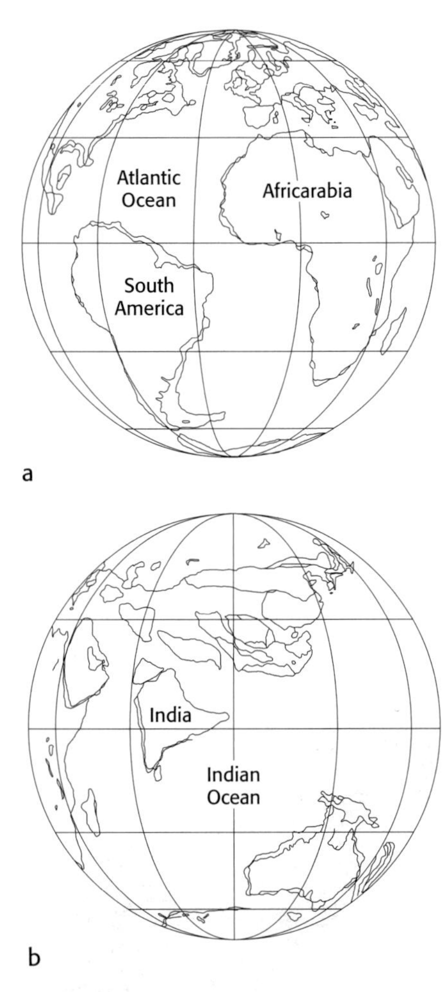

Figure 17.2 Eocene world geography at about 50 Ma, showing the Atlantic Ocean (a) and Indian Ocean (b). The northern land masses were close together, and land animals could walk freely from one to the other through the comparatively mild climates of the Eocene Arctic. India, Australasia, Africarabia, and South America could all be considered as island continents, because Australasia, South America, and Antarctica were linked only by a difficult passage through polar regions.

South American fauna were already isolated geographically, and can be recognized there.

Among all this diversity, the dominant group of Paleocene mammals was a set of generalized, rapidly evolving early "ungulates," most of them were herbivores of various sizes. But the **arctocyonids** had low, long skulls with canines and primitive molars, and were probably raccoon-like omnivores. *Chriacus* (Figure 17.1) had much the same size and body plan as the tree-climbing coati, but *Arctocyon* itself was the size of a bear and probably had much the same omnivorous ecology. **Mesonychids** were probably otter-like carnivores or scavengers, but some of them were good running predators on land. For those interested in the largest of anything, the mesonychid *Andrewsarchus* from the Eocene of Mongolia was the largest terrestrial carnivore/scavenger among mammals, with a skull nearly 1 m (3 ft) long.

Paleocene mammals are generally primitive in their structure, but after a drastic turnover at the end of the epoch, many new groups appeared in the Eocene that survive to the present.

The Eocene

The turnover at the end of the Paleocene is partly related to a chance event: climatic change briefly allowed free migration of mammals across the northern continents of Eurasia and North America. Roughly the same fossil faunas are found across the Northern Hemisphere in North America and Eurasia. In contrast, South America, Africa + Arabia, India, and Australasia were island continents to the south of this great northern land area (Figure 17.2), and their faunal evolution is discussed separately in Chapter 18.

Many modern groups of mammals appeared very early in the Eocene, including rodents, advanced primates, and modern artiodactyls and perissodactyls. There are some disputes about molecular evidence linked with this radiation. For example, molecular evidence suggests that whales form a clade with artiodactyls (antelope, cattle, deer, pigs, and so on), while carnivores are linked with perissodactyls (horses, rhinos, tapirs, and so on). In contrast, zoologists and paleontologists have always linked artiodactyls and perissodactyls within a large group of herbivores, the ungulates. (If the molecular evidence holds, then "ungulates" are perhaps an ecological group, but not an evolutionary one.)

Some of the problem may lie in the fact that living mammals are merely the survivors of a massive radiation that included extinct mammals that have left us fossil skeletons but which cannot be sampled for molecular studies. These results need a lot of further analysis and debate.

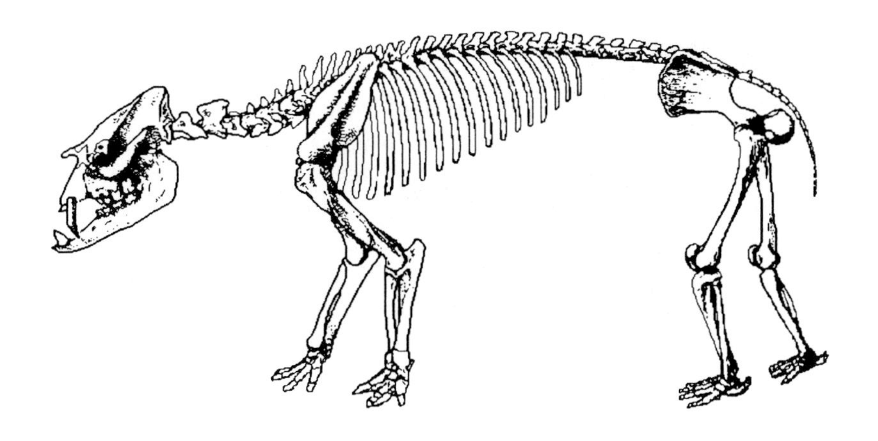

Figure 17.3 Many groups of mammalian vegetarians have evolved to about 5 tons body weight at different times and on different continents. This is *Astrapotherium* from the Early Cenozoic of South America. (After Riggs.)

By the end of the Early Eocene, digging, running, climbing, leaping, and flying mammals were well established at all available body sizes. Above all, Eocene faunas record the evolution of many different groups of mammals into herbivores of all sizes. Many of these early herbivores were small or medium-sized, including the earliest known horse, *Hyracotherium*, but soon there were large-bodied herbivores that ranged up to 5 tons. In North America the large herbivores were uintatheres, followed by titanotheres; in South America they were astrapotheres (Figure 17.3); and in the Old World, especially in Africa, they were arsinoitheres.

Perissodactyls and artiodactyls appeared abruptly at the base of the Eocene in North America. Perissodactyls may have invaded (from Asia?), but it is likely that artiodactyls evolved in North America from an arctocyonid such as *Chriacus* (Figure 17.1). Small at first, both groups evolved long, slim, stiff legs and other adaptations for fast running.

Proboscideans (elephants and related groups) and sea cows, which belong to the Afrotheria, evolved along the African shores of the tropical ocean that spread east–west between Africa and Eurasia (Figure 17.2). Many other herbivores evolved in isolation in South America. Whales were evolving from land mammals, possibly along the southern coasts of Eurasia. *Ambulocetus* was a swimming creature that can confidently be identified as a primitive cetacean, and probably had an ecology like a large sea lion. It swam in otter-like fashion, with a flexible spine, but it still had fairly effective limbs for moving about on shore, and had not yet evolved the tail flukes of later whales.

Mammals did not evolve quickly into large carnivores. Some early carnivorous mammals, the mesonychids, arctocyonids, and creodonts, were probably the equivalents in size and ecology of hyenas, coyotes, and dogs. Larger mammals seems to have been omnivores or herbivores.

On land, the carnivorous mammals were outclassed in body size in the northern continents by large, flightless birds with massive heads and tearing and stabbing beaks, the **diatrymas** (Figure 17.4). Similar carnivorous birds called **phorusrhacids** (Figure 13.22) evolved independently in South America, and both groups were the dominant predators in their respective ecosystems for some time. At the same time, some crocodiles became important predators on land: for example, the pristichampsid crocodiles of Europe and North America evolved the high skulls, serrated teeth, and rounded tails of terrestrial carnivorous reptiles. One Eocene crocodile evolved hooves!

Figure 17.4 The giant carnivorous bird *Diatryma*. (Reconstruction by Bob Giuliani. © Dover Publications Inc., New York. Reproduced by permission.)

La Grande Coupure

Toward the end of the Eocene, many families on land and in the sea became extinct and were replaced by others. Naturally, the mammals that became extinct have come to be called archaic and the survivors modern, but this does not necessarily imply that there were major functional differences between them. The event has been called *La Grande Coupure,* "the great cut-off," and it has been well documented in Europe and Asia. Even so, the extinction was much less abrupt than the K–T event. Because it was gradual rather than catastrophic and was accompanied by changes in climate and ocean currents, agents here on Earth were probably responsible.

The Oligocene

As Antarctica became isolated and began to refrigerate, Earth's climate began to cool on a global scale. It seems that the cooling took place in sharp steps, occasionally reversing for a while, so that there may have been a series of climatic events, each of which set up stresses on the ecosystems of the various continents. For example, a rapid cooling in southern climates in the Mid Oligocene seems to have had global effects, and there were some abrupt extinctions among North American mammals. Later events were even more severe, however.

The Later Cenozoic

In the Miocene the refrigeration of the Antarctic deepened, and its ice cap grew to huge size, affecting the climate of the world. Vegetation patterns changed, creating more open country, and a major innovation in plant evolution produced many species of grasses that colonized the open plains. The mammals in turn responded, and a grassland ecosystem evolved on many continents, continuing with changes to the present. The Savanna Story receives separate treatment later in this chapter.

Climatic and geographical changes allowed exchanges of mammals between continents, often in pulses as opportunities occurred. A favorite example is *Hipparion,* a horse that migrated out of North America, where horses had originally evolved and spent most of their evolutionary history. It spread across the plains of Eurasia about 11 Ma, leaving its fossils as markers of a spectacular event in mammalian history.

By the end of the Miocene, the mammalian fauna of the world was essentially modern. Two further events demand special attention: the great series of ice ages that have affected Earth over the last few million years (Chapter 21), and the rise to dominance of animals that greatly changed the faunas and floras of Earth—humans (Chapters 20 and 21).

ECOLOGICAL REPLACEMENT: THE GUILD CONCEPT

Although ancient mammal communities may have included some strange-looking animals, nevertheless certain ways of life are always present in a fully evolved tropical ecosystem. Plant life is abundant and varied, and provides food for browsers and grazers, usually medium to large in size. Small animals feed on high-calorie fruits, seeds, and nuts. Pollen and nectar feeding is more likely to support really tiny animals. Carnivores range from very small consumers of insects and other invertebrates to medium-sized predators on herbivorous mammals; scavengers can be any size up to medium. There may be a few rather more specialized creatures, such as anteaters, arboreal or flying fruit-eaters, or fishing mammals.

Easily categorized ways of life that have evolved again and again among different groups of organisms are called **guilds**, and their recognition helps to make sense of some of the complexity of mammalian evolution on several continents over more than 60 million years.

For example, the woodpecker guild includes many creatures that eat insects living under tree bark. Woodpeckers do this on most continents. They have specially adapted heads and beaks for drilling holes through bark, and very long tongues for probing after insects. But there are no woodpeckers on Madagascar, where the little lemur *Daubentonia,* the aye-aye, occupies the same guild. It has ever-growing incisor teeth, like a rodent, and instead of using beak and tongue like a woodpecker, it gnaws with its teeth and probes for insects with an extremely long finger.

On New Guinea, where there are no primates and no woodpeckers, the marsupial *Dactylopsila* has evolved specialized teeth and a very long finger for the same reasons. Because these three species all belong to the same guild, understanding the adaptations of any one of them helps us to interpret other members. In the Galápagos Islands, the woodpecker finch *Camarhynchus* does not have a long beak but uses a tool, usually a cactus spine, to probe into crevices. In Australia, some cockatoos fill the woodpecker niche, but they rely on the brute strength of their beaks to rip off bark, and they have not evolved the sophisticated probing devices of the others. Another small mammal evolved woodpecker devices 50 million years ago. *Heterohyus,* from the Eocene of Germany, had powerful triangular incisor teeth, and the second and third fingers of each hand were very long, with sharp claws on the ends.

Some guilds are unexpected (to me). For example, there is a recognizable guild of small mammals that live among rocks. From marmots and rock hyraxes to chinchillas, pikas, and rock wallabies, small rock-dwelling mammals on several different continents look alike, behave alike, and even sound alike.

Of course, there is no guarantee that a guild will be occupied by only one major group. In the tropics today, small arboreal animals that feed at night are almost all mammals, but the daytime feeders are almost all birds. Most medium sized predators and scavengers are mammals, but raptors are very effective at smaller body weights.

Cenozoic Mammals in Dinosaur Guilds

All Mesozoic mammals were small, and small body size implies that animals can play only a limited number of ecological roles, mainly insectivores and omnivores. When dinosaurs disappeared at the end of the Cretaceous, some of the Early Paleocene mammals quickly evolved to take over many of their ecological roles, particularly as omnivores and vegetarians. Others continued to occupy the same small-bodied guilds that the Mesozoic mammals had occupied for a 100 million years. Even today, 90% of all mammal species weigh less than 5 kg (11 lb).

Dinosaurs dominated many guilds in the Cretaceous, including that of large browsers. Most of them, such as the ceratopsians, hadrosaurs, and iguanodonts, weighed about 5–7 tons as adults. The K–T extinction wiped out all these creatures, and it was not until the Late Paleocene that the guild was occupied again, by large mammals.

Although some birds are large herbivores (ostriches are omnivorous, but much of their food is browse), mammals are the dominant browsers and grazers today. Even at the very beginning of the Paleocene, the mammals were dominated not by insectivores but by the largely herbivorous early ungulates. Very late in the Cretaceous, some mammals evolved molars even more complex than tribosphenic molars. The new teeth permitted or even required complex jaw motions, but they allowed much more shearing and grinding than before. The capacity for grinding more and tougher food allowed mammals to turn to low-calorie vegetarian diets.

There seems to be something special about the 5- to 7-ton range for the largest land herbivores. This limit applied to all dinosaurs except for sauropods, and it has apparently applied to all mammals since, including living elephants and rhinos. Presumably there is some metabolic reason for this limit, associated with the fact that browse and forage is low in calories. The 5- to 7-ton size was approached by different mammalian groups in the different continents of the Paleocene and Eocene (Figure 17.3). The best record is in North America, where uintatheres and titanotheres followed the dinosaurs.

Uintatheres (Figure 17.5) were most successful in the northern continents, but they may have managed to cross from an original home in South America in the Paleocene. They had massive skeletons and gradually increased in size through the Paleocene and Eocene. They had large canine teeth modified into cutting sabers, but they were not carnivores. The large flattened molar teeth were used for grinding vegetation. (The sabers were probably for fighting between adults: compare the last set of large-bodied synapsids, Figure 10.14.) *Uintatherium* itself (Figure 17.5) was as large as a rhino.

These large herbivores were replaced in the large herbivore guild in North America, and later in Asia, by perissodactyls called **titanotheres** (or brontotheres). These were small in the Early Eocene, by the Middle Eocene they were large, and at the end of the Eocene they were very large indeed (Figures 17.6; 17.7). They evolved massive blunt horns as they evolved to larger body size. The horns have been interpreted as ramming devices, but most of them have a shape and a position on the head that would have been much better designed for pushing and wrestling (Figure 17.7). Titanotheres became extinct at the end of the Eocene, and their guild was filled by the modern-looking rhinos in Eurasia and North America. Later,

Figure 17.5 *Uintatherium*, one member of a major group of large-bodied Paleocene and Eocene North American vegetarian mammals. (Reconstruction by Bob Giulani. © Dover Publications Inc., New York. Reproduced by permission.)

Figure 17.6 The Eocene titanothere *Brontops*. (Reconstruction by Bob Giulani. © Dover Publications Inc., New York. Reproduced by permission.)

rhinos were joined in the guild by elephants, which had evolved in Africa but did not leave that continent until the Miocene.

Creodonts and Carnivores: Ecological Replacement or Displacement?

As we have seen, the larger-bodied carnivorous guild vacated by dinosaurs was occupied in the Paleocene by crocodiles and flightless birds. Many early mammals were insectivores and only became carnivores with the evolution of larger size. The success of the new early ungulates and their young meant prey for early carnivorous mammals, recognized in fossils by their biting and slashing teeth, powerful jaws, and strong, clawed feet.

Most of these early hunters were **creodonts**. Creodonts included oxyaenids, which were probably ambush predators, and hyaenodonts, which were runners. Some mesonychids were also carnivores or scavengers. These mammals were the dominant small- and medium-sized carnivorous animals in Paleocene and Eocene ecosystems.

Some new small carnivorous mammals, the **miacids**, appeared in the Middle Paleocene but for 20 m.y. remained rather few in number and small in size. They lived alongside the larger carnivorous groups of creodonts and mesonychids. At the end of the Eocene all the larger predators except a few hyaenodonts became extinct, and miacids underwent an evolutionary radiation that produced all the modern types of carnivores: the mustelids (otters, weasels, and badgers), viverrids (mongooses and civets), canids (dogs), felids (cats), ursids (bears), and nimravids (sabertooths).

Did miacids suddenly evolve some character or characters that allowed them to outcompete the mesonychids and creodonts? Or did the disappearance of their competitors allow miacids to evolve into larger-bodied carnivores? In other words, was this change in the carnivore guild the result of replacement or competition? The extinction of creodonts has been attributed at various times to increased intelligence of modern carnivores over creodonts and/or to better running adaptations.

Leonard Radinsky attacked this question in 1982 by studying the skull structure of living carnivores. The four largest carnivore groups—viverrids, canids (dogs), felids (cats), and mustelids (weasels)—have distinctly different skulls that reflect the different ways they bite and kill. Cats and weasels have short, powerful jaws, and

Figure 17.7 Titanotheres evolved to very large body size between the Early Eocene and the Early Oligocene. (a) Shows (i) an early small titanothere, *Eotitanops*, and (ii) a gigantic late one, *Brontotherium*, in the act of displaying. (b) Shows *Brontotherium* walking naturally with the head held low. The huge double horns look to me like wrestling structures rather than ramming devices. (From Osborn.)

they kill small prey with a powerful bite at the back of the neck. Large cats kill larger prey with a powerful neck bite that strangles the prey. Dogs, on the other hand, have long, slender jaws, and they kill small prey with a head shake. They can kill large prey only by tearing at them in packs.

Radinsky hoped to be able to distinguish a new killing capability among miacids that would have given them a competitive edge over creodonts, but he could not see one. In short, there is no obvious key innovation behind the radiation of the modern Carnivora.

Without a key innovation, then, modern carnivores must have succeeded because their larger competitors disappeared for unknown reasons, along with other mammal groups, at the end of the Eocene at *La Grande Coupure*. This is an example of guild replacement, not of displacement.

A final piece of evidence that supports this argument deals with the latest surviving creodonts, the hyaenodonts. The large running mesonychids disappeared at the end of the Eocene, and thereafter a surviving group of hyaenodonts became very large, replacing them. Modern carnivores did not take on the role of large predators until later, in the Miocene, with much larger cats and dogs.

THE SAVANNA STORY

Modern Savannas

Research by Samuel McNaughton and his colleagues on the savanna grazing ecosystem of East Africa revealed patterns that may also be true for other ecosystems in space and time (see McNaughton, 1985).

Herbivores tend to graze off the tops of any plants they can reach, because the top of the plant contains the most tender, juicy parts, and is less well protected by any mineral or chemical compounds the plant produces. Grazing thus promotes the survival and evolution of plants that tend to grow sideways rather than upward. If grazing is continuous, such plants are selected because they lose less of their foliage. They are not shaded out by competitors that grow upward, because the grazing animals remove those competitors. Low plants tend to occupy a smaller area than high plants, so there is space for more plants in a grazed environment. This may often translate into more species as well as more individuals of one species.

McNaughton fenced off savanna areas to protect them from grazing. It turned out that grazed areas actually had much more available vegetation per cubic centimeter than fenced areas. Plants that are not grazed grow tall and airy, not low and bunched. This happens on lawns too, where mowing is artificial grazing. In areas that are grazed, therefore, food resources are densely packed. A grazing animal can get more food per bite than in ungrazed areas, and it feeds more efficiently.

For example, a cow needs a certain level of nutrition per mouthful in order to survive, considering the energy that is required to move, bite, chew, and digest that mouthful. If the cow lived on the Serengeti Plains of East Africa, it could not survive if it had to crop grassland that had grown more than about 40 cm (16 in) high, but the same environment, already grazed down to 10 cm (4 in) high, would provide a very rich food supply.

Grazed plants react in more sophisticated ways than by simply altering their growth habit. After some time, they coevolve with the grazers to produce different reproductive patterns and structures. For example, plants that can regrow from the base rather than the growing tip will be favored, as will plants that reproduce by runners.

All this has important consequences. It implies that a grazing ecosystem is balanced evolutionarily so that the herbivores are controlling the type and density of their food resources, but at the same time the response of the plants forces certain behavioral patterns and perhaps social structures on the herbivores. The ecosystem will tilt out of balance unless the grazing pressure is maintained at a minimum level to keep the low-growing plants at an advantage over possible competitors.

Grazing animals probably can't do this if they are solitary. Solitary grazers have two problems: they have to spend energy to defend a territory, and in open country they are liable to predation from running carnivores. Living in herds is an efficient solution to this problem, because it removes the need to spend energy on defense of a territory, it increases the chance of early warning of the approach of a predator, it allows group defense, and it provides a better guarantee of the heavy and continuous grazing that maintains a healthy ecosystem.

Furthermore, with a seasonal and local variation in food supply, it is easy to envisage the evolution of a set of grazing species, each specializing in a different part of the available food. In the Serengeti, for example, three different grazers eat grass and herbs. Zebra eat the upper parts of the blades of grass and the herbs, wildebeest follow up and eat the middle parts, and the Thomson's gazelle eats the lower portions. The teeth and digestive systems of each animal are specialized for its particular diet. Thus, a succession of animals grazes the plain at different times, each species modifying the plants in a way that permits its successor to graze more efficiently. A great diversity of grazers is encouraged: today there are ten separate tribes of bovid antelopes on the savannas of Africa.

Because these principles are so general, they have probably operated at least since grasslands spread widely in the Miocene. Furthermore, there were low-plant ecosystems even before the evolution of the first grasses at the end of the Oligocene. The first horses seem to have grazed in open country in the Paleocene, for example. If dinosaurs were warmblooded, they probably faced similar problems related to feeding requirements per mouthful. Even if dinosaurs were cold-blooded, with lower metabolic requirements, they would still have faced similar problems.

Similar principles probably apply to browsers too. Obviously, the rules will be rather different, because the defense of many plants against browsing is to grow tall quickly. And finally, herbivores, whether they are grazers or browsers, are a food resource for predators and scavengers. The animals of the African savannas are in a delicate and interwoven ecological network.

McNaughton's work, which explicitly defined principles that many workers had guessed at previously, is likely to be a breakthrough not only in the understanding of modern savanna ecosystems but in the interpretation of past ones too.

Savannas in the Fossil Record

A major climatic change in the Miocene was apparently triggered by the refrigeration of the Antarctic and the growth of its huge ice cap. The cooler climate encouraged the spread of open woodland in subtropical latitudes, at the expense of thicker forests and woods. In California, for example, this occurred around 12 Ma, when the climate changed from wet summers to dry summers. There had been open woodland on Earth ever since the Permian, but the plants that grew in the open had been ferns and shrubs. The new feature of Miocene open country was the spread of grasses, with their high productivity.

Savanna ecosystems produce a great deal of edible vegetation, even though grasses have high fiber and low protein. Grasses are adapted to withstand severe grazing; they recover quickly after being cropped because they grow throughout the blade instead of mainly at its growing tip. They have evolved tiny silica fragments, or phytoliths, that make them tough to chew and cause significant tooth wear in grazing animals.

The spread of grasses was perhaps encouraged at first by intense grazing pressure, but the whole savanna ecosystem quickly stabilized, no doubt through mechanisms like those suggested by McNaughton. There was a rapid and spectacular evolutionary response, especially from the mammals, which evolved many different grazing forms. This event and its continuation into plains ecosystems today is called "The Savanna Story" by David Webb, who has done the most to document it (for example, Webb, 1977). The change in vegetation was worldwide, and although the North American evidence is most complete, similar trends can be traced on all the continents with subtropical land areas. On each continent, the savanna fauna evolved from animals that lived there before the major climatic change.

The animals that were particularly successful in savanna ecosystems were grazers or browsers on these open plains with only scattered woodland patches. Deer and antelope evolved to great diversity. Often their teeth evolved to become very long for their height, or **hypsodont**, with greatly increased enamel surfaces. Elephants, rodents, horses, camels, and rhinos, for example, all evolved jaws and teeth with adaptations for better grinding. Presumably, hypsodont teeth wore longer and permitted a grazer to chew tough fibers and resist the abrasion of phytoliths.

Figure 17.8 The gigantic rhinoceros *Baluchitherium*. (Reconstruction by Bob Giuliani. © Dover Publications Inc., New York. Reproduced by permission.)

The larger savanna animals also showed changes in size and locomotion consistent with life in open country where there was nowhere to hide. They became taller and longer-legged, well adapted for running fast. Some of these Miocene plains animals were gigantic, and they included the largest land mammal that has ever lived, the Eurasian rhino *Baluchitherium* (Figure 17.8), as well as the tallest camel.

In the Miocene of North America, the grazers were at first native ruminants such as camelids and horses. Eurasian deer arrived and radiated during the Miocene. The Late Miocene savanna fauna of North America was very rich, peaking at about 50 genera of ungulates and large carnivores, dominated by hypsodont horses, camelids, and pronghorns.

Starting about 9 Ma, this North American savanna ecosystem suffered a series of shocks, including a great extinction that began about 6 Ma. New genera evolved and new immigrants arrived, but they did not come close to replacing the losses.

Figure 17.9 The giant Miocene camel *Aepycamelus*. (Reconstruction by Bob Giuliani. © Dover Publications Inc., New York. Reproduced by permission.)

The extinction patterns are interesting. All nonruminant artiodactyls disappeared except for one peccary, *Platygonus*, which evolved shearing teeth and shifted toward a coarser, more fibrous diet. Grazing horses flourished, but browsing horses disappeared. Only hypsodont camelids survived, while short-toothed forms became extinct. The casualties included a giant camel *Aepycamelus*, 3.5 m (12 ft) high (Figure 17.9), which was probably a giraffe-like browser. Among the ruminants, the major survivors were the pronghorns, which are hypsodont.

The common ecological pattern is adaptation to coarse fodder and more open country, and it presumably reflects a change from savanna to steppe grassland. Perhaps rain shadow effects were produced by major uplift in western North America; but overall, global climates became colder in the Late Cenozoic.

EVOLUTION BY IMPROVEMENT

The fossil record of mammals is so good that we can trace related groups of mammals through long time periods, and often across large areas and across geographic and climatic barriers. In many cases, we can see considerable evolutionary change in the groups, and because we understand the biology of living mammals rather well, we can interpret the changes confidently. Often the changes can be linked with specific biological functions and can be seen as allowing the animals to perform those functions in a more effective way.

Many people do not like the concept of improvement, or of evolutionary progress, which is another way of saying the same thing. However, there are parameters that we can measure that show clearly that many clades of organisms do get better through time as doing what they do. The easiest way to show this is to use simple mechanics (biomechanics, if applied to animals or plants). So, for example, dinosaurs moved in a mechanically better way than their ancestors, and so do living mammals. Horses in particular have evolved clever mechanical couplings between bone, muscle, and tendon that gives living horses, including racehorses, much better running performance than their predecessors. The examples are too numerous to list here, and one could make the same arguments about physiology, biochemistry, reproduction, and so on. I simply want to make the point that it is legitimate to write about "progress" as applied to evolution. I have space to deal here with only one example: horses; but I will post others on the Web site.

Horses

Horses are only one of a group of ungulates called perissodactyls (they usually have an odd number of toes on each foot). Tapirs and rhinos are also living perissodactyls (Figure 17.10), and titanotheres (Figure 17.6) are extinct perissodactyls. While rhinos have had an interesting, eventful, and complex history, tapirs have evolved slowly enough to qualify as living fossils. This is especially interesting because horses have evolved so radically and so rapidly.

Perissodactyls appeared in numbers in the Late Paleocene, replacing some of the earlier ungulates as small-bodied browsers. They already show good adaptations for running, and their success can be explained in two ways: either the better running ability of the earliest perissodactyls made them more predator-proof than the earlier mammals they replaced, and/or better running ability was needed as Paleocene forest gave way to more open woodland in the Early Eocene. The first perissodactyls had larger and more complex brains, too, and they began a radiation into different and often larger body types.

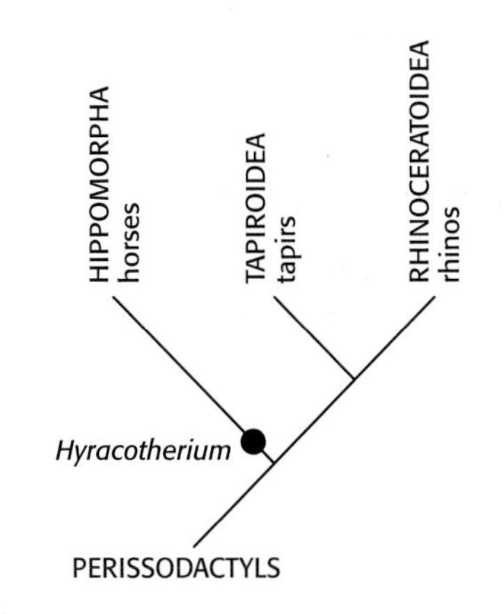

Figure 17.10 Phylogram of the major groups of living perissodactyls.

Hyracotherium, the first known horse and one of the first perissodactyls (Figure 17.11), was a browsing animal: the smallest species was about the size of a cat, but other species weighed up to about 35 kg (75 lb). *Hyracotherium* first appeared in the Early Eocene of North America and Europe. Males were larger than females, with wider faces and longer canine teeth; presumably this was related to fighting among males. Large horses today fight with teeth and hooves, but at cat size or even at 35 kg, a kick does not have much impact, so *Hyracotherium* must have had a quite different fighting style. It was once thought that *Hyracotherium* lived in forests, but paleobotanical evidence shows that horses lived in open woodland right from the start.

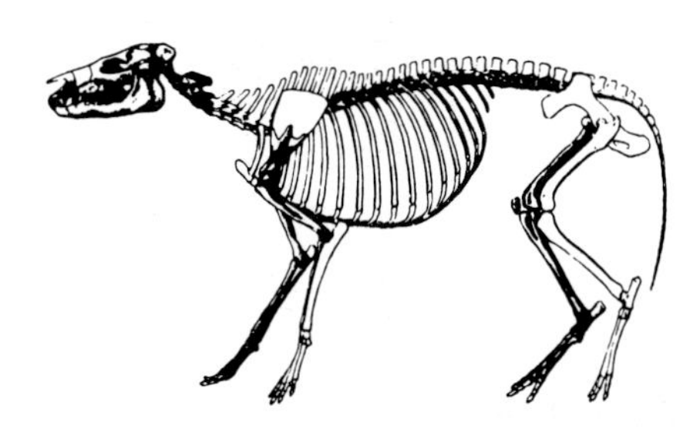

Figure 17.11 *Hyracotherium* was the earliest horse, but was also close to the ancestor of rhinos and tapirs (see Figure 17.10). (After Cope.)

The history of horses is perhaps the best known of any fossil vertebrate group. It was one of the earliest evolutionary sequences to be worked out well: paradoxically, the first good version was published in 1851 by Darwin's rival Sir Richard Owen, who named the earliest horse *Hyracotherium* and traced its lineage to the living *Equus*. Twenty years later, Darwin's friend Thomas Huxley used an improved version of the horse story in his lectures to help promote Darwin's theory of evolution

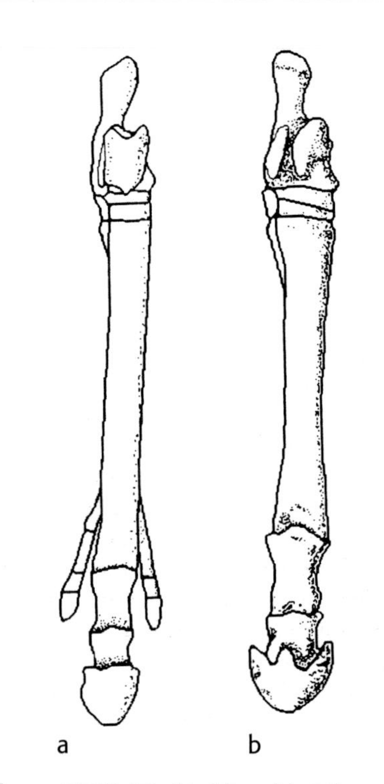

Figure 17.12 The hind feet of the Miocene horse *Merychippus* (a) and the modern horse *Equus* (b). Fast running requires that there should be as little mass as possible at the end of a limb, and *Merychippus* had evolved close to that ideal. *Equus* finally lost the side toes. (After Scott.)

by natural selection. In 1876, however, Huxley visited the United States and immediately had to rewrite his lectures to incorporate the much richer North American fossil record of horses.

Hyracotherium had rather generalized teeth and probably a vegetarian diet to match. The teeth of *Mesohippus*, in the Early Oligocene, evolved to accentuate a combination of shearing and crushing, perhaps associated with succulent plants and seeds in its diet. Oligocene horses were rather bigger than their Eocene ancestors (perhaps 40–50 kg), and they had longer legs and feet. The number of toes dropped to three on each foot (*Hyracotherium* had four toes on its front feet), and the body frame of ribs and spine grew stronger with greater size.

With the cooling and climatic change in the Miocene, the decline in forested area, and the evolution of grasses as a major component in the larger open areas between woodland patches, there was more food for grazers as well as browsers. A number of Miocene horses survived as browsers, but others began to exploit the new food resource in earnest. Grazing exposes teeth to more abrasive plant material, and a grazer is more likely to get grit and sand in its teeth than a browser is; both conditions lead to greater tooth wear. The height of the molars of *Parahippus* was double that of earlier horses, the enamel was more complex, and the teeth contained more cement. Body size also doubled in *Parahippus.*

An explosive breakthrough of grazing horses came around 17–18 Ma, in the Middle Miocene, as many new species appeared. Species of *Parahippus* and *Merychippus* are good examples: they were the size of small ponies, with three toes on each foot (Figure 17.12a).

A whole set of evolutionary changes in *Merychippus* were adaptations for savanna grazing rather than savanna browsing. The teeth grew larger and longer, with deep roots and still more complex enamel, and they were supported by cement so they would last longer under the impact of a silica-rich diet. They were set along a longer, larger jaw with a wide, flat muzzle, in a longer, deeper, heavier skull. Sophisticated analysis of jaw motions, revealed by wear patterns on its teeth, show that *Merychippus* chewed with a jaw action that differed distinctly from that of its predecessors, accentuating sideways slicing and shearing rather than shearing with downward crushing and compression.

Many of the new grazing horses were larger than their predecessors (100–50 kg), as low-quality forage encouraged larger gut capacity. The limbs became relatively longer and better designed for faster running, as safety came to depend more on speed than on stealth and camouflage. The gait became truly ungulate. Instead of running on a padded foot, as a tapir does and as earlier horses had done, *Merychippus* ran on a hoofed foot raised onto the toes, and the side toes were reduced in size and importance (Figure 17.12). Fast running requires mass to be reduced as much as possible at the ends of the limbs. The feet had powerful ligaments that could flex to reduce the stress of impact in running.

The new savanna habitats encouraged an increased diversity among horses. Between 18 Ma and 15 Ma there was a dramatic increase in the number of grazing horses in North America; in fact, species numbers increased until there were 18 grazing species altogether in North America. Some of the new horses were large, others small, but all retained the improvements in skeletal and tooth characters that allowed them to graze in open grassland. A group of small, three-toed North American horses called **hipparions** evolved perhaps about 15 Ma in the Middle Miocene. *Cormohipparion* is the horse that found its way out of North America and invaded savannas throughout the Old World, leaving fossils so abundant that they form a valuable marker at about 11 Ma, the so-called *Hipparion* Event (Figure 17.13).

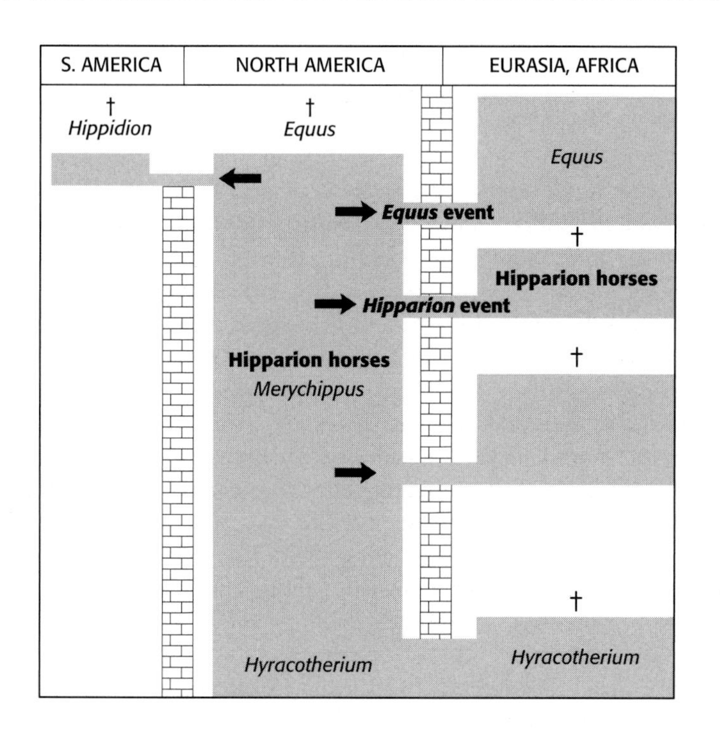

Figure 17.13 The biogeography of horses through time. Diagrams like this have been drawn since the 1880s. This version is based on the work of Bruce MacFadden (1992). Time is not to scale. The diagram shows how Huxley could tell a reasonable but incomplete story of horse evolution from evidence in Eurasia, and why he had to rewrite his lectures when he saw the North American evidence. At times, horses migrated out of North America across geographic barriers (shown as brick walls). Finally, New World horses became extinct, to be reintroduced from the Old World in 1492.

Later hipparion species also found their way out of North America to create other, less dramatic markers in Old World fossil collections.

The Late Miocene North American horse *Pseudhipparion* holds the record for extreme tooth evolution. Although it was tiny, it had molar teeth that are proportionately higher than in any other horse known. Normally, teeth stop growing when their roots form. In living horses this happens when the molars are about half worn down. But in *Pseudhipparion* the tooth root did not form until very late in life, so the molars were ever-growing for most of its life. The same extended tooth development has evolved independently in rodents, rabbits, and pronghorns, and in some extinct grazing animals, for the same reason: to prolong the life of the molars and of the animal that uses them.

Teeth have been used as direct indicators of diet, for the reasons specified above. Hypsodonty may well have evolved for chewing resistant grasses, but a hypsodont horse could browse soft vegetation. Grasses run a different photosynthetic pathway (called C4) than do many shrubs and trees (which run C3). For our purposes, it's important to note that C4 grasses set up a carbon isotope anomaly (Chapter 2) of $\partial^{13}C$ around −25 per mil in their leaves, whereas C3 shrubs and trees set up a $\partial^{13}C$ of around −12. The difference is large enough to show up in the carbon isotopes in the teeth of the horses that eat the plants (though the actual numbers are less). Here is a cautionary tale: there were six species of horses living in Florida about 5 Ma, all of them hypsodont. But analysis of their teeth revealed that the largest one, *Dinohippus*, was browsing on shrubs rather than grazing on grasses! Others were eating a mixed diet, including the extremely hypsodont *Pseudohipparion* (see above). It's not easy to see what all this means, except that animals will eat what's available. Deer are perfectly happy to eat grape leaves and my wife's garden instead of grazing as they are supposed to; there are flies that eat toads, and deer that eat baby birds.

In North America, the adaptation of horses to open country was accelerated by the rise of the western mountains. The High Plains became steppe grassland rather than savanna. In the Pliocene, modern horses evolved from hipparions by reducing their toes to one (Figure 17.12, right), and *Equus* itself evolved perhaps 5 Ma:

paradoxically enough, the closest relative of the first *Equus* is the browsing *Dinohippus* (see above). The *Equus* event marks its invasion of the Old World (Figure 17.13). It reached India by 3 Ma and Europe by 2.6 Ma; zebras reached Africa at about the same time.

Horses today are reduced to only seven species worldwide, all of them native to the Old World. Przewalski's horse of Central Asia and the domesticated horse survive only under human protection. All so-called "wild" horses have escaped or have been released from domestication. One endangered species of wild ass survives in Africa and one in Asia, and three species of zebra remain in Africa.

But this low diversity is a result of catastrophic extinctions in the last 2 m.y. At the end of the Pliocene, horses were abundant and diverse on all the continents except Australasia and Antarctica. Viewed at that time by a zoologist, their body plan and their adaptations would have been seen as remarkably successful. The most ironic twist to the horse story is, of course, that horses became extinct in the Americas in the Late Pleistocene (Figure 17.13 and Chapter 21) and were reintroduced only in 1492.

Further Reading

Archibald, J. D., et al. 2001. Late Cretaceous relatives of rabbits, rodents, and other extant eutherian mammals. *Nature* 414: 62–5. [The branching of northern placentals was under way in the Cretaceous, but not as intensively as molecular evidence suggests.]

Bromham, L., et al. 1999. Growing up with dinosaurs: molecular data and the mammalian radiation. *Trends in Ecology and Evolution* 14: 113–18. [Review of this controversy.]

Foote, M., et al. 1999. Evolutionary and preservational constraints on origins of biological groups: divergence times of eutherian mammals. *Science* 283: 1310–14. [Argument against deep Cretaceous roots for many mammal lineages.]

Gaeth, A. P., et al. 1999. The developing renal, reproductive, and respiratory systems of the African elephant suggest an aquatic ancestry. *Proceedings of the National Academy of Sciences* 96: 5555–8.

Gingerich, P. D., et al. 2001. Origin of whales from early artiodactyls: hands and feet of Eocene Protocetidae from Pakistan. *Science* 293: 2239–42, and comment 2216–17.

Luo, Z-X. et al. 2001. Dual origin of tribosphenic mammals. *Nature* 409: 53–7, and comment 28–31. Comment in *Science* 291, 26.

MacFadden, B. J., et al. 1999. Ancient diets, ecology, and extinction of 5-million year-old horses from Florida. *Science* 283: 824–7, and comment 773. [Trying to work out the diets of fossil horses is not as simple as we thought!]

MacFadden, B. J. 1992. *Fossil Horses.* Cambridge: Cambridge University Press.

Madsen, O., et al. 2001. Parallel adaptive radiations in two major clades of placental mammals. *Nature* 409: 610–14.

McNaughton, S. J. 1985. Ecology of a grazing ecosystem: the Serengeti. *Ecological Monographs* 55: 259–94.

Murphy, W. J., et al. 2001a. Molecular phylogenetics and the origins of placental mammals. *Nature* 409: 614–18.

Murphy, W. J., et al. 2001b. Resolution of the early placental mammal radiation using Bayesian phylogenetics. Science 294: 2348–51, and comment 2266–9.

O'Leary, M. A., and M. D. Uhen. 1999. The time of origin of whales and the role of behavioral changes in the terrestrial–aquatic transition. *Paleobiology* 25: 534–56.

Radinsky, L. 1982. Evolution of skull shape in carnivores. 3. The origin and early radiation of the modern carnivore families. Paleobiology 8: 177–95.

Rose, K. D. 1996. On the origin of the order Artiodactyla. *Proceedings of the National Academy of Sciences* 93: 1705–9.

Savage, R. J. G., and M. R. Long. 1987. *Mammal Evolution: An Illustrated Guide.* London: Natural History Museum.

Springer, M. S., and W. W. de Jong. 2001. Which mammalian supertree to bark up? *Science* 291: 1709–11.

Springer, M. S., et al. 2003. Placental mammal diversification and the Cretaceous–Tertiary boundary. *Proceedings of the National Academy of Sciences* 100: 1056–61. [The latest synthesis of molecular data: a model study, in my opinion.]

Thewissen, J. G. M., et al. 2001. Skeletons of terrestrial cetaceans and the relationship of whales to artiodactyls. *Nature* 413: 277–81, and comment 259–60.

Van Valkenburgh, B. 1999. Major patterns in the history of carnivorous mammals. *Annual Reviews of Earth & Planetary Sciences* 27: 463–93.

Webb, S. D. 1977. A history of savanna vertebrates in the New World. Part 1, North America. *Annual Review of Ecology and Systematics* 8: 355–80.

CHAPTER EIGHTEEN

Geography and Evolution

Natural selection operates on individual organisms partly by their response to their environment. On a larger scale, the evolution of larger groups of organisms is strongly affected by major geographic effects. In this chapter I discuss some aspects of Cenozoic evolution that were affected by geography.

AUSTRALIA

Australia is linked in people's minds with exotic creatures such as kangaroos and jillaroos, but they are only a part of the story of evolution on this isolated continent. Australian plants, insects, amphibians, reptiles, birds, and mammals are all unusual. Australia and New Zealand were part of Gondwana in Cretaceous times, joined to Antarctica in high latitudes. The climate was mild, however, and pterosaurs, dinosaurs, and marine reptiles have been found there. In Early Cenozoic times the two land masses broke away from Antarctica and began to drift northward and diverge. In the process, both Australia and New Zealand became isolated geographically and ecologically from other land masses, and evolution among their faunas and floras led to interesting parallels with other continents.

Among amphibians, Australia has (or had) at least two species of frogs that brood young in their stomachs. Instead of the colubrid snakes and vipers that are abundant elsewhere, Australia has had a radiation of elapid snakes (cobras and their relatives) into 75 species, all of them virulently poisonous. The largest Australian predators are the salt-water crocodiles (the world's largest surviving reptiles), which lurk along northern rivers and shorelines, and large monitor lizards related to the Komodo dragon of Indonesia. Monitors are ambush predators, the largest living Australian monitor being 2 m (over 6 ft) long. Smaller monitors dig for prey like the badgers of larger continents. In contrast, most Australian mammals are herbivores.

Extinct Australian reptiles included giant horned tortoises (Figure 18.1) that weighed up to 200 kg (450 lb), a monitor 7 m (23 ft) long that weighed perhaps a ton, and competed with large terrestrial crocodiles of about the same size and weight. The giant snake *Wonambi* was 6 m (19 ft) long and must have weighed 100 kg (220 lb). Extinct Australian birds included *Dromornis*, the heaviest bird that has ever evolved (Chapter 13).

Australia is the only continent with living monotremes. They have been in Australia since the Early Cretaceous (Chapter 15), and only one Early Cenozoic

Figure 18.1 *Meiolania*, a very large extinct tortoise with horns and a clubbed tail, about 2 m (6 ft) long, from Lord Howe Island off the east coast of Australia. (After Gaffney.)

monotreme tooth from Argentina shows that they once ranged more widely over Gondwana. The surviving monotremes are egg-laying mammals, including the duck-billed platypus and the spiny echidna of Australia and New Guinea. Many aspects of monotreme biology are bizarre: for example, the platypus swims in muddy water with its eyes, ears, and nostrils tightly shut, searching for its crustacean prey with electrical sensors in its beak. Since monotremes have evolved to include the specialized platypus and ant-eating echidnas, it's likely that their fossil record will eventually show us many other surprises.

Marsupials had originally evolved in the northern continents, but as we shall see later in the chapter, they reached South America and radiated there in the Cenozoic. Marsupial fossils have now been discovered in Eocene rocks in Antarctica and Australia, so it is likely they reached Australia from South America across Antarctica when the region was much warmer than it is now, well before the refrigeration of Antarctica in Oligocene times (Figure 18.2).

By the Late Cenozoic, marsupials had evolved to fill most of the ecological roles in Australia that are performed by placental mammals on other continents (Figure 18.3). Wallabies and kangaroos are grazers comparable with antelope and deer, wombats are large burrowing "rodents" rather like marmots, and koalas are slow-moving browsers like sloths. The cuscus is like a lemur, and the numbat is a marsupial anteater. There are marsupial cats, marsupial moles, and marsupial mice, and at least six gliding marsupials can be compared with flying squirrels. The honey possum *Tarsipes* is the only nonflying mammal that lives entirely on nectar and pollen, which it gathers with a furry tongue. The small marsupial *Dactylopsila* of New Guinea has evolved specialized teeth and a very long finger to become a marsupial woodpecker (Chapter 17). The Tasmanian wolf and the Tasmanian devil are marsupial carnivores comparable in size and ecology to wolf and wolverine. They once ranged over the main continent of Australia. The Tasmanian devil is now confined to Tasmania, and the Tasmanian wolf is probably extinct.

The fossil record of extinct Australian marsupials is even more impressive. Entire families of marsupials are now extinct. Many were very large, including giant kangaroos and giant wombats that each weighed 200 kg or so (450 lb). *Thylacoleo* was a Pleistocene carnivore whose name means the marsupial lion. It was indeed the size of a leopard, and had efficient stabbing and cutting teeth. It was better adapted for cutting off chunks of flesh than any living carnivore is (Figure 18.4). Diprotodonts

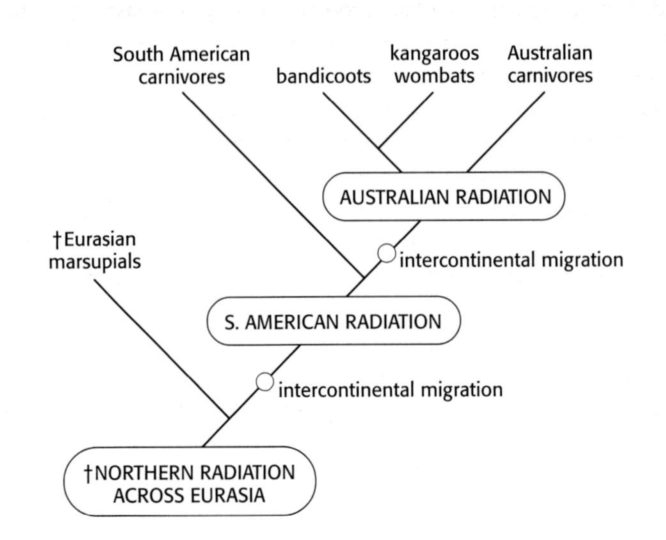

Figure 18.2 The biogeographic evolution of marsupials. An Early (Cretaceous) radiation in Asia and North America was followed by a Late Cretaceous and Cenozoic radiation in South America, and Cenozoic dispersal through Antarctica to Australia, where a spectacular radiation occurred.

kangaroos	antelope
wallabies	rabbits
wombats	marmots
phalangers	squirrels
koala	sloths
"mice"	cats and weasels
"moles"	moles
numbats	anteaters
Tasmanian devil	wolverine
†diprotodonts	rhinos, tapirs
†marsupial "lion"	large cats
†Tasmanian "wolf"	dogs

Figure 18.3 A gallery of Australian marsupials, all of which have placental ecological counterparts on other continents.

were quadrupedal Pleistocene marsupials about the size of tapirs and rhinoceroses (Figure 18.5). They were the largest marsupials ever: the largest diprotodont was the size of a small elephant, almost 3 m (10 ft) long and 2 m (over 6 ft) high at the shoulder, weighing probably close to 3 tons. Discoveries of enormous numbers of Miocene bats and marsupials at Riversleigh, in Queensland, will soon allow a better description of the radiation of these Australian mammals.

People often talk of marsupials as primitive and inferior to placentals, and it's true that today they are outclassed in diversity and range by placentals. But marsupials do not always have inferior adaptations (Chapter 15). For example, a kangaroo is rather clumsy as it hops slowly around on the ground, using its tail as an extra limb in what is really a five-footed movement. It does use more energy than a placental at this speed. But at high speed a kangaroo is not only very fast (up to 60 kph, or 40 mph), but its incredibly long leaps are much more efficient than the full stride of a four-footed runner of the same weight.

It was surprising to find in 1992 that a mysterious primitive placental mammal had also reached Australia by Eocene times. *Tingamarra* or its ancestors must also have walked across Antarctica to reach Australia and drop a right upper molar tooth into a billabong in southeast Queensland, but that's all the evidence we have right now. *Tingamarra's* tooth shows us that marsupials did not come to dominate Australian mammalian faunas because they were isolated from placental competition there.

Dromornithids (mihirung in aboriginal legend) are giant extinct Australian birds that must have evolved flightlessness and large body size (Figure 18.6). *Dromornis* was probably as large as *Aepyornis*, the elephant bird of Madagascar, and rivals it for the heaviest bird of all time. The living Australasian emu and cassowary are large ground-running ratites. Ratites are distributed on the southern continents that are remnants of Gondwana (Figure 13.23).

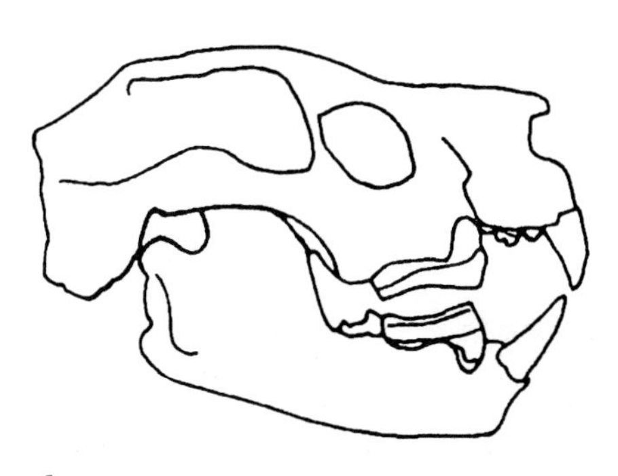

Figure 18.4 *Thylacoleo*, an extinct Australian marsupial carnivore the size of a leopard. (After Lydekker.)

Figure 18.5 A diprotodont, one of several large extinct quadrupedal marsupials in Australia. Diprotodonts weighed up to 3 tons. (Reconstruction by Bob Giulani. © Dover Publications Inc., New York. Reproduced by permission.)

The isolated position of Australia has meant that only very mobile birds and placental mammals (bats and humans) have reached it. Humans brought with them a host of other invaders, such as rats, cats, dogs, sheep, cattle, rabbits, cactus, fish, and cane frogs, with serious results for the Australian ecosystem. Captain Cook's first reaction to a kangaroo was to set his dog on it! More recently, other bizarre introductions have helped to restore a little of the damage—for example, the organism that causes the rabbit disease myxomatosis, and the dung beetles that keep Australian grasslands from being buried in cattle dung. The biogeographic story of Australia is still in an active phase.

NEW ZEALAND

New Zealand was part of Gondwana until the Cretaceous, and it had a normal fauna at that time. But it had no land mammals until humans arrived, and the rest of its prehistoric fauna suggests that migration into the region was difficult. The native fauna includes only four amphibian species, primitive frogs that hatch as miniature adults from the egg with no tadpole stage. There are only a few native reptile species: 11 geckos that all have live birth, 18 skinks, of which 17 have live birth, and the tuatara, which is an ancient and primitive reptile. New Zealand has no snakes and no normal lizards. The only native mammals are two species of bats.

The dominant prehistoric creatures of New Zealand were birds. The kiwis survive as nocturnal insectivores, but the major vegetarians were moas, very large ratites. The largest moa (females were much larger than males) was 3.5 m (11 ft) in height (Figure 13.23).

Moas coevolved with New Zealand plants so that 10% of the native woody plants have a peculiar branching pattern called divarication—they branch at a high angle to form a densely growing plant with interlaced branches that are difficult to pull out or break. There are few leaves on the outside, and the largest, most succulent leaves are on the inside. But nine species of divaricating plants that grow more than 3 m (10 ft) tall look more like normal trees once they reach that height, and other divaricating species grow more normally on small offshore islands. The only reasonable explanation of divarication is that it evolved as a defense against browsing moas, the largest of which was about 3 m tall.

Figure 18.6 A dromornithid, or mihirung, a giant extinct bird from Australia, shown with a living emu for scale.

Other vegetarian guilds that were filled by small mammals on other land masses were partly occupied by moas and other birds and partly by huge flightless insects—

enormous weevils and wetas (giant grasshoppers). It's not easy to identify the major prehistoric predators, but they were present. The largest surviving New Zealand birds (the kiwi, for example) are well camouflaged, although there is no obvious surviving predator on them. But extinct New Zealand raptors include a bird that was the largest goshawk that ever evolved (3 kg or 7 lb in weight) and a huge extinct eagle that weighed about 13 kg (30 lb).

LIMERICK 18.1
In New Zealand, evolution's been slower,
Vegetarian diversity's lower.
Giant weevils eat seeds,
The wetas eat weeds,
But the grasses are cut by the moa.

SOUTH AMERICA

South America is in many ways more interesting than Australia for mammalian evolution because we know its history in more detail. South America split away from Africa in the Late Cretaceous (around 80 Ma) to become an island continent (Figure 18.7).

Figure 18.7 South America drifted west and then west–northwest during the Cenozoic, and for most of that time was an island continent accessible only to lucky or mobile immigrants.

In Cretaceous times the South American mammals and dinosaurs included unique forms belonging to primitive Jurassic groups that had become extinct everywhere else but continued to evolve in South America. Examples include the giant dinosaur *Megaraptor*, a large sphenodont, and early mammals. Triconodonts, symmetrodonts, and multituberculates (Chapter 15) have all been collected from Cretaceous rocks in South America, yet therian mammals (marsupials and placentals) are not found.

Around the end of the Cretaceous, marsupials and placental herbivores arrived in South America, presumably from North America, and South America probably provided the gateway to a route across Antarctica for marsupials to reach Australia.

The climatic changes at the end of the Eocene seem to coincide with the arrival of a further few immigrants into South America: rodents and monkeys, tortoises, and colubrid snakes. The same climatic changes led to the spread of Oligocene grasslands over much of South America, and the early expansion of the South American placentals into a guild of open-country grazers.

Apart from these brief periods of immigration, Cenozoic evolution in South America took place in isolation for over 60 m.y. The strange South American mammals in particular are well known, and they divided up available ecological roles in the usual way. Charles Darwin noticed peculiar fossil mammals in Argentina during his voyage on the Beagle, and later expeditions to Argentina have found hundreds of beautifully preserved Cenozoic fossils.

From Early Cenozoic times, the South American marsupials took on the roles of small insectivores (and still do). There is a living aquatic marsupial with webbed feet and a watertight pouch. *Argyrolagus* was a rabbit-sized marsupial that looked like a giant kangaroo rat. It hopped and had ever-growing molars for grazing coarse vegetation. The arrival of the placental rodents did not affect these small marsupials. One of the most successful marsupials in the world, even in the face of intense competition from placentals, is the small omnivorous opossum, *Didelphis*.

The placental grazers had evolved by the Miocene into a bewildering variety of forms ranging from rhino-sized to rabbit-sized. *Thoatherium* and *Diadiaphorus* (Figure 18.8) had an uncanny resemblance to horses, with long faces, horse-like front teeth, grinding molars, straight backs, and slender legs ending in one or three toes. Some of their relatives looked like camels. Large vegetarians such as *Toxodon* had large grinding molars that grew through most of the life of the animal (Figure 18.9).

Armadillos, sloths, and anteaters are also characteristic South American mammals. Armadillos and their relatives evolved heavy body armor for protection and became highly successful opportunistic insectivores and scavengers. The

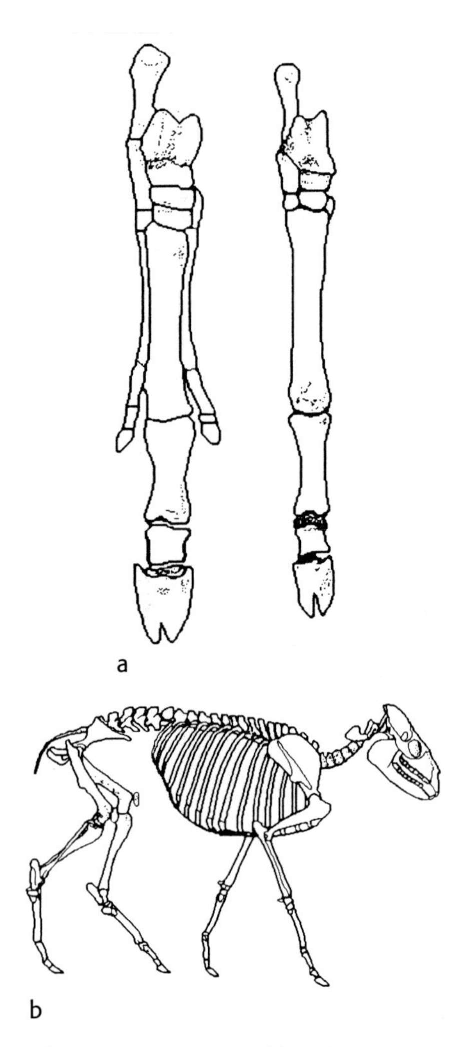

Figure 18.8 (a) The hind feet of two Cenozoic mammals from South America, *Diadiaphorus* (left), and *Thoatherium* (right), the thoat. Compare the hind feet of horses in Figure 17.12. These strikingly similar structures evolved in parallel in these two separate lineages of savanna running animals. (b) The skeleton of *Diadiaphorus*, showing that the resemblance to horses involved the entire body structure. (After Scott.)

Pleistocene armadillo *Glyptodon* was very large, probably a vegetarian, 1.5 m (5 ft) long. It had a thick armored skullcap as well as body armor, and some glyptodont species had a spiked knob at the end of the tail (Figure 18.10). Glyptodonts were certainly too big to burrow like the smaller armadillos, and they had to be heavily armored and armed to survive out on the surface. Naturally, their skeleton was very strong to support all the weight of the armor.

Sloths now live in trees, eating leaves and moving with painful slowness. But remains of huge ground sloths have been found in South America, including one that must have been almost as large as an elephant. Anteaters evolved from the same group of ancestors but are now specialized to an amazing extent for eating termites, beginning by tearing apart their nests with tremendously powerful clawed forearms.

The most impressive South American creatures were the larger carnivores. None of them were placental mammals, and most were marsupials. This is not surprising, considering how savage the surviving little marsupial insectivores are, but it is unusual compared with other continents. Borhyaenids were basically like wolves, but were generally larger. *Proborhyaena* was as big as a bear and probably had a similar way of life. *Borhyaena* itself was a wolf-sized Miocene marsupial with canine teeth adapted for stabbing and molars that had evolved into meat-slicing teeth (Figure 18.11). It was a successful medium-sized carnivore, but it was the last of the large borhyaenid carnivores. They were replaced by invading placentals from the north and by giant predatory birds.

Thylacosmilids looked like large cats. *Thylacosmilus* was a marsupial sabertooth, but its savage stabbing canines were better designed than those of the placental sabertooth cats of North America. In *Thylacosmilus* the sabers were longer, slimmer, more securely anchored in huge, recessed tooth cavities extending far up the face; thus, they were better protected from damage than those of true cats (Figure 18.12). The sabers were ever-growing and self-sharpening, and they were backed by more powerful neck and head muscles. Presumably they were adapted to killing large (placental) herbivores by stabbing and slashing deep into the soft tissues of throat or belly. The cheek teeth were not as powerful as those of placental cats, however.

These amazing marsupials had unusual competitors for mastery of the carnivorous guild, the phorusrhacids: flightless, ostrich-sized birds equipped with very powerful tearing beaks as well as foot talons (Figure 13.22). It seems that the phorusrhacids eventually gained the upper hand over the carnivorous marsupials.

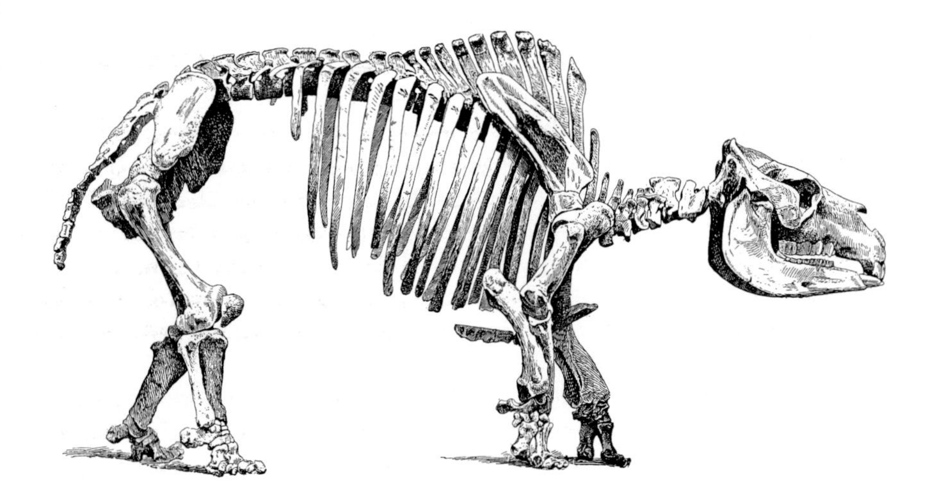

Figure 18.9 *Toxodon*, a large vegetarian mammal from the Cenozoic of South America. (After Lydekker.)

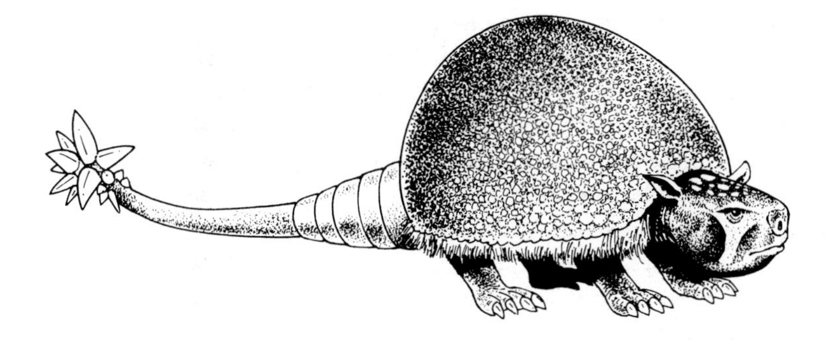

Figure 18.10 A glyptodont, a giant, heavily armored extinct relative of living armadillos. This one was close to 3 m (9 ft) long. (Reconstruction by Bob Giuliani. © Dover Publications Inc., New York. Reproduced by permission.)

South America had its own group of crocodiles, the sebecids. They apparently evolved in Gondwana in the Cretaceous, survived the K–T extinction, and radiated in the Early Cenozoic in South America to become powerful terrestrial predators. Unlike aquatic crocodiles, they had high, deep skulls and snouts. Other crocodilians in South America also evolved into unusual morphologies; for example, a duck-billed caiman is known from the Miocene of Colombia.

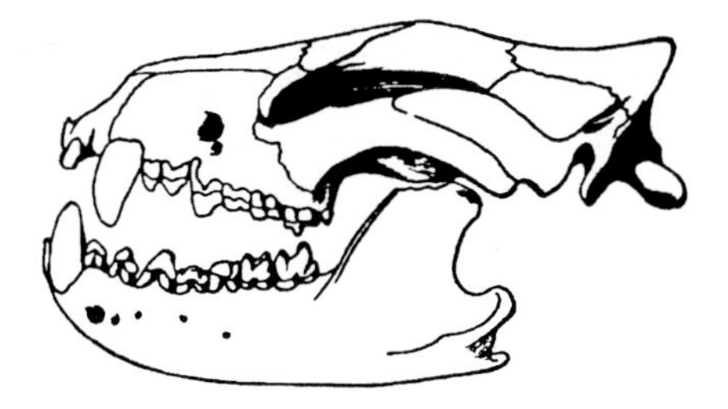

Figure 18.11 *Borhyaena,* a wolf-sized marsupial carnivore from the Cenozoic of South America. (After Sinclair.)

The South American ecosystem gained new immigrants in Oligocene times, around 25 Ma, with the arrival of rodents and primates, probably from Africa by way of islands in the widening Atlantic Ocean (Figure 18.7). Both groups radiated widely. The primates radiated into the distinctive New World monkeys, evolving habits and characters in parallel with gibbons and Old World monkeys. The rodents evolved into forms that include the world's largest rodent (by far): *Phoberomys* from the Miocene of Venezuela weighed 700 kilograms (about 1500 lb)! Other members of the Cenozoic South American fauna included more giants, the largest flying birds of all time, the teratorns (Chapter 13). The largest turtle of all time, *Stupendemys,* lived along the north coast close to *Phoberomys.*

This unique ecosystem suffered four tremendous shocks in ten million years and has almost completely disappeared. First, Antarctica froze up, with the result that the Humboldt Current, flowing most of the way up the west coast of South America, became much colder and stronger. Second, tectonic activity along the Pacific coast raised the Andes as a major mountain chain. Together, these two events drastically lowered rainfall over most of the continent, and much of the area turned from forest and well-watered plain to dry steppe. This led, in the Later Miocene, to the extinction of many animals, including the terrestrial crocodiles and especially the large-bodied savanna herbivores.

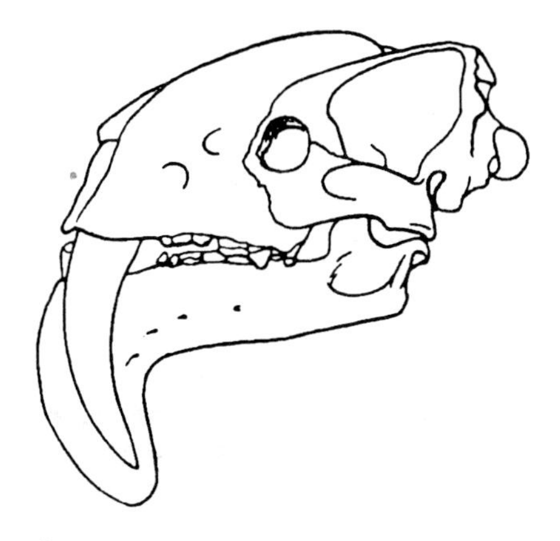

Figure 18.12 The skull, jaws, and teeth of the South American marsupial sabertooth *Thylacosmilus.* (From Riggs.)

Third, South America drifted northward towards Central and North America (Figure 18.7). By about 6 Ma, the gap was small enough to allow a few animals to cross it, more or less by accident. North American raccoons and some mice and rats crossed to the south, while two kinds of sloths crossed to the north. The effect of the competition was seen almost immediately. Many borhyaenids were replaced by raccoons, and the largest of them, the bear-like *Proborhyaena,* was replaced by a bear-sized raccoon. Finally, at about 3 Ma, the last important sea barrier was bridged, and animals could walk from one continent to the other.

Ecological principles suggest what should happen when an exchange of animals takes place. A larger continent such as North America should contain a larger diversity of animals than its smaller counterpart, and the fossil record confirms that this was true just before the exchange. Therefore, if the same proportion of animals from each continent migrated to the other, one would expect more North American animals to go south than the reverse. If a continent can hold only so many families or genera of animals, then one would predict extinctions on each continent, but more in South America than in North America. The effect would be accentuated because

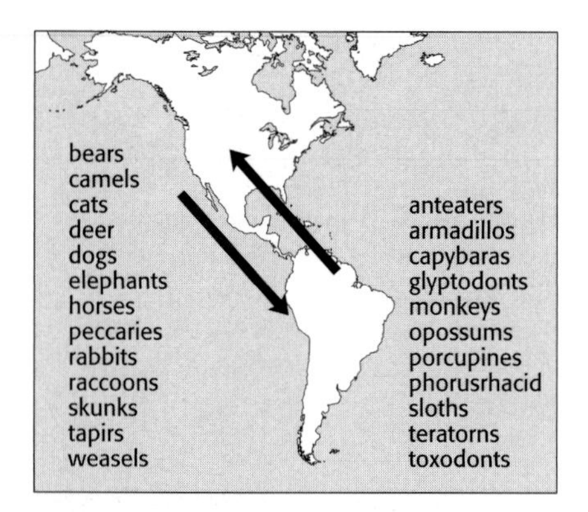

Figure 18.13 The Great American Interchange.

North America was at least intermittently connected with Eurasia, and altogether this huge northern area of temperate open country held a great variety of savanna animals. In contrast, the area of savanna in South America was not as large as one would think, because the continent is widest in equatorial latitudes and narrows significantly to the north and south. South American savanna faunas might have been very vulnerable to invasion from the north.

The major exchange happened after 3 Ma (Figure 18.13). Camels, elephants, bears, deer, peccaries, horses, tapirs, skunks, rabbits, cats, dogs, kangaroo rats, and shrews entered South America. Monkeys, opossums, anteaters, sloths, armadillos, capybaras, toxodonts, porcupines, and glyptodonts migrated north, with the giant birds—a phorusrhacid and a few teratorns.

The South American immigrants to North America flourished there, and so did the successful North American immigrants that moved south. Overall, however, there was a net major extinction of South American groups. The large, native marsupial carnivores and most of the phorusrhacids seem to have been outcompeted by the cats and dogs from the North, and the remaining savanna browsers and grazers were almost all wiped out, perhaps outcompeted by the northern horses and camels, perhaps hunted out by the new predators. The sabertooth marsupial *Thylacosmilus* was replaced by a real sabertooth cat. Even earlier invaders suffered: the bear-sized raccoon was replaced by a true bear.

Overall, the result was nearly as expected, with the South American animals coming out much the losers. North American invaders survive in strength today in South America, including all the South American cats, the llamas, and dozens of rodents.

The geographical changes that had permitted the interchange also changed the climate of the Atlantic Ocean, and this in turn caused drastic changes in the land ecology of North and South America as the northern ice ages began in earnest around 2.5 Ma. South American faunas suffered another catastrophic extinction in the Late Pleistocene. This time, similar extinctions took place in North America too, and we shall examine this in Chapter 21.

AFRICA

Africa (plus Arabia, so perhaps I should write Africarabia) was part of Gondwana until the Cretaceous, when it broke away from South America on the west and Antarctica and India on the east (Figure 6.4). From Late Cretaceous times onward, Africa and South America, and the animals and plants living there, had increasingly different histories.

Africa had dinosaurs much like those of the rest of the world during the Late Cretaceous, but there is no record of any African Cretaceous mammal. This may change soon, because molecular evidence suggests that there must have been spectacular evolution among the Afrotheria (Chapter 15) on the isolated continent of Africarabia, which had split from South America but was not close to Europe or Asia. The continent lay south of its present position, bounded on the north by the tropical Tethys Ocean.

Our first look at the fossil Cenozoic life of Africa comes from the Eocene rocks of Egypt, laid down on the northern edge of the continent. Shallow warm seas teemed with microorganisms whose shells formed the limestones from which the Pyramids and the Sphinx were carved.

Here we find early whales and sea cows, which probably evolved adaptations for marine life in swamps and deltas along the shores of Tethys. Moeritheres were amphibious animals related to sea cows and to elephants. *Moerithium* itself is an Eocene animal from Africa, and looked like a small, fat elephant with the ecology of a hippo. Other Eocene fossils from Egypt include some early primitive carnivores, the creodonts (also known from other continents).

By Oligocene times, Egypt was the site of lush deltas where luxuriant forest growth housed rodents, primates, and bats, all recent Eurasian immigrants. Pig-like anthracotheres had crossed from Eurasia, but there were African groups too. Hyraxes are small- to medium-sized vegetarians that look much like rodents. *Arsinoitherium* was a large-bodied browser (Figure 18.14).

Eocene and Oligocene African mammals are a mixture of native African groups and a few successful immigrants from Eurasia. Even in the Late Oligocene, the large African mammals were still arsinioitheres and a diverse set of elephants. But in the Miocene, Africarabia drifted far enough north to bring it close to Eurasia, and finally the two continental edges collided around 24 Ma (Figure 18.15). There were important times of uninterrupted migration between the two land masses. The interchanges affected animal life throughout the Old World, almost on the same scale as the Great American Interchange (Figure 18.16).

Twelve families of small mammals appeared in Africa in the Early Miocene, mostly insectivores and rodents from Eurasia. Early deer, cattle, antelope, and pigs largely replaced the hyraxes at medium sizes, and rhinos and the first giraffes were large invaders. Cats arrived and began to replace the older creodonts. Going the other way, elephants walked out of Africa into Eurasia, in at least two major adaptive groups, mastodons and true elephants. Some large creodonts even reinvaded Eurasia from Africa.

In a second exchange around 15 Ma, a new set of African animals, including apes, quickly spread over the woodlands and forests of Eurasia. Hyenas and shrews migrated into Africa.

It is not clear whether the continental collision itself altered world climate, or whether climate was affected more by major events in the Southern Hemisphere. Whatever the cause or causes, the Miocene change from forest to savanna was partly responsible for the success of the large number of grazing animals listed above.

By the end of the Miocene, more immigrants had appeared in Africa: small animals, including many bats, and the three-toed horse *Hipparion.* Meanwhile, hippos evolved in Africa, and the antelope and cattle that had arrived earlier evolved into something close to the incredible diversity we see today in the last few game reserves.

Africa and Eurasia have been connected by land since the Miocene, but this does not automatically imply free exchange of animals. For example, the Mediterranean Sea dried into a huge salty desert like a giant version of Death Valley around 6 Ma. Only a few animals could have crossed this barrier. Later, the development of desert conditions in the Sahara formed another fearsome barrier to animal migration for most of the past few million years. Today, North African animals are more like those of Eurasia than those of sub-Saharan Africa.

The Early Pleistocene saw a large extinction in Africa, with one-third of the mammals becoming extinct. But they were replaced by newly evolving species, so total diversity remained high. Africa apparently did not feel the effects of the ice ages too drastically. In contrast, when European animals were pushed southward by the advancing cold and ice, they could not cross the Mediterranean and the Sahara, and many became extinct. Human hunting activities have affected Africa less than other continents, perhaps because humans evolved gradually there and the animals had time to adjust to them. On other continents human impact was much more sudden and severe. Animals formerly widespread over the world are now confined to Africa

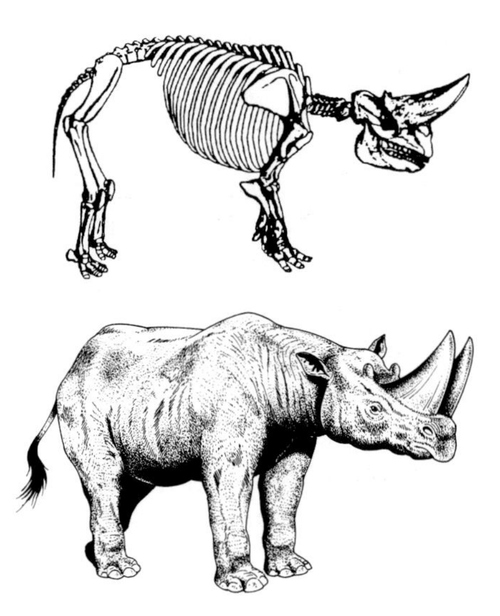

Figure 18.14 *Arsinoitherium*, a large browsing mammal weighing close to 5 tons, from the Oligocene of Africa. (Skeleton from Andrews. Reconstruction by Bob Giulani, © Dover Publications Inc., New York. Reproduced by permission.)

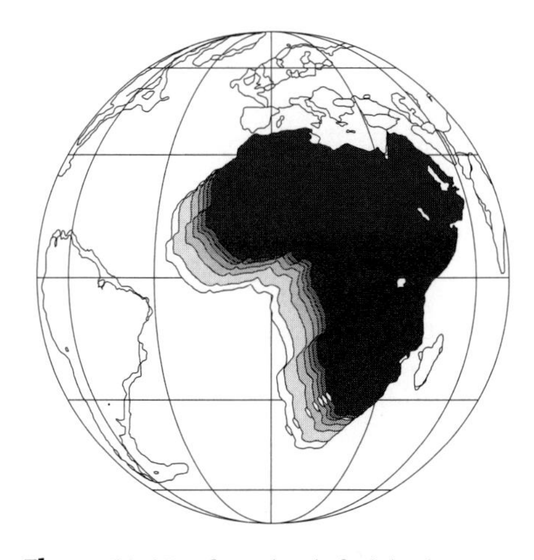

Figure 18.15 Africarabia drifted slowly northeast during the Cenozoic. Finally it collided with western Asia in the Early Miocene along a line that is now the Zagros Mountains. The African continent then rotated slightly clockwise, splitting away from Arabia. At the end of the Miocene the northwest corner of Africa collided with western Europe to close off the Mediterranean Sea as a vast lake that quickly dried up. Meanwhile the Red Sea opened up as Africa swung away from Arabia, and the great African Rift Valley was formed.

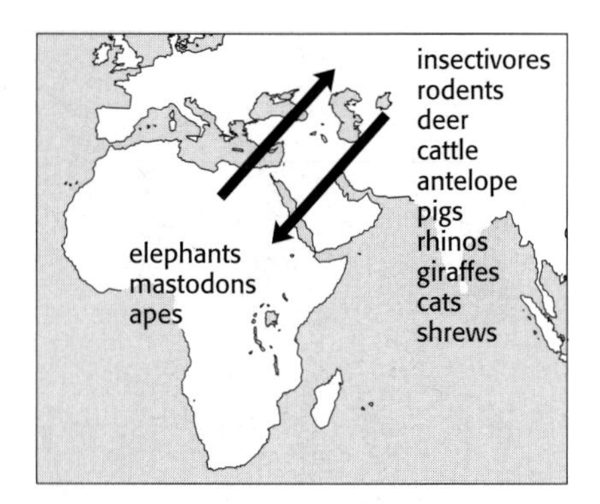

Figure 18.16 The Great Old World Interchange of animals between Africarabia and Eurasia in the Miocene.

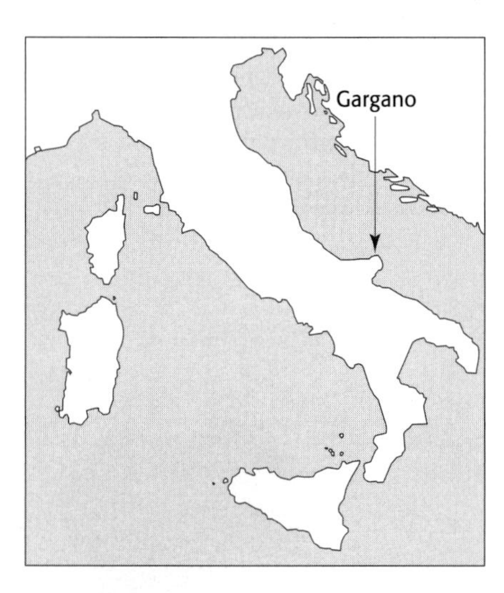

Figure 18.17 The geographic setting of the Gargano Peninsula, in Italy, which formed an island in Miocene times.

or nearly so (rhinos, lions, cheetahs, hyenas, and wild horse-like species, zebras). Protected there by the geographical, climatic, and historical events of the Late Cenozoic, many creatures survived relatively successfully in Africa until this century.

ISLANDS AND BIOGEOGRAPHY

Strict geographic barriers prevent land plants and animals living on islands from moving easily to other land areas, and potential invaders also must cross barriers. This means that island faunas and floras tend to evolve in greater isolation than those with wider and more variable habitats. Of course, this is true at any scale, whether we look at small islands or continent-sized ones. Islands past and present can teach us a great deal about evolution. It is no accident that Darwin was particularly enlightened by his visit to islands like the Galápagos, and Wallace by his years in Indonesia.

We have seen some of the vagaries of continental faunas over a time scale of tens of millions of years, but it is worth looking at cases where smaller-scale events on smaller islands over smaller lengths of time show the rapidity and power of natural selection in isolated populations.

The Raptors of Gargano

In 1969, three Dutch geologists were exploring the Mesozoic limestones of the Gargano Peninsula in southern Italy (Figure 18.17). Sometime in the Early Cenozoic, this block of land was raised above sea level and caves and fissures formed in the limestone. In Early Miocene times the Gargano area was cut off from the mainland by a rise in sea level to form an island in the Mediterranean Sea of the time. Land animals living there were isolated on the island as the sea rose. Over only a few million years, animals occasionally fell into fissures in the limestone, where they were covered by thin layers of soil and preserved as fossils. Today, the limestones are quarried for marble, and the bones can be found in the pockets of ancient soils exposed in the quarry walls.

No large animals were isolated on Gargano as it was cut off. The only large reptiles were swimmers (turtles and crocodiles) and the only mammalian carnivore was also a swimmer, a large otter with rather blunt teeth that probably ate shellfish most of the time and would not have hunted on land.

Because there were no land carnivores, small mammals evolved quickly into spectacular forms. Small rabbit-like pikas were abundant, and gigantic dormice evolved on the island. Giant hamsters were eventually outcompeted by true rats and mice. Some of the Gargano mice grew to giant size, with skulls 10 cm (4 in) long, and many evolved fast-growing teeth as complex as those of beavers. They probably chewed very tough material. *Hoplitomeryx* is a small deer which evolved horns instead of antlers (Figure 18.18).

If there were no cats, dogs, or other terrestrial carnivores, how were the rodents kept under control? By disease and starvation? And why did *Hoplitomeryx* evolve spectacular horns, if there were no carnivores to fight off? The horns were too lethal to have been used for fighting between individuals of the species.

The answer to these questions seems to have been raptors—birds of prey. A giant buzzard, *Garganoaetus*, was as large or larger than a golden eagle. Presumably it hunted by day. It would have been perfectly capable of taking a small or young

Hoplitomeryx, and the horns may have evolved to protect the back of the neck against raptors. Normally, small deer hide in vegetation, but Gargano was a bare, limestone island, with no cover by day. At night, the owls took over: the largest barn owl of all time evolved on Gargano.

Figure 18.18 A deer with horns instead of antlers, *Hoplitomeryx*, evolved in geographic isolation on the Gargano Peninsula in Miocene times. (After Leinders.)

Giant Pleistocene Birds on Cuba

Cuba had a strange set of animals isolated on it during the ice ages. Pleistocene mammals have been found in enormous numbers in limestone cave deposits, and we have a reasonable idea of the unusual ecology the island must have had. In particular, there were enormous numbers of ground sloths and rodents, and insectivores were very common. Tens of thousands of mouse jaws have been found in one cave, and another site yielded over 200 ground sloths. In addition, large numbers of fossil vampire bats imply that there were large numbers of warmblooded animals for them to prey on. Similar but less spectacular fossils have also been found on Puerto Rico and Hispaniola.

There are practically no carnivorous mammals in these deposits, and as at Gargano, we are forced to wonder what kept the animal populations in check. The answer here too seems to be raptors. In the caves with the animal bones there are also great numbers of the bones of small birds. This suggests that the cave deposits are mainly the accumulations of owl pellets and bat colonies. But the size of the bones indicates that the owls were producing pellets much larger than normal owls do.

In 1954 a gigantic fossil owl was discovered, large enough to have preyed upon juvenile ground sloths. Later that year, a fossil eagle bigger than any living species was found. A fossil vulture as large as a condor, and a fossil barn owl as large as the species at Gargano, fill out a picture of a set of predators quite alien to our experience today.

The gigantic owl *Ornimegalonyx* must have stood a meter high. It may not have been a powerful flyer, because its breastbone looks weak relative to the rest of the skeleton. But with its tremendous beak and claws, it could have preyed successfully on rodents and young sloths. By day the giant eagle would have performed the same function—it is larger than the monkey-eating eagle of the tropical forest today. Presumably the giant vulture fed from the carcasses of giant ground sloths, and the other large owls added to the flying nocturnal predators.

The whole ecosystem became extinct towards the end of the Pleistocene on Cuba and on all the other Caribbean islands. We don't know enough of the geological history of Cuba to suggest that human intervention caused these extinctions.

Other Biogeographic Islands

Islands in the biogeographic sense do not have to be small pieces of land separated by water. Mountain-dwelling species can be isolated by stretches of plains country around them, so that in some parts of Africa each mountain system has its own spider fauna. Lake faunas are separated by watersheds. Woodland faunas can be separated by open stretches of prairie. Even in the ocean, shallow-water animals find deep ocean basins just as much of a barrier to them as land masses are. Marine organisms can often be thought of as living on water islands surrounded by seas of land.

For example, in the Late Oligocene, around 30 Ma, walruses and sea lions evolved around the edges of the North Pacific, while seals evolved around the edges of the North Atlantic. The two groups did not intermingle until much later in the Cenozoic, presumably because they were separated by the barrier of the north–south land masses of the Americas.

Further Reading

Archer, M., et al. 1989. Fossil mammals of Riversleigh, Northwestern Queensland: preliminary overview of biostratigraphy, correlation and environmental change. *Australian Zoologist* 25: 29–65.

Diamond, J. M. 1990. Biological effects of ghosts. *Nature* 345: 769–70. [The prehistoric ecology of New Zealand.]

Flannery, T. F. 1995. *The Future Eaters: An Ecological History of the Australasian Lands and People.* New York: George Braziller. [Part I is the story of Australasian life before the arrival of humans.]

Flynn, J. J., and A. R. Wyss. 1998. Recent advances in South American mammalian paleontology. *Trends in Ecology & Evolution* 13: 449–54.

Godthelp, H., et al. 1992. Earliest known Australian Tertiary mammal fauna. *Nature* 356: 514–16.

Kappelman, J., et al. 2003. Oligocene mammals from Ethiopia and faunal exchange between Afro-Arabia and Eurasia. *Nature* 426: 549–52, and comment 509–11.

Marshall, L. G. 1988. Land mammals and the Great American Interchange. *American Scientist* 76: 380–8.

Marshall, L. G. 1994. The terror birds of South America. *Scientific American* 270 (2): 90–5.

Pascual, R., et al. 1992. First discovery of monotremes in South America. *Nature* 356: 704–6.

Rasmussen, D. T., and E. L. Simons. 1988. New Oligocene hyracoids from Egypt. *Journal of Vertebrate Paleontology* 8: 67–83. [Hyraxes formed an Oligocene mini-radiation.]

Rich, P. V., and G. F. van Tets (eds). 1985. *Kadimakara.* Victoria, Australia: Pioneer Design Studio. [The extinct animals of Australia.]

Richardson, K. C., et al. 1986. Adaptations to a diet of nectar and pollen in the marsupial *Tarsipes rostratus* (Marsupialia: Tarsipedidae). *Journal of Zoology, London A* 208: 285–97.

Sânchez-Villagra, M. R., et al. 2003. The anatomy of the world's largest extinct rodent. *Science* 301: 1708–10, and comment 1678–9. [*Phoberomys*, 600 kg (that's my guess: the authors claim 700 kg).]

Savage, R. J. G., and M. R. Long. 1987. *Mammal Evolution: An Illustrated Guide.* London: British Museum (Natural History).

Simpson, G. G. 1980. *Splendid Isolation.* New Haven: Yale University Press. [South American mammals.]

Springer, M. S., et al. 1997. Endemic African mammals shake the phylogenetic tree. *Nature* 368: 61–4.

Stanhope, M. J., et al. 1998. Molecular evidence for multiple origins of Insectivora and for a new order of endemic African insectivore animals. *Proceedings of the National Academy of Sciences* 95: 9967–72.

Stehli, F. G., and S. D. Webb (eds). 1985. *The Great American Biotic Interchange.* New York: Plenum.

Webb, S. D. 1991. Ecogeography and the Great American Interchange. *Paleobiology* 17: 266–80.

CHAPTER NINETEEN

Primates

We are particularly interested in our own ancestry. After all, the recent evolution of primates has produced humans, the most widespread, powerful, and potentially destructive biological agent on Earth.

Most living primates are small, tropical, tree-dwelling animals that eat high-calorie food, mainly insects. This is particularly true of groups that retain the most primitive primate characters. Taken at face value, this suggests that primate ancestors searched for insects, fruit, seeds, or nectar on small branches, high in trees, and in smaller bushes. Evolutionary evidence supports this scenario, because primates are most closely related to three other mammal groups that also live in trees—tree shrews, dermopterans or "flying lemurs," and bats (Figure 19.1).

This group of small arboreal mammals had invaded forest habitats by the end of the Cretaceous, and several lines must have survived the K–T extinction. The surviving members of that clade are bats, primates, tree shrews, and the one surviving species of dermopteran, the colugo or "flying lemur" of the Indonesian rain forest (Figure 19.1). Because they were such close relatives, it has been difficult for paleontologists to distinguish early small primates from early members of the other groups. All these animals were small and probably ate nectar, gum, pollen, seeds, insects, and fruit in the canopy forest.

Living primates have large eyes, turned forward to give excellent stereoscopic vision. The combination of large eyes and stereoscopic vision may have evolved in primates—as in cats, owls, and fruit bats—to help search for food by sight rather than smell, because it allows the animal to judge the distance of a food item without moving its head. Stereoscopic vision promotes agility and coordination, especially when an animal has hands and feet adapted for grasping and fine manipulation, with pads and nails rather than paws and claws. Grasping feet and hands allow primates to forage along narrow branches, and live prey or other food can be reached or seized by a hand or hands rather than by a lunge with the whole body and head. Compare the coiled strike of snakes and the tongue strike of chameleons (Figure 19.2), frogs, and toads, which all do the same thing in different ways.

Primate fetuses show rapid growth of the brain relative to the body, so they are born with relatively larger brains than other mammals. Gestation time is long for body size, and primates have small litters of young that develop slowly and live a long time. Primates evolved high learning capacity, complex social interactions, and unusual curiosity. The evolution of curiosity is useful in searching for food, and

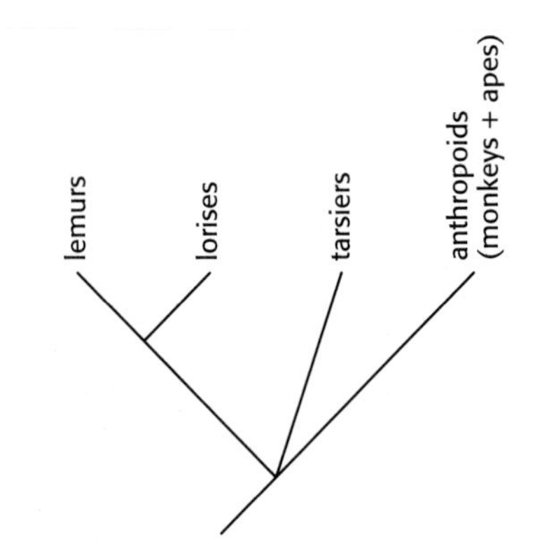

Figure 19.1 Cladogram of living primates. The branches between the groups are very deep in time, so the branching pattern close to the base is disputed.

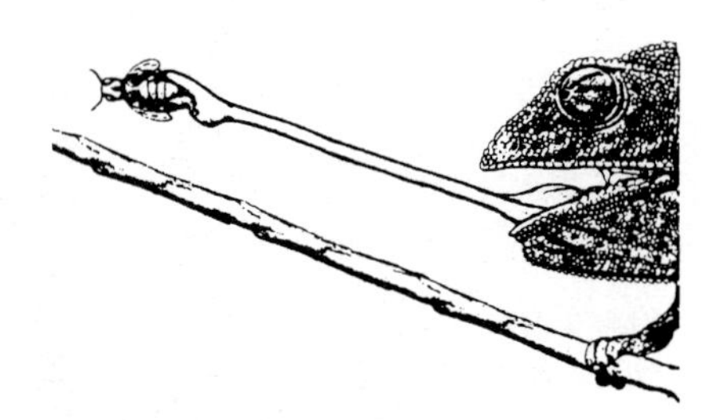

Figure 19.2 Catching small, agile prey without moving the entire body.

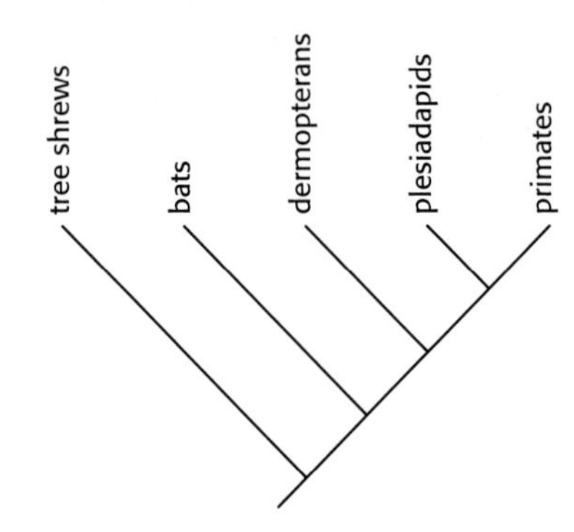

Figure 19.3 Cladogram of some tree-dwelling mammal groups: primates and their nearest relatives. This cladogram, one of several possibilities, shows primates and plesiadapids as closest allies. Primates might *include* plesiadapids. If so, there has to be a name for those primates that *aren't* plesiadapids. Euprimates has been suggested for this.

high learning capacity, memory, and intelligence help individuals to make correct responses in a complex, ever-changing environment.

Living primates are often divided into two groups: small-brained, small-bodied animals called **prosimians**, and the relatively large-brained **anthropoids** (monkeys and apes). However, prosimians contain two clades, each with a long evolutionary history: the **tarsiers** of Southeast Asia on the one hand, and the **lorises** of Africa and the **lemurs** of Madagascar on the other hand (Figure 19.3). It is most likely that the lorises and lemurs are an African group of primates that forms one clade, while tarsiers form another. Anthropoids are a clade. But there are many early primates that muddy our picture of early primate evolution. In particular, it is probable but not yet clear that tarsiers are more closely linked with anthropoids than the others, and it is quite unclear how many early primate groups fit into the evolutionary scheme.

THE LIVING PROSIMIANS

Living **lemurs** are confined to Madagascar, and must have reached that island from Africa. Molecular evidence suggests that the ancestors of lemurs reached Madagascar in Paleocene times, but no other primates did so until humans did so about 2000 years ago. Lemurs flourished in their island refuge. The actual fossil record of lemurs in Madagascar goes back only as far as the Miocene, but that is enough to document a startling radiation into at least 45 species of lemurs on the island, adapted to a great variety of life styles.

Living lemurs (Figure 19.4) are specialists at vertical clinging and leaping, in which the front limbs are used for manipulating, grasping, and swinging, while the hind limbs are powerful for pushing off. Most lemurs are medium-sized (weighing a few pounds), and are omnivorous, eating fruits and leaves. A few lemurs are small: the mouse lemur weighs only 50 g or so (about 2 ounces). The largest lemur, *Archaeoindris*, reached around 200 kg, the size of a gorilla, and became extinct only recently; as an adult it must have been a ground dweller. The recently discovered extinct lemur *Palaeopropithecus* was adapted for moving slowly in the forest canopy in the same way as the South American sloth, while *Megaladapis* was probably rather like the Australian koala in its ecology.

Lorises and **bushbabies** are small, slow-moving, nocturnal hunters of insects. Lorises live in African and Southeast Asian tropical forests, while bushbabies are exclusively African.

Tarsiers, in comparison, are small, agile primates, adapted to eating small animals and insects. Today they survive only in Southeast Asia, but were much more widespread in the past. Essentially, tarsiers are living fossils, with possible ancestors in the Early Cenozoic that seem to have had much the same anatomy and way of life. They diverged from anthropoids so long ago that the two groups share little similarity today.

EARLIEST PRIMATES

The best-known early primates are **plesiadapids**, an important group of animals found mainly in the Paleocene of North America and Europe (Figure 19.5). Larger plesiadapids were rather heavy in build, ecologically like squirrels or marmots, small brained and rather small-eyed, with teeth adapted for cropping vegetation (Figure 19.4). Because plesiadapids looked and probably lived like large rodents, they must have competed to some extent with Paleocene multituberculates (Chapter 15).

Figure 19.4 The black lemur, a living prosimian primate from Madagascar. (From Meyers.)

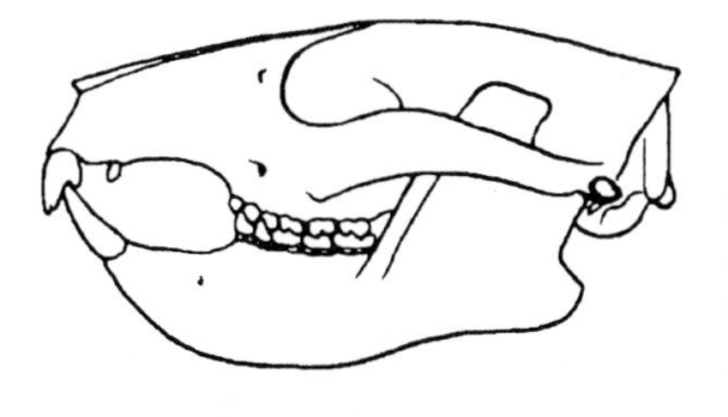

Figure 19.5 The skull of a typical plesiadapid, one of a radiation of early primate-like animals that included animals with an ecology like rodents. (After Gingerich and Krause.)

Figure 19.6 *Notharctus*, an adapid from the Eocene of North America. It probably had an ecology like that of its living relatives, the lemurs (Figure 19.4). Like other vegetarian primates, *Notharctus* was rather small-brained (for a primate). (Negative 319565. Courtesy of the Department of Library Services, American Museum of Natural History.)

Most people think that plesiadapids did not give rise to any descendant groups. However, some of the last plesiadapids evolved some characters that are found also in euprimates: the plesiadapid *Carpolestes* had grasping hands and feet, and the big toe had a nail (like a euprimate) while the other toes had claws (like other plesiadapids). But *Carpolestes* did not have stereoscopic vision, and apparently did not leap. If the carpolestids were the plesiadapids that did evolve into euprimates, they did it in the Paleocene and in Asia, because *Carpolestes* and all other North American plesiadapids became extinct at the end of the Paleocene, apparently in competition with rodents and the euprimates that arrived from Asia.

Recently research has begun to focus on China in the search for the early primate radiation, the euprimate lineage, and the ancestry of anthropoids. Anthropoids may have evolved from euprimates in East Asia.

As we have seen (Chapter 17), a warm event at the end of the Paleocene allowed Asian mammals to reach North America. Among the primates that arrived were omomyids and adapids. **Adapids** include *Diacronus*, from Paleocene rocks in south China, a plausible ancestor for the Early Eocene *Cantius* from western North America. Adapids look like small lemurs in limb structure, and are related to them. They presumably moved in the same way (Figure 19.6). Many adapids evolved toward larger body size and turned from branch-stalking for insects to a swinging and leaping mode of movement, and to a diet that included much more plant material as well as animal prey, with some eating fruit and others leaves. **Omomyids** were

Figure 19.7 Life restoration of the small Eocene omomyid *Tetonius*, from North America, as an alert tarsier-like animal. By L. Kibiuk, under the supervision of K. D. Rose. (Courtesy of Kenneth D. Rose of The Johns Hopkins University.)

small, alert, active nocturnal insect eaters in the forest (Figure 19.7). Ecologically they were probably like tarsiers, and in terms of evolution they were probably near the base of the anthropoid/tarsier clade (Figures 19.1, 19.8). *Eosimias* ("the dawn monkey") from the Middle Eocene of China, is tiny, about the same size as the smallest living anthropoid, the pygmy marmoset of South America. From incomplete specimens, *Eosimias* has been claimed as the earliest, most primitive anthropoid yet discovered; yet strong counterarguments suggest that it is a tarsier or close tarsier relative, and therefore not relevant to anthropoid ancestry. Nothing like *Eosimias* reached North America, as far as we know.

The northern continents slowly cooled as they drifted northward during the Eocene. Finally, at the end of the Eocene, primates disappeared from northern latitudes. Refugee adapids reached Southeast Asia in the Late Eocene, but "anthropoid" characters that some of them have are likely evolved independently in this region, and they seem to have died out without descendants. By Oligocene times there were practically no primates left in northern continents.

THE ORIGIN OF ANTHROPOIDS

The living higher primates, or anthropoids (monkeys and apes), have evolved into a variety of life styles and habitats that extends from the huge herbivorous gorillas to the tiny gum-chewing marmosets of South America. Various Eocene primates were most likely adept at four-footed climbing and leaping from branch to branch in three dimensions, using the full grasp of hands and feet for catching and holding small branches (Figures 19.6, 19.7). All the different ways in which lemurs, monkeys, gibbons, great apes, and humans move could have evolved from this generalized style shared by early primates. Of course, that does not help us find which Eocene primates were the anthropod ancestors.

The arm-swinging or brachiating of gibbons could have arisen by emphasizing the arms in movement. The careful, multilimbed climbing of orangutans in trees, the agility of monkeys, the four-footed scrambling and shambling of heavy apes, and the trotting of baboons on the ground could each have evolved by using all the limbs equally. The bipedal walking and running of australopithecines and humans could have been achieved by accentuating the role of the hind limbs in powerful pushing and of the forelimbs in grasping and handling.

Living anthropoids are divided into three evolutionary groups: **cercopithecoids** (Old World monkeys), **ceboids** (New World monkeys), and **hominoids** (Figure 19.8), which include gibbons, apes, and humans. Apart from *Eosimias* and some potential descendants in Southeast Asia, all other evidence points to the Eocene of Africa as the time and place for the radiation of anthropoids, and may imply they originated there too.

The Late Eocene Primates of Egypt

In Late Eocene and Oligocene times, the Fayum district of Egypt, not far from Cairo, lay on the northern shore of Africarabia as it drifted slowly northeastward (Chapter 18). Thousands of fossilized tree trunks, some of them more than 30 m

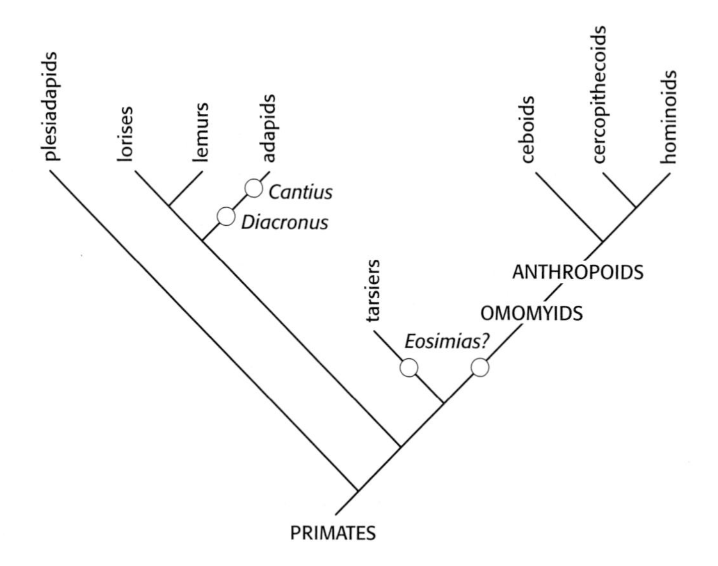

Figure 19.8 One of several possible phylograms of primate evolution. Radiations of omomyids and adapids in northern continents have provided much of the evidence for early primate evolution, and we may expect revisions as more evidence turns up in Africa and Asia.

(100 ft) long, show that tropical forests of mangroves, palms, and lianas grew along the levees of a lush, swampy delta. Water birds such as storks, cormorants, ospreys, and herons were abundant, as they are today around the big lakes of central Africa. Fishes, turtles, sea snakes, and crocodiles lived in or around the water, and early relatives of elephants and hyraxes foraged among the rich vegetation. The primates presumably ate fruit in the trees. The same fauna has been discovered as far south as Angola, so the Fayum animals were widespread around the coasts of Africa in Eocene and Oligocene times.

More than 2000 specimens of 19 species of fossil primates have now been collected from the Fayum deposits, most of them on expeditions led by Elwyn Simons. The primates include tarsiers, lorises, and bushbabies. But the others are anthropoids, all of which look like tree-climbing fruit and insect eaters.

Many of the Fayum primates have some advanced characters, but do not look like the direct ancestors of monkeys or apes. They are placed into a basal stem-anthropoid group called parapithecids (Figure 19.9). Parapithecids are small, weighing only up to 3 kg (7 lb). Their skulls are rather like those of Old World monkeys, but the rest of the skeleton looks primitive, more like that of South American monkeys. This miscellaneous group probably had a basic style of primate ecology, eating fruit in the trees. *Apidium*, for example, seems to have been adapted for leaping and grasping in trees.

Those Fayum anthropoids that are well enough known to compare individual sizes are sexually dimorphic. Males are larger and had much larger canine teeth than females, implying that males displayed or fought for rank, and that the animals had a complex social life that included groups of females dominated by a single male. Among living primates, it is generally the larger-bodied species that have these characters, especially in the Old World. However, the Fayum anthropoids show that size is not important in evolving these sex-linked characters, and they also suggest that these traits may well be basic to anthropoids. Simons and his colleagues suggest that they arose when anthropoids became active in daylight: group defense may be linked with the social structure.

Aegyptopithecus (Figure 19.10) is the best known of the Fayum anthropoids. It is a larger, monkey-sized primate with an adult weight of 3–6 kg (7–14 lb). Its heavy limb bones suggest that it was a powerfully muscled, slow-moving tree climber, ecologically like the living howler monkey of South America. It had many primitive characters, but its advanced features were more like those of apes than monkeys. Its

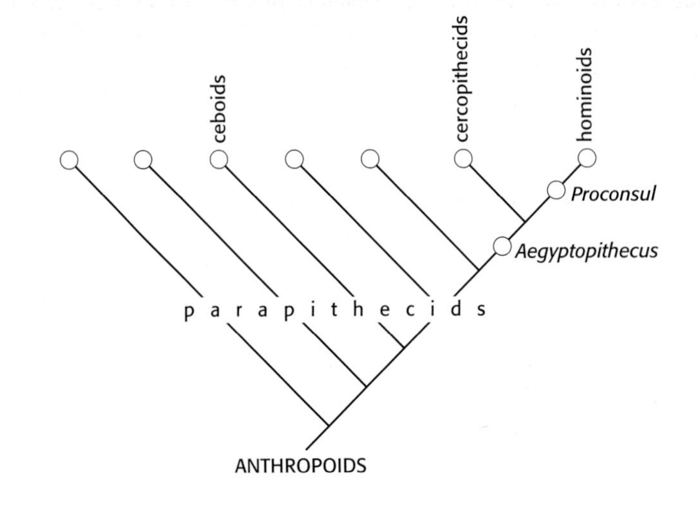

Figure 19.9 Phylogram of the higher primates, including evidence from the fossils of the Fayum. A miscellaneous group of Fayum anthropoids, the parapithecids, include the ancestor of ceboids (New World monkeys), and of *Aegyptopithecus*, which has characters that allow it to have been the ancestor of both Old World monkeys (cercopithecids) and the hominoids. *Proconsul* is a possible ancestor of all hominoids. This scheme is consistent with biogeographic evidence that suggests the ceboids diverged from Old World primates around the end of the Eocene.

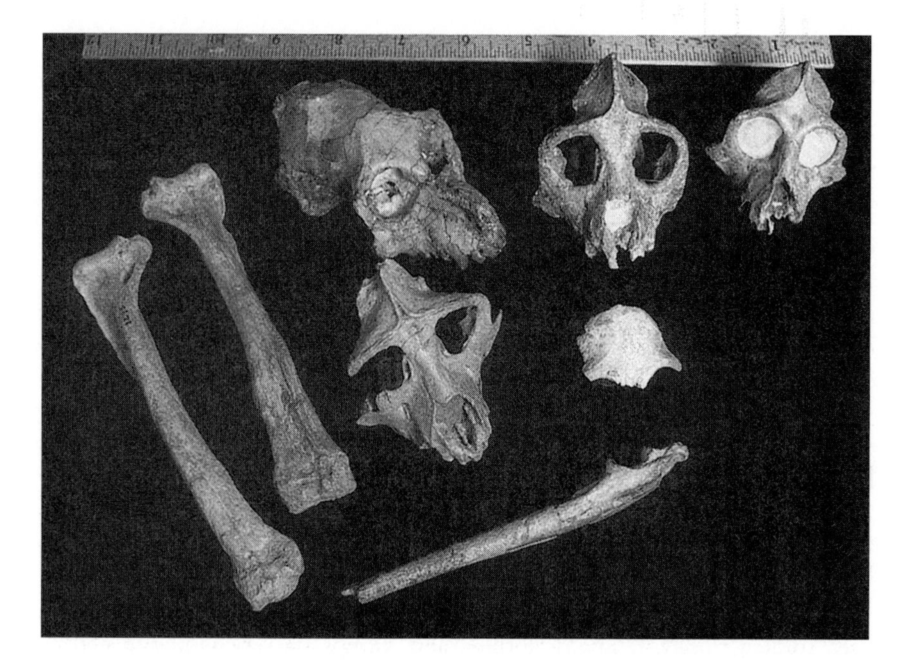

Figure 19.10 *Aegyptopithecus*, a little anthropoid from the Late Eocene of Egypt. It is the closest species we have yet found to the common ancestor of Old World primates. (Courtesy of Elwyn L. Simons of Duke University.)

brain was large for its body size, for example, and its foot bones were like those of Miocene apes. It had powerful jaws for its size, too. It may well be the common ancestor of all higher primates in the Old World: the cercopithecoids, or Old World monkeys, and *Proconsul* and the line leading to hominids (Figure 19.9).

The New World Monkeys

Primates reached South America by Oligocene times, and evolved there in isolation, never again influenced by exchange and contact with other primate groups. No prosimian or ape-like primate has ever been found in South America. Instead, the New World primates evolved to fill the ecological niches that monkeys and gibbons occupy in Old World forests.

New World primates, the **ceboids**, probably evolved from African immigrants that crossed the widening Atlantic in Early Oligocene times. For want of better information I have shown them as diverging from early Fayum parapithecids (Figure 19.9). But ceboids have some unique characters: prehensile tails that can be used as a fifth limb, and four more teeth than Old World primates, **cercopithecoids**. Ceboids are related to cercopithecoids only in that they are both anthropoids (Figure 19.9); many of their monkey-like characters evolved independently. Ceboid color vision uses different nerve pathways than that of cercopithecoids and apes, for example, and it evolved once only, very early.

The earliest ceboids, *Dolichocebus* from Patagonia and *Branisella* from Bolivia, are both Late Oligocene in age, perhaps 26 or 27 Ma. *Dolichocebus* is very much like the living squirrel monkey, and represents an ideal ancestor for it. In the same way, some Miocene South American monkeys are much like living spider monkeys and howler monkeys. A genuinely modern-looking owl monkey is known from Miocene rocks of Colombia at 12–15 Ma. Many of today's South American monkeys therefore qualify as living fossils. Either they evolved early and rapidly, or they have a longer fossil record still to be discovered.

All of them are tree dwellers. South American primates did not evolve into

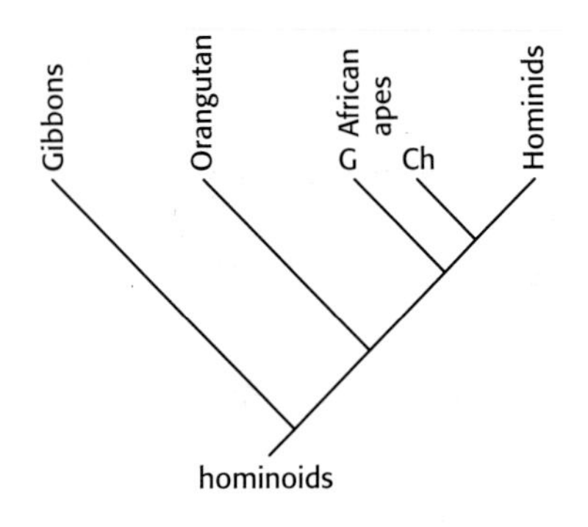

Figure 19.11 The cladogram of living hominoids that is suggested by molecular and fossil evidence. G, gorillas, Ch, chimps. The timing of the branching points is uncertain, and is not likely to be settled soon. Extinct groups are not shown on this cladogram.

terrestrial ways of life as Old World primates did, even though there have been extensive savannas in South America since the Miocene.

The Old World Monkeys

We know more about the emergence of Old World monkeys. The small fossil *Victoriapithecus* from the Miocene of East Africa is an ideal ancestor for this group. Interestingly, it is small even for a monkey at 3–5 kg (7–11 lb), and seems to have had a semiterrestrial ecology rather than the tree-dwelling habit that one might have expected.

EMERGENCE OF THE HOMINOIDS

Living hominoids include hylobatids (gibbons); pongids or Asian apes (only orangutans survive); the African apes (chimps and gorillas); and hominids (only humans survive). The physical, molecular, and genetic structure of living hominoids has been studied closely. Humans, gorillas, and chimps are very similar in genetic makeup and in protein chemistry, much closer than they are in body structure, but the orangutan differs significantly, and gibbons even more.

Hominoids almost certainly evolved from some African genus like Aegyptopithecus. The DNA clock suggests that the common ancestor of all hominoids split from monkeys about 33 Ma. The gibbons split off about 22 Ma, followed by the orangutan lineage at about 16 Ma. Finally the various living lineages of panids and hominids diverged from one another between 10 and 6 Ma (Figure 19.11). Protein clocks suggest more recent branching points.

Miocene Hominoids

About 20 Ma, Africarabia formed a single land mass that lay south of Eurasia and was separated from it by the last remnant of the Tethys Ocean. African animals were evolving largely in isolation from the rest of the world, and some groups, including the hominoids, were confined to Africarabia at this time, though they were widespread across it.

Early Miocene faunas of Africa were dominated by elephants and rhinos at large body size, primitive deer and hyraxes at medium size, and insectivores common at small sizes. The environment was forest, broken by open grassland and woodland. Primates of all kinds flourished, although it is difficult to describe their ecology and habits because body skeletons are not as well known as skulls. But prosimians and monkeys were rare, while hominoids were diverse and abundant. We have over 1000 hominoid fossils from the Early Miocene of Africa, most dating from 19 to 17 Ma and most from East Africa.

The dominant hominoids were the ape-like **dryomorphs**. Like living African apes, they had relatively small cheek teeth with thin enamel, implying a soft diet of fruits and leaves, and a way of life foraging and browsing in trees like most living monkeys. (True monkeys were scarce at this time, remember.) Dryomorphs varied in weight from a large species of *Proconsul* in which males weighed about 37 kg (80 lb) down to *Micropithecus* at about 4 kg (9 lb), and their locomotion varied accordingly. *Micropithecus* and *Dendropithecus* were not as well adapted for arm-swinging as living gibbons, but they were lightly built and relied more on brachiating than did other Miocene primates.

The best-known and most important form is *Proconsul* itself (Figure 19.12). There were several species of this animal by 18 Ma. The most complete specimens are from a small species that weighed only about 9 kg (20 lb) but had a baboon-sized brain; that is, its brain was larger relative to body size than that of living monkeys. Its skeleton was a mixture of primitive characters that are also found in monkeys, gibbons, and chimps; altogether, these characters indicate a rather basic quadrupedal, tree-climbing, fruit-eating primate that could be the ancestor of all later hominoids. *Proconsul* had advanced hominoid characters of the head and jaws, though most of its body skeleton remained unspecialized. A large *Proconsul* may have spent a lot of time in deliberate climbing or on the ground, like a living chimp. Like them, it was probably versatile in its movements, and capable of occasional upright behavior.

Morotopithecus is a large ape from Uganda, probably as old as 20 Ma. Although we do not have a skeleton as complete as *Proconsul*, *Morotopithecus* is clearly large (40–50 kg, or 100 lb), and the pieces we have are more advanced than the same pieces of *Proconsul*. In other words, *Morotopithecus* is probably close to the direct ancestry of all later hominoids. It was probably a rather heavy slow climber, hanging in trees and eating fruit.

Africarabia drifted northward during the Miocene (Chapter 18) and finally collided with Eurasia to form an irregular mountain belt from Iran to Turkey. The collision interrupted tropical oceanic circulation and set off climatic changes. Temperatures cooled in East Africa, and almost all the northern continents experienced dramatic changes in faunas and floras. Forests became much more open, and grasses evolved to form wide expanses of savanna.

An exchange of animals between Africarabia and Eurasia added to the ecological turmoil of the times (Chapter 18). In that process, African hominoids successfully invaded Eurasian plains and woodlands.

In Africa the dryomorphs remained in the forests, which were thinned or diminished by cooling temperatures. They came under increasing pressure from the evolving monkeys. Monkeys have increased in abundance and diversity so that today they rather than apes dominate the remaining forests of Africa and Asia.

Some late dryomorphs reached Eurasia, but they were apparently numerous only in Europe. *Dryopithecus* is a European fossil, known from Spain to Hungary. Perhaps because it lived in a cool region, *Dryopithecus* was bigger and stronger than most dryomorphs. It shows adaptations for branch-swinging with the trunk more or less vertical, which gave it some of the characters of the living orang. Its skull is not like that of orangutans, however, and the position of *Dryopithecus* is closer to African hominoids.

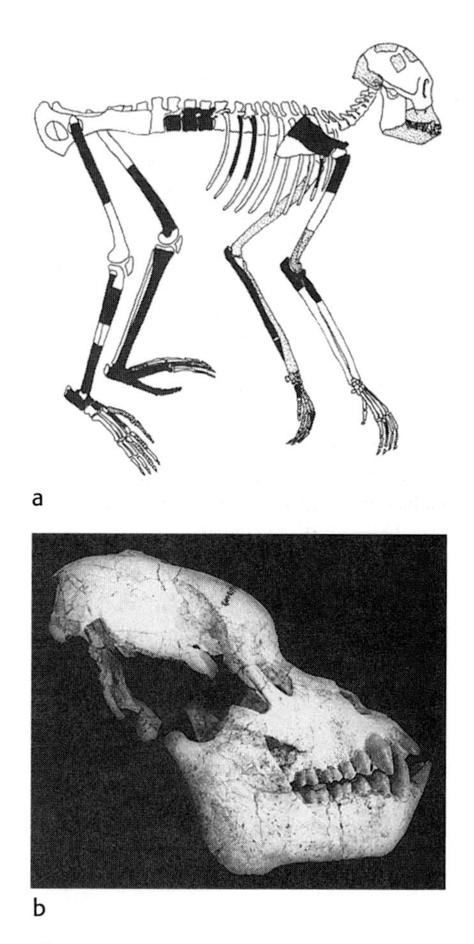

Figure 19.12 (a) Reconstruction of *Proconsul*, the Miocene dryomorph from East Africa that could be the common ancestor of later hominids. Parts in black are reconstructed; other bones are known. (b) The best-known skull of *Proconsul*. The jaw does not project so much if the skull is tilted to the position it had in a quadrupedal pose in life (compare top picture). (Courtesy of Alan Walker of Pennsylvania State University.)

Sivapithecids

At about 14 Ma, new hominoids appeared alongside the dryomorphs: the **sivapithecids** were the dominant group in East Africa. The earliest sivapithecid is *Kenyapithecus*, dated about 14 Ma. It is generalized enough to be a descendant of *Proconsul* and/or *Afropithecus*, and to be the ancestor of all later large apes: sivapithecids + pongids on one hand, and panids + hominids on the other.

Sivapithecids are apes, known from a wide area that stretches from East Africa to Central Europe, and eastward as far as China (Figure 19.13). *Sivapithecus* itself was an Asian ape. We have a great number of sivapithecid fossils, but they are mostly jaws, skulls, and isolated teeth; few body or limb bones are well known. Thus we can reconstruct sivapithecid heads rather well, but we know little about their body anatomy, posture, or locomotion.

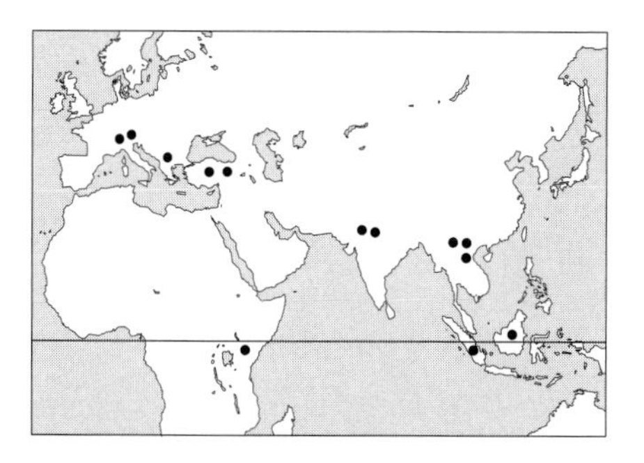

Figure 19.13 The geographical distribution of sivapithecids. Miocene specimens from Kenya, Hungary, Greece, Turkey, India, and China have all been given different names. The giant Pleistocene form *Gigantopithecus* and the living orangutan, *Pongo*, are later sivapithecid descendants.

Many different names have been applied to sivapithecids in their various

Figure 19.14 Reconstruction of the giant Asian sivapithecid ape *Gigantopithecus*. (© Stephen Nash and Russell Ciochon. Courtesy of Russell Ciochon, University of Iowa.)

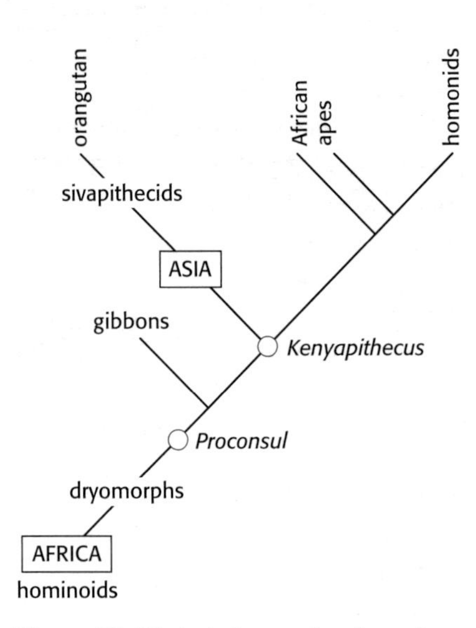

Figure 19.15 A phylogram that shows the sivapithecids fitting into hominoid evolution as ancestors of the orangutan.

countries of discovery. Hungarian, Turkish, Kenyan, Indian, Chinese, and Greek specimens were all given different names, for example. Part of the problem of naming sivapithecids is that there is a good deal of variation between individuals. As in orangutans, the skulls of males are much larger and broader than those of females.

All sivapithecids had thick tooth enamel and powerful jaws, suggesting that their diet required prolonged chewing and great compressive forces on the teeth. In living primates with thick enamel, such as orangutans or mangabeys, teeth and jaws like these are correlated with a diet of nuts, or fruits with hard rinds. One can hear an orangutan cracking nuts a hundred meters away! Perhaps sivapithecids diverged from the dryomorph diet of soft leaves and fruits to exploit a food source that had so far been available only to pigs, rodents, and bears.

Nut eating can be an activity of tree or ground dwellers, or creatures making the evolutionary and ecological transition from woodland to open ground. We cannot yet tell whether sivapithecids were foraging for fruit and nuts in the trees (with an arboreal life like that of orangutans) or under the trees (with adaptations for ground living).

One late sivapithecid was adapted to live entirely on the ground. The huge ape *Gigantopithecus* lived in southern and eastern Asia from about 7 Ma well into the Pleistocene. It had huge grinding teeth and weighed several hundred pounds (Figure 19.14). It probably lived on very coarse vegetation, as an ecological equivalent of the giant ground sloth of the American Pleistocene, or the Asian giant panda, or the African mountain gorilla. *Gigantopithecus* survived in Asia as recently as 300,000 or 250,000 years ago. It was certainly contemporaneous with *Homo* in eastern Asia, and its bones, teeth, and jaws may be responsible for Himalayan folklore about the abominable snowman, or yeti.

It is clear now that sivapithecids have nothing to do with human ancestry but are instead ancestors of the living Asian ape, the orangutan. In Africa, dryomorphs evolved toward hominids (Figure 19.15). The molecular clock suggests 17 Ma for the divergence, and the new Kenyan fossils seem to agree with that estimate.

After about 11 Ma, migration between Africa and Eurasia was essentially cut off. The hominoid groups evolved independently in Eurasia and Africa, eventually leaving the sivapithecids in Asia and the hominids in Africa. The African fossil record of hominids is horribly incomplete during the critical time after 11 Ma when they radiated to become separate lineages: we simply haven't found these fossils yet.

Between 8 and 5 Ma, the climate of Eurasia slowly changed to encourage even more open grasslands instead of woodland and forest. Then the history of Eurasian apes became one of struggling survival rather than innovation and evolution. European dryomorphs disappeared around 8 Ma, and the only remaining sivapithecids were the East Asian animals that led to *Gigantopithecus* and the orangutan. This means that Eurasia is not the continent in which to search for direct human ancestry (Figure 19.15). It is the African story that we must now follow.

By 7–5 Ma, the African forest had become dominated by monkeys, who displaced the dryomorphs ecologically and presumably restricted all the surviving forest apes. This is the time interval in which we can look for dramatic finds in the near future.

Further Reading

Benefit, B. R. 1999. *Victoriapithecus*: the key to Old World monkey and catarrhine origins. *Evolutionary Anthropology* 7: 155–74.

Bloch, J. I., and D. M. Boyer. 2002. Grasping primate origins. *Science* 298: 1606–10, and comment 1564–5; arguments, *Science* 300: 741. [The plesiadapid *Carpolestes* evolved some euprimate-like characters, probably independently.]

Boissinot, S., et al. 1998. Origins and antiquity of X-linked triallelic color vision systems in New World monkeys. *Proceedings of the National Academy of Sciences* 95: 13749–54. [The unusual color vision of New World monkeys evolved only once. See Sumner and Mollon, 2000.]

Chaimanee, Y., et al. 2000. A lower jaw of *Pondaungia cotteri* from the Late Middle Eocene Pondaung Formation (Myanmar) confirms its anthropoid status. *Proceedings of the National Academy of Sciences* 97: 4102–5. [The title says it all: more evidence for a very early anthropoid radiation in Asia.]

Ciochon, R., et al. 1990. *Other Origins: The Search for the Giant Ape in Human Prehistory*. New York: Bantam Books. [*Gigantopithecus*.]

Ciochon, R., et al. 1996. Dated co-occurrence of *Homo erectus* and *Gigantopithecus* from Tham Khuyen cave, Vietnam. *Proceedings of the National Academy of Sciences* 93: 3016–20.

Collura, R. V. and C.-B. Stewart. 1995. Insertions and duplications of mtDNA in the nuclear genomes of Old World monkeys and hominoids. *Nature* 378: 485–4. [Some problems in molecular clocks.]

Dean, D., and E. Delson. 1992. Second gorilla or third chimp? *Nature* 359: 676–7.

Fleagle, J. G., and R. F. Kay. (eds). 1994. *Anthropoid Origins*. New York: Plenum.

Gebo, D. L., et al. 1997. A hominoid genus from the Early Miocene of Uganda. Science 276: 401–4. [*Morotopithecus*.]

Gebo, D. L., et al. 2000. The oldest known anthropoid postcranial fossils and the early evolution of higher primates. *Nature* 404: 276–8. [New material of *Eosimias*.]

Godinot, M., and M. Mahboubi. 1992. Earliest known simian primate found in Algeria. *Nature* 357: 324–6.

Heesy, C. P. 2001. Rethinking anthropoid origins. *Evolutionary Anthropology* 10: 119–21. [Some of the issues being discussed in 2001: astonishingly fundamental ones!]

Jaeger, J.-J., et al. 1999. A new primate from the Middle Eocene of Myanmar and the Asian early origin of anthropoids. *Science* 286: 528–30. [*Bahinia*.]

Kay, R. F., et al. 1997. Anthropoid origins. *Science* 275: 797–804, and discussion, v. 278, 2134–6. [Strange methods were used to get from the data to the cladogram].

Kelley, J., and Q. Xu. 1991. Extreme sexual dimorphism in a Miocene hominoid. Nature 351: 151–3, and comment 111–12. [*Lufengpithecus*.]

Krause, D. W. 1986. Competitive exclusion and taxonomic displacement in the fossil record: the case of rodents and multituberculates in North America. *Contributions in Geology of the University of Wyoming, Special Paper* 3: 95–117.

Martin, R. D. 1990. *Primate Origins and Evolution*. London: Chapman and Hall.

Moyá-Solá, S., and M. Köhler. 1996. A *Dryopithecus* skeleton and the origins of great-ape locomotion. *Nature* 379: 156–9, and comment 123–4.

Olson, S. L., and D. T. Rasmussen. 1986. Paleoenvironment of the earliest hominoids: new evidence from the Oligocene avifauna of Egypt. *Science* 233: 1202–4.

Rasmussen, D. T., et al. 1998. Tarsier-like locomotor specializations in the Oligocene primate *Afrotarsier*. *Proceedings of the National Academy of Sciences* 95: 14848–50.

Rook, L., et al. 1999. *Oreopithecus* was a bipedal ape after all: evidence from the iliac cancellous architecture. *Proceedings of the National Academy of Sciences* 96: 8795–9.

Rose, K. D. 1990. Postcranial skeletal remains and adaptations in early Eocene mammals from the Willwood Formation, Bighorn Basin, Wyoming. *Geological Society of America Special Paper* 243: 107–33.

Ross, C. F. 2000. Into the light: the origin of Anthropoidea. *Annual Reviews of Anthropology* 29:147–94.

Schwartz, J. H. 1990. *Lufengpithecus* and its potential relationship to an orangutan clade. *Journal of Human Evolution* 19: 591–605.

Seiffert, E. R., et al. 2003. Fossil evidence for an ancient divergence of lorises and galagos. *Nature* 422: 421–4. [By the Eocene: evidence from the Fayum of Egypt.]

Setoguchi, T., and A. L. Rosenberger. 1987. A fossil owl monkey from La Venta, Colombia. *Nature* 326: 692–4.

Simons, E. L. 1987. New faces of *Aegyptopithecus* from the Oligocene of Egypt. *Journal of Human Evolution* 16: 273–89.

Simons, E. L., et al. 1999. Canine sexual dimorphism in Egyptian Eocene anthropoid primates: *Catopithecus* and *Proteopithecus*. *Proceedings of the National Academy of Sciences* 96: 2559–62.

Simons, E. L. 1995. Skulls and anterior teeth of *Catopithecus* (Primates: Anthropoidea) from the Eocene and anthropoid origins. *Science* 268: 1885–8, and comment 1851.

Stewart, C.-B., et al. 1987. Adaptive evolution in the stomach lysozymes of foregut fermenters. *Nature* 330: 401–4, and comment 315.

Sumner, P. and J. D. Mollon 2000. Catarrhine photopigments are optimised for detecting targets against a foliage background. *Journal of Experimental Biology* 203: 1963–86.

Tattersall, I. 1993. Madagascar's lemurs. *Scientific American* 268 (1): 110–17.

Walker, A., and M. Teaford. 1989. The hunt for *Proconsul*. *Scientific American* 260 (1): 76–82.

Ward, S., et al. 1999. *Equatorius*: a new hominoid genus from the Middle Miocene of Kenya. *Science* 285: 1382–6, and comment 1335–7.

Yoder, A. D., et al. 1996. Ancient single origin for Malagasy primates. *Proceedings of the National Academy of Sciences* 93: 5122–6.

CHAPTER TWENTY

Evolving Toward Humans

We know practically nothing of the evolution of the hominoid lineages that led to gorillas and chimps. Molecular and genetic evidence suggests that our closest living relatives are chimps, with gorillas a little further away. Our own lineage, the **hominids**, probably separated from that of chimps around 5 or 6 Ma. Even so, our DNA is more than 95% identical to that of chimps. Obviously the 5% that is different reflects very important evolutionary changes in our bodies, brains, and behavior.

Over time, there have been perhaps a dozen species of hominids, but we, as *Homo sapiens*, are the only surviving one. As many as six earlier species of *Homo*, ranging back to about 2 Ma, have become extinct, and another six hominid species have usually been placed in the genus *Australopithecus*, which ranges back to about 4 Ma and is generally accepted as containing the ancestor of *Homo*.

This picture is being rapidly changed by the discovery of earlier hominids. Be warned that almost everything I have written in this chapter is being argued over by paleoanthropologists. I have tried, as usual, to select what I think are the most likely hypotheses.

Sahelanthropus may be the oldest hominid yet found, at 6 Ma or 7 Ma. It is from Chad, far from the "classical" East African sites (Figure 20.1). It is known from skull pieces, and shows a puzzling combination of very "primitive" characters (small brain, for example) with "advanced" characters such as eyebrow ridges. It may force us to reevaluate what characters are really primitive for hominids. It opens up the possibility that more early hominids will be discovered in the Sahel rather than East Africa. However, the single skull was badly crushed, and it's possible that a different reconstruction would allow a different interpretation. As usual, we need more fossils!

Orrorin, dates from perhaps 6 Ma in East Africa, and is much closer to the split between gorillas, chimps, and hominids. It is known mostly from a few pieces of limb bones, so is difficult to place. Rivals of the discovery team are dropping surreptitious hints that it may in fact be an ancestral gorilla or chimp rather than a hominid.

Since we cannot yet give a reasonable story of very early hominid evolution, we move to the australopithecines, the hominid species that are not *Homo*. These hominids lived in Africa, south of the Sahara Desert, from perhaps 4.4–4.3 Ma to about 1.4 Ma or a little later. Overall, we are reasonably sure of the position of australopithecines in hominid evolution (Figure 20.2) and have enough evidence to reconstruct a vivid picture of australopithecine life.

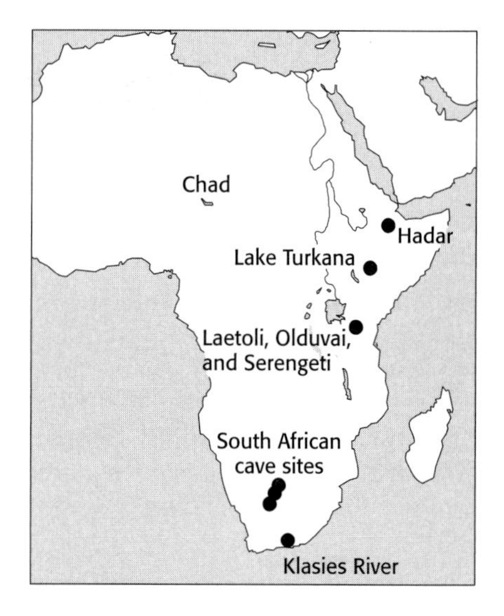

Figure 20.1 Some important hominid-bearing regions and localities in Africa.

AUSTRALOPITHECINES

Early Australopithecines

The earliest two species of australopithecines are *Ardipithecus ramidus*, described from rocks in Ethiopia dated at 4.3–4.4 Ma, and *Australopithecus anamensis*, described from rocks in Kenya dated at 4.1–4.2 Ma. *Ardipithecus ramidus* is the most primitive (i.e., ape-like) australopithecine yet found, and was probably a forest dweller. The slightly younger *Australopithecus anamensis* has a jaw that is even more ape-like, but its arm and leg bones suggest upright (bipedal) posture and locomotion. It would make a good ancestor for later *Australopithecus*: in fact some specimens currently identified as *A. afarensis* may actually belong to *A. anamensis*. *A. anamensis* is rather large, perhaps 50 kg, 110 lb. New specimens are still being cleaned, and new information will pour out in the next few years. However, it is fair to say that there are no major evolutionary surprises (yet) in these new fossils.

Footprints at Laetoli

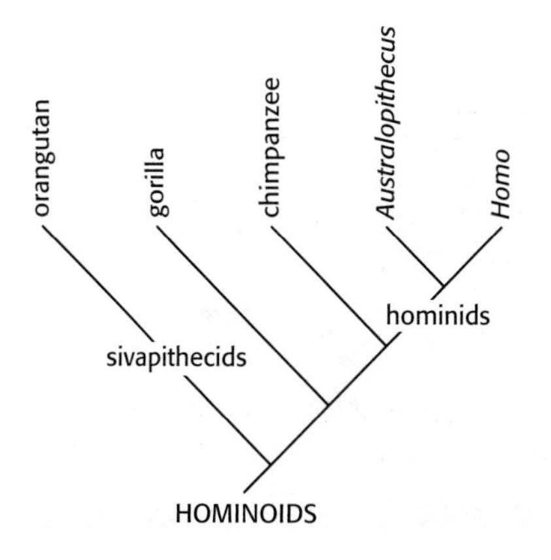

Figure 20.2 The separation of hominids from Asian and African apes. The origin of the hominids is also the point at which bipedalism evolved. The gorilla and chimp do not form a clade in this hypothesis, which proposes that chimps are closer to hominids than gorillas.

The East African Rift splits East Africa from Ethiopia to Zambia and Malawi. Among its unusual geological features are volcanoes that sometimes erupt carbonatite ash, which is composed largely of a bizarre mixture of calcium carbonate and sodium carbonate. One of these volcanoes, Sadiman, stood near the Serengeti Plain, in northern Tanzania (Figure 20.1). After carbonatite ash is erupted, the sodium carbonate in it dissolves in the next rain, and as it dries out the ash sets as a natural cement. Any animals moving over the damp surface in the critical few hours while it is drying will leave footprints that can be preserved very well. As long as the footprints are covered up quickly (for example, by another ash fall), rainwater percolating through the ash will react with the carbonate to make a permanent record.

Sadiman erupted one day about 3.6 Ma, towards the end of the dry season. Ash fell on the plains near Laetoli, 35 km (20 miles) away, and hominids walked across it, leaving their footprints along with those of other creatures. The vital point about the tracks is that the hominids were walking fully erect, long before hominid jaws, teeth, skull, and brain reached human proportions, shape, or function.

Why would a hominid become bipedal? Most suggestions are related to carrying things with the hands and arms (infants, weapons, tools, food), to food gathering (seeing longer distances, foraging over greater ranges, climbing vertically, reaching high without climbing at all), to defense (seeing longer distances, throwing stones, carrying weapons), to better resistance to heat stress (less sweat loss and better cooling), or to staying within reach of rich food resources by migrating with the great plains animals (carrying helpless young over long distances). These are all reasonable suggestions, but all are difficult to test.

It's certain that the footprints at Laetoli were made by australopithecines that were walking upright. All australopithecines are similar below the neck, apart from size differences, so they all probably moved in much the same way. Their movements were probably not exactly like ours, but their leg and hip bones indicate that they walked and ran well. At the same time, the limb joints and toes suggest that

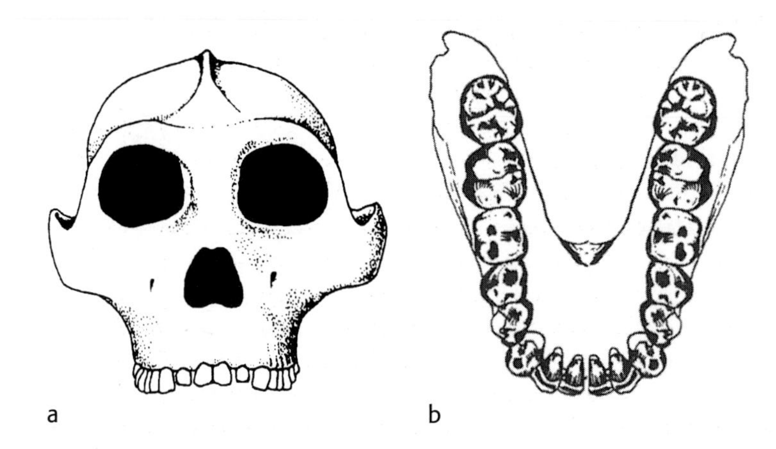

Figure 20.3 The skull, jaw, and teeth of australopithecines. (a) The skull of *Australopithecus robustus*. (After Howell.) (b) The teeth and jaw of *A. afarensis*, idealized after the specimen AL–400. (Adapted from Johanson and Maitland.)

they spent a lot of time climbing in trees as well as walking upright on the ground.

Probably the trend toward the use of the forelimbs for gathering food and the hind limbs for locomotion began among tree-dwelling primates long before *Australopithecus*. But this thought is based mainly on my own experience in picking and eating fruit, and the realization that forelimbs are more effective for that job than teeth and jaws alone. The final achievement of erect bipedality on the ground was probably an extension of previous locomotion and behavior, rather than something completely different.

Australopithecines were smaller than most modern people. They varied around 40 kg (90 lb) as adults, but their bones were strongly built for their size. The skull was even stronger, and very different from ours (Figure 20.3a). The relative brain size was about half of ours, even allowing for the smaller body size of *Australopithecus*, but the jaw was heavy and the teeth, especially the cheek teeth, were enormous for the body size. The canine teeth were large and projecting. The whole structure of the jaws and teeth suggests strength (Figure 20.3b).

The small size of the brain and the thickness of the skull may be linked with another feature that separates us from *Australopithecus*. The birth canal in the pelvis of australopithecines is wide from side to side, but narrow from front to back, so that there may have been a special mode of delivery for even the small-brained babies that australopithecines had (Figure 20.4). In *Homo* the birth canal is rounder (Figure 20.4), presumably to accommodate the passage of a baby with a very large head (and a very large brain). If so, a larger brain was important enough that this visible difference in skeletal anatomy was evolved in *Homo*.

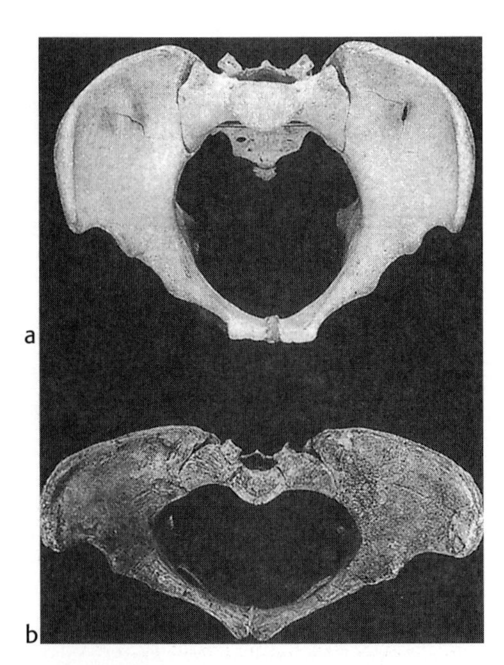

Figure 20.4 The difference in structure between the pelvis of a modern woman (*Homo sapiens*, photograph a), and the reconstructed pelvis of the female australopithecine Lucy (*Australopithecus afarensis*, photograph b). The modern pelvic canal is much rounder, presumably to accommodate the birth of a larger-brained baby. (Courtesy of Owen Lovejoy, Kent State University. © C. Owen Lovejoy.)

Like ourselves, australopithecines were built for trotting endurance rather than blinding speed, an adaptation that would be better suited to foraging widely in open woodland than to skulking in forests, provided that *Australopithecus* was not easily picked off by large sprinting carnivores. (Reasons may have been related to group defense, which allows baboon troops to roam freely on the ground, or to early possession and use of defensive tools.) *Australopithecus* had long arms and fingers that were capable of sensitive motor control. In a tall biped walking upright, the arms would have been free for carrying, throwing, and manipulating.

Australopithecus afarensis

The best-known collections of early australopithecines are from Laetoli and from Hadar in Ethiopia (Figure 20.1). Each district has produced spectacular finds. At Laetoli there are the footprints, plus remains of at least 22 individuals; at Hadar, bone fragments from at least 35 individuals are preserved rather better. All the specimens belong to one species, *Australopithecus afarensis,* which was closely related to *A. anamensis,* and was probably ancestral to all the later species of *Australopithecus* and to *Homo* as well (Figure 20.5).

Hadar lies in the Afar depression, a vast arid wilderness in northeast Ethiopia (Figure 20.1). At 3–4 Ma it was the site of a lake fed by rivers tumbling out of winter snowfields on the plateau of Ethiopia. The australopithecines lived and are fossilized along the lake edges. Delicately preserved fossils such as crab claws and turtle and crocodile eggs suggest that the australopithecines had rich protein foods available to them, and skeletons of hippos and elephants suggest that there was rich vegetation in and around the lake edges. All the Hadar specimens are dated at about 3.2 Ma, so they are considerably later than the Laetoli australopithecines.

The best-preserved Hadar skeleton is the famous Lucy. Lucy was small by our standards, a little over 1 m (42 in) in height. She was full-grown, old enough to have had arthritis. Her brain was small at about 385 cc, compared with 1300 cc for an average human. Her large molar teeth suggest that *A. afarensis* was a forager and collector eating tough fibrous material.

New data suggest that *A. afarensis* was no more dimorphic than modern humans: males were bigger than females, but not to the extent seen in baboons, for example. Across primates, extreme male size is correlated with intense physical competition between males for females; monogamy is associated with low levels of dimorphism. It would be interesting if it turns out that all hominids (australopithecines and *Homo*) have always had this sort of social structure.

Baboons sleep in high places — trees or high rocks — and are great opportunists in taking whatever food is available. They live and forage in troops and have a cohesive social structure that gives them effective protection from predators even though they are fairly small as individuals. But *Australopithecus* walked upright, whereas baboons trot on four limbs. Ecologically (but perhaps not socially!), *Australopithecus* may have been a super-baboon. Walking upright, with its arms free for carrying, it may have been a more effective forager than a baboon, which can carry only what it can put into its mouth and stomach. Perhaps the requirements and advantages of efficient troop foraging and defense encouraged tight social cohesion among australopithecines, long before tools permitted technological advances.

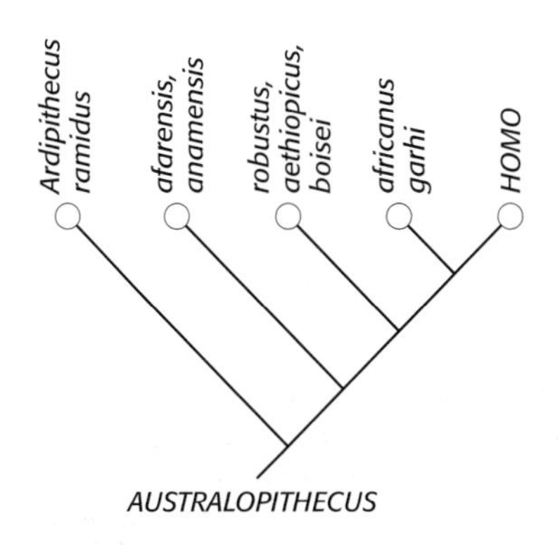

Figure 20.5 Simplified cladogram of australopithecines. I have grouped rather similar species together into clades: the early primitive species *afarensis* and *anamensis,* for example; the "robust" australopithecines (they may belong to only one species, which would have the name *robustus*); and the "gracile" *africanus* and *garhi. Ardipithecus* is more primitive than any of them. Note that if this cladogram is correct, *Australopithecus* is not a clade unless you include *Homo* in it. This "problem" is not really a problem: the aim of cladograms is to portray evolution, the naming schemes are simply convenient.

Australopithecus in South Africa

Isolated caves scattered over the high plains of South Africa are mined for limestone, and hominid fossils have been found encased in the limestone (Figure 20.6). But cave deposits are difficult to interpret and date accurately. Roof falls and mineralization by percolating water have disturbed the original sediments, and few of the radioactive minerals in cave deposits allow absolute dating. Thus there have been problems in relating South African hominid fossils to their well-dated East African counterparts.

New research will soon change that. A new find at Sterkfontein in 1998 has been claimed to be about 4 Ma in age. This is not outrageous, but it is very early, and the claim will no doubt be examined very carefully.

To date, the best-known early australopithecine from South Africa is *Australopithecus africanus* (Figure 20.6). Although it was about the same body weight as *A. afarensis, A. africanus* was taller but more lightly built and had a larger brain, perhaps 450 cc. The teeth and jaws continued to be large and strong, with molars twice as large as chimpanzee molars, suggesting that the diet remained mainly vegetarian. However, new evidence from isotopes in the teeth suggest that *A. africanus* ate vegetarian animals as well, possibly catching small animals, or scavenging meat from

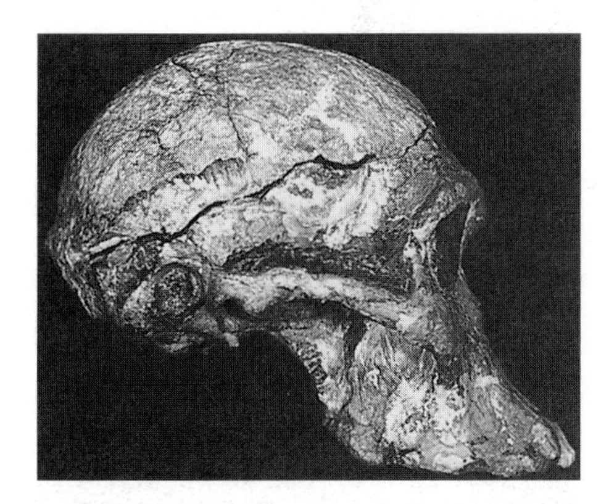

Figure 20.6 A skull of *Australopithecus africanus* from the cave deposits at Sterkfontein in South Africa. This specimen was discovered by deliberately dynamiting the cave limestone, which is very solid. When the smoke cleared after one blast, the skull was found blown almost in two, showing the brain cavity still empty and lined with lime crystals. The skull has been rejoined along the line of breakage. (Negative 2A 333. Courtesy of the Department of Library Services, American Museum of Natural History.)

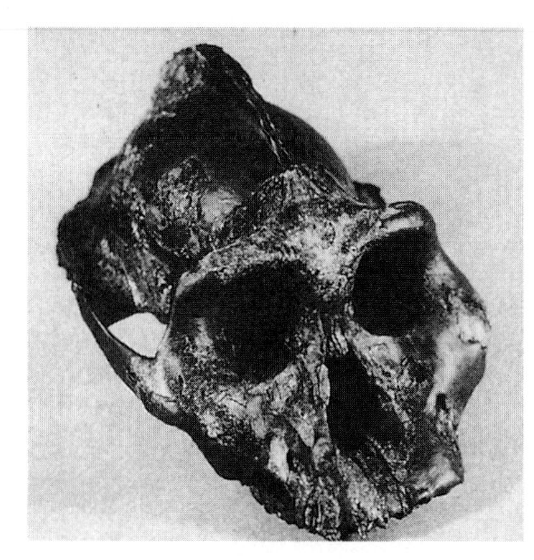

Figure 20.7 Alan Walker found this robust australopithecine skull in northern Kenya. First called the "Black Skull," it is now usually called *Australopithecus aethiopicus.* (Courtesy of Alan Walker of Pennsylvania State University.)

carcasses. Tooth wear suggests an average life span for *A. africanus* of perhaps 20 years, maybe with a maximum of 40 years, about the same as a gorilla or chimpanzee. The arms were relatively long compared with *A. afarensis*, suggesting that *A. africanus*, though perfectly erect and able to walk and run on the ground, spent a good deal of time in trees.

Robust Australopithecines

Australopithecines with heavily built skulls are called robust to distinguish them from those with lightly built skulls (such as *A. africanus*) which are called gracile. The best example of a robust skull is the oldest one, the so-called Black Skull (Figure 20.7) from the Turkana Basin of northern Kenya (Figure 20.1) dating from about 2.5 Ma. The Black Skull is usually called *Australopithecus aethiopicus.* It has a skull much heavier and stronger than *A. afarensis*, although the brain was no larger and the body was not very different. The jaw extended further forward, the face was broad and dish-shaped, and there was a large crest on the top of the skull for attaching very strong jaw muscles (Figure 20.7). The molar teeth of the Black Skull are as large as any hominid teeth known, about four or five times the size of ours. Yet the front teeth of robust australopithecines are small.

Later robust forms have been found all over East and South Africa between 2.5 Ma and 1.4 Ma. In East Africa they are usually called *Australopithecus boisei* (the famous Zinj of Louis Leakey), and in South Africa they are called *A. robustus* (Figure 20.3a). There are enough fossils to suggest that robust australopithecines changed over the million years of their history, evolving a larger brain (perhaps 500 cc rather than 400 cc) and a flatter face.

The robust australopithecines are certainly linked ecologically. The large jaw weight and the huge molars, with their very thick tooth enamel, were adaptations that indicate great chewing power and a diet of coarse fiber. However, almost all the characters that are used to define robust australopithecines are connected with the huge teeth, and the modifications of the jaws and the face during growth that are required to accommodate the teeth. So any australopithecine population that evolved huge teeth would have come to look "robust." Therefore the robust australopithecines may not be an evolutionary group. They may be three separately evolved species; they may be three related species; or they may be variants of the same species (which would have to be called *robustus*). Some specialists prefer to give robust australopithecines their own generic name, *Paranthropus*, but I have not used this name (Figure 20.5). The robust australopithecines could easily have evolved from *A. afarensis.*

Australopithecus garhi, and Butchering Tools

An astonishing find was reported in 1999. Rocks in Ethiopia dated at 2.5 Ma yielded enough pieces of two or three skeletons to allow the description of a new species, *Australopithecus garhi.* Since then, beds of the same age have yielded evidence of the use of stone tools for butchering meat and smashing bones.

Australopithecus garhi is a normal gracile australopithecine, except that it has very large teeth for the size of its jaw and skull. The skull is far too primitive for it to belong to *Homo*, and its brain size is only about 450 cc. But given its age, location, and the features of its skeleton, *A. garhi* would be a good ancestor for *Homo.*

Some of the animal bones in the same rock bed had been sliced and hammered in ways that betray intelligent butchering. Most likely, the butchers used their tools carefully, because there were no suitable rocks nearby, and all tools had to be carried in (and carried out for further use).

Before 1999, it had generally been thought that the defining characters of *Homo* versus *Australopithecus* included a larger brain and the use of tools. The new evidence suggests that *A. garhi* was making, carrying, and using tools effectively. Perhaps the great ecological advantages gained by the invention of butchering tools encouraged exactly those changes in the *Australopithecus garhi* lineage that led quickly to increased brain size, reduced tooth size, and the status of first *Homo.*

Once again, apparently major transitions disappear as we collect more fossils: we have seen this for the transition between birds and dinosaurs, between cynodonts and mammals, and now between australopithecines and *Homo.*

LIMERICK 20.1
He butchered and hammered the dead,
We assume he was very well fed,
More astonishing still,
He learned this new skill
With a very small brain in his head.

THE APPEARANCE (OR NOT) OF *HOMO*

Perhaps as early as 2.4 Ma, hominids with increased brain size and reduced teeth and jaws appeared in Africa. They are sufficiently like ourselves in jaws, teeth, skull, and brain size to be classed as *Homo.* But because one genus always evolves from another, there is always room to argue just where to draw the line, and this is happening as we try to decide which species actually was the first *Homo.* Increasingly, we realize that there is a great difference between early, transitional forms, and later species that everyone agrees as belonging to *Homo.* I will continue to call the transitional forms *Homo* until there is more of a consensus.

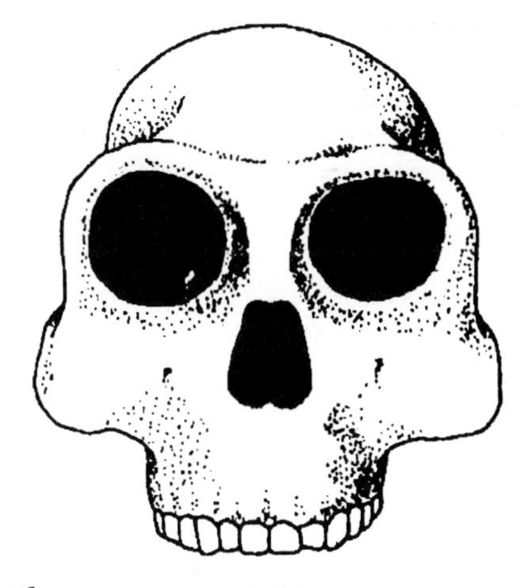

Figure 20.8 *Homo habilis*: a reconstruction of specimen KNM-ER1470. (Redrawn after Howell.)

Transitional Species That May or May Not Remain in Homo

The earliest specimen claimed to be *Homo,* with an age of at least 2.33 Ma, consists only of an upper jaw, so has not been given a species name. The most familiar transitional species is *Homo habilis* (Figure 20.8). *Homo rudolfensis* is known from East Africa around 2 Ma, largely from skulls, and has been given a separate name. A new find at Olduvai Gorge (Figure 20.1) seems to suggest that *habilis* and *rudolfensis* are the same species. This story is bound to change as we find more fossils. Meanwhile I will call them "early *Homo*" or *Homo habilis.*

Early *Homo* was small by modern standards, perhaps just over 1 m (4 ft) tall, but was at least as heavy as contemporary robust australopithecines at about 30–50 kg (65–110 lb). The difference in brain size is striking, however. The brain size was about 650 cc, considerably larger than the brain of an australopithecine. Perhaps, then, early *Homo* is marked by a new level of brain organization.

We have only a few sets of bones of *H. habilis,* but there is enough evidence from hands, legs, and feet to suggest that *H. habilis* spent a lot of its time climbing in trees.

We have a good record of the tools that were used by *Homo habilis* (and probably *Australopithecus garhi* before it). They are called Oldowan tools because they were first identified by the Leakeys in Olduvai Gorge. They are often large and clumsy-looking objects with simple shapes, and not all of them were useful tools in themselves. Instead, many objects may be the discarded centers (cores) of larger stones from which useful scraping and cutting flakes had been removed by hammering with other stones (Figure 20.9).

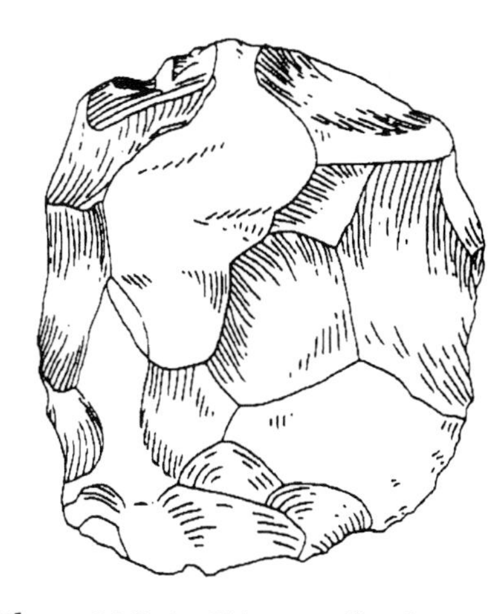

Figure 20.9 An Oldowan artifact shown about half-size. Perhaps it was some sort of tool in itself, but maybe it was used as a core for chipping off more useful smaller tools. (After Gowlett.)

Oldowan tools demonstrate the use of stone in a deliberate, intelligent way, and the flakes were probably made and used for cutting up food items. For example, an excavation in the Turkana Basin turned up the skeleton of a hippopotamus lying near an ancient river bed. Cobbles naturally occurring close by on a gravel bank in the river had been broken to produce simple tools. Marks on the hippo bones showed that they had been scraped, and that the tendons and ligaments had been cut, to allow meat to be taken from the carcass. There was no indication that the hippo had been killed by the tools.

Nicholas Toth has reproduced and used Oldowan-style artifacts from East African rock types (see Schick and Toth, 1993). He showed that the toolmaker was sophisticated in selecting appropriate rocks and making the most of them. Toth's experiments on fresh carcasses of East African animals show that Oldowan axes, flakes, and cores are excellent tools for slitting hides, butchering carcasses, and breaking bones for marrow. Toth was also able to determine that *habilis* was right-handed!

Some Oldowan sites were visited many times. They contain accumulations of bones, stones, and tools, brought to the site over periods of years. Flakes were made on site from stones that had been carried there. This may not indicate a systematic return to a homesite, but it does indicate an intelligent return to sites that perhaps were particularly suitable for food processing and tool making.

From Super-Baboon to Super-Jackal

Was early *Homo* a hunter or a scavenger? This may be a nonquestion, because all hunters will eat a fresh carcass, and all scavengers will cheerfully kill a helpless prey if they can. Evidence from Turkana and Olduvai suggests that early *Homo* was a scavenger on large carcasses but hunted small- and medium-sized prey. Thus early *Homo* may have had the ecology of a super-jackal, foraging in groups over long distances in search of large, fresh carcasses killed by other predators. Rhinos, hippos, and elephants have thick and leathery hides, difficult for vultures, jackals, and hyenas to pierce, but stone tools allowed *Homo* to make short work of dismembering a large carcass. Between carcass finds of large animals, early *Homo* may have foraged for leopard kills of medium-sized animals, left hanging in trees. Early *Homo* may also have been an opportunistic hunter of small- to medium-sized prey that was brought to central sites for butchering, and also a forager searching for fruits, berries, grains, roots, grubs, locusts, and lizards. *Australopithecus africanus* may have eaten small animals as well. Early *Homo* (and possibly *A. garhi*) used tools to make that opportunistic way of life more efficient.

The concept is exciting. A new ecological niche opened up, or became much more profitable, with the invention of tools and the ability to use them intelligently. Visiting American anthropologists with no previous experience in the African bush were able to learn quickly how to find large carcasses of animals killed in woodlands and smaller carcasses cached in trees by leopards (see Blumenschine and Cavallo, 1992); it is perfectly reasonable to expect that early *Homo* could have done so too. Simultaneous or consequent changes in diet, brain size, and possibly even social structure are consistent with the apparently rapid advances in skull characters, but not body anatomy, in early *Homo*, and its replacement of gracile australopithecines. Perhaps early *Homo* did not compete ecologically with the surviving australopithecines (the robust forms). Certainly robust australopithecines and *Homo* coexisted for over a million years in the same environments.

One can imagine how early *Homo* could have improved its competitive ability by exploring and exploiting the possibilities of tool use. Weaponry would naturally follow from tool use during scavenging. Food and infants could be transported safely from place to place with carrying devices. Increasing behavioral complexity would probably act to increase the value of brain growth and learning ability, and perhaps we may speculate (but not too wildly) about the increasing value of, or need for, sophisticated communication within and among social groups of humans.

Hominids and Cats in South Africa

Most hominid fossils found in South Africa have come from caves, many of which had steep or vertical entrances. It is unlikely that the hominids lived in the caves. Instead, the piles of bones there probably fell into the caves from above. In addition to hominid skeletons (mostly australopithecines), the fossils include bones of rodents, hyraxes, antelopes, baboons, two species of hyenas, leopards, and three extinct species of stalking sabertooth cats, one as big as a lion and two the size of a leopard.

C. K. Brain realized that the hyrax skulls in the cave deposits are all damaged in a peculiar way (see Brain, 1981). Leopards always eat hyraxes completely, except for the fur, the gut, and the skull and jaws, and as they get at the brain and tongue they leave characteristic tooth marks on the skull, just like those on the fossil hyraxes. Cheetahs today can eat the backbones of baboons but not those of antelopes. Fossils from Swartkrans cave include many antelope vertebrae but none from baboons. Only baboon skulls are found. Furthermore, it looks as if the cave fossils were selected by size. There are very few juvenile baboon skeletons at Swartkrans, but many juvenile australopithecines. Some of the primate skulls show teeth marks that look exactly like those made today by leopard canines. Some of the fossil antelopes are bigger than those killed today by leopards, and sabertooth cats may have been responsible for them.

Leopards today like to carry their prey up trees. On the bare plains of South Africa, a cave entrance is one of the few places that seedlings can find safe rooting away from browsers, fire, and winter frost. Brain suggests that Pliocene and Pleistocene sabertooth cats killed prey and carried them to safe places to eat, undisturbed by jackals and hyenas, in trees growing at the entrances of caves such as Swartkrans and Sterkfontein. Uneaten parts of the carcasses fell into the caves, away from the hyenas, and were buried and preserved as soil, debris, and limestone deposits filled the cave.

Hominoids may have been a preferred meal for sabertooths for a long time. Many well-preserved fossils of sivapithecids and gibbons in south China, at about 6 Ma, consist almost entirely of skulls and skull fragments, with few other bones of the skeleton, and there are large and impressive sabertooth canines in the same beds.

The large cats were the dominant carnivores in South Africa when the cave deposits were formed, and we can imagine them stalking and killing fairly large prey animals, including *Australopithecus robustus.* But there are relatively few fossils of early *Homo* at Swartkrans, Sterkfontein, or among other early cave deposits, suggesting that early *Homo* was either rare or comparatively safe from big cats by virtue of habits, intelligence, or defensive methods and weapons. Early *Homo* did not have to be immune from big cat predation, just well defended enough that the big cats hunted other prey most of the time. *Homo,* of course, eventually replaced *Australopithecus* in South Africa.

There are many bones of *Australopithecus africanus* in the rock bed Member 4 at Sterkfontein, but no tools. Member 5, which overlies it, contains many tools, including choppers and diggers, animal bones with cut marks on them, and a few

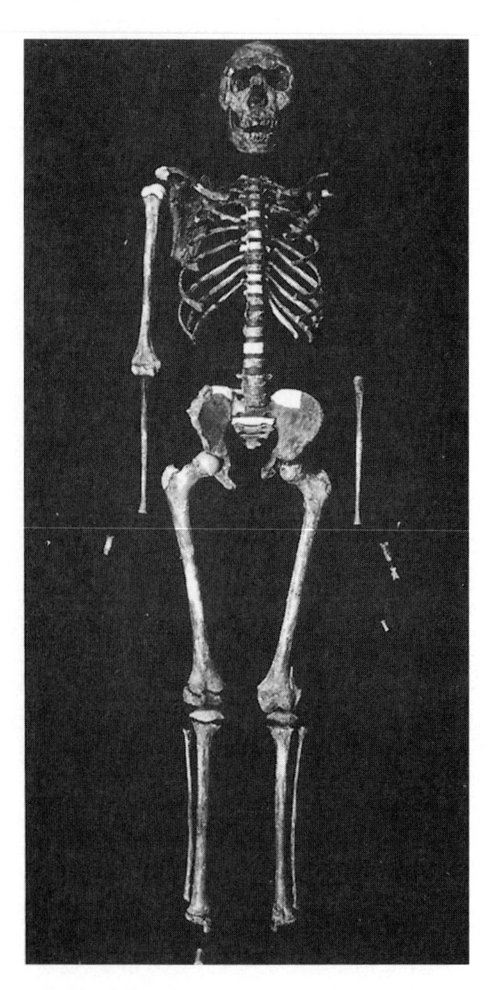

Figure 20.10 The "Nariokotome boy," KNM–ER–15000, the best specimen of early *H. erectus*, found in 1984 near Lake Turkana in Kenya. (Courtesy of Alan Walker of Pennsylvania State University.)

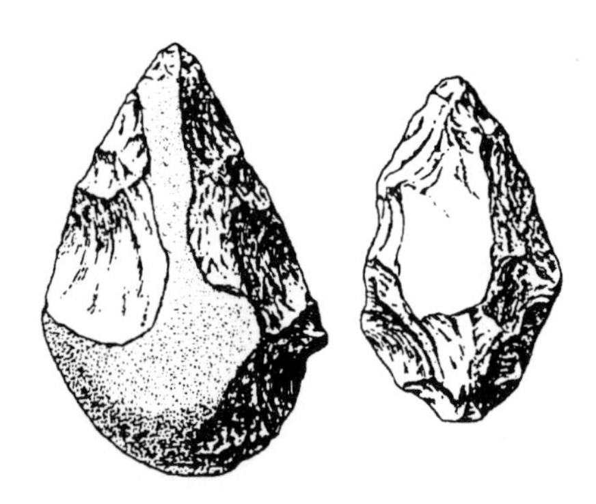

Figure 20.11 An Acheulian tool kit, well designed for heavy-duty butchering. Shown about half size. (Redrawn after Mary Leakey: I have "freshened up the edges.")

fossils of *Homo habilis*. The contrast between these two levels is striking in every aspect of their fossil record. As Brain sees it, the replacement of big cats by *Homo* as the dominant predators in South Africa was a major step toward human control over nature, and the beginning of our rise to dominance over the planet.

HOMO ERECTUS: THE FIRST "REAL" *HOMO*?

Some extraordinary changes took place in the African plains ecosystem, beginning about 1.5 Ma. It is tempting to associate them with the appearance of new species of human, *Homo erectus*. An excellent specimen of *H. erectus* was discovered in 1984 west of Lake Turkana in Kenya, in sediment dated about 1.5 Ma. Although the body had been trampled by animals, so that the bones were broken and spread over 6 or 7 m, careful collecting recovered an almost complete skeleton. The skeleton came from a boy 11 or 12 years old who stood 1.6 m (65 in) high (Figure 20.10). It is therefore possible that adult males stood close to 1.8 m (6 ft) tall by 1.5 Ma. The nose was enlarged and projected, as in modern humans but unlike australopithecines or *H. habilis*. This character suggests that *H. erectus* was adapting to greater exposure to dry air, for longer times and during greater activity.

Homo erectus was strongly built, and was a specialized walker and runner with large hip and back joints capable of taking the stresses of a full running stride. There is less evidence of tree-climbing ability than there is in early *Homo*, though *H. erectus* would have been no worse at it than we are. *H. erectus* is also advanced in skull characters. The skull is thick and heavy by our standards, but brain size had increased to around 900 cc.

Quite suddenly, at about 1.4–1.5 Ma, all over East Africa, *Homo erectus* is found associated with a completely new set of stone tools. The **Acheulean tool kit** is much more effective than the older Oldowan, but experiments by Nicholas Toth show that Acheulean tools required much greater strength and precision to make and use than Oldowan tools. Acheulean craftsmen shaped their stone cores into heavy axes and cleavers at the same time as they flaked off smaller cutting and scraping tools. Most Acheulean tools are well explained as heavy-duty butchering tools (Figure 20.11). And around this time, robust australopithecines, *A. africanus*, and sabertooth cats all became extinct in South Africa. All other species of early *Homo* were already gone from Africa, and by 1 Ma, the last robust australopithecines and the last two species of sabertooth cats were gone too.

It is tempting to correlate all these events with the achievement of some dramatically new level of intellectual, physical, and technical ability in *Homo erectus*. *H. erectus* was much bigger than any preceding human. Most paleontologists believe that the evidence from anatomy, from tools, and from animal remains found with *H. erectus* suggests that this was the first effective human hunter of large animals. Alan Walker has suggested that the entire ecosystem of the African savanna was reorganized as *Homo erectus* came to be a dominant predator instead of a forager, scavenger, and small-scale hunter (see Walker and Shipman, 1996). African kill sites with butchered animals suggest a sophisticated level of achievement.

The physical stature and ecological impact of this new species of *Homo* is the reason some experts suggest we should redefine the origin of *Homo* to the appearance of *Homo erectus*, perhaps placing *Homo habilis* and the other transitional forms into a genus of its own that would not be *Homo*.

The first fossils to be named *Homo erectus* were in fact collected a hundred years ago on the island of Java, in present-day Indonesia. The earliest of the Java specimens of *H. erectus* may be as old as 1.8 Ma, though this date is contested. However, specimens of *H. erectus* have been discovered in the southern Caucasus, in Georgia, and they date to around 1.7–1.8 Ma. It seems increasingly likely that the emigration of *H. erectus* from Africa to Asia occurred almost as soon as the species evolved. It was rapid, and it extended across the warm regions of southern Asia from the Middle East to Indonesia.

A few, very early African fossils can be assigned anatomically to *Homo erectus*. They also date from about 1.8 Ma, well before the invention of the Acheulean tool kit in Africa. So we have two populations of *Homo erectus*, one in Africa and one in east Asia. Neither had the Acheulean tool kit, which appeared in Africa around 1.5 Ma, as we have seen.

Should we call all these fossils *Homo erectus*, implying that they are the same species? Several scholars argue that genes could not have been transferred between these separate populations, and that Acheulean tools weren't either. Therefore, the argument goes, since the Asian ones carry the name *Homo erectus*, the African ones need a new name (*Homo ergaster* is usually used).

However the names turn out, the evolutionary reality is that the ancestors of *Homo erectus* left Africa almost as soon as *Homo* evolved there, before the African branch invented Acheulean tools.

Other specimens of *Homo erectus* from China are compatible with this story. The Chinese specimens have an age around 1.0 Ma and younger. *H. erectus* may have reached as far east as the island of Flores, in Indonesia, before 750,000 BP (years before present), a feat which involved two sea crossings, of 15 and 12 miles. There are no fossils, only a few tools, but the story fits with the fact that three major animals became extinct quite suddenly on Flores around 900,000 BP: a pygmy elephant, a giant tortoise, and a giant lizard related to the Komodo dragon.

The Asian specimens of *H. erectus* had their own versions of stone tool making styles. Specimens from Java had a brain size just under 1000 cc, but brain size had reached 1100 cc by the time of "Peking Man," who occupied caves outside Beijing between 500,000 and 300,000 BP. The successful long-term occupation of north China by these people indicates that they had solved the problems of surviving a challenging seasonal climate. Some Asian *erectus* made tools out of rhino teeth, since they were living in an area without good tool-making stone.

Here is the problem with the separation of these humans into *H. erectus* and *H. ergaster*. If hominids were spread widely but thinly over the Old World, the widely separated populations would tend to diverge in some characters, both anatomical and cultural. But would they become totally separated groups, geographically and genetically, long enough to evolve into separate species? One could look at a modern analog: leopards extend over a range today from southern Africa to the Pacific coast of Asia. No-one would argue that they were anything but one species.

It's difficult to separate reality from theory, and it's impossible to use a naming scheme that will make everyone happy. Even readers of *National Geographic* have to deal with messy alternative schemes, each of which has enthusiastic and often intolerant proponents.

The migration of *Homo erectus* left a corridor of humanity that stretched from South Africa to eastern Asia. All these populations evolved larger body size and more advanced skull characters, and all made new tool kits. There is no explicit reason (or evidence) to claim that they diverged into two or more species. Skulls dating from 1 Ma are well within the range of a single species (*Homo erectus*), after close to a million years of supposed separation. Given the mobility of humans, there was no necessary dramatic or long-lived separation between these pan-tropical populations. There was one founding, and dominant human species, *Homo erectus*, with locally variable anatomy and culture, just like *Homo sapiens* today. At least, that's a hypothesis that can stand until more new evidence turns up.

It looks as if *Homo erectus* was the first species to control fire. There is good evidence for camp fires in a South African cave at Swartkrans, dating from at least 1 Ma, and *H. erectus* lived in the cold north of China around 1.4 Ma.

We know from the shape of the pelvis that *H. erectus* babies were born as helpless as modern human babies are, and it is clear that the brain grew a lot after birth, as our brains do. This implies a long period of care for a baby that probably could not walk for several months. That is an enormous price to pay for a larger brain, and would only have been evolutionarily worthwhile (selected for) if there was a large pay-off for learning and intelligence.

All these lines of evidence imply a complex and stable social structure for *Homo erectus*, though details are certainly not available. I will make one comment of my own. The cooperation required to build, start, control, maintain, and transport a fire is very high. It is difficult (for me) to imagine a campfire without conversation. But as soon as any hominid evolved language, that would have begun the novel process of transferring abstract information and knowledge directly and immediately from individual to individual, replacing indirect methods such as taking and showing, demonstrating and copying, or sharing the same real experience. The ability to short-circuit the processes of teaching and learning must have allowed much more knowledge to be transmitted, absorbed, and retained in a society, with obvious advantages to all its members.

Later modifications in educational techniques have served mainly to accelerate the transfer of information and to make it possible at a distance, by the invention of writing and reading, numbers and alphabets, schools, printing, telephones, broadcasting, and so on.

Many experts believe that language is a very recent invention, essentially by *Homo sapiens*. That may be true for the complex activity that all modern humans are so good at. But language must have evolved, like every other characteristic of modern humans. I suspect that the information explosion set off by modern electronics is only the latest in a series that started around a campfire a million years ago. This book and the Macintosh I wrote it on are direct continuations of that tradition.

After *Homo erectus*

After *Homo erectus*, the current story of human evolution becomes very messy as anthropologists argue about the origin of modern humans. As fossil and molecular evidence has accumulated, anthropologists have accepted finer and finer subdivisions for species. In 2003 we have a majority opinion that several species of *Homo* have evolved during the last million years, with all but one becoming extinct. As always, you are free to construct your own interpretation of the story, and as always, new data will cause us all to rethink and reinterpret in the future.

According to the current majority story, some local populations of *Homo erectus* and/or *Homo ergaster* evolved separately and significantly. A separate species, named *Homo antecessor*, has been proposed for a group of well-preserved and distinct specimens found in Spain, dated close to a million years ago, and living on what was then the fringes of humanity. Obviously, *H. antecessor* must have had ancestors somewhere in Africa. Some *antecessor* left a beautiful pink quartzite hand ax on the top of a pile of skeletons in a Spanish cave. (Obviously, we are free to speculate about the meaning of the act.)

Perhaps dating as far back as 500,000 BP, a population, perhaps descended from *H. antecessor* and best known from central Europe, has been called *Homo heidelbergensis*. Around 400,000 BP, *H. heidelbergensis* was making beautifully crafted hunting spears in Germany. They are throwing spears, up to 3.2 m long (10 ft), carved to angle through the air like modern javelins, and the spears are associated with butchered horses and other bones from elephant, rhino, deer and bear.

Homo heidelbergensis in turn seems to have evolved into the Neanderthal people, usually now called *Homo neanderthalensis*. They were strongly adapted for life in cold-climates along the fringes of the Ice-Age tundra from Spain to Central Asia.

We have only patchy evidence from Africa during this time, and we assume that *Homo erectus* (*ergaster*, if you like) continued to thrive across that continent. Meanwhile, *Homo erectus* thrived in eastern Asia, and we have populations from Indonesia ("Java Man") and China ("Peking Man") in the age range 1 Ma to 500,000 Ma.

THE ORIGIN OF *HOMO SAPIENS*

For some years now, the majority story has been that around 300,000 BP an African *Homo* species (*erectus/ergaster/heidelbergensis/rhodesiensis/antecessor*/archaic *sapiens*) (all these names have been used, so we need more specimens to decide precisely what to call it) gradually developed a distinctive new type of stone tool technology which we call Middle Stone Age/Middle Paleolithic technology: MSA for short. And gradually, one of these African populations evolved into fully modern *Homo sapiens*.

This story was strengthened when a new set of fossils was described from Ethiopia in 2003. Three skulls, dating to about 160,000 BP, were identified as *Homo sapiens*. They still have some "archaic" features such as big eyebrows that would allow us to recognize them as "different" even if they were wearing blue jeans and T-shirts, so they are called *Homo sapiens idaltu*, a separate subspecies. Their tools were on the boundary between MSA and Acheulean. This time was in the middle of a very cold Ice Age, when other populations of *Homo* outside the tropics may have been stressed and fragmented. It seems that *Homo sapiens* had some feature or features that allowed it to expand and compete effectively against other *Homo* species.

During the last interglacial (around 120,000 BP) modern humans, *Homo sapiens sapiens*, are found in South Africa. And sometime after that, a small population of pioneering or refugee *Homo sapiens* left Africa for the first time and expanded into and over the Old World, presumably through the Middle East, which was not such a severe desert as it is now. Modern humans may have traversed around the shores of the Indian Ocean to Southeast Asia, but they were apparently not yet able to sweep aside other species easily. *Homo neanderthalensis* was firmly entrenched in the colder areas of Europe and western Asia, and *Homo erectus* in eastern Asia. As the interglacial ended and the climate turned cool again, Neanderthals reoccupied the Middle East around 70,000 BP, replacing modern humans.

All this changed, beginning around 45,000 BP. Modern humans now swept all competing species from the Old World: the Neanderthals from Europe, and *Homo erectus* from Asia. This was overwhelming competition, perhaps even ethnic cleansing: the fossil record suggests there were no survivors. And, of course, *Homo sapiens* occupied also spread to parts of the world previously not occupied by any *Homo*: Australia, the Americas, and Polynesia. All living humans are thus descended from what must have been a comparatively small original population of *H. sapiens*.

This fundamental scenario is fiercely debated, and it is important. A few anthropologists still argue that a widespread *Homo erectus* evolved into a widespread *Homo sapiens* across the entire Old World, and the other names simply represent subspecies or variants that do not imply an inability to interbreed. There have been claims of transitional skeletons between *erectus* and *sapiens* in Indonesia, China, Africa, and the Middle East.

All these events are so recent that their effects can still be seen in, and interpreted from, the genetics of *Homo sapiens*. And those genetic data overwhelmingly support the out-of-Africa model. Even if genetic clocks do not run accurately, the story would not change: only the dates would.

The DNA structure of living humans shows a distinct division into African and non-African types. Furthermore, the variation in modern human DNA is very restricted. A single breeding group of chimpanzees in the Taï in Africa has more variability than does the entire human race today. This suggests very strongly that all modern humans are descended from an ancestral population that was not only small—say 10,000 or so—but was small for a long time.

If this estimate is correct, there is no way that 10,000 people could have interbred and inhabited more than a relatively small area, even if they were wandering hunter-gatherers. The only possible conclusion (if the assumptions and calculations are correct) is that indeed all living humans are exclusively descended from a small ancestral *sapiens* population who evolved in a restricted region in Africa, and spread from there throughout the Old World.

The difference that remains today between "African" and "non-African" DNA is explained if a small founder population left Africa, carrying with them only a small sample of the genetic variation that had by then evolved across Africa. These founder populations expanded as they occupied Eurasia, growing into a large population with a distinctly non-African DNA structure. Once again, a small subset of east Asian humans crossed the Bering Strait and populated the Americas with people who had even less genetic variation.

This scenario was first favored in the late 1980s, when it was called the Mitochondrial Eve or Out-of-Africa hypothesis. The original presentation of the hypothesis was flawed (it was based on poor use of the computer program that processed the data), but it is now strongly supported by masses of new data. Pressing questions involve the timing, and what happened to the human populations of the rest of the Old World.

THE NEANDERTHALS

The people we call Neanderthals lived between 90,000 and 30,000 BP in Europe and along the mountain slopes on the northern edges of the Middle East, as far east as Iraq. They were named after a site in the Neander Valley in Germany. Neanderthals had a way of life that was distinctly sophisticated in living sites, tools, and behavior.

Neanderthals differ from living humans in having big faces with large noses, large front teeth, and little or no chin (Figure 20.12). These characters are connected: Neanderthal front teeth show heavy wear, as if they used their incisors for something that demanded constant powerful pressure (softening hides by chewing, as Eskimos used to do?). Human faces are plastic, especially in early growth, and either by use or by genetic fixation, the frontal chewing of Neanderthals seems to have encouraged growth of the facial bones to support the front teeth against the skull. The nasal area was essentially swung outward from the face, so that the nose was

a

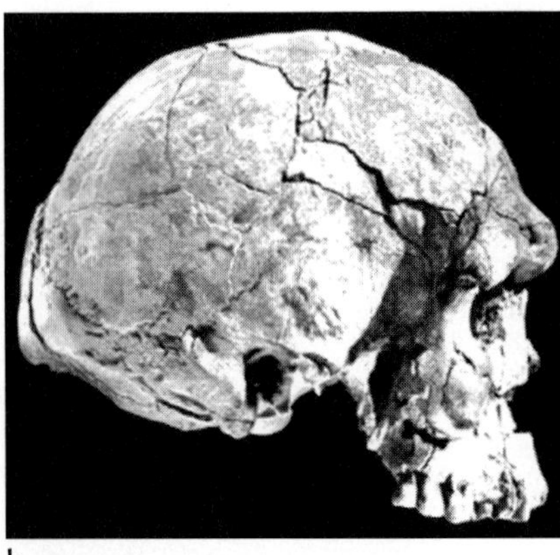

b

Figure 20.12 (a) Skull of a Neanderthal. (Redrawn, idealized, and simplified from Trinkaus). (b) Skull of *Homo sapiens idaltu*, from Ethiopia, the oldest skull yet that has been assigned to *Homo sapiens* (age about 160,000 BP). Research published by White et al., 2003. (Photo donated by Tim White of the University of California, Berkeley.)

even bigger and more projecting than it is in most hominids. Neanderthal brain size, at 1450–1500 cc, was at least equal to that of living humans and sometimes greater. Another special Neanderthal character was a very strong, stocky body with very robust bones, which may have helped conserve body heat in a cold climate and/or may reflect a lifestyle that required great physical strength. Most Neanderthal fossils are found in deposits laid down in the harsh climates of the next-to-last ice age.

Most Neanderthal tools are made in a style called Mousterian. They include scrapers, spear points, and cutting and boring tools (Figure 20.13) made from flakes carefully chipped off a stone core. Marks on Neanderthal teeth suggest that they stripped animal sinews to make useful fibers by passing them through clenched teeth, just as Australian aboriginals do. But perhaps the most enlightening Neanderthal finds are their ceremonial burials. Bodies were carefully buried, with grave offerings of tools and food. Enormous quantities of pollen were found with the body of Shanidar IV, a Neanderthal man buried in Iraq. The pollen came from seven plant species in particular. All seven have brightly colored flowers, all seven bloom together in the area in late April, and all have powerful medicinal properties. It is difficult to avoid the conclusion that Shanidar IV was carefully buried with garlands of healing herbs chosen from early summer flowers, suggesting an intense concern for the abstract world.

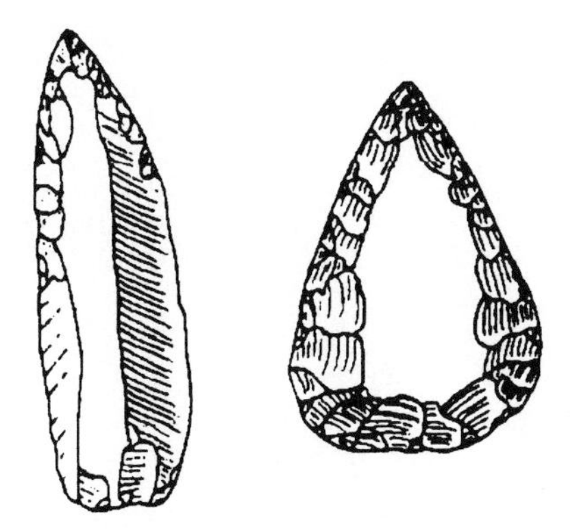

Figure 20.13 A Neanderthal tool kit, shown about half size. (After Bordes.)

Neanderthals became adapted to life in the cold climate along the edges of the ice sheets from western Europe to central Asia, by evolving characters of their own. The more geographically isolated they were, the more extreme their Neanderthal characters became, until they became visibly different from both *heidelbergensis* and from the *sapiens* populations that were evolving in Africa. The reasonable interpretation of these facts is that Neanderthals were not *Homo sapiens* but a separate and extinct species, *Homo neanderthalensis.*

In the Middle East, Neanderthals seem to have alternated with *Homo sapiens,* with a fluctuating border between them. Both peoples in the region were making the same Mousterian tools, which have been identified as far south as the Sudan, but none of the fossil skulls are intermediate, suggesting that the two did not interbreed. Neanderthals lived in the Middle East in cooler, wetter times, while *Homo sapiens* lived there in hotter, drier times. Each was fitted to a particular climatic zone in which the other could not compete; neither was "superior" during those tens of thousands of years.

Neanderthals disappeared from the Middle East about 45,000 BP, then from eastern and central Europe, and finally from northwest Europe (France and northern Spain). The last dated surviving Neanderthals held out in upland France until about 38,000 BP and in southern Spain and Portugal until about 30,000 BP.

European Neanderthal sites typically contain less standardized tools, made only from local stone and flint, but the last Neanderthals in western Europe showed a distinctly more advanced culture. Their tools are called Châtelperronian. These tools show clear evidence of the old Mousterian style, but they also have some similarities to the Aurignacian tools that the newly arrived, fully modern humans (the CroMagnon people), were using at about the same time in western Europe. The last Neanderthal sites in France also contain simple ornaments, and it is tempting to suggest that Neanderthals may have copied some of the CroMagnon technology and art.

This suggestion is controversial, because Neanderthal artifacts are not found at CroMagnon sites, or the reverse. Some anthropologists think that the late Neanderthals were evolving their own advanced culture just before their demise (perhaps in the same way that Aztecs and Incas were evolving new political systems

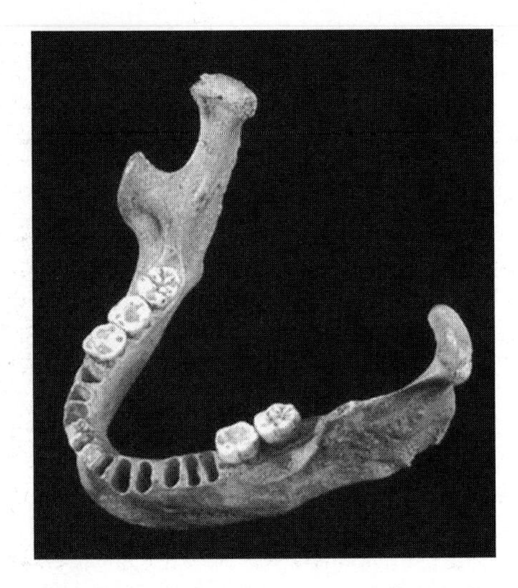

Figure 20.14 A *Homo sapiens* jaw from Romania, old enough (around 35,000 BP) to have lived alongside Neanderthals. Its molar teeth are astonishingly large (Neanderthal-sized, to be perfectly blunt about it). (Courtesy Erik Trinkaus of Washington University, St. Louis.)

Figure 20.15 Portraits of a woolly mammoth and a woolly rhinoceros, by unknown CroMagnon artists. (After drawings made by de Breuil from cave paintings in France.)

just as the Spanish arrived). Neanderthal and CroMagnon peoples overlapped in Europe for 10,000 years or so, a very long time in historical terms.

The skeletons at Châtelperronian sites are Neanderthal. A beautifully preserved skull of a late Neanderthal baby from France shows a structure of the inner ear that is quite different from that in modern humans. Neanderthal DNA (recovered from the original fossil from the Neander Valley) is distinctly different from that of any living human population, and also different from two samples of CroMagnon DNA.

As I wrote this chapter, new evidence suggested that there had been some interbreeding between *Homo sapiens* and Neanderthals. An extremely early *sapiens* skull turned up in Romania, early enough to be contemporary with Neanderthals. That skull was totally *sapiens* in morphology, except that its molar teeth were astonishingly large (Figure 20.14). They were larger than any *sapiens* teeth ever discovered, but about average for Neanderthal. Was this a hybrid? As usual, we need more evidence!

CroMagnons and Neanderthals were quite different socially. There is not only the evidence of art and symbolism; CroMagnons may well have been better big-game hunters, and some people even argue that CroMagnons had true language, while Neanderthals did not. But Californians are quite different socially from Papua New Guineans: that means nothing genetically. We need a better explanation. What superiority did *Homo sapiens* have? Was it weaponry? or social cohesion? It was not language as such: Neanderthals could certainly speak.

As CroMagnon people took over the cold and forbidding European peninsula, they were only a local population on the northwest fringe of the human species, but they are important because they have yielded us the best-studied set of fossils, tools, and works of art from the depths of the last Ice Age (Figure 20.15).

CroMagnon sites yield richer and more sophisticated art and sets of tools, and more complex structures and burials than Neanderthals. In particular, CroMagnons made stone tips for projectiles. Neanderthals may have had arrows and spears, but if they did, they were not stone tipped. Mousterian tools are more frequently wood-working tools than those of CroMagnons, who worked more with bone, antler, and stone.

CroMagnons and their contemporaries in eastern Europe, the Gravettians, left evidence of a capacity for habitat destruction that is typically modern in style. From Russia to France, sites contain the remains of thousands of horses and hundreds of woolly mammoths. CroMagnons were also responsible for the magnificent cave paintings of extinct Ice Age animals, drawn by people who saw them alive (Figure 20.15), and they made (and presumably played) bone flutes. CroMagnons were painting on cave walls at 30,000 BP, and Gravettians were firing small terracotta figurines in kilns more than 25,000 years ago. CroMagnons and their contemporaries had a tremendous ecological impact on the world (Chapter 21).

So is *Homo sapiens* collectively guilty of ethnic cleansing? All other species of *Homo* disappeared. It is unlikely that some natural catastrophe affected *Homo* but not the other surviving apes. It is much more likely that the branches of human descent were pruned by other humans.

Migration-with-ethnic-cleansing may not be one's favorite image of the human race, but we have to deal with evidence. Certainly genocide occurred in the past (read the Bible or the newspaper). Events in Sierra Leone and Liberia come to mind as I write, but Bosnia, Rwanda, Kurdistan, Uganda, the Congo, and Nazi Germany should not be forgotten either, and the history of any continent has horrific examples. Within living memory, cannibal villagers in Papua New Guinea would try to kill off all members of a target community, because survivors would be likely to retaliate.

EVOLUTION AMONG HUMANS TODAY

Given the vastly different biological and ecological environments of the species of *Homo* since 2.4 Ma, it's likely that the selective pressures on soft-part anatomy and behavior have been as intense as those on skeletal features. There is clear evidence among living humans of regional evolution to suit the particular environment; for example, some of the characters of soft anatomy, body proportions, and even parts of the mitochondrial genome are strongly linked to climate in many human groups. Nose shape is strongly correlated with humidity. Eskimos are endomorphic to resist body cooling. Peoples living at high altitude in Asia, Ethiopia, and South America have adapted physiologically to low oxygen levels. There are extremely ectomorphic, dark-skinned tropical people, pink-skinned people in northern Europe, and so on. Testis size varies markedly among human groups, as does the frequency of twinning.

Such features must have evolved under intense regional selection, combined with the slow spread and mixing of genes at a time when human groups could not travel great distances. Such differences are visibly diminishing in certain modern populations (Hawaii, Brazil, California, and London come to mind).

The evolution of behavior cannot be assessed very well from the fossil record, but the variation in social structure within hominoid species is very great and suggests radical behavioral evolution, at least over the past 10 m.y. New research on mating patterns among animals suggests that much of human sexual anatomy and sexual behavior may be linked with the evolutionary breakthrough that began with early *Homo* and the use of tools to achieve ecological dominance.

It is sometimes claimed that natural selection no longer acts on modern humans because our surroundings are so artificial. Most people are now more insulated against diseases, environmental fluctuations, and accidents than humans were only a few centuries ago. Yet selection still operates strongly even in the most "advanced" societies. The genes for sickle-cell anemia are now generally harmful, instead of being favored in malarial regions. The genes that predispose non-Europeans to diabetes and gall-bladder cancer are more easily triggered into action on a westernized ("coca-colonized") diet, whereas they were beneficial in their original cultural environment. Among the Micronesian population of Nauru, nearly two-thirds of the population have diet-induced diabetes by the age of 55, and 50% of all Pima Indians of Arizona have diabetes.

Finally, it appears that although the maximum human lifespan has not changed very much, average life expectancy has. At least some human societies now depend strongly on physical, social, and intellectual nurturing of children long past physical maturity. Characteristics that are normally juvenile attributes in primates, such as imagination, curiosity, play, and learning, are now encouraged in early adult years. The trend is presently social. There is no evidence yet of any evolutionary feedback creating delayed physical maturity or increased mental capacity. It is not yet clear whether this will occur and alter human biology as well as human culture. Potentially, an increased learning period could have enormous consequences for us and for every living thing on Earth.

Further Reading

Abbate, E., et al. 1998. A one-million-year-old *Homo* cranium from the Danakil (Afar) depression of Eritrea. *Nature* 393: 458–60.

Adcock, G. J., et al. 2001. Mitochondrial DNA sequences in ancient Australians: Implications for modern human origins. *Proceedings of the National Academy of Sciences* 98: 537–42; and comment in *Science* 291: 230–1.

Agnew, N., and M. Demas. 1998. Preserving the Laetoli footprints. *Scientific American* 279 (3): 44–55.

Aiello, L. C., and J. C. K. Wells. 2002. Energetics and the evolution of the genus *Homo. Annual Reviews in Anthropology* 31: 323–38.

Ambrose, S. 2001. Paleolithic technology and human evolution. *Science* 291: 1748–53.

Appenzeller, T. 1998. Art: evolution or revolution? Science 282: 1451–4.

Asfaw, B., et al. 1999. *Australopithecus garhi*: a new species of early hominid from Ethiopia *Science* 284: 629–5, and comment 572–3. See also companion paper by J. de Heinzelin et al. on environment and behavior of *A. garhi*, 625–9.

Bahn, P. G. 1998. Neanderthals emancipated. *Nature* 394: 719–20. [Neanderthals at Arcy-sur-Cure evolved their culture independently.]

Bahn, P. G., and J. Vertut. 1998. *Journey Through the Ice Age.* Berkeley, California: University of California Press. [The art work of the Cro-Magnons.]

Bahn, P. 2003. A child in time. *Nature* 424: 490. [Review of the "Neanderthal hybrid" monograph.]

Balter, M. 1996. Cave structure boosts Neandertal image. *Science* 271: 449.

Balter, M. 1999. New light on the oldest art. *Science* 283: 920–2.

Balter, M. 2001. Scientists spar over claims of earliest human ancestor. *Science* 291: 1460–1. [*Orrorin*, at 6 Ma.]

Balter, M. 2001. In search of the first Europeans. *Science* 291: 1722–5.

Balter, M. 2001. Paleontological rift in the Rift Valley. *Science* 292: 198–201. [Grown-ups behaving badly: the viciousness among paleoanthropologists.]

Balter, M. 2001. What, or who, did in the Neandertals? *Science* 293: 1980–1. [Disputes, no answers.]

Balter, M. 2002. What made humans modern? *Science* 295: 1219–25. [News summary.]

Bermúdez de Castro, J. M., et al. 1997. A hominid from the Lower Pleistocene of Atapuerca, Spain: possible ancestor to Neandertals and modern humans. *Science* 276: 1392–5, and comment 1331–3. [*Homo antecessor.*]

Bermúdez de Castro, J. M., et al. 1999. A modern human pattern of dental development in Lower Pleistocene hominids from Atapuerca-TD6 (Spain). *Proceedings of the National Academy of Sciences* 96: 4210–13. [*Homo antecessor.*]

Blumenschine, R. J., and J. A. Cavallo. 1992. Scavenging and human evolution. *Scientific American* 267 (4): 90–6.

Blumenschine, R. J., et al. 2003. Late Pliocene *Homo* and hominid land use from western Olduvai Gorge, Tanzania. *Science* 299: 1217–21, and comment 1193–4. [Casts doubt on the validity of *Homo rudolfensis.*]

Brain, C. K. 1981. *The Hunters or the Hunted? An Introduction to African Cave Taphonomy.* Chicago: University of Chicago Press.

Brantingham, P. J. 1998. Hominid-carnivore coevolution and invasion of a predatory guild. *Journal of Anthropological Archaeology* 17: 327–53.

Brunet, M., et al. 2002. A new hominid from the Upper Miocene of Chad, Central Africa. *Nature* 418: 145–51, and comment 133–5. [*Sahelanthropus.* See also companion paper on the age and environment 152–5.]

Cann, R. L. 2001. Genetic clues to dispersal in human populations: retracing the past from the present. *Science* 291: 1742–8. [Very nice state-of-the-art review paper.]

Caramelli, D., et al. 2003. Evidence for a genetic discontinuity between Neandertals and 24,000-year-old anatomically modern Europeans. *Proceedings of the National Academy of Sciences* 100: 6593–7, but see comment in *Nature* 423: 468. [Neanderthal and CroMagnon DNA is different.]

Cartmill, M. 1998. The gift of gab. *Discover* 19 (11): 56–64. [Origin of language.]

Churchill, S. E. 1998. Cold adaptation, heterochrony, and Neandertals. *Evolutionary Anthropology* 7: 46–61.

Clarke, R. J. 1998. First ever discovery of a well-preserved skull and associated skeleton of an *Australopithecus. South African Journal of Science* 94: 460–3.

Defleur, A., et al. 1999. Neanderthal cannibalism at Moula-Guercy, Ardèche, France. *Science* 286: 128–31.

Diamond, J. M. 1992. *The Third Chimpanzee.* New York: HarperCollins.

Diamond, J. 2001. Unwritten knowledge. *Nature* 410: 521. [Why did humans evolve to live so long?]

Diamond, J. M. 2003. The double puzzle of diabetes. *Nature* 423: 599–602. [Why are many people predisposed to get diabetes, and why are Europeans not among them?]

Diamond, J. M., and J. I. Rotter. 1987. Observing the founder effect in human evolution. *Nature* 329: 105–6. [South African Afrikaner genetics.]

Fernández-Jalvo, Y., et al. 1999. Human cannibalism in the Early Pleistocene of Europe (Gran Dolina, Sierra de Atapuerca, Burgos, Spain). *Journal of Human Evolution* 37: 591–622. [Someone ate *Homo antecessor*, and it looks as if it was *Homo antecessor.*]

Fifer, F. C. 1987. The adoption of bipedalism by the hominids: a new hypothesis. *Human Evolution* 2: 135–47. [Stone throwing.]

Gabunia, L., et al. 2001. Dmanisi and dispersal. *Evolutionary Anthropology* 10: 158–70. [Very early *Homo erectus* in Georgia.]

Gibbons, A. 1993. Pleistocene population explosions. *Science* 262: 27–8.

Gibbons, A. 1998. Ancient island tools suggest *Homo erectus* was a seafarer. Science 279: 1635–7. [In Indonesia.]

Gibbons, A. 1998. Which of our genes makes us human? *Science* 281: 1432–4.

Gibbons, A. 2001. The riddle of coexistence. *Science* 291: 1725–9. [Between humans and Neanderthals.]

Gibbons, A. 2002. In search of the first hominids. *Science* 295: 1214–19. [News summary.]

Gore, R. 1997. The dawn of humans. *National Geographic* 191 (2): 72–99 [February]; 191 (5): 84–109 [May]; 192 (1): 96–113 [July]; and 192 (3): 92–9 [September].

Haile-Selassie, Y. 2001. Late Miocene hominids from the Middle Awash, Ethiopia. *Nature* 412: 178–81, with companion paper on the age and paleoenvironment 175–8; also see comment in Science 293: 187–9. [*Ardipithecus* at over 5 Ma.]

Harpending, H. C., et al. 1998. Genetic traces of ancient demography. *Proceedings of the National Academy of Sciences* 95: 1961–7.

Harris, E. E., and J. Hey. 1999. X chromosome evidence for ancient human histories. *Proceedings of the National Academy of Sciences* 96: 3320–4, and comment in *Science* 283: 1828.

Holden, C. 1998. No last word on language origins. *Science* 282: 1455–8.

Holden, C. 1999. A new look into Neandertals' noses. *Science* 285: 31–3.

Holden, C. 2001. Oldest human DNA reveals Aussie oddity. *Science* 291: 230–1. [DNA from Mungo Man.]

Hou, Y., et al. 2000. Mid-Pleistocene Acheulean-like stone technology of the Bose Basin, South China. *Science* 287: 1622–6, and comment 1566. [Beautiful stone tools, presumably made by *Homo erectus*, at 800,000 BP.]

Ingman, M., et al. 2000. Mitochondrial genome variation and the origin of modern humans. *Nature* 408: 708–13, and comment 552–3. [More evidence for out of-Africa.]

Johanson, D., and B. Edgar. 1996. *From Lucy to Language.* New York: Simon & Schuster.

Klein, R. G. 1999. *The Human Career: Human Biological and Cultural Origins.* 2d ed. Chicago: University of Chicago Press. [Comprehensive text with hundreds of references.]

Kramer, P. A. 1999. Modelling the locomotor energetics of extinct hominids. *Journal of Experimental Biology* 202: 2807–18. [How Lucy walked (efficiently!)].

Kunzig, R. 1997. The face of an ancestral child. *Discover* 18 (12): 88–101. [*Homo antecessor* from Spain.]

Kunzig, R. 1999. Learning to love Neanderthals. *Discover* 20 (8): 68–75. [A Neanderthal/*sapiens* "hybrid" child, a sympathetic article.]

Larick, R., and R. L. Ciochon. 1996. The African emergence and early Asian dispersals of the genus *Homo. American Scientist* 84: 538–51.

Leakey, M., and A. Walker. 1997. Early hominid fossils from Africa. *Scientific American* 276 (6): 74–9. [*A. anamensis*.]

Lewin, R. A. 1998. *The Origin of Modern Humans.* 2nd edn New York: Scientific American Library.

Li, T., and D. A. Etler. 1992. New Middle Pleistocene hominid crania from Yunxian in China. *Nature* 357: 404–7. [Apparent transition between *erectus* and *sapiens* in China.]

Lieberman, D. E. 1998. Sphenoid shortening and the evolution of modern human cranial shape. *Nature* 393: 158–62.

Lovejoy, C. O. 1988. Evolution of human walking. *Scientific American* 259 (5): 118–25.

McCollum, M. A. 1999. The robust australopithecine face: a morphogenetic perspective. *Science* 284: 301–5, and comment, 230–1.

McHenry, H. M., and K. Coffing. 2000. *Australopithecus* to *Homo*: transformations in body and mind. *Annual Reviews of Anthropology* 29: 125–46. [Excellent review of the transition as seen in 2000].

Mishmar, D., et al. 2003. Natural selection shaped regional mtDNA variation in humans. *Proceedings of the National Academy of Sciences* 100: 171–6.

Nesse, R. M., and G. C. Williams. 1995. *Why We Get Sick: An Introduction to Darwinian Medicine.* New York: Times Books. [Read this, and then ask whether natural selection has ended! Paperback edition, Vintage Books, 1999.]

Ovchinnikov, I. V., et al. 2000. Molecular analysis of Neanderthal DNA from the northern Caucasus. *Nature* 404: 490–3, and comment 453–4. [It differs from modern human DNA.]

Partridge, T. C., et al. 2003. Lower Pliocene hominid remains from Sterkfontein. *Science* 300: 607–12, and comment 562. [The date, around 4 Ma, seems very early!]

Pennisi, E. 1999. Did cooked tubers spur the evolution of big brains? *Science* 283: 2004–5. [See Wrangham et al., 2001 for the long professional version.]

Pinker, S. 1994. *The Language Instinct.* New York: William Morrow and Company. [Chapter 11 deals with the origin of language among humans.]

Rand, D. 2001. Mitochondrial genomics flies high. *Trends in Ecology and Evolution* 16: 2–4. [New research that shows mtDNA changes are NOT neutral or clock-like.]

Relethford, J. H. 1998. Genetics of modern human origins and diversity. *Annual Reviews of Anthropology* 27: 1–23. [Very important and accessible review of what genetics can and cannot say about the evolution of modern humans. Genetic data are not as powerful as many people think!]

Reno, P. L., et al. 2003. Sexual dimorphism in *Australopithecus afarensis* was similar to that of modern humans. *Proceedings of the National Academy of Sciences* 100: 9404–9, and comment 9103–4. [This could be important because social structure such as monogamy is associated in primate biology with low levels of dimorphism. No proof, but reasonable inference.]

Richmond, B. G., and D. S. Strait. 2000. Evidence that humans evolved from a knuckle-walking ancestor. *Nature* 404: 382–5; comment in *Science* 287: 2131–2.

Schick, K. D., and N. Toth. 1993. *Making Silent Stones Speak.* New York: Simon and Schuster.

Schlieckelman, P., et al. 2001. Natural selection and resistance to HIV. *Nature* 411: 545–6.

Semaw, S., et al. 2003. 2.6-million-year-old stone tools and associated bones from OGS-6 and OGS-7, Gona, Afar, Ethiopia. *Journal of Human Evolution* 45: 169–77.

Semino, O., et al. 2000. The genetic legacy of Paleolithic *Homo sapiens sapiens* in extant Europeans: a Y chromosome perspective. *Science* 290: 1155–9, and comment 1080–1. [Where did Europeans come from, and when?]

Senut, B., et al. 2001. First hominid from the Miocene (Lukeino formation, Kenya). *Comptes Rendus de l'Académie des Sciences, Paris* 332: 137–44, and companion paper on the paleoenvironment, 145–52; see comment in *Nature* 410: 526–7. [*Orrorin.*]

Shreeve, J. 1995. *The Neanderthal Enigma: Solving the Mystery of Modern Human Origins.* New York: William Morrow and Company. [A masterpiece of scientific writing. Discusses human origins, not just Neanderthals. Paperback edition, Avon Books, 1996.]

Shreeve, J. 1995. The Neanderthal peace. *Discover* 16 (9): 71–81.

Sponheimer, M., and J. A. Lee-Thorp. 1999. Isotopic evidence for the diet of an early hominid, *Australopithecus africanus. Science* 283: 368–70, and comment 303. [Meat?]

Strauss, E. 1999. Can mitochondrial clocks keep time? *Science* 283: 1435–8.

Stringer, C., and R. McKie. 1997. *African Exodus: the Origins of Modern Humanity.* New York: Henry Holt. [Very readable; stresses Stringer's own views, of course.]

Suwa, G., et al. 1997. The first skull of *Australopithecus boisei. Nature* 389: 489–92, and comment 445–6. [*A. boisei* and *A. robustus* may be the same.]

Swisher, C. C., et al. 1994. Age of the earliest known hominids in Java, Indonesia. *Science* 263: 1118–21, and comment 1087–8.

Swisher, C. C., et al. 1996. Latest *Homo erectus* of Java: potential contemporaneity with *Homo sapiens* in Southeast Asia. *Science* 274: 1870–4.

Tattersall, I. 1999. *The Last Neanderthal.* 2nd edn Westview Press. [Neanderthals as a separate species.]

Tattersall, I. 2000. Once we were not alone. *Scientific American* 282 (1): 56–62. [Mostly, this is a vehicle for Jay Matternes paintings of hominid species. But it is also a good summary of the 2000 view of hominid evolution.]

Thieme, H. 1997. Lower Palaeolithic hunting spears from Germany. *Nature* 385: 807–10, and comment 767–8. [Throwing spears from 400,000 BP.]

Toth, N. 1987. The first technology. *Scientific American* 256 (4): 112–21.

Toth, N., et al. 1992. The last stone ax makers. *Scientific American* 267 (1): 88–93.

Treves, A., and L. Naughton-Treves. 1999. Risk and opportunity for humans coexisting with large carnivores. *Journal of Human Evolution* 36: 275–82.

Trinkaus, E., et al. 2003. An early modern human from the Pestera cu Oase, Romania. *Proceedings of the National Academy of Sciences* 100: 11231–6. [*Homo sapiens*, but with HUGE molars.]

Vekua, A., et al. 2002. A new skull of early *Homo* from Dmanisi, Georgia. *Science* 297: 85–9, and comment 26–7.

Walker, A., and P. Shipman. 1996. *The Wisdom of the Bones: in Search of Human Origins.* New York: Alfred A. Knopf. [Superb account of the way an outstanding scientist works and thinks, with a vivid reconstruction of *H. erectus.*]

Wallace, D. C., et al. 1985. Dramatic founder effects in Amerindian mitochondrial DNAs. *American Journal of Physical Anthropology* 68: 149–55.

Wallace, D. C. 1999. Mitochondrial diseases in man and mouse. *Science* 283: 1482–8.

Weiss, K. M., et al. 1984. A New World Syndrome of metabolic diseases with a genetic and evolutionary basis. *Yearbook of Physical Anthropology* 27: 153–78. [Diabetes etc.]

Wheeler, P. E. 1991. The influence of bipedalism on the energy and water budget of early hominids. *Journal of Human Evolution* 21: 117–37.

White, T. 2003. Early hominids: diversity or distortion? *Science* 299: 1994–6. [*Kenyanthropus* is just a smashed *Australopithecus afarensis.*]

White, T. D., et al. 2003. Pleistocene *Homo sapiens* from Middle Awash, Ethiopia. *Nature* 423: 742–7; accompanying paper 747–52; comment 692–5; and comment in Science 300: 1641. [Skulls from 160,000 BP.]

Wong, K. 2000. Who were the Neanderthals? *Scientific American* 282 (4): 98–107. [The question in early 2000, with cameo comments by some of the major researchers.]

Wrangham, R. W., et al. 2001. The raw and the stolen: cooking and the ecology of human origins. *Current Anthropology* 40: 567–94. [Includes comments by dissenters.]

Wyckoff, G. J., et al. 2000. Rapid evolution of male reproductive genes in the descent of man. *Nature* 403: 304–9.

Zollikofer, C. P. E., et al. 2002. Evidence for interpersonal violence in the St. Césaire Neanderthal. *Proceedings of the National Academy of Sciences* 99: 6444–8. [St. Césaire survived a vicious blow to the head, so he must have been cared for.]

CHAPTER TWENTY-ONE

Life in the Ice Age

Climate is one of the most important environmental factors for all organisms, and climatic changes have almost certainly been major factors affecting the evolution of life. Plate tectonic movements can change oceanic and continental geography, and those geographic changes can modify seasonal climatic patterns and affect the ecology and evolution of organisms in major global events (Chapter 6). There were major effects on life as world geography changed with the breakup of Pangea in the Late Mesozoic and Early Cenozoic. Some effects resulted directly from geographic isolation (Chapter 18), but others were mediated through the indirect effects of geography on climate. Many puzzles of Mesozoic evolution may be resolved when we can reconstruct paleoclimates more accurately. Certainly this has been the result of concentrated research programs on Cenozoic paleoclimates.

We are living through an ice age now, and have been for the past 2.5 million years or so. We happen to live during a warm stage in it, but there is no sign that it is over. Great ice sheets expanded and covered much of the northern continents then retreated again. They have done so at least 17 times in the past two million years. Yet ice ages have been rare during Earth's history. How did the present ice age affect life?

ICE AGES AND CLIMATIC CHANGE

Ice ages are not common events in Earth's history. There was a widespread ice age toward the end of the Pre-Cambrian around 600 Ma (Chapter 4). In Late Ordovician times, when northern Gondwana was over the South Pole, a great ice sheet spread over most of North Africa and probably further, triggering enough changes in marine life to mark the end of the Ordovician Period and the beginning of the Silurian. Gondwana drifted across the South Pole during the rest of the Paleozoic Era, with a particularly important glacial period in South America at the end of the Devonian. A small ice sheet lay over South Africa in the Early Carboniferous, but large-scale glaciation once again spread over most of Gondwana in the Late Carboniferous and Early Permian. Traces of this event, in the form of scratched rock surfaces and piles of glacial rock debris, are widespread in South Africa, South America, Australia, India, and Antarctica (Chapter 10). A Northern Hemisphere glaciation occurred in Siberia late in the Permian. But afterward there was no major ice age for 250 m.y., until the present one began. Paleoclimatic evidence suggests that Earth's surface cooled over the past 60 m.y., until finally the planet dropped into the present ice age.

The only external factors that could generate major climate change are astronomical processes—**changes in Earth's orbit** or **changes in solar radiation**. Such changes occur, but they are probably too small to generate major climate change by themselves. They cause fluctuations in climate, however. An **asteroid impact** could conceivably trigger a climate change, but only for a short time and only if conditions were already just right to start *and maintain* a change over considerable time.

It seems that we must look for mechanisms here on Earth for major climatic changes. Two processes can affect the amount of solar radiation that Earth retains. Some solar radiation is reflected back into space (the **albedo effect**), and a change in the amount of heat reflected would cool or warm Earth. Gases in the atmosphere, especially carbon dioxide and methane, are very effective in absorbing solar radiation (the **greenhouse effect**), and changes in the amounts of these gases could strengthen or weaken the effect of solar radiation, or override it completely.

The basic preconditions for climate change on Earth are simple. For an ice age, there must be a lot of snowfall in areas where it will build up rather than remelt. Such a situation can occur if Earth's global geography is arranged in the right way. An ice age, or any other climate change, can be encouraged or discouraged by the geographic changes that result from plate tectonic movements. But geography changes through the action of plate tectonics. Changes in geography also act to vary the albedo of Earth, the scale and activity of ocean currents, and the distribution of heat to different regions, all of which affect climate.

In general, ice ages require large areas of land in high latitudes. The poles should be isolated from warm water. Finally, to lock Earth into a long glacial period, there must be room for large continental ice sheets to spread out and provide high reflectivity to large regions. Geography thus controls whether or not Earth's heat is well distributed, and whether or not polar ice sheets can form. Plate tectonics controls continental distributions, and the necessary conditions for generating an ice age may arise from time to time just by the motions of the plates.

Earth's major climate changes contain distinct fluctuations. Dramatic advances and retreats of ice can occur even while Earth is locked into an ice age. Huge areas of the northern continents are still covered by debris dropped by ice sheets during 40 or so glacial advances and retreats during the past two million years. Large temperature fluctuations are recorded by microfossils in seafloor sediments. World sea level has fluctuated up and down by more than 70 m (220 ft) as 5% of Earth's water has alternately been frozen into ice sheets and melted away. Such changes in sea level are recorded worldwide in sedimentary deposits far from the ice sheets. For example, islands and atolls in the Atlantic and Pacific Oceans have been repeatedly exposed and reflooded.

Ancient rocks also show evidence of frequent and important change in sea level. For example, regular cycles of limestone, sandstone, and coal formation in the Carboniferous rocks of North America and Europe resulted from the cyclic rise and fall of sea level, and those rocks were deposited in tropical latitudes far from the glaciations of Gondwana that indirectly generated them. Many other cases of climatic cycling have been identified, even at times when Earth had no ice sheets, so we should look for a general cause for them, unconnected with ice sheets as such.

The **astronomical theory of ice ages** was suggested more than a century ago. It was worked out by hand by Milutin Milankovitch in the 1920s, and refined by computer calculations in the 1970s. It has been confirmed by evidence from microfossils that record temperature fluctuations in the oceans. The Milankovitch theory

BOX 21.1 The Components of The Milankovitch Theory

Tilt. Increased or decreased tilting of the axis (Figure 21.1a), which varies between 22° and 24.5°, increases or decreases the effect of the seasons in a cycle of about 41,000 years (Figure 21.2).

Precession. Earth's orbit is not a circle but an ellipse, with the sun at one focus (Figure 21.2). One pole is closer to the sun in its winter, while the other is closer in its summer. Thus, at any time, one pole has warm winters and cool summers, while the other pole has warm summers and cool winters. However, the slow rotation or precession of Earth's orbit around the sun alternates the effect between the two poles in cycles of 19,000 or 23,000 years (Figure 21.1c).

Eccentricity. Earth's orbit varies so that it is more elliptical at some times than at others, strengthening or weakening the precession effect (Figure 21.1b). This variation in orbital eccentricity affects climate in cycles of about 100,000 years and about 400,000 years. Of course, when eccentricity is low (when the orbit is closer to being circular), the precession effect is much lessened.

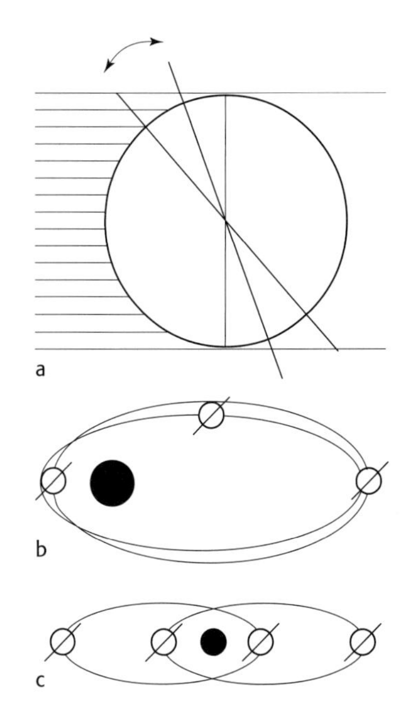

Figure 21.1 Some parameters of Earth's orbit can change over time, and as they do so they affect Earth's climate. (a) Greater or lesser tilting of Earth's axis causes stronger or weaker seasons. (b) The eccentricity of Earth's elliptical orbit means that one pole almost always feels greater seasonal effects than the other; changes in eccentricity weaken or strengthen that effect. (c) The precession of the elliptical orbit alternates the eccentricity effect between the poles.

suggests that slight variations in Earth's orbit around the sun and in the tilt of Earth's axis make significant differences to climate (Box 21.1).

Most important, Milankovitch cycles can trigger the advance and retreat of ice sheets, if conditions for an ice age are already present. Computer models of ice advances and retreats agree well with data from the geological record. The models suggest that the present mild climate on Earth is very unusual for our geography. Interglacial periods with reduced northern ice sheets are very short in comparison with glacial periods with large ice sheets.

Now it is time to apply all this theory to Earth's present ice age, the one in which we are living.

THE PRESENT ICE AGE

Earth has been locked into an ice age since about 2.5 Ma, but its effects have been most marked in the Northern Hemisphere. Thus the northern glaciations that began in the Late Pliocene and continued through the Pleistocene were centered on huge new ice sheets, mostly around the North Atlantic Ocean (Figure 21.3). At the same time there was severe cooling in the Southern Hemisphere.

Once ice sheets built up, they altered climatic patterns in the North Pacific and North Atlantic. Sea surfaces in the North Atlantic froze as far south as New York and Spain (Figure 21.3). Warm Gulf Stream waters were diverted eastward toward North Africa, instead of bringing warm, moist climate to western Europe as far north as Scandinavia.

At its maximum somewhere around 20,000 ± 2000 BP, Canadian ice advanced as far south as New York, St. Louis, and Oregon. Ice scour removed great blocks of rock and transported them for hundreds of miles. The North American ice sheets diverted the jet stream and the main storm track southward (Figure 21.3). The western United States became much wetter than it is today, so that great freshwater

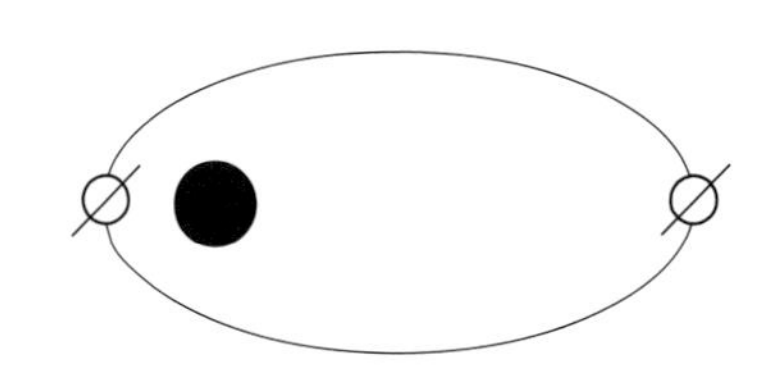

Figure 21.2 Earth's axis is tilted. As it orbits around the sun, solar radiation is concentrated on one hemisphere and then the other, giving Earth seasons. As the seasons go by, we see the Sun gradually moving higher in the sky, then lower.

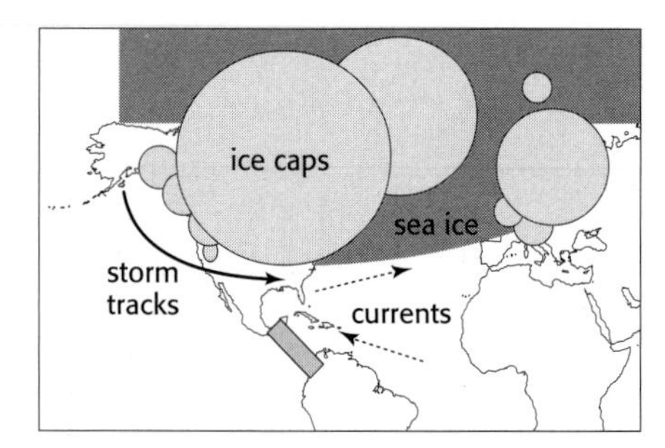

Figure 21.3 The winter climate of the North Atlantic 18,000 years ago. The fully developed ice sheets cooled the North Atlantic so that sea ice extended as far south as New York and Spain. Winter storm tracks were diverted far south in North America, forming large lakes in what is now the western and southwestern desert country. Warm winds and currents in the Atlantic were also far south of their present latitude. They brought moisture to what is now the Sahara Desert, leaving much of Europe as ice sheet and tundra.

lakes formed from increased rainfall and from meltwater along the front of the ice sheet. River channels were blocked by ice to the north, and at the southern edges of the ice sheets, almost all of the melt water drained south to the Gulf of Mexico down a giant Mississippi River.

As the North American ice sheets began to melt and retreat, water flow down the Mississippi to the Gulf of Mexico must have increased enormously. The water draining from the melting North American ice sheet changed the seawater composition of the Gulf of Mexico as it poured southward down the Mississippi in enormous quantities beginning about 14,000 BP, perhaps at ten times its current flow. Finally, as the edge of the ice sheet retreated, the Great Lakes began to drain to the Atlantic instead, first down the Hudson and then the St. Lawrence.

More subtle effects occurred in warmer latitudes. Increased rainfall in the Sahara during ice retreat formed great rivers flowing to the Nile from the central Sahara; they were inhabited by crocodiles and turtles, and rich savanna faunas lived along their banks.

LIFE AND CLIMATE IN THE ICE AGES

Amazingly, the severe fluctuations of climate do not appear to have affected ice-age plants or animals very much. Glacial advances and retreats, though they were rapid on a geological time scale, were slow enough to allow communities to migrate north and south with the ice sheets and the climatic zones and weather patterns affected by them. Communities close to mountain glaciers were able to adjust to advances and retreats by simply moving up and down in altitude. Tropical rain forests were very much reduced, but the habitat did not disappear, and their fauna and flora survived well. Tropical savannas were more extensive during the drier times that accompanied glaciations.

The most interesting effects were controlled by changes of sea level that occurred with every glacial advance and retreat. Each major glaciation dropped world sea level by 120 m or so (about 400 ft), exposing much more land area and joining land masses together. Each new melting episode reflooded lowlands to recreate islands.

Most continents carry examples of creatures stranded by flooding and the warming that occurred during and after the last ice retreat. In the Sahara Desert, for example, there are cypress trees perhaps 2000 years old. They set seed that never germinates because the climate is now much drier than it used to be. Ancient rock paintings of giraffes and antelope confirm the evidence of the cypress trees. Giraffes migrated south to the savannas; cypresses are confined to the north around the Mediterranean; and the Sahara Desert is a fearsome barrier to biological exchange.

A few creatures were trapped in geographical cul-de-sacs and wiped out. Advancing ice sheets, not St. Patrick, wiped out snakes from Ireland, and snakes have not yet been able to cross the Irish Sea to recolonize the island. The Loch Ness Monster is impossible because Loch Ness was frozen under a mile of ice at 18,000 BP. The forests of western Europe were trapped between ice sheets from Scandinavia and wiped out. After the ice sheets melted, western Europe was recolonized by deciduous hardwoods; elsewhere, in North America, Scandinavia, and Siberia, the great boreal forests are dominated by conifers.

In the seas, many species of warm-water molluscs that retreated southward with the first great cooling episode were trapped in marginal seas around the North Atlantic: in the Caribbean, the Mediterranean, and the North Sea. They could not

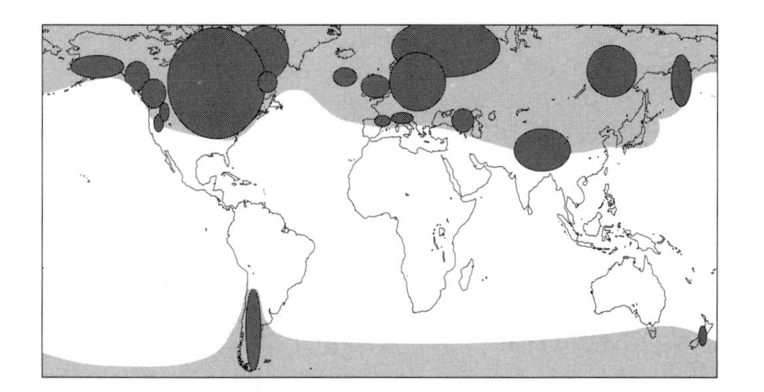

Figure 21.4 Cartoon of world geography at 18,000 BP. The largest ice sheets are shown in darker gray. Areas of land and sea, covered with tundra and sea ice respectively, are shown in lighter gray. I have indicated some important geographic changes to coastlines. For example, much of the seafloor off Southeast Asia, Australia, and the West Indies was dry land at the time.

BOX 21.2 Major Land Masses, Now and At 18,000 BP

MAJOR LAND MASSES NOW		MAJOR ICE AGE LAND MASSES
1. Eurasia plus Africa	6. New Guinea	1. One major world continent
2. The Americas	7. Borneo	2. Antarctica
3. Antarctica	8. Madagascar	3. Meganesia: Australia + New Guinea
4. Australia	9. Baffin Island	4. Madagascar
5. Greenland	10. Sumatra	5. New Zealand

find a clear way of retreat southward, and they suffered great extinctions. But most of these events occurred in early ice advances, and the creatures that survived the first few glacial episodes were well adapted to Pleistocene conditions and suffered little extinction during subsequent glaciations.

We have good evidence of the plant and animal life of the Pleistocene. Enormous bone deposits in Alaska and Siberia and fossils found in caves, sinkholes, and tar seeps have provided excellent evidence of rich and well-adapted ecosystems on all continents.

There have been much greater changes in terrestrial animals and plants during and after the last glaciation than in any previous one, and the effects have often varied with the size of the land area. The land areas sometimes showed dramatic changes. For example, Alaska and Siberia were joined across what is now the Bering Strait, and Greenland was joined to North America, to form one giant northern continent. Australia was joined to New Guinea, and Indonesian seas were drained to form a great peninsula jutting from Asia (Figure 21.4). Box 21.2 compares the major land masses now and at 18,000 BP at the height of the last glaciation; it shows how drastically seawater barriers were removed to join land masses together.

CONTINENTAL CHANGES

On major continents, the larger birds and mammals of the Pleistocene were most unlike their modern counterparts. Just before the last ice advance, North America had mastodons, mammoths, giant bison, ground sloths, sabertooth cats, horses, camels, and dozens of other large mammals. Eurasia had most of these, plus giant deer and woolly rhinos. The giant ape *Gigantopithecus* roamed the Himalayan

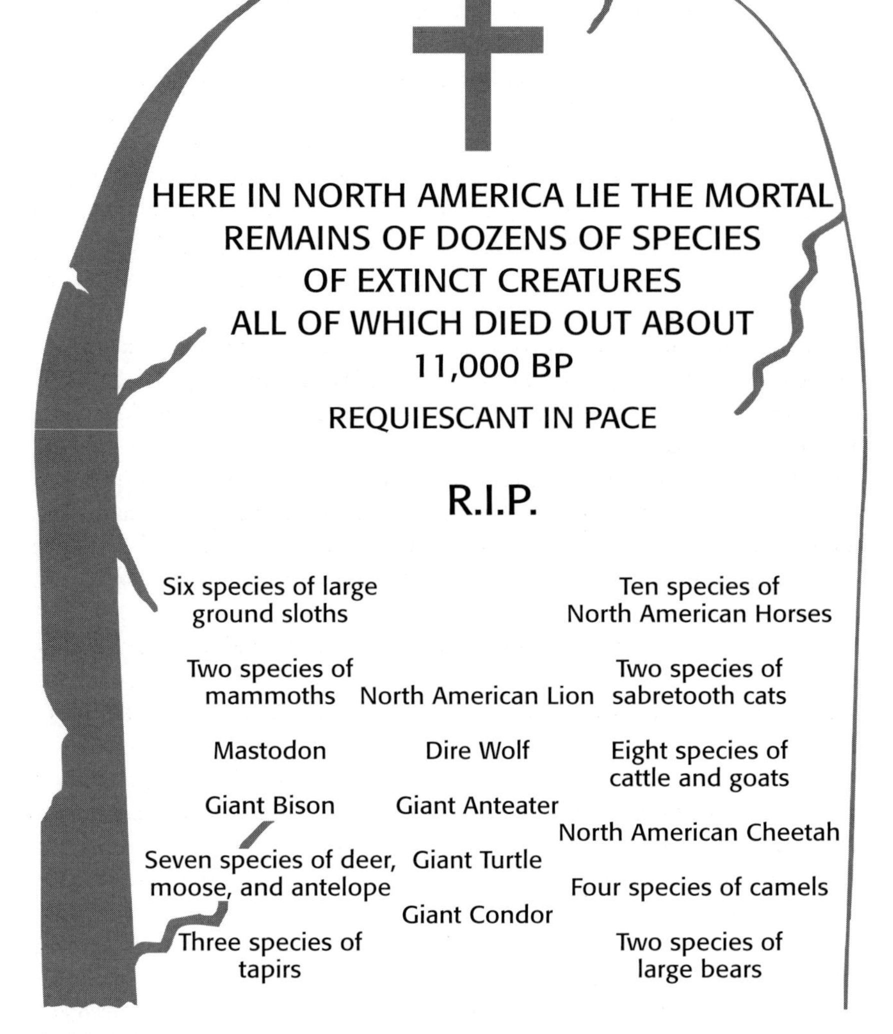

BOX 21.3

slopes (Chapter 19). The moas of New Zealand and the elephant birds of Madagascar are well known (Chapter 18), but Australia had giant ground birds as large as these and a dozen giant marsupials. All these creatures are now extinct.

The catastrophic extinctions occurred at different times on different continents. In each case, the mammals and birds were part of flourishing ecosystems. For example, North America has a very good fossil record of large Pleistocene mammals. Twenty genera became extinct in the 2 m.y. before the last ice sheet melted, then 35 genera were lost in less than 3000 years! Radiocarbon dates for the extinction cluster around 11,000 BP; where the record is good, the extinctions look sudden. North American extinctions are listed (in part) in Box 21.3.

Figure 21.5 The discovery of the famous frozen Beresovka mammoth in Siberia in 1900. (From an old magic-lantern slide.)

Some ice-age animals, such as the woolly mammoth and the woolly rhinoceros, were specifically adapted to life in cold climates. They were much hairier than their living relatives, and they have been found in areas that were very close to the ice sheets at the time. Woolly mammoths were sometimes killed by falling into ice crevasses. Their bodies have been found still frozen in permafrost in Siberia, preserved well enough to tell us quite a lot about their way of life (Figure 21.5). Gallons of frozen stomach contents show that woolly mammoths ate sedges and grasses and browsed tundra trees such as alder, birch, and willow. The tusks of adults were well shaped for clearing snow from forage in winter. Woolly mammoths were adapted specifically to high latitudes, and we have evidence of their reaction to ice advances and retreats: the only change was in size. Siberian woolly mammoths were about 20% larger during the warm interglacials than they were at the coldest times.

Large Pleistocene mammals were well able to withstand climatic change as well as climatic severity. Their large body sizes gave them low metabolic rates, so they could live on rather poor-quality food. As adults, they were largely free from the danger of predation by carnivores. Yet the large mammals and birds became extinct, while smaller species did not suffer as much. The plants the large mammals ate are still living, and so are the small birds, mammals, and insects that lived with them. In the oceans, nothing happened to large marine mammals.

In North and South America, the extinctions took place in a short time toward the end of the ice age, very close to 11,000 BP. This was an unusually cold, dry time, so it has been easy for North American geologists to argue that the extinction was related to climate change.

But that explanation, even if true, covers only some of the American extinctions and does not apply at all to the rest of the world. For example, the giant ground sloths of Arizona were browsers and ate semidesert scrub that was available in the area before, during, and after they died out. Other things being equal, we should prefer another hypothesis if it explains more data more simply.

There is no question that climatic change around 11,000 BP was rapid. Yet the very same species of animals had already survived a dozen or more similar events. There is nothing climatically unique about the last ice retreat. The previous ice retreat, about 125,000 BP, was just as sudden but caused no extinctions. So if climatic change did not result in the extinctions, what did? The problem of Pleistocene extinctions has been debated ever since ice-age animals were discovered, and there is still continuing major disagreement.

The strongest evidence supports an idea put forward in its current form mainly by Paul Martin. Martin's **overkill hypothesis** (see Martin, 1990) gets its name because he stresses one human behavior in particular: hunting. In every case, invading humans were skilled hunters, encountering animals that had never seen humans before. Martin listed seven major lines of evidence (Box 21.4). All the pieces of evidence, argues Martin, are consistent with the idea that the sudden arrival of human

BOX 21.4 Paul Martin's Evidence in Favor of the Overkill Hypothesis

1 Large mammals and ground-living birds were affected most. North America lost 35 genera, and South America lost even more.
2 Extinctions occurred in different areas at very different times.
3 Extinct animals were not replaced.
4 Extinctions were closely linked in time and space with human arrival.
5 Large mammals survived best in Africa and Asia. Extinctions were much more severe in the New World (Australasia and the Americas).
6 Where extinctions are well dated, they were sudden: North America and New Zealand are the best examples.
7 There are very few places where mammal remains occur with human remains or human artifacts. This implies that coexistence was brief.

invaders in an ecosystem was responsible for the extinctions. Other corollaries of human arrival may play an important part, so Martin's idea should not be judged entirely on the hunting overkill that he stresses most. To test his idea, we can draw on data from the only three major continents that were colonized suddenly by humans: North and South America, and Australia.

THE AMERICAS

Human Arrival

Humans crossed into North America from Siberia at a time when the Bering Strait region was a dry land area, Beringia. In the depths of the Ice Ages, Beringia was a frigid plain swept by violent winds blowing dust and sand from the edge of the ice sheet. Yet a varied Arctic vegetation supported a fauna of large ice-age mammals, including woolly mammoths, horses, camels, sheep, deer, musk-oxen, and ground sloths.

Even then, Beringia was separated from the rest of North America by the ice sheets of the Canadian Shield and the Rocky Mountains, which flowed together in what is now Alberta (Figure 21.4). An ice-free corridor to the south opened up into the rest of the Americas only as the main Canadian ice sheet retreated eastward. The important event in human migration is not when people reached Beringia, but when they broke past the ice barriers to the temperate and tropical Americas to the south.

Did humans reach the Americas only as the last glacial period ended, or had they done so long before? There is now compelling evidence that humans were living in Monte Verde, in southern Chile, at 12,500 BP or before. Most likely, these people had arrived by boat along the western American coast: there is scattered evidence of very early American fisherfolk along the west coast at sites in British Columbia, southern California, and Peru. These people seem to have eaten shellfish, seabirds, and small fish. As far as we can tell, they had very little effect on American continental ecosystems, and may not even have ventured inland across the mountain barriers that lie behind the entire west coast. American continental ecosystems did not receive full human impact until around 11,000 BP, with the arrival of inland big game hunters.

Surviving Native Americans have a distinctive, fairly homogeneous DNA structure, probably because they are all descendants of a small group of immigrants who managed to enter the Americas through some kind of physical bottleneck. Eventually Native Americans paid a terrible price for this. When Europeans arrived with new diseases in 1492, any susceptibility to a particular disease was almost universal among the population. Imported diseases decimated Native Americans, where a population with a more varied gene pool would have suffered less catastrophically.

But when did the great wave of continental colonists arrive (after the coastal fisherfolk)? A distinctive and short-lived tool and weapon culture, the Clovis culture, spread rapidly across North America from Washington to Mexico. All the radiocarbon dates for Clovis sites in the western United States cluster around 11,000 BP. The invaders were already skillful hunters of large mammals across the far northern plains of Asia and Beringia.

Soon after 11,000 BP, there is evidence of human occupation throughout the Americas. By then, the Clovis artifacts had been replaced by more advanced and more regionally varied sets of tools, and people had adopted varied life styles appropriate to the regions.

Large Animals

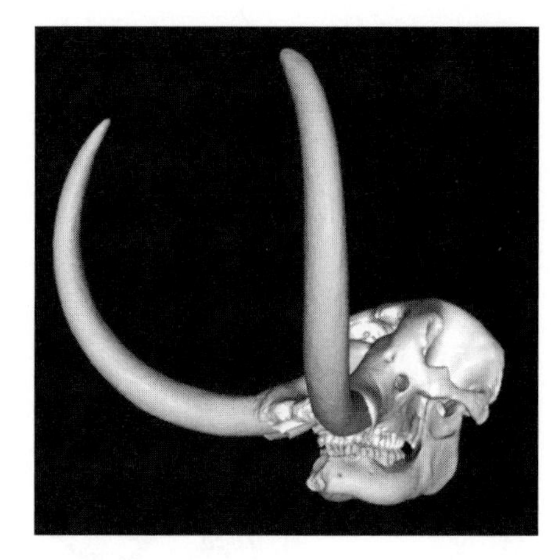

Figure 21.6 3-D computer image of the skull of a large male American mastodon, as reconstructed by Daniel Fisher and Gretchen Moeser. (Courtesy Daniel Fisher of the University of Michigan. © Daniel Fisher.)

Along with the arrival of Clovis culture in the Americas, we see evidence of large animal hunting. These PaleoIndians hunted mammoths and mastodons. There are cut marks on mastodon bones found close to the edge of the ice sheet near the Great Lakes, and it seems that humans butchered the carcasses into large chunks and cached them for the winter in the frigid waters under shallow, ice-covered lakes, just as Inuit do today in similar environments. We can tell that the PaleoIndians' favorite hunting season for mastodons was late summer and fall, whereas natural deaths occurred mainly at the end of winter when the animals were in poor condition. A mammoth skeleton from Naco had eight Clovis points in it. Two juvenile mammoths and seven adults were killed with Clovis tools near Colby, Wyoming, and the way the bones are piled suggests meat-caching there too.

It cannot have been easy to kill these elephants in any direct attack. Mastodons were lethally effective using their tusks against one another (Figures 21.6; 21.7), and that skill would easily have carried over into effective defense against their natural predators. But Clovis was another story: armed with formidable weapons, traps, poisons, and intelligent group hunting tactics, the American megafauna was devastated. The last mammoths in the Great Lakes area, the last native North American horses, and ground sloths and mountain goats in the southwest all died out around 11,200 BP. In west Texas, mammoths, horses, and bison are common before 11,000 BP, but after that we find only bison bones. The skeleton of a giant extinct turtle found in Little Salt Spring in southwest Florida had a sharpened wooden stake jammed between the shell and the breastbone. The turtle had apparently been killed and cooked on the spot in its shell. There are ground sloths, bison, and a young elephant at about the same level, but afterward the PaleoIndians ate white-tailed deer.

Megaherbivores and Medium-Sized Animals

There was extinction among medium-sized mammals, although one would expect some of them (camels, horses, and deer, for example) to have been resistant to extinction because of their speed, agility, and rapid reproduction, even in the face of expert hunters.

The answer to this puzzle may be found in ecosystems that include megaherbi-

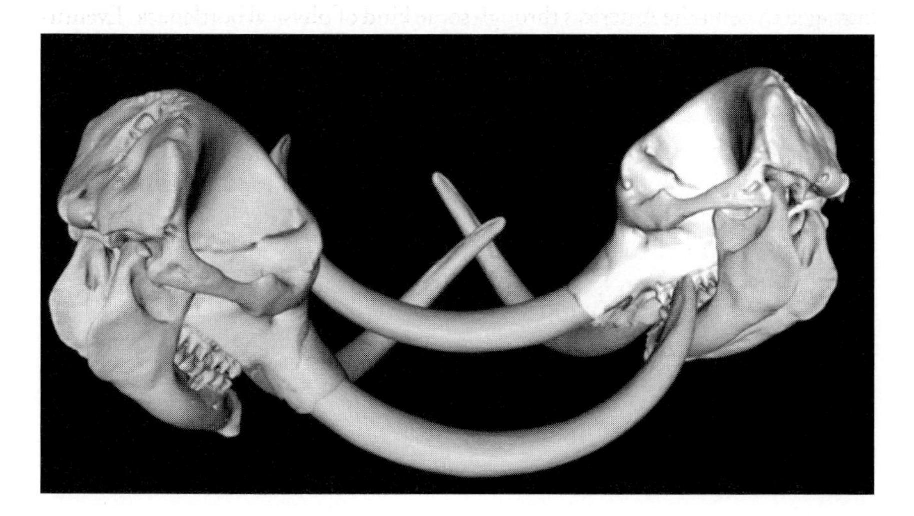

Figure 21.7 Reconstruction of fighting between two adult male American mastodons. The reconstruction is based on massive damage to some fossil male skulls, which implies lethal or sublethal strikes on a target area in which the tusk would smash under the cheekbone toward the braincase. (Courtesy Daniel Fisher of the University of Michigan. © Daniel Fisher.)

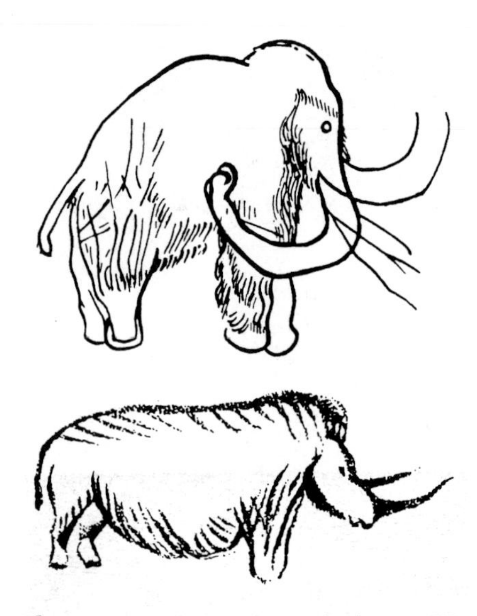

Figure 21.8 Pleistocene megaherbivores. We don't have to rely on the inferences of paleontologists to tell us that there were megaherbivores in Pleistocene ecosystems. They were observed and illustrated by competent ecologists of the time. (Same as Figure 20.15.)

LIMERICK 21.1
They fall from the branches to wait,
But they're 12,000 summers too late.
You can smell them for miles,
They're rotting in piles,
The fruits that the gomphotheres ate.

vores (very large herbivorous mammals more than 2000 kg [2 tons] in weight). On the plains of Africa, for example, the largest animals, elephant and rhino, can have drastic effects on vegetation. Elephants destroy trees and turn dense forest into open woodland by opening up clearings in which smaller browsing animals multiply. Eventually elephants turn any local habitat into grassland. They then migrate to another woodland habitat, leaving the trees to recover in a long term ecological cycle that can take decades to complete. White rhinos graze high grass so effectively that they open up large areas of short grassland for smaller grazing animals.

Thus, in the long run, megaherbivores keep open habitats in which smaller plains animals can maintain large populations. Where elephants have been extinct for decades, the growth of dense forest is closing off browsing and grazing areas, and smaller animals are also becoming locally extinct. Many of the problems in African national parks today occur because they are not big enough to allow these cycles of destruction and migration to take place naturally.

But what would happen if megaherbivores were completely removed from an ecosystem—by hunting, for example? Megaherbivores breed slowly and cannot hide. They would be particularly vulnerable to skillful hunters. Norman Owen-Smith proposed that the disappearance of Pleistocene megaherbivores (Figure 21.8) soon led to the overgrowth of many habitats, reducing their populations of smaller animals too. Thus, even if early hunters hunted or drove out only a few species of megaherbivores, they could have forced ecosystems so far out of balance that extinctions would then have occurred among medium-sized herbivores too, especially if hunters were forced to turn to the latter as prey when the megaherbivores had gone.

There may be more subtle effects of removing large herbivores. Plants sometimes coevolve with herbivores that disperse their seeds. Very large herbivores are likely to encourage the evolution of large, thick-skinned fruits, and a sudden extinction would leave the fruits without dispersers. Even today, guanacaste trees of Central America produce huge crops of large fruits, most of which lie and rot. Daniel Janzen suggested that these fruits coevolved with large elephants (gomphotheres), which became extinct with the other large American mammals (see his paper in Martin and Klein, 1984).

Predators and Scavengers

Predator species such as the sabertooth cats (Figure 21.9) and the North American lion could have been reduced to dangerously low levels by the removal of their prey by overkill; there is no need to think in terms of the direct, systematic overkill of predator species that modern humans often carry out. In turn, scavengers may also depend on populations of large mammals to provide the carcasses they feed on. Thus, the giant teratorn known from the La Brea tar pits (Figure 13.2) is extinct, and the so-called "California" condor once nested from the Pacific coast to Florida. Pleistocene caves high on vertical cliffs in the Grand Canyon of Arizona contain bones, feathers, and eggshells of this condor, along with the bones of horses, camels, mammoths, and an extinct mountain goat. The condor vanished from this area at the same time as the large mammals did, presumably because its food supply largely disappeared.

Survivors

What about the surviving large mammals in North America? It turns out that many of them were originally Eurasian and crossed into the Americas late in the Pleistocene. Thus, bear, moose, musk-oxen, and caribou had been exposed to human hunting in Eurasia before 11,000 BP. There were no North American extinctions after 8000 BP at the latest, presumably when a new stable balance had evolved. There were separate regional cultures in the Americas by this time, but there were no new significant extinctions even though tools and weapons had improved.

Bison were a special case. They were American natives, and although the immense long-horned bison (Figure 21.10) became extinct, the smaller bison survived in great numbers. Perhaps the removal of larger competitors encouraged this success. Moreover, bison survived in the face of intense and wasteful hunting by PaleoIndians, whose methods were by no means as ecologically sound as their descendants sometimes claim. A well-studied site in Canada reveals that PaleoIndians, hunting on foot with stone weapons, would stampede whole herds of bison along preplanned routes that led to a cliff edge or buffalo-jump. The bison would then be finished off and butchered at the cliff base. The method naturally resulted in the deaths of many young animals; only about a quarter of the animals killed were full-grown. The site, appropriately named Head-Smashed-In, had been used for more than 5000 years when it became obsolete as firearms reached the tribes in the nineteenth century. Given that there are dozens more buffalo-jump sites stretching across the Great Plains from Alberta to Texas, most of them still to be investigated, it seems less surprising that humans would be capable of the overkill that the fossil record suggests, and more surprising that the bison adjusted so successfully for so long to human predation pressure.

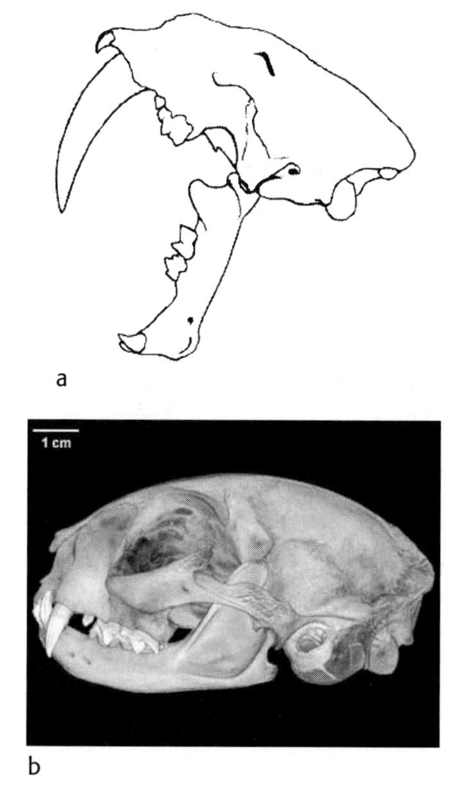

Figure 21.9 (a) *Smilodon*. This large sabertooth cat occurs in large numbers among the Pleistocene fossils of the La Brea tar pits in Los Angeles. It was a major predator in the Pleistocene ecosystems of North America, and is now the State Fossil of California. (b) CAT scan of a cat (sorry, but it's true!). Compare carefully and you will see that proportionately, these two cats differ only in the extravagant size of the upper canine tooth in *Smilodon*. (Courtesy Timothy Rowe and DigiMorph Project at the University of Texas at Austin.)

AUSTRALIA

Australia suffered more severe extinctions than any other continental sized land mass. It lost every terrestrial vertebrate larger than a human. It lost a giant horned turtle as big as a car, and its giant birds, the dromornithids (Chapter 18). It lost its top predators, including *Megalania*, the largest terrestrial lizard that ever evolved, 7 m (24 ft) long. *Megalania* was related to the living Komodo dragon but weighed more than eight times as much. Other predators were a huge terrestrial crocodile, a carnivorous kangaroo, and a 5-m (16-ft) python. Australia lost about 20 large marsupials, including all the diprotodonts, huge four-footed vegetarians the size of tapirs; a wombat the size of a cow; and the largest kangaroo of all time, *Procoptodon*, a browser 3 m high that was the ecological equivalent of a tapir or ground sloth. Only a few large animals survived in Australia, but small animals were less affected.

The extinctions are dated to around 45,000–50,000 BP, a time that coincides, as far as we can tell, with human arrival in Australia. The slow-running giant dromornithid bird *Genyornis* disappeared from habitats where fast-running emus survived, but there seems to be a memory of the dromornithids in aboriginal legend as the mihirung.

The extinctions coincide roughly with a change in vegetation associated with increased burning. This was not a time of climatic change, so the increased burning may have been generated by the early Australians. They were migrating into a dry country ecosystem that was unfamiliar to them because they came from the moister tropical ecosystems of New Guinea. They had to learn slowly how to adapt to drier Australian conditions, just as Europeans had to do tens of thousands of years later.

One of the easiest ways of clearing Australian vegetation is to burn it: burning makes game easier to see and hunt, and Australian aborigines today have complex

Figure 21.10 The huge long-horned bison of the Pleistocene North American plains, *Bison latifrons*. (Reconstruction by Bob Giuliani. © Dover Publications Inc. Reproduced by permission.)

timetables for extensive seasonal brush burning that has dramatic effects on regional ecology. Most likely, then, the Australian extinctions were the direct result of human invasion, through the introduction of large-scale burning as well as hunting.

On the basis of this evidence from three continents, it looks as if Martin's hypothesis of human impact is stronger than any other. Now we will look at smaller land masses that were subjected to much the same human impact and see how they fared rather differently.

ISLAND EXTINCTIONS

Island animals can often evolve into unique sets of creatures, and geographical changes that connect previously isolated areas can have dramatic and damaging effects on species and communities (Chapter 18). Human arrival has often had a catastrophic effect on island faunas. The world may have 25% fewer species than it did a few thousand years ago, and most of those extinctions took place on islands. For example, native Tasmanians killed off a unique penguin sometime after the thirteenth century, 600 years before they in turn were wiped out by European settlers.

On Madagascar, large lemurs, giant land tortoises, and the huge flightless elephant birds (Chapter 13) disappeared after the arrival of humans somewhere between 0 and 500 AD. Here too, large forest areas were cut back and burned off to become grassland or eroded, barren wasteland. No native terrestrial vertebrate heavier than 12 kg (25 pounds) survived after 1000 BP. Humans took a long time to penetrate the forest of this large island, and the extinction may have taken 1000 years instead of being sudden. It's clear that human arrival was part of a "recipe for disaster" (as David Burney called it in 1993). The desperate erosion and poverty of much of the countryside of Madagascar today underlines the fact that humans are still involved in self-destructive deforestation, in spite of the evidence all around them of its horrific after-effects.

A panda-sized marsupial lived in New Guinea in the Late Pleistocene, and although it is now extinct, the plants that it ate are still flourishing.

New discoveries of extinct flightless birds in Hawaii suggest that devastating extinctions followed the arrival of Polynesians. The Hawaiian Islands are famous for honeycreepers, which evolved there into many species like Darwin's finches on the Galápagos Islands. But there were 15 more species of honeycreepers before humans arrived. Two-thirds of the land birds on Maui were wiped out by the Polynesians, probably by a combination of hunting, burning, and the arrival of rats. As in New Zealand (Box 21.5), the extinctions that followed the European arrival were severe but not as drastic, probably because the bird fauna was already so depleted.

The same process is recorded on almost all the Pacific islands in Melanesia, Polynesia, and Micronesia. Almost all of them, apparently, had species of flightless birds that were killed off by the arriving humans and/or their accompanying rats, dogs, pigs, and fires. As many as 2000 bird species may have been killed off as human migration spread across the ocean before the arrival of Europeans. It may not be an accident that Darwin was inspired by the diversity of the Galápagos Islands: these were never occupied before European discovery in 1535, and human impact was relatively slight until whalers arrived in strength around 1800.

Several islands in the Mediterranean Sea (Cyprus and Crete are good examples) held fascinating evolutionary experiments after the ice ages. There were pigmy elephants and pigmy hippos, giant rodents, and dwarf deer. These mammals had

BOX 21.5 Case Study: New Zealand

A thousand years ago, New Zealand was an isolated set of islands without land mammals (except for two species of bats). Birds were the dominant vertebrates, and the largest were the moas, giant flightless browsing birds the size of ostriches (Chapter 18) (Figure 21.11). The moas and other native creatures survived as glacial periods came and went, yet they became extinct within a few hundred years of the arrival of the Polynesian Maori people after 1000 AD.

There seem to have been two main reasons for the extinctions, and all of them are connected with human arrival. First, evidence of hunting is clear and appalling. Midden sites that extend for acres are piled with moa bones, with abundant evidence of wasteful butchering. The bones are so concentrated in some places that they were later mined to be ground up for fertilizer. The middens contain bones of 11 of the 12 extinct species of moas, and they also contain bones of tuataras, very primitive reptiles. Second, the Maori brought rats with them, which ate insects directly, killed off reptiles by eating their young, and exterminated birds by robbing their nests. The tuataras (Chapter 10), the giant flightless wetas (insects that had been the small-bodied vegetarians of New Zealand [Chapter 18]), and many flightless birds including the only flightless parrot, the kakapo, were practically wiped out by rats. There were many other more subtle ecological changes. A giant eagle that probably preyed on moas died out with them, for example. And when the moa were gone, the Maori took up serious cannibalism, because humans were the largest remaining protein packets.

Half of the original number of bird species in New Zealand were extinct before Europeans arrived, and the new settlers only acted to increase the changes in New Zealand's landscape and biology. Forests were cleared even faster, and new mammals were introduced. European rats were the worst offenders against the native birds, but cats, dogs, and pigs were also destructive, rabbits destroyed much of their habitat, and deer competed with browsing birds. The tuatara now lives only on a few small, rat-free islands, and the kakapo survives precariously in remote areas where it is threatened by wild cats. Bird populations are still dropping in spite of efforts to save them.

As a microcosm of the problem, consider the Stephen Island wren, the only flightless songbird that has ever evolved. This species had already been exterminated from New Zealand by the Polynesian rat before European arrival. The entire remaining population of this species, which was by then confined to one island, was caught and killed by Tibbles, a cat brought to the island in 1894 by the keeper of a new lighthouse.

A convict colony established by the British wiped out an endemic seabird on Norfolk Island, between Australia and New Zealand. Several small, unique native birds fared better for a while on Lord Howe Island, further north: they lived alongside the early settlers until a shipwreck allowed rats to reach the island in 1918. Within a few years five species had completely disappeared.

evolved on these isolated islands in much the same way as did the fauna of Gargano during Miocene times (Chapter 18). Many of the island animals disappeared as Neolithic peoples discovered how to cross wide stretches of sea and colonized the islands several thousand years ago.

Europeans killed off the dodo on Mauritius (Figure 21.12) and several species of giant tortoises there and in the Galápagos; deliberate burning, and the goats, pigs, and rats they brought, completed a great deal of destruction of native plants, birds, and animals. They killed off the great auk of the North Atlantic, the huia and other small birds of New Zealand (Box 21.5), and an unknown number of species of birds of paradise in New Guinea, all to satisfy the greed of egg and feather collectors. They drove fur-bearing mammals close to extinction worldwide.

LIMERICK 21.2
The morning was hardly propitious
When sailors discovered Mauritius.
They killed off the lot,
Stewed them up in a pot,
And pronounced them extinct, but
delicious.

Irrespective of race, color, and creed, it seems, human arrival in the midst of a fauna and flora unused to hunting pressure, to extensive burning, or to rats, cats, and pigs, has spelled disaster. One common factor among the victims is naiveté. Charles Darwin described the complete lack of fear of humans of the Galápagos animals and birds, an almost universal feature of creatures never exposed to human hunting; and modern observers like Tim Flannery have recorded the same behavior. New Zealand's inhabitants had never seen a land mammal before the Polynesians arrived.

LIMERICK 21.3

The Du is a bird that's now long lost
It laid eggs in a pile of warm compost
But the chicks that did hatch
Were too easy to catch
And became Melanesian pot-roast.

The Du

Very large bird bones were discovered a few years ago on the Isle of Pines, off New Caledonia. The fossil was named *Sylviornis*, and although it was large, it was not a ratite, but a very large flightless megapod. Megapods are a family of birds that includes the mallee fowl of Australia and ranges through eastern Indonesia and Australasia. No bird anywhere near that size now lives on New Caledonia. *Sylviornis* is therefore extinct, and its remains are dated at about 3500 BP, when humans had already reached its island.

Melanesian folk tales from the Isle of Pines describe a giant red bird, the Du (Figure 21.13), which did not sit on its single egg to hatch it. Although the Melanesians did not know it at the time, this behavior is unique to megapod birds, which lay their eggs and cover them with rotting vegetation to keep them at an even warm temperature. The male keeps close control of the egg temperature by adjusting the compost heap on an almost hourly basis for weeks at a time. The Du, then, was *Sylviornis*, and the legend shows that it was known to early humans.

The Isle of Pines has large areas covered by large and mysterious mounds, never associated with original human artifacts. The mounds are the right size to be Du hatching mounds, still preserved in enormous numbers. They give some idea of the numbers of the Du, and they illustrate the massive disaster that overtook the bird at a time when there was no significant climatic change in its environment.

Figure 21.11 One of the moas of New Zealand. (After Frohawk.)

EXPERIENCED FAUNAS

We have already seen that many survivors in North America had been used to hunting pressure in Eurasia. Humans developed their hunting skills in the Old World, and although there were extinctions of large mammals there, they were spread out over longer times than the New World extinctions were.

For example, in Africa *Homo erectus/ergaster/antecessor* butchered giant baboons, hippos, and the extinct elephant *Deinotherium*. The remains of 80 giant baboons have been found at one site dated at about 400,000 BP. At Torralba, in Spain, the remains of 30 elephants, 25 horses, 10 wild oxen, 6 rhinos, and 25 deer were found on one site.

It therefore fits the overkill or human-impact hypothesis that the most important local extinctions in the Old World took place in habitats that modern humans were invading in strength for the first time. These invasions took place along the edges of the ice sheets, and even then humans are implicated in the disappearance of only a few large mammals of the northern Eurasian tundra, especially the woolly mammoths and the woolly rhinoceros.

Figure 21.12 The dodo of Mauritius. This giant flightless pigeon was hunted out by meat-hungry European sailors. We now have only some dried-out museum scraps and a few drawings. (From a nineteenth-century drawing.)

It looks as if mammoths became extinct as advanced hunting techniques allowed humans to range closer to the ice sheets. For example, the advance of ice sheets toward the peak of the last glaciation seems to have driven the Gravettian people (Chapter 20) out of the northern Carpathian Mountains of Central Europe toward the south and the east, where they discovered and invaded the mammoth steppe of Ukraine for the first time. The Gravettians were already using mammoth bones as resources. At Predmost in the Czech Republic, a site dating from just before the coldest period of the last glaciation (28,000–22,000 BP) contains the bones of at least 1000 woolly mammoths. These people routinely buried their dead with mammoth shoulder bones for tombstones.

The pattern in these extinctions was always the same. The large mammals were hunted out of the optimum part of their range, and then the last survivors hung on in the inhospitable (usually northern) parts of their range until newly invading humans or climatic fluctuations killed them off. For example, woolly mammoths, woolly rhinoceroses, and giant deer, along with horses, elk, and reindeer, reinvaded Britain from Europe after the ice sheets began to retreat and birch woodland and parkland spread northward. Mammoths flourished in Britain until 12,800 BP at least, but then human artifacts appeared at 12,000 BP, and the largest animals of the tundra fauna quickly disappeared.

The giant deer is sometimes called the Irish elk, partly because it is best known from Ireland. It was not an elk but a deer the size of a moose, with the largest antlers ever evolved, more than 3 m (10 ft) in span (Figure 21.14). It was adapted for long-range migration and open-country running, and its diet was the high-protein willow vegetation on the edges of the northern tundra. It once ranged from Japan to France, but it did not reach North America, where the moose is an approximate ecological counterpart.

Figure 21.13 *Sylviornis*, the extinct Du of New Caledonia. The bird stood about 1 m (3 ft) high. (Redrawn from a reconstruction by Poplin and Mourer-Chauviré; the head ornament is based on oral legend of the Melanesians of New Caledonia.)

The giant deer disappeared from Eurasia in a sequence that started in eastern Siberia and proceeded westward. It survived longest in western Europe and finally, after the ice sheets melted and sea level rose, it was confined to the island of Ireland. The giant deer flourished there in a warm period until about 11,000 BP, but it then died out in a cold period, possibly because it was unable to retreat southward to a better climate. Humans did not reach Ireland from Europe and Britain until the climate warmed again, around 9000 BP.

Mammoths, which had lived much farther south, were confined to the tundra north of the Black Sea by 20,000 BP. We have an interesting record of life around 15,000 BP on the plains of Russia and Ukraine. Several dozen living sites were built on low river terraces by people from the Gravettian culture. Each major site contains the remains of several large buildings whose foundations and lower walls were made entirely from mammoth bones. The buildings were large, 4–7 m (13–23 ft) across and up to 24 sq. m (240 sq. ft) in area. The foundation was made of the heaviest bones, carefully aligned. Skulls at the base were followed by jaws and then long bones, with the resulting pattern providing an aesthetic geometric arrangement as well as sound architecture. The roofs were probably lighter structures made from branches, hides, or sod. (Only a thousand years ago, the Inuit of Greenland were using whale skulls and whale ribs in the same way, roofing the dwellings with sod.)

Figure 21.14 The giant Pleistocene deer of Eurasia, *Megoceros*, usually called the Irish elk. Males carried the largest antlers ever evolved. (Reconstruction by Bob Giuliani. © Dover Publications Inc. Reproduced by permission.)

One group of four buildings was built using bones from at least 149 mammoths. It's not clear whether the mammoths had been killed, or whether bones from old skeletons had been collected. Zoia Abramova has remarked that it would have been easier to obtain bones from living mammoths than to dig them out of permafrost (see Soffer and Praslov, 1993). Others disagree, arguing that dwelling sites may even have been chosen because they were close to massive mammoth bone accumulations. Since no one involved in the arguments has completed even one of these tasks, we are not likely to get any agreement soon. Certainly, the sites in this unique area give some idea either of the numbers of mammoths that once roamed the plains, or of the hunting efficiency of these stone-age peoples, or both.

LIMERICK 21.4

The Ukrainian tundra was bleak
Till they found a solution unique,
But lack of supplies
Soon caused the demise
Of their mammoth construction
technique.

Inside, the Gravettians laid down clear river sand as a floor, built hearths (fueled by mammoth bone), and remained secure and warm during the winter. They made finishing touches to stone tools (leaving behind their antler hammers and chipped flakes), skinned animals (leaving behind the bones), ground ochre for dye, and sewed with ivory needles.

The mammoth-bone dwellings are surrounded by pits dug into the permafrost that were probably used to store meat long-term. Lewis Binford has made a close and vivid comparison between the whale- and caribou-hunting Inuit of Arctic

North America today and the tundra dwellers of the mammoth steppe (see Soffer and Praslov, 1993). Many features of their buildings and the food storage pits are closely similar, implying that the ancient Gravettians were effective hunters even if they also used and re-used old bones.

Older generations of Inuit did not hesitate to attack a 25-ton bowhead whale from flimsy boats, though their modern descendants prefer motor boats and assault rifles. Ice-age hunters may or may not have attacked an 8-ton mammoth directly, but it doesn't take a great deal of imagination to see why man and mammoth could not have coexisted for long in this open steppe country that had been the main range of the species. (The same Gravettian people had built mammoth-bone dwellings in Central Europe several thousand years before, when mammoths still lived there.)

Some woolly mammoths survived in the permafrost areas of northern Siberia until perhaps 10,000 BP, in an area where humans arrived late. Even then, there was still one mammoth refuge left, in the Wrangel Islands, a small group of low lying islands off the north coast of Siberia. Forage was poor, and the last mammoths were small, perhaps 2 tons instead of the 6 tons of their ancestors. The last woolly mammoths died out in the Wrangel Islands only 3000–4000 years ago, at a time when there were large cities in the ancient civilizations of Eurasia and the Egyptian pyramids were already old. We are not sure what killed off these last pitiful survivors of the great mammoths, but humans reached the Wrangel Islands about that time.

There is a myth that primitive peoples live in ecological harmony with the plants and animals around them, and that it has been only with the arrival of modern civilization that major ecological imbalances have arisen. We have seen several examples that explode this myth, and there are many more.

Ancient peoples have destroyed their own civilizations on islands with delicate ecosystems. The Easter Islanders who built their famous enormous stone statues on the island also deforested their fertile, productive land until it became a barren waste and they became a wretched band of refugees surviving by shoreline scavenging and primitive fishing. But sophisticated peoples on large continents have harmed themselves too. The Anasazi Indians, who built a complex civilization on the Colorado Plateau, stripped their environment of trees until the erosion and siltation that followed ruined their irrigation projects and they disappeared as a significant people. (Tim Flannery calls these sorts of self-destructive societies *The Future Eaters.*) But are we doing any better?

THE WORLD TODAY

The Spanish introduced cattle to Argentina in 1556; by around 1700 there were about 48 million head of wild cattle on the plains. By 1750 they had been all but exterminated by a comparatively sparse human population with primitive firearms. This is even more incredible than the North American slaughter of about 60 million bison a century later with much more effective rifles, and it is more evidence in support of the plausibility of the overkill hypothesis.

Stripping tropical forest from hillsides not only removes the plants and animals that are best adapted to life there, but it results in erosion that removes the few nutrients left in the soil, destroying any agricultural value the land may have. It also results in much greater run-off and downstream flooding, which destroys or silts up rivers, irrigation channels, and fields downstream, harming ecosystems and productivity there too. This scenario has been played out now in Ethiopia, Madagascar, and Haiti in horrific proportions; it is happening throughout the rain forests of

Brazil and Indonesia, it is destroying the reservoirs that provide water for the Panama Canal, and yet we do not seem to have learned the lesson.

One can argue that humans at 11,000 BP, perhaps even at 500 BP, did not know enough ecology, did not have enough recorded history, did not know enough archeology or paleontology, and did not have enough of a global perspective to realize the consequences of their impact on an ecosystem. But that is not true today. We have the theory and the data to know exactly what we are doing. We transport species to new continents and islands without proper ecological analysis of their possible impact. We approach our environment sometimes with stupidity, sometimes with greed, but usually with both.

We know very well that the tropical regions of the world are a treasure house of species, many of them valuable to us and many of them undescribed. Yet we deliberately introduce alien carnivorous fish into tropical lakes, ruining fisheries that have been stable for centuries.

We know that clearing tropical forests will quickly destroy the low level of nutrients in the soil and will render those areas useless for plant growth. Yet we go ahead anyway, sometimes for a quick profit on irreplaceable timber, sometimes to achieve a few years' agricultural cropping before the land is exhausted.

We poach gorillas and shoot animals for trophies. Indonesians and Malaysians clear tropical forests to supply the wood for the 11 billion disposable chopsticks used each year in Japan. Africans destroy rhinos to supply Asians with useless medicines and Yemenis with dagger handles; Africans poach elephants for ivory that ends up as ornaments and trivia in Japan, Europe, and North America.

Fishing grounds have been plundered worldwide. Tanzanian and Filipino fishermen use dynamite sticks, killing off the reefs their fish depend on and any hope they have of catching anything next year. Filipinos capture tropical fish for American aquariums and Chinese restaurants and "medicine" shops by dosing their reefs with cyanide and catching the few survivors. Japanese, Norwegians, and Icelanders catch whales under the guise of "research" though they sell the meat: after all, they've already paid for the ships and want to get their money back. Giant clams are poached from marine preserves all over the South Pacific to feed the greed of Chinese "gourmets."

North Americans complain about the destruction of tropical rain forests for export to Japan, while their own lumber companies are felling the last of the old Douglas fir and redwood forests of the northwest for export to Japan. Italians and French shoot millions of little songbirds each year for "sport." All industrial nations continue to destroy forests and lakes with acid rain, though we know how to prevent the pollution that causes it. Ignorance is not the problem in any of these cases: poverty, greed, and arrogance are to blame.

We could live perfectly well—in fact, we could have a vastly increased quality of life—without disturbing the equilibrium of marmosets, gorillas, orangutans, chimps, whales, and all the other endangered species, if we took a grip on our own biology. What we need is a sense of collective responsibility and enlightened self-interest. It's a difficult message to get across because evolution and society, and simple principles of economics, all favor the short-term goals of individuals rather than the long-term welfare of communities or societies.

It is in the interest of everyone, for themselves and for their children, to make our future secure not just for survival but for quality of life. If we don't solve our problems by our own voluntary actions, natural selection will do it for us. If we can learn anything from the fossil record, it is that extinction is the fate of almost every species that has ever lived on this planet. There is no automatic guarantee of success. Every individual in every generation is tested against the environment. We have the power and the knowledge to control our environment on a scale that no other species has ever done. So far, we have used those abilities to remove thousands of other species from the planet. If we destroy our environment to the point where the human species fails the test, becoming either extinct or less than human, it will serve us right.

But the greater tragedy would be our legacy, because we'll destroy much of the world's life along with ourselves. I believe that any rational God would have intervened long ago to prevent the wholesale destruction of so many of His creatures. We have only ourselves and one another to blame and to rely on.

The anthropologist David Pilbeam once wrote that we have only just begun to tap the potential of the human brain. He had better be right.

Further Reading

Climate and ice ages

Broecker, W. S., and G. H. Denton. 1989. What drives glacial cycles? *Scientific American* 262 (1): 49–56.

Broecker, W. S., et al. 1989. Routing of meltwater from the Laurentide Ice Sheet during the Younger Dryas cold episode. *Nature* 341: 318–21.

COHMAP members. 1988. Climatic changes of the last 18,000 years: observations and model simulations. *Science* 241: 1043–52.

Jones, P. D., et al. 2001. The evolution of climate over the past millennium. *Science* 292: 662–7.

Raymo, M. E. 1998. Glacial puzzles. *Science* 281: 1467–8. [Subtleties in current research.]

Ruddiman, W. F. 2001. Earth's climate: past and future. New York: W. H. Freeman.

Zachos, J., et al. 2001. Trends, rhythms, and aberrations in global climate 65 Ma to present. *Science* 292: 686–93. [Review of Cenozoic climate as seen in 2001.]

Ice age life and the great extinction

Barnosky, A. D. 1986. "Big game" extinction caused by late Pleistocene climatic change: Irish Elk (*Megaloceros giganteus*) in Ireland. *Quaternary Research* 25: 128–35.

Dillehay, T. D. 1999. The Late Pleistocene cultures of South America. *Evolutionary Anthropology* 7: 206–16.

Fedje, D. W., and H. Josenhaus. 2000. Drowned forests and archaeology on the continental shelf of British Columbia. *Geology* 28: 99–102. [Migration route for first Americans?]

Fisher, D. C. 1984. Mastodon butchery by North American PaleoIndians. *Nature* 308: 271–2.

Frison, G. C. 1989. Experimental use of Clovis weaponry and tools on African elephants. *American Antiquity* 54: 766–83.

Frison, G. C. 1998. Paleoindian large mammal hunters on the plains of North America. *Proceedings of the National Academy of Sciences* 95: 14576–83.

Geist, V. 1986. The paradox of the giant Irish stags. *Natural History* 95 (3): 54–65.

Gladkih, M. I., et al. 1984. Mammoth-bone dwellings on the Russian Plain. *Scientific American* 251 (5): 164–75.

Guthrie, R. D. 1990. *Frozen Fauna of the Mammoth Steppe.* Chicago: University of Chicago Press. [Paperback, excellent reading and excellent science.]

Harris, J. M., and G. T. Jefferson (eds). 1985. *Rancho La Brea.* Los Angeles: Los Angeles County Natural History Museum.

Kerr, R. A. 2003. Megafauna died from big kill, not big chill. *Science* 300: 885.

Kurtén, B. 1976. *The Cave Bear Story.* New York: Columbia University Press.

Kurtén, B., and E. Anderson. 1980. *Pleistocene Mammals of North America.* New York: Columbia University Press.

Lister, A., and P. G. Bahn. 1994. *Mammoths.* New York: Macmillan.

Marshall, E. 2001. Pre-Clovis sites fight for acceptance. *Science* 291: 1730–2.

Martin, P. S. 1990. 40,000 years of extinction on the "planet of doom." *Palaeogeography, Palaeoclimatology, Palaeoecology* 82: 187–201.

Martin, P. S., and R. G. Klein (eds). 1984. *Pleistocene Extinctions.* Tucson: University of Arizona Press.

Meltzer, D. J. 1993. Coming to America. *Discover* 14 (10): 90–7. [Monte Verde.]

Nemecek, S. 2000. A taste for science. *Scientific American* 282 (4): 24. [Dan Fisher tests his ideas on the refrigeration of mammoths by Clovis people.]

Owen-Smith, N. 1987. Pleistocene extinctions: the pivotal role of megaherbivores. *Paleobiology* 13: 351–62.

Pielou, E. C. 1991. *After the Ice Age: The Return of Life to Glaciated North America.* Chicago: University of Chicago Press. Paperback.

Sandweiss, D. H., et al. 1998. Quebrada Jaguay: early South American maritime adaptations. *Science* 281: 1830–2, and comment 1775–7.

Schurr, T. G. 2000. Mitochondrial DNA and the peopling of the New World. *American Scientist* 88: 246–53.

Soffer, O., and N. D. Praslov (eds). 1993. *From Kostenki to Clovis: Upper Paleolithic—PaleoIndian Adaptations.* New York: Plenum Press. [Chapters by Olga Soffer and Lewis Binford.]

Sutcliffe, A. J. 1985. *On the Track of Ice Age Mammals.* London: British Museum (Natural History).

Vartanyan, S. L., et al. 1993. Holocene dwarf mammoths from Wrangel Island in the Siberian Arctic. *Nature* 362: 337–40, and comment 283–4.

The modern world (the last 5000 years or so)

Bahn, P. G. and J. Flenley. 1992. *Easter Island, Earth Island.* London: Thames and Hudson. [See also summary of this work by Diamond, 1995.]

Black, F. L. 1992. Why did they die? *Science* 258: 1739–40. [Diseases that decimated Native Americans.]

Burney, D. A. 1993. Recent animal extinctions: recipes for disaster. *American Scientist* 81: 530–41.

Burney, D. A., et al. 2003. *Sporormiella* and the late Holocene extinctions in Madagascar. *Proceedings of the National Academy of Sciences* 100: 10800–5.

Diamond, J. M. 1986. The environmentalist myth. *Nature* 324: 19–20.

Diamond, J. M. 1989. The present, past and future of human-caused extinctions. *Philosophical Transactions of the Royal Society of London B* 325: 469–77.

Diamond, J. M. 1991. Twilight of Hawaiian birds. *Nature* 353: 505–6.

Diamond, J. M. 1992. Twilight of the pygmy hippos. *Nature* 359: 15.

Diamond, J. M. 1994. The last people alive. *Nature* 370: 331–2. [Societies that self-destructed.]

Diamond, J. M. 1995. Easter's end. *Discover* 16 (8): 62–9. [The destruction of Easter Island by its own inhabitants. Required reading!]

Diamond, J. M. 2001. Australia's last giants. *Nature* 411: 755–7. [Oz megafaunal extinctions tied more firmly to human arrival.]

Flannery, T. F. 1995. *The Future Eaters.* New York: George Braziller. [The history of Meganesia (Australasia and associated islands). This book has much deeper significance than simply a regional history.]

Flannery, T. F., 1996. *Throwim Way Leg: Adventures in the Jungles of New Guinea.* London: Weidenfeld and Nicolson. [Enthralling account by an Australian zoologist in New Guinea, showing how easy it is for people to see species go extinct without either noticing or caring very much.]

Goldschmidt, T. 1996. *Darwin's Dreampond: Drama in Lake Victoria.* Cambridge, Mass.: MIT Press. [The devastation of the astounding biology of Lake Victoria.]

Gould, S. J. 1996. The dodo in the caucus race. *Discover* 17 (11): 22–33.

Holdaway, R. N., and C. Jacomb. 2000. Rapid extinction of the moas (Aves: Dinornithiformes): model, test, and implications. *Science* 287: 2250–4, and comment by Jared Diamond, 2170–1.

Jackson, J. B. C., et al. 2001. Historical overfishing and the recent collapse of coastal ecosystems. *Science* 293: 629–37. [Required reading, showing that humans began to wreck marine ecosystems as soon as they learned how to fish.]

Montevecchi, B. 1994. The great auk cemetery. *Natural History* 103 (8): 6–9.

Quammen, D. 1996. *The Song of the Dodo.* New York: Scribner. [Highly recommended.]

Reeves, B. O. K. 1983. Six milleniums of buffalo kills. *Scientific American* 249 (4): 120–35.

Steadman, D. W. 1995. Prehistoric extinctions of Pacific island birds: biodiversity meets zooarchaeology. *Science* 267: 1123–31.

Steadman, D. W., et al. 2002. Rapid prehistoric extinctions of iguanas and birds in Polynesia. *Proceedings of the National Academy of Sciences* 99: 3673–7. [You can reach other papers from the references.]

Stearns, B.P., and S. C. Stearns. 1999. *Watching, From the Edge of Extinction.* New Haven: Yale University Press. [Cases of imminent extinction.]

Stiner, M. C., et al. 1999. Paleolithic population growth pulses evidenced by small animal exploitation. *Science* 283: 190–4. [Overhunting around the ancient Mediterranean.]

Worthy, T. H., and Holdaway, R. N. 2002. *The Lost World of the Moa: Prehistoric Life of New Zealand.* Bloomington, IN: Indiana University Press.

Index

Page numbers in *italic* refer to figures.